BIOTECHNOLOGY IN AGRICULTURE, INDUSTRY AND MEDICINE SERIES

BIOCHEMICAL ENGINEERING

BIOTECHNOLOGY IN AGRICULTURE, INDUSTRY AND MEDICINE SERIES

Agricultural Biotechnology: An Economic Perspective
Margriet F. Caswell, Keith O. Fuglie, and Cassandra A. Klotz
2003. ISBN: 1-59033-624-0

Biotechnology in Agriculture and the Food Industry
G.E. Zaikov (Editor)
2004. ISBN: 1-59454-119-1

Governing Risk in the 21st Century: Lessons from the World of Biotechnology
Peter W.B. Phillips (Editor)
2006. ISBN: 1-59454-818-8

Biotechnology and Industry
G.E. Zaikov (Editor)
2007. ISBN: 1-59454-116-7

Research Progress in Biotechnology
G.E. Zaikov (Editor)
2008. ISBN: 978-1-60456-000-8

Biotechnology and Bioengineering
William G. Flynne (Editor)
2008. ISBN: 978-1-60456-067-1

Biotechnology: Research, Technology and Applications
Felix W. Richter (Editor)
2008. ISBN: 978-1-60456-901-8

Biotechnology: Research, Technology and Applications
Felix W. Richter (Editor)
2008. ISBN: 978-1-60876-369-6
(Online Book)

Biotechnology, Biodegradation, Water and Foodstuffs
G.E. Zaikov and Larisa Petrivna Krylova (Editors)
2009. ISBN: 978-1-60692-097-8

Industrial Biotechnology
Shara L. Aranoff, Daniel R. Pearson, Deanna Tanner Okun, Irving A. Williamson, Dean A. Pinkert, Robert A. Rogowsky, and Karen Laney-Cummings
2009. ISBN: 978-1-60692-256-9

Industrial Biotechnology: Patenting Trends and Innovation
Katherine Linton, Philip Stone, Jeremy Wise, Alexander Bamiagis, Shannon Gaffney, Elizabeth Nesbitt, Matthew Potts, Robert Feinberg, Laura Polly, Sharon Greenfield, Monica Reed, Wanda Tolson, and Karen Laney-Cummings
2009. ISBN: 978-1-60741-032-4

Biosensors: Properties, Materials and Applications
Rafael Comeaux and Pablo Novotny (Editors)
2009. ISBN: 978-1-60741-617-3

Perspectives on Lipase Enzyme Technology
J. Geraldine Sandana Mala and Satoru Takeuchi
2009. ISBN: 978-1-61741-977-8

Biochemical Engineering
Fabian E. Dumont and Jack A. Sacco (Editors)
2009. ISBN: 978-1-60741-257-1

BIOTECHNOLOGY IN AGRICULTURE, INDUSTRY AND MEDICINE SERIES

BIOCHEMICAL ENGINEERING

FABIAN E. DUMONT
AND
JACK A. SACCO
EDITORS

nova publishers
New York

For permission to use material from this book please contact us:
Telephone 631-231-7269; Fax 631-231-8175
Web Site: http://www.novapublishers.com

LIBRARY OF CONGRESS CATALOGING-IN-PUBLICATION DATA

Dumont, Fabian E.
Biochemical engineering / Fabian E. Dumont and Jack A. Sacco.
p. cm.
Includes index.
ISBN 978-1-60741-257-1 (hardcover)
1. Biochemical engineering. I. Sacco, Jack A. II. Title.
TP248.3.D86 2009
660.6'3--dc22

2009016575

Published by Nova Science Publishers, Inc. ✢ New York

CONTENTS

PREFACE

Biochemical engineering is the application of engineering principles to conceive, design, develop, operate, and/or use processes and products based on biological and biochemical phenomena. Biochemical engineering influences a broad range of industries, including health care, agriculture, food, enzymes, chemicals, waste treatment, and energy, among others. Historically, biochemical engineering has been distinguished from biomedical engineering by its emphasis on biochemistry and microbiology and by the lack of a health care focus. This is no longer the case. There is increasing participation of biochemical engineers in the direct development of pharmaceuticals and other therapeutic products. Biochemical engineering has been central to the development of the biotechnology industry, given the need to generate prospective products on scales sufficient for testing, regulatory evaluation, and subsequent sale. This book begins with a review of biodiesel processing technology, the use of varied biodiesel in diesel engines and an analysis of economic scale and ecological impact of biodiesel fuel. Other areas of research include the application of biochemical engineering in the fishery industry, algae growth, and waste water management.

Increasing demand and price of fossil fuel has been a challenge for world scientific researchers and governments which results in a huge impact upon economic development. Biodiesel, as an alternative diesel fuel that can be generated from renewable sources such as animal fat, vegetable oils, and recycled cooking oil, seems to be a promising solution for future in a sustainable manner with respect to energy security and reduction of green house gas (GHG) emission. Chapter 1 is a review of development of biodiesel processing technology, use of varied biodiesel in diesel engines and analysis of economic scale and ecological impact of biodiesel fuel.

Biodiesel can be produced either by chemical (pyrolysis, microemulsification, solid-liquid phase conversion, and transetherification) or biochemical (lipase) methods. Some scientists also demonstrated the potential of employing microwave irradiation or supercritical fluid for derivation of biodiesel. However, the most common process for commercial biodiesel production is to apply alkali as catalyst and mix with methanol for the formation of fatty acid methyl ester (FAME). It has been produced more than 10 millions tonnes of biodiesel and applied as B5 or B20 product in the market.

Most of conclusions from research reports of emission test of different biodiesel resources indicated a significant decrease in particular matter (PM), hydrocarbons, SOx and CO_{2equ} at global level but slightly increase in NOx and CO or CO_2. A research investigated the characteristics of mutagenic species, trans,trans-2,4-decadienal (tt-DDE), and polycyclic aromatic hydrocarbons (PAHs) in the exhaust of diesel engines operated with biodiesel blend fuels made from recycled cooking oil. It showed that tt-DDE and PAHs tend to accumulate in particulate for cold-start driving. Despite of its advantages on environmental protection, the lubricant properties of the biodiesel are able to extend the engine life but oxidation of biodiesel fuel may cause the maintenance problem and result in damage on engine in short-term duration.

However, the biodiesel employed as a renewable energy has also forced the change in food price and supply chain. Therefore, to establish an integral infrastructure of combining energy, economics, environment and agriculture becomes a major issue for the biodiesel application.

In Chapter 2, acyl ascorbates were synthesized through the condensation of various fatty acids with L-ascorbic acid using immobilized lipase in a water-soluble organic solvent, and their properties as food additive were examined. The optimal conditions, which were the type of organic solvent, reaction temperature, the initial concentrations of substrates and the molar ratio of fatty acid to ascorbic acid, for the enzymatic synthesis in a batch reaction were determined. The continuous production of acyl ascorbate was carried out using a continuous stirred tank reactor (CSTR) and plug flow reactor (PFR) at 50°C, and each productivity was *ca.* 6.0 x 10 for CSTR and 1.9 x 10^3 g/(L-reactor·d) for PFR for at least 11 days, respectively. The temperature dependences of the solubility of acyl ascorbate in both soybean oil and water could be expressed by the van't Hoff equation, and the dissolution enthalpy, ΔH, values for the soybean oil and water were *ca.* 20 and 90 kJ/mol, respectively, irrespective of the acyl chain length. The decomposition kinetics of saturated acyl ascorbate in an aqueous solution and air was empirically expressed by the Weibull equation, and the rate constant, k, was estimated. The activation energy, E, for the rate constant for the decomposition in both systems depended on the acyl chain length. The surface tensions of acyl ascorbates in an aqueous solution were measured by the Wilhelmy method, and the critical micelle concentration (CMC) and the residual area per molecule were calculated. The CMC values were independent of temperature but dependent on the pH. The effect of pH of aqueous phase on the stability of O/W emulsion prepared using acyl ascorbate as an emulsifier was examined, and the high stability at pHs 5 and 6 was ascribed to the largely negative surface-charge of droplets in the emulsion. The addition of saturated acyl ascorbate, whose acyl chain length was from 8 to 16, lengthened the induction period for the oxidation of linoleic acid in a bulk and microcapsule with maltodextrin as a wall material. The oxidative stability in bulk system increased with increasing the acyl chain length, whereas that in the microcapsule was the highest at the acyl chain length of 10. The esterification of various polyunsaturated fatty acids, such as linoleic, α- and γ-linolenic, dihomo-γ-linolenic, arachidonic, eicosapentaenoic, docosahexaenoic and conjugated linoleic acids with ascorbic acid and subsequent microencapsulation significantly improved their oxidative stability.

Poly(γ-glutamic acid) (γ-PGA), a novel polyanionic and multifunctional macromolecule synthesized by *Bacillus* species, has attracted considerable attention because of its eco-friendly, biodegradable and biocompatible characteristics. Recently, its application in a wide range of fields such as food, agriculture, medicine, hygiene, cosmetics and environment has been explored. Chapter 3 reviews the literature reports on the application of γ-PGA as a flocculating agent, and adsorbent for cationic dyes and chemical mutagens, affected by several process parameters including pH, temperature, contact time, metal cations, concentration and molecular weight of γ-PGA.

Molecular imprinting has proved to be an effective technique for generating specific recognition sites in synthetic polymers. These sites are tailor-made in situ by copolymerization of functional monomers and cross-linked around the template molecules. The print molecules are subsequently extracted from the polymer, leaving accessible complementary binding sites in the polymer network. Despite significant growth within the field, the majority of template molecules studied thus far are low molecular weight compounds and generally insoluble in aqueous systems. In biological systems, molecular recognition occurs in aqueous media. So, in order to create molecular imprinted polymers capable of mimicking biological processes, it is necessary to synthesize artificial receptors which can selectively recognize the target biological macromolecules such as peptides and proteins in aqueous media. Actually, the synthesis of molecular imprinted polymers specific for biomacromolecules has been a focus for many scientists working in the area of molecular recognition, since the creation of synthetic polymers that can specifically recognize biomacromolecules is a very challenging but potentially extremely rewarding work. The resulting molecular imprinted polymers with specificity for biological macromolecules have considerable potential for applications in the areas of solid phase extraction, catalysis, medicine, clinical analysis, drug delivery, environmental monitoring, and sensors. In Chapter 4, the authors discuss the challenges associated with the imprinting of peptides and proteins, and provide an overview of the significant progress achieved within this field. This review offers a comparative analysis of different approaches developed, focusing on their relative advantages and disadvantages, highlighting trends and possible future directions.

The activated sludge process is a mature and widely-adopted biotechnology for treating both municipal and industrial wastewater for more than one century. However, a large quantity of excess sludge is inevitably generated as a byproduct of biological conversion of organic matters during the process. Treatment and disposal of this byproduct usually accounts for up to 60% of the total capital and operation cost; thus it poses a great challenge in the field of environmental biotechnology. In order to solve this problem, some strategies for minimizing sludge production have been explored and developed, e.g. lysis-cryptic growth, bacteriovoric metabolism, maintenance metabolism and uncoupled energy metabolism-associated sludge reduction, etc. Lysis-cryptic growth technique is using either physical or chemical forces (e.g. heat treatment, ozonation, chlorination, etc.) to disintegrate and mineralize sludge. However, this method is difficult to control, expensive to implement and have a low efficiency. Such drawbacks weaken its capability in practice. Bacteriovoric metabolism method highly depends on the properties of predators and requires strict control of growth conditions to promote specific predator to bloom. Uncoupled energy metabolism of activated sludge is an alternative to reduce excess sludge generation in the activated sludge process. Microbial metabolism is basically includes interrelated catabolic and anabolic reactions. Under normal conditions, catabolism of microbes is tightly coupled with anabolism in the light of energy requirements. However, energy uncoupling can be triggered when some abnormal conditions are present, such as excess carbon source and nutrients limitation; high temperature; alternative aerobic-anaerobic cycle; and presence of metabolic inhibitors. Under such conditions, energy generation from catabolizing substrate is in excess with respect to the anabolism requirement, resulting in dissipation of part of energy through futile cycles (e.g. energy becomes heat). As a result, the biomass yield would be reduced significantly.

Chapter 5 aims to offer an overview of different excess sludge reduction methods with special focus on uncoupled energy metabolism-associated sludge reduction and particularly on addressing the mechanisms behind it. Meanwhile, potential interactions and genetic behaviors of microbes under uncoupled growth conditions, and some advanced microbiological tools are also discussed.

A large amount of raw material is processed with the demand of fishery products. Membrane technology and bioprocessing have increasingly involved in the fishery industry, particularly the fishery waste utilization and the treatment of fishery waste water. A comprehensive review on application of membrane technology in the fishery industry is addressed in Chapter 6. Fishery wastes in forms of solid and liquid contain high content of organic compounds which may cause the pollution to the environment. Membrane technology can recover valuable compounds from these fishery wastes; therefore, it not only reduces the risk of pollution but also improves economical benefit of the fishery industry. Comparing with other competing methods, membrane technology can serve as a mild temperature, simple and large-scaled method to achieve both high efficiency and maximum preservation of natural properties of recovered compounds. A series of valuable compounds such as protein, enzymes, collagen and marine flavor could be recovered from fishery by-products by membrane technology. Pressure-driven membrane processes which are microfiltration, ultrafiltration, reverse osmosis and nanofiltration have been employed. These membrane processes could work individually or be combined with other biochemical reactions to develop a hybrid multistage membrane process or a membrane reactor for the desirable recovery rate, high purity of recovered compounds and development of new products (e.g. bioactive peptides). In addition to recovery of valuable compounds, membrane process has been also applied for treatment of fish pond and fishing process water to achieve water recycle. In the future, more success of membrane technology in the aspect of the fishery industry is to be expected.

In Chapter 7, synthesis of amylase by *Aspergillus oryzae* strain FQB-01 was followed in submerged liquid and solid state fermentations. The submerged cultures were carried out in media prepared with brewery (BW) and meat processing (MPW) wastewaters supplemented with different starch concentrations (10, 20 30 and 40 g/L).

Amylase productions (116 and 111 EU/mL) in the BW and MPW media supplemented with 40 g of starch/L of medium were slightly higher than those obtained in the same media supplemented with 30 g of starch/L (113 and 107 EU/mL) after 84 h of fermentation. In addition, the initial chemical oxygen demand in both wastes was reduced by at least 95%. Optimal pH and temperature for amylase activity were estimated at 5.8 and 46.4°C, respectively. In the optimal conditions, the enzyme showed a high stability at 40 or 50°C (pH = 5.8) or at pH values of 5.0 and 6.0 (T = 46.4°C) in the absence of starch.

The optimum conditions for high amylase production (539 EU/g of dry bagasse) under solid stated fermentation were particle size of bagasse in the range of 5-10 mm, incubation temperature of 32.5°C, pH of 5.9, moisture content of bagasse of 75%, starch concentration of 70.5 mg/g of dry bagasse and inoculum size of 1.4×10^7 spores/g of dry bagasse.

Mammalian cell-enclosing microcapsules have been investigated as devices for bioproduction, cell therapy and stem cell research. Reduction in the diameter of the vehicles is an important issue as it induces beneficial effects such as higher molecular exchangeability between the enclosed cells and the ambient environment, as well as higher mechanical stability and biocompatibility. In Chapter 8 we describe the effectiveness of using a jetting process involving the formation of a stretched thin jet of aqueous polymer solution and its subsequent breakup into droplets in a co-flowing water-immiscible liquid for obtaining droplets of about 100 μm in diameter. The droplet production process and the processes for obtaining gelated microcapsules through a thermal and peroxidase-catalyzed gelation process are also described. In addition, we introduce the production of cell-enclosing hydrogel fibers using the same device developed for the production of cell-enclosing microcapsules.

Wastewater treatment systems must be operated under hydrodynamic conditions that allow maintaining the biomass in suspension and promoting intimate contact between substrates and biomass. The systems used to maintain the mixture (mechanical stirring, aeration, etc.) exert shear stress on the biomass which can affect its physical properties (density and diameter) and specific activity.

When biomass is subjected to moderate shear stress, stable and dense structures can be formed, improving its retention, and the substrate transfer rates are also favoured. However, high shear stress generally leads to the loss of biomass activity and to the formation of particles with low diameters, which are washed-out of the system. Therefore, it is very important to control shear stress acting on biomass particles in order to optimize the performance of wastewater treatment systems.

The effects of impact stress and hydraulic stress by gas or liquid on the efficiency of different biological systems for carbon and nitrogen removal are discussed in Chapter 9.

As explained in Chapter 10, biological wastewater treatment modelling has become and important tool in process engineering. There are state of the art activated sludge models (ASMs) available, which have found wide application in the engineering community. Biofilm models have found less application in engineering practice so far, and a gap has developed between biofilm research and engineering practice in the biofilm modelling community. In this context biofilm and floc structure have played different roles in biological wastewater treatment modelling. Activated sludge models (ASMs) do not explicitly take floc structure into account. In contrast biofilm structure has been strongly emphasized in biofilm models over the past decades. Biofilm models have as a result evolved with increasing complexity from one- to two- to three-dimensional models. One reason for this is that biofilm structure is crucially linked to diffusion by Fick's laws of diffusion, since it is known that diffusion is an important process in biofilm systems. The application of Fick's laws of diffusion has thus been a driving force towards the development of multidimensional biofilm models with increased model complexity, because biofilms have a complex, heterogenous three-dimensional structure. The increasing complexity has not led, however, to increased application of biofilm models in engineering practice, and there is a trend towards simplified (e.g. zero-dimensional) models for this purpose. Further it has been shown that diffusion and structure play an important role in activated sludge systems. The role of activated sludge structure has recently led to the development of multidimensional activated sludge models in activated sludge research, whilst the state of the art ASMs for engineering practice do not take floc structure into account.

In Chapter 11, microalgae *Chlorella fusca* ACOI 621, *Chlorella vulgaris* ACOI 879, *Scenedesmus acutus* ACOI 538 and *Scenedesmus obliquus* ACOI 550, all native from Portugal, were characterized in terms of specific growth rate. The effect of pH and the presence of Cr(VI) in concentrations up to 25 mg l^{-1} (50 mg l^{-1} for *Chlorella fusca*) has been evaluated. The logistic equation of population growth $n = 1/\left[\left(1/n_0 - 1/K\right) e^{-\mu t} + 1/K\right]$

adequately describes the cellular growth. Experiments at pH = 6.5 and temperature around 24.5 °C, in the absence of Cr(VI), led to specific growth rates (μ) of 0.0370, 0.0284, 0.0359 and 0.0162 h^{-1} and maximum biomass concentrations (K) of 403.3, 369.2, 542.9 and 604.1 mg l^{-1} for *C. fusca*, *C. vulgaris*, *S. acutus* and *S. obliquus*, respectively. Experiments carried out with the same algae at approximately 21 °C, also in the absence of Cr(VI), gave μ values of 0.0241, 0.0357, 0.0272 and 0.0289 h^{-1} and K values of 292.6, 169.9, 263.1 and 327.8 mg l^{-1} for initial pH = 6.5 and μ values of 0.0115, 0.0177, 0.0137 and 0.0158 h^{-1} and K values of 35.9, 3.0, 32.8 and 54.7 mg l^{-1} for initial pH = 7.9. Higher pH results in a significantly lower growth rate and *C. vulgaris* seems to be the less resistant microalgae to changes in the environmental conditions. Looking simultaneously at μ and K values, the best performance in terms of growth kinetics was obtained for *S. acutus* and *C. fusca*. Growth inhibition is visible for Cr(VI) ≥ 5 mg l^{-1} but concentrations up to 1 mg l^{-1} seem not to seriously affect algal growth, even increasing the *C. fusca* specific growth rate. For Cr(VI) < 1 mg l^{-1}, μ varies between 0.08 and 0.17 h^{-1}, depending on the algal species. The growth of *C. vulgaris* is severely inhibited by Cr(VI) = 5 mg l^{-1}. The production of metabolites is small compared with biomass production, for all Cr(VI) concentrations. The organic carbon content of algae is about 40%-50% (dry basis), except for *S. obliquus* (around 30%). The biomass of *C. fusca* and *S. acutus* presents the greatest sedimentation rates. The presence of high Cr(VI) concentrations negatively affects the sedimentation.

In Chapter 12, the short-term effects of glucose addition on nitrification and activated sludge settlement were investigated in two laboratory-scale sequencing batch reactors (SBRs): one with the addition of glucose (G-Reactor) and the other without the addition of glucose (N-Reactor). The characteristics of nitrification activity, nitrite accumulation, and activated sludge settlement were examined. A high specific nitrification rate was obtained in the N-Reactor, while a high volumetric nitrification rate was obtained in the G-Reactor. Nitrite accumulation occurred in both reactors, and the nitrite/total oxidized nitrogen ratio in both reactors was over 67%. Nitrite accumulation in both reactors was due to low pH caused by the processing of nitrification. In the G-Reactor, the biomass concentration did not change much; in the N-Reactor, the biomass concentration decreased with time. The reason for decreasing biomass concentration in the N-Reactor was as follows: (1) high extracellular polymeric substances (EPS) produced in the N-Reactor due to shortage of organic carbon substrate, resulting in poor settlement of activated sludge flocs; (2) poor settlement of activated sludge flocs causing activated sludge wash out of the system, and, consequently, a low sludge retention time occurred; and, finally, (3) the low sludge retention time further encouraged the poor settlement of activated sludge flocs.

It is known that the proteases have applications in several industrial processes such us leather processing, laundry detergents, producing of protein hydrolysates and food processing, as well as in the peptide synthesis in non conventional media. The application of proteases as catalyst of short oligopeptides in aqueous-organic media, have received a great deal attention as a viable alternative to chemical approach because of their remarkable characteristics. On the other hand, alkaline proteases have also been used to improve the cleaning efficiency of detergents. Detergent enzymes account for about 30% of the total worldwide enzyme production and represent one of the largest and most successful applications of modern industrial biotechnology. The aim of Chapter 13 was to study the performance of proteolytic enzymes of *Acacia caven* (Mol.) Molina pollen for its potential application as an additive in various laundry detergents formulations and as catalyst of the peptide synthesis in aqueous-organic media. Pollen grains (35 mg/ml) were suspended in 0.1M Tris-HCl buffer pH 7.4 and slowly shaken for 2 h at 25° C. Then, the slurry was centrifugated for 30 min at 8000 rpm and the supernatant (crude enzyme extract, CE) was tested in protein content (Bradford's method) and proteolytic activity (using BAPNA and Z-Ala-pNO as substrates). A partial characterization of *Acacia caven* CE was carried out: enzyme extract displayed maximum proteolytic activity at pH 8 and 35-40° C; it showed remarkable thermal stability after 1.5 h at 25-40° C but it decreased as long as temperature increased to 60° C. On the other hand, the enzyme extract was incubated with different surfactants and commercial laundry detergents at 25-60° C during 30 min and 1h; and it showed high stability and compatibility with them. The peptide synthesis catalyzed by *Acacia caven* CE was carried out in a mixture of 0.1M Tris-HCl buffer pH 8.5 and ethyl acetate (50:50 ratio) at 37° C using 2-mercaptoethanol as activator and TEA as neutralizing agent of the amino component (Phe-OMe.HCl). Carboxylic components were selected in base of the highest preference of CE. The identification of synthesized peptide products was carried out by HPLC-MS. According to the obtained results, this work contributes with a new variety of phytoprotease useful as catalyst of the peptide synthesis and as additive of laundry detergents.

The parallel AN/AO process, first proposed by the authors to efficiently use denitrifying phosphorus removing bacteria, was briefly introduced in Chapter 14. Deactivation of phosphorus-accumulating organisms (PAO) occurred in the process when its SRT (Sludge Retention Time) and HRT (Hydraulic Retention Time) were too long, i.e. SRT and HRT were 30d and 18h respectively. PAO deactivation was observed also in three anaerobic-anoxic SBR reactors fed with different COD/NO_3^--N synthetic wastewater using seed sludge from the parallel AN/AO process. Possible factors that could cause PAO deactivation such as pH, temperature, internal/external return ratio, SRT, HRT, DOC (Dissolved Organic Carbon) at the beginning of anoxic stages and NO_3^--N concentration at the beginning of anaerobic stages were studied. Results showed that SRT and HRT were main factors accounting for PAO deactivation occurrence in the parallel AN/AO process while DOC concentration at the beginning of anoxic stages and NO_3^--N concentration at the beginning of anaerobic stages were main factors influencing PAO activity in the anaerobic-anoxic SBR reactors. PAO rejuvenation occurred in both configurations shortly after main influencing factors were reset to right values: PAO was rejuvenated by adjusting SRT and HRT to 15d and 9h respectively for the parallel AN/AO process; by controlling DOC at the beginning of anoxic stages and NO_3^--N concentration at the beginning of anaerobic stages lower than 3 mgL-1 and 2.3 mgL-1 respectively could rejuvenate PAO in anaerobic-anoxic SBR reactors.

Dialysis and adsorption units are commonly used in liver support devices for the removal of albumin-bound toxins such as bilirubin. In Chapter 15, an engineering approach to the analysis of a liver support device implementing these units is presented. Starting from the physico-chemical description of the basic phenomena involved in the detoxification process, a mathematical model of a recirculating albumin dialysis liver support device was built and used to calculate bilirubin clearances obtained by the device with different operating conditions.

The results highlight the possible existence of an optimum dialysate albumin concentration; furthermore, the overall bilirubin clearances obtained in the simulations did not exceed 4% of the blood flow-rate fed to the device, this poor performance being limited by the slow bilirubin mass transfer across the membrane. The information presented in this chapter can be helpful for the optimization of existing liver support devices and for the design of new ones; nevertheless, for a complete assessment of the device performance, a similar analysis should be extended to the clearance of other toxins and some of the model parameters should be also checked against clinical data.

In: Biochemical Engineering
Editors: F.E. Dumont and J.A. Sacco, pp. 1-40

ISBN: 978-1-60741-257-1

Chapter 1

A REVIEW OF BIODIESEL AS RENEWABLE ENERGY

John Chi-Wei Lan[1], Amy Tsui[2], Shaw S. Wang[1,2] and Ho-Shing Wu[1]

[1]Department of Chemical Engineering and Materials Science, Yuan Ze University, ChungLi, Taiwan

[2]Departement of Chemical and Biochemical Engineering, Rutgers University, The State University of New Jersey, United States

Abstract

Increasing demand and price of fossil fuel has been a challenge for world scientific researchers and governments which results in a huge impact upon economic development. Biodiesel, as an alternative diesel fuel that can be generated from renewable sources such as animal fat, vegetable oils, and recycled cooking oil, seems to be a promising solution for future in a sustainable manner with respect to energy security and reduction of green house gas (GHG) emission. This article is a review of development of biodiesel processing technology, use of varied biodiesel in diesel engines and analysis of economic scale and ecological impact of biodiesel fuel.

Biodiesel can be produced either by chemical (pyrolysis, microemulsification, solid-liquid phase conversion, and transetherification) or biochemical (lipase) methods. Some scientists also demonstrated the potential of employing microwave irradiation or supercritical fluid for derivation of biodiesel. However, the most common process for commercial biodiesel production is to apply alkali as catalyst and mix with methanol for the formation of fatty acid methyl ester (FAME). It has been produced more than 10 millions tonnes of biodiesel and applied as B5 or B20 product in the market.

Most of conclusions from research reports of emission test of different biodiesel resources indicated a significant decrease in particular matter (PM), hydrocarbons, SOx and CO_{2equ} at global level but slightly increase in NOx and CO or CO_2. A research investigated the characteristics of mutagenic species, trans,trans-2,4-decadienal (tt-DDE), and polycyclic aromatic hydrocarbons (PAHs) in the exhaust of diesel engines operated with biodiesel blend fuels made from recycled cooking oil. It showed that tt-DDE and PAHs tend to accumulate in particulate for cold-start driving. Despite of its advantages on environmental protection, the lubricant properties of the biodiesel are able to extend the engine life but oxidation of biodiesel fuel may cause the maintenance problem and result in damage on engine in short-term duration.

However, the biodiesel employed as a renewable energy has also forced the change in food price and supply chain. Therefore, to establish an integral infrastructure of combining energy, economics, environment and agriculture becomes a major issue for the biodiesel application.

1. Introduction

Improving energy security, decreasing vehicle contribution to air pollution and achieving reduction or even eliminating greenhouse gas (GHG) emissions are primary goals compelling governments to identify and commercialise alternatives to the petroleum fuels. Over the past two decades, several candidate fuels have emerged such as compressed natural gas (CNG), liquefied petroleum gas (LPG) and electricity power. These fuels feature a number of benefits over petroleum fuel, however, they also exhibit a number of drawbacks like requirement of costly modifications on applied engines and the development of separate fuel distribution that limit their ability to capture a significant share of the market.

Biofuels like bioethonal and biodiesel have the potential to overcome those disadvantages of replacing traditional fuels. Biodiesel, as an alternative and renewable fuel consisting of the alkyl esters of fatty acids, can be derived from animal fats, vegetable oil and waste cooking oil. It has been receiving a lot of attention lately due to its impacts upon energy security, offering prospect of reduction of air-pollutants emissions as well as economic and sustainable development compared to fossil fuel. In its principal use, biodiesel is a potential replacement for conventional diesel, which in this instance, is the term used to describe diesel generated from crude oil. Most research studies have depicted no appreciable difference between biodiesel and diesel in engine durability or in carbon deposits.

The biodiesel has been in commercial use as an alternative fuel since 1988 in many European countries. It can be produced from a great variety of feedstocks including vegetable oil and animal fat as well as waste cooking oils. The choice of feedstocks depends largely upon geography. The biodiesel from Europe is primary produced from rapeseed oil while in the United States both rapeseed and soybean oil are used and in Taiwan as well as Japan waste cooking oil is employed. Biodiesel has several distinct advantages compared with diesel fuel in addition to being fully competitive with diesel in most technical aspects.

Biodiesel fuel is reliable, renewable, biodegradable and non-toxic. It is less harmful to the environment for it contains practically no sulfur and substantially reduced emissions of unburned hydrocarbon (HC), carbon monoxide, sulfates, polycyclic aromatic HC (PAH) and particulate matter. It has fuel properties comparable to mineral diesel and because of great similarity; it can be mixed with mineral oil and used in standard diesel engines with minor or no modifications at all. Biodiesel works well with new technologies such as catalysts (which can reduce the soluble fraction of diesel particulates but not the solid carbon fraction), particulate traps and exhaust gas re-circulation. Being an agricultural product, all countries have the ability to produce and control this energy source which is a situation very different to the crude oil business. This work discusses the benefits of biodiesel, its reaction chemistry, and the various sources and components involved in the production of biodiesel. Certain components will be chosen based on optimum characteristics to be observed more closely when detailing the kinetics and process design for a selection of process systems.

1.1. Benefits of Biodiesel: Economics

As of 2007, the United States had biodiesel production capacity of 1.85 billion gallons from 165 commercial biodiesel plants [1]. It is calculated that 1.16 jobs would be created per million liters of annual production in a biofuels plant [2]. This number would be higher in more labor intensive regions. Biodiesel can be produced worldwide, and in a study done by Johnston and Holloway, its production has the potential to improve economies [2]. The study determined that Malaysia, Indonesia, Argentina, the US, and Brazil are the top five largest potential producers of biodiesel due to current their current production of palm and soybean. Developing countries with the highest profit potential include Malaysia, Indonesia, Phillipines, Papua New Guinea, and Thailand. Developing countries with highest profitable biodiesel export potential are Malaysia, Thailand, Colombia, Uruguay, and Ghana [36]. Biodiesel can thus create hundreds of jobs and contribute millions of dollars to a country's GDP [2].

1.2. Benefits of Biodiesel: Politics

The United States consumes 0.53 billion cubic meters of diesel annually. [1] Producing more biodiesel domestically also lowers dependence on foreign crude oil. A significant factor holding back large scale production of biodiesel is consumer demand. Europe created this demand by making alternative fuel use mandatory [2,9]. In Europe, biodiesel production has surpassed 2.0 billion litres as of 2004 [2]. This is primarily due to the legislation passed in the 1990s making use of alternative fuels mandatory [2]. Because diesel fuel comprised 66% of on-road, liquid fuel demand, biodiesel saw rapid popularity there [2]. As a result of increased demand, capacity for biodiesel production has increased significantly in Europe [3], rising from almost none in 1991 to over 5000 million liters in 2008 as shown in Figure 1.

Legislation under consideration would require motor vehicle fuel sold in the United States from 2002 onward to contain a minimum quantity of renewable fuel. 2 Renewable fuels include biodiesel, ethanol or any other liquid fuel produced from biomass or biogas. Precise estimates of the minimum quantity guidelines are a current topic of discussion. It is assumed that the minimum percentage by volume of renewable fuel content will increase from 1.2 percent in 2002 to four percent by 2016.

Using current long-term U.S. Department of Energy projections for highway energy use as a baseline, 3 renewable fuel use in the United States would increase from current levels of about 1.9 billion gallons to more than 8.8 billion gallons by 2016. As shown in Figure 2, the majority of renewable fuel would be accounted for by ethanol produced from grain, however biodiesel is expected to account for about 15 percent of total renewable fuel use by 2016.

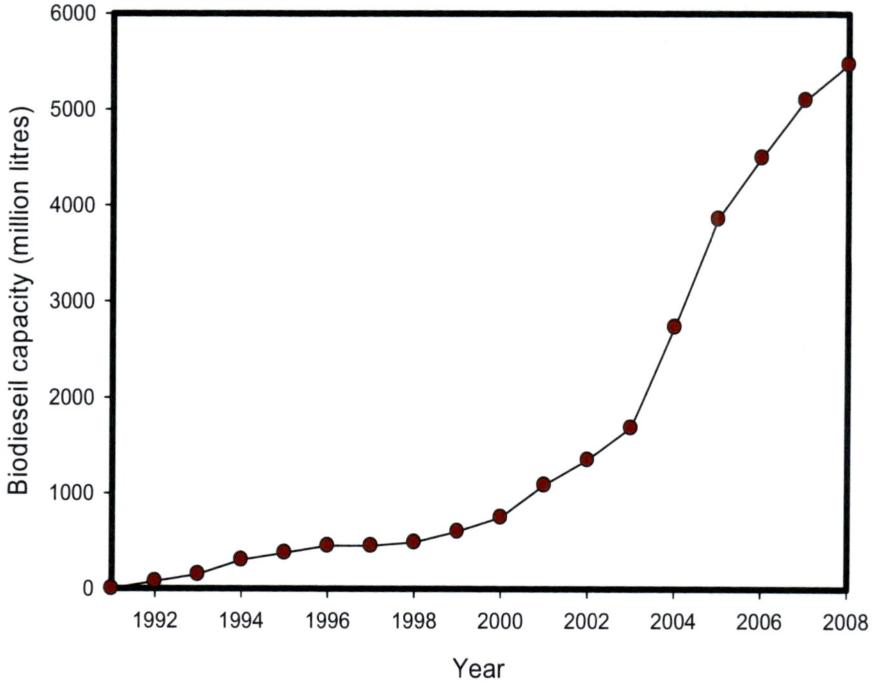

Figure 1. Changes in biodiesel capacity from 1991 to 2008 in Europe.

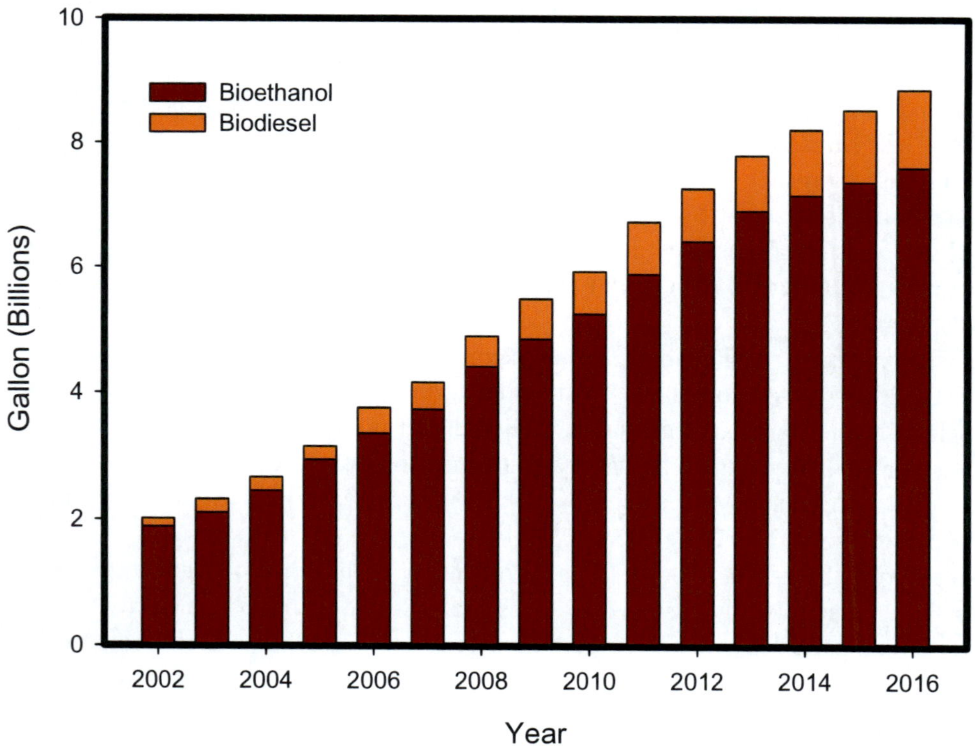

Figure 2. Renewable fuel demand in United state.

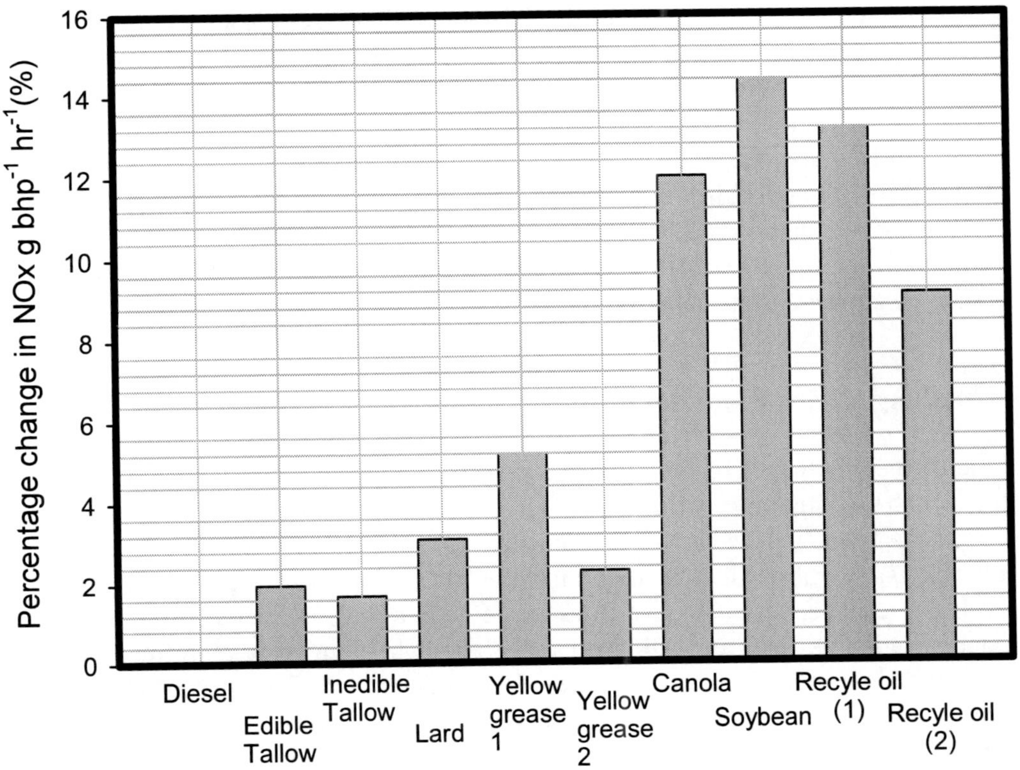

Figure 3. Increase in NOx emissions from CI engines using various B100 fuels.

1.3. Benefits of Biodiesel: Environment

As stated earlier, biodiesel is derived from renewable sources such as vegetable oils, animal fats, and waste cooking oils. Once produced and used, the byproducts of its combustion in automobile engines are carbon dioxide and water only. The source of biodiesel, such as soybeans, will absorb CO_2 during its lifetime through photosynthesis. In this way, biodiesel is considered carbon neutral. However, when production is considered, biodiesel is not neutral. Fossil fuels are still required to create the steam, electricity, and methanol needed for manufacturing, and to fuel the equipment for farming and transportation and materials. Even with all this fossil fuel input, biodiesel is still an energy efficient fuel. From one unit of fossil fuel energy used to produce the biodiesel, 3.2 units of energy are created as biodiesel fuel [4]. It is also estimated that biodiesel still has 41% less carbon dioxide emissions than petroleum based diesel [2]. Biodiesel also reduces other emissions such as particulate matter, hydrocarbons, and carbon monoxide. This is due to the 11% oxygen by weight content that allows for more complete combustion. [4] Furthermore, biodiesel is a natural substance and therefore is biodegradable if spilled. It is also comparably better than other popular renewable energies. Soybean based biodiesel has a 93% energy gain compared to only 25% for corn derived ethanol [5]. An 80000-km durability test was performed on two engines using diesel and biodiesel (methyl ester of waste cooking oil) as fuel in order to examine emissions resulting from the use of biodiesel by Yang et al (2007). The test biodiesel (B20) was blended with 80% diesel and 20% methyl ester derived from waste cooking oil. The results presented that the average total PAH emission factors were 1097 and 1437 μg bhp-h-1 for B20 and diesel, respectively. For most ringed-PAHs and total-PAHs, B20 has lower PAH emission levels than that of diesel fuel. For both B20 and diesel, total PAH emission levels decreased as the driving mileage accumulated [6]. Some studies have shown also, that biodiesel may actually produce an increase in NOx emissions as shown in Figure 3. However, this larely varies due to composition of the fuel. In actuality, some fuels decrease emissions, while others are seen to increase generalizing statement [4].

1.4. Challenges with Biodiesel

One of the challenges of the biodiesel industry is improving efficiency to make the production cost-competitive with diesel [1]. In 2003, biodiesel cost over $0.50/l while diesel cost $0.35/l [7]. This high cost is mostly due to the usage of virgin vegetable oil as a feedstock [7]. Soybean oil cost $0.36/l in June 2002 [7]. This is already over the cost of diesel. Using cheaper feedstocks, such as waste cooking oil, is seen as a promising way of reducing cost, as they are estimated to be about half the cost of refined oils. The obstacle with cheaper feedstocks is the higher content of FFA and other unwanted ingredients. Although biodiesel has been proven profitable, but if there are more lucrative alternatives, actors will not pursue it. The analysis focuses on quantifiable economic costs and benefits driven by markets, since few production decisions in competitive agricultural and fuels markets are driven by non-market logic. Situations in which non-economic considerations might influence production decisions are noted. Table 1 considers the regional actors required to realize biodiesel production [8]. It shows that the development and economic analysis of biodiesel industry can be influenced by several of factors. Those factors are the major challenges for considerations of which shall be taken priority.

2. Biodiesel Production

Methods such as pyrolysis, microemulsification, solid-liquid phase conversion, and transetherification applied to reduce the high viscosity of vegetable oils to enable their use in general diesel engines without operational problems have been investigated. Transesterification is the most common technique used for biodiesel production. The most commonly prepared esters ate methyl esters due to methanol is the least expensive alcohol, although there are exceptions in some countries. Although those fresh or used oils and fats can be suitable for biodiesel production; however, changes in the reaction procedure frequently have to be made because the presence of water of free fatty acid (FFA) in feedstocks. This section discusses the reaction based on transesterification technologies.

Table 1. Regional Actors in Biodiesel sector

Actor	Need	Benefits	Problems	Alternatives
Farmer	Crops with value added use	Good rotation crops: breaks disease cycles	Market price below breakeven cost	Growing barley, peas, lentils
Crusher	Oil and meal market	n/a	Difficult to find local crushers	Involvement in another ag enterprise
Meal user	Regional alternative for livestock feed	Good for dairies	More than 12% canola mean in feed is not applicable	Importing alternative meals
Biodiesel producer	Low cost of regional feedstocks	Oil feedstock; low pour point	Oil extraction efficiency	Biolubricants as potentially candidates
Blender /Distributor	Meet market demand for biodiesel	Tax credit; easily blended	Minimal economic incentive to expand storage facilites	Synthetic lubricity additives probably cheaper
End user	Warranties for fuel and engines	Regional alternative fuel source	Engine warranties, fuel price	Petroleum diesel

2.1. Biodiesel Reaction Chemistry

Biodiesel is produced from the catalytic transesterification, a type of alcoholysis of vegetable oils, animal fats, or waste cooking oils with an alkyl alcohol group. During transesterification, an alkoxy is exchanged between an ester compound and an alcohol to produce a different ester and alcohol. As shown in Figure 4, the transesterification of biodiesel produces three moles of fatty acid methyl esters (FAMEs) from one mole of triglyceride and methanol. This reaction actually occurs in three steps as shown in Figure 5. In the reaction mechanism, the methanol forms a tetrahedral intermediate at an ester group on the triglyceride (TG) and then detaches to form a diglyceride (DG) and a FAME. [10] This is repeated stepwise until the monoglyceride (MG) is converted to glycerine (GL). Excess alcohol should be used to drive the reaction forward, for which a 6:1 molar ratio of alcohol to oil is most commonly the case. [10-12]

2.2. Rate Law

The differential equations governing the reactions are described by Nourredini and Zhu [12] as shown in equations 1 through 5, where E and A are ester and alcohol respectively.

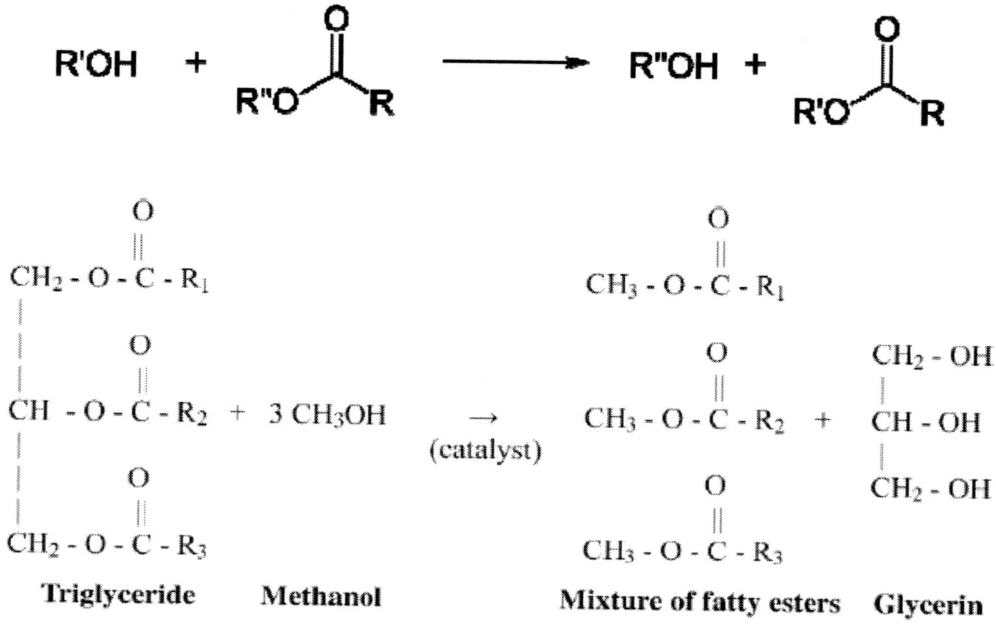

Figure 4. Above, generic transesterification process. Below, biodiesel transesterification process where R1-3 are long hydrocarbon chains.

$$TG + CH_3OH \underset{K_2}{\overset{K_1}{\Leftrightarrow}} DG + R_1COOCH_3$$

$$DG + CH_3OH \underset{K_4}{\overset{K_3}{\Leftrightarrow}} MG + R_2COOCH_3$$

$$MG + CH_3OH \underset{K_6}{\overset{K_5}{\Leftrightarrow}} DG + R_3COOCH_3$$

overall reaction

$$TG + 3CH_3OH \underset{K_8}{\overset{K_7}{\Leftrightarrow}} 3RCOOCH_3 + \text{glycerol}$$

Figure 5. Reaction mechanism for transesterifcation of biodiesel (TG: triglyceride; DG: diglyceride and MG: monoglyceride).

$$\frac{d[\mathrm{TG}]}{dt} = -k_1[\mathrm{TG}][A] + k_2[\mathrm{DG}][A] - k_7[\mathrm{TG}][A]^3 + k_8[A][\mathrm{GL}]^3 \quad (1)$$

$$\frac{d[\mathrm{DG}]}{dt} = k_1[\mathrm{TG}][A] - k_2[\mathrm{DG}][E] - k_3[\mathrm{DG}][A] + k_4[\mathrm{MG}][E] \quad (2)$$

$$\frac{d[\mathrm{MG}]}{dt} = k_3[\mathrm{DG}][A] - k_4[\mathrm{MG}][E] - k_5[\mathrm{MG}][A] + k_6[\mathrm{GL}][E] \quad (3)$$

$$\frac{d[E]}{dt} = k_1[\mathrm{TG}][A] - k_2[\mathrm{DG}][E] + k_3[\mathrm{DG}][A] - k_4[\mathrm{MG}][E] + k_5[\mathrm{MG}][A] - k_6[\mathrm{GL}][E] + k_7[\mathrm{TG}][A]^3 - k_8[\mathrm{GL}][E]^3 \quad (4)$$

$$\frac{d[A]}{dt} = \frac{-d[E]}{dt}$$

$$\frac{d[\mathrm{GL}]}{dt} = k_5[\mathrm{MG}][A] - k_6[\mathrm{GL}][E] + k_7[\mathrm{TG}][A]^3 - k_8[\mathrm{GL}][E]^3 \quad (5)$$

Darnoko determined the reaction rate constants for the reaction mechanism, and the results are shown in Table 2. The rate constant for triglyceride to diglyceride is the lowest and is therefore the rate determining step, identifying k1 as the rate constant for the rate law. According to Nourredini and Zhu [12], when using a 6:1 ratio of methanol to soybean oil, a 2nd ordermechanism with a 4th order shunt mechanism is the most appropriate kinetics.

Table 2. Reaction rate constant k for triglyceride, diglyceride, and monoglyceride hydrolysis over a temperature range of 50 to 65° C

Glyceide	Temperatutr (℃)	Reaction rate constant, k (wt% min^{-1})	R^2
TG→DG	50	0.018	0.9865
	55	0.024	0.9966
	60	0.036	0.9822
	65	0.048	0.9903
DG→MG	50	0.036	0.9940
	55	0.051	0.9974
	60	0.070	0.9860
	65	0.098	0.9678
MG→Glycerol	50	0.112	0.9733
	55	0.158	0.9619
	60	0.141	0.9862
	65	0.191	0.9843

2.2. Reaction Temperature

As shown in Table 2, reaction rate constants increased with increasing temperature, indicating a quicker reaction. In another work, the effect of temperature on conversion was analyzed. Similarly, with increasing temperature, the overall conversion increased. In both cases, there is not a concrete optimal temperature, but instead, the highest temperature in the investigated range was considered the best; 65 and 70° C respectively. Still, a more common reaction temperature used is 60 °C. [11, 14] This is still reasonable since the conversion differences between 60 and 70° C is very small. In general, using a range of 50 to 70° C is recommended. While higher temperatures result in slightly better conversions, using a lower temperature would save in operating costs. Newer processes propose utilizing supercritical conditions where temperatures are above 280°C. These proceses will be discussed later.

2.3. Improving Miscibility

A constant issue with transesterification is that fat and oil do not mix with alcohol. The reaction is thus in two-phases and is mass transfer limited. [9] There are three methods for improving miscibility: adding a cosolvent, mixing during the reaction, and using a membrane reactor. A common cosolvent used it tetrahydrofuran (THF).

2.4. Alcohol Reagent

Methanol, isopropanol, butanol, and ethanol are some of the alcohols that can be used for the transesterification. Methanol is the most frequently used alcohol because of its lower cost and smaller molecular mass, which means less material is used up in relation to the amount of esters produced. Because three moles of alcohol is required for the reaction, regardless of which alcohol, it ends up being cheaper to use methanol (as shown in Table 3)

Furthermore, FAMEs, produced using methanol, are considered to have the proper characteristics as an ideal replacement of petrodiesel fuel, including viscosity, boiling point, and cetane number. [15] The importance of these characteristics is explained later. Alcohol choice can depend on the type of catalyst used. Tamalampudi et al. showed that methanol is best for use in enzyme catalyzed production of biodiesel with the Jatropha oil feedstock [16]. It was hypothesized that low molecular weight and high polarity of methanol may allow it to more easily diffuse and access the enzyme [16].

Table 3. Comparison of the cost of methanol and ethanol (based on price of 2004)

Alcohol	Cost/gallon	Gram moles/gallon	Cost/gram-mole
Methanol	US$ 0.61	93.56	US$ 0.0065
Ethanol	US$ 1.45	64.82	US$ 0.0224

However, a common downside of biodiesel from methanol is the high cloud point, which makes it less desirable in colder climates. Work has shown that cloud point and pour point can be lowered when longer chain alcohols are used instead of methanol [17], as shown in Table 4. In conventional base-catalyzed methods, absolute ethanol (99.9%) must be used to limit water content that would neutralize the catalyst. Absolute ethanol is also very hygroscopic, or absorbs and retains water, so is not commonly used in industry [18]. Ethanol forms an azeotrope with water, so it is difficult to remove and recycle [19]. Therefore, though methanol is more toxic, it is considered better overall.

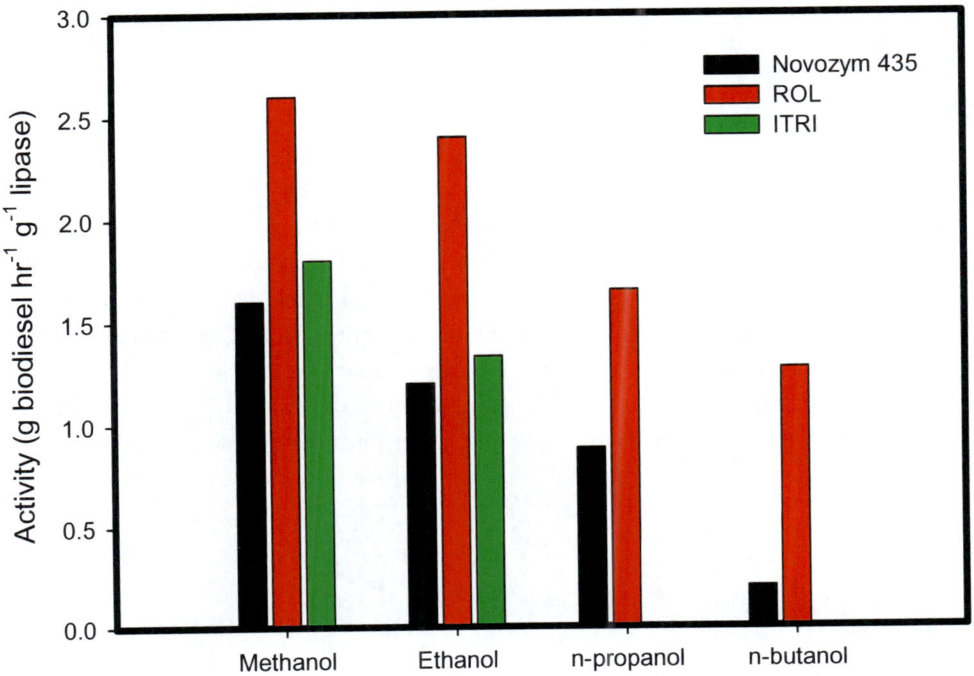

Figure 6. Effect of different alcohols on lipase activities. Reaction condition: Jatropha oil 5g; alcohol-oil molar ratio 1:1; lipases 0.2g; reaction temperature 30°C; reaction time 60 min. [16;32]

Table 4. Cloud point and pour point of the bio-diesels from various oils

Oils	Cloud point				Pour point			
	Methyl	Ethyl	2-Propyl	Butyl	Methyl	Ethyl	2-Propyl	Butyl
Linseed	0	-2	3	-10	-9	-6	-12	-13
Canola	1	-1	7	-6	-9	-6	-12	-16
Sunflower	1	-1	n.d.	n.d.	-8	-5	n.d.	n.d.
Rapseed	0	-2	n.d.	n.d.	-15	-15	n.d.	n.d.
Soybean	5	3	n.d.	n.d.	6	n.d.	n.d.	n.d
Recycle	12	9	n.d.	n.d.	n.d.	n.d.	n.d	n.d.

Also, the usage of supercritical alcohol has been shown to be able to perform esterification and transesterification simultaneously. This means that feedstocks with high fatty acid content can also be used. Typically, esterification is performed as an expensive pretreatment to the feedstock to reduce free fatty acid content [20]. Afterwards, steps for catalyst and soap removal are also necessary, but not when using super critical methanol, which requires no catalyst and also esterifies FFAs [20].

2.5. Catalyst

Because the reaction does not heavily favor one side or the other, a catalyst is essential for an efficient and higher yield reaction. The common types of catalyst used are acid, base, enzyme, solid metal catalyst, and solid super base. Each type has its advantages and disadvantages which are discussed below.

Acid Catalysis

The acid catalyst acts by donating a proton to the carbonyl group which increases the ester activity [21]. This protonation makes the oxygen in the carbonyl group positively charged and open to receiving electrons from the alcohol. The advantages of using acid are that it does not produce the saponification side reaction that base does. The soap produced from saponification causes difficult separation downstream. Acid's downside is that the reaction rate is much slower than using base [22, 23]. Conversion is also highly controlled by water and fatty acid content; less of both lead to high conversions [18]. Therefore, acid catalysts are generally only used for pretreatment in an esterification reaction [19] as shown below:

$$\underset{\textbf{Fatty Acid}}{HO-\overset{\overset{\displaystyle O}{\|}}{C}-R} \quad + \quad \underset{\textbf{Methanol}}{CH_3OH} \quad \xrightarrow{(H_2SO_4)} \quad \underset{\textbf{Methyl ester}}{CH_3-O-\overset{\overset{\displaystyle O}{\|}}{C}-R} \quad + \quad \underset{\textbf{Water}}{H_2O}$$

Common acids used in this process are sulfuric acid and phosphoric acid [19], but acids are not used commercially as of 2003 [7].

Base Catalysis

Conversely, base catalysts act on the alcohol by deprotonating it. This allows it to more readily react with the feedstock [21]. Base-catalyzed reactions occur up to 4000 times faster than acid-catalyzed and the base is less corrosive to materials [22, 16]. In base-catalyzed reactions, its typical to use 6:1 mole ratio of alcohol to triglyceride in order to drive the reaction forward [19]. Since bases will react with water and FFA in a saponification reaction as shown below, base catalysts are frequently used when vegetable oil is the feedstock as there is less FFA and water content [19]. If there is greater than 1% FFA content, pretreatment is required [19]. Because of the saponification reaction, it is important to carefully analyze how much base to use. Base catalysis has the same conversion dependence as acid [18]. To use the best of both types of catalysis, acid-base catalysis is often used. This takes advantage of the very fast reaction rate of base-catalysis while reducing saponification with acid. This occurs in a two step process. First, the acid catalyst is used to convert the free fatty acids to esters which decreases the free fatty acid content (preferably to 1%)[21]. Free fatty acids are what react with the base to produce soaps, so reduction in its content will likewise reduce the amount of soap produced. The feedstock is then base-catalyzed for transesterification.

$$HO-\overset{\overset{\displaystyle O}{\|}}{C}-R \quad + \quad KOH \quad \rightarrow \quad K^{+}\,{}^{-}O-\overset{\overset{\displaystyle O}{\|}}{C}-R \quad + \quad H_2O$$

Fatty Acid **Potassium Hydroxide** **Potassium soap** **Water**

There are two types of base catalysts: homogenous and heterogeneous. Common homogenous bases are sodium hydroxide (NaOH) and potassium hydroxide (KOH). The amount of additional catalyst required to neutralize FFAs can be calculated according to equations derived from [19] at a limit of 5% FFA:

$$\text{NaOH [\%FFA](0.144) + 1\%} \tag{8}$$

$$\text{KOH [\%FFA](0.197)/(purity) + 1\%} \tag{9}$$

$$\text{Na(C}_3\text{HO) [\%FFA](0.190) + 0.25\%} \tag{10}$$

Heterogeneous Base Catalysts

Heterogeneous examples are alkali metal compounds like alumina and zeolites [21]. Solid bases are proposed to eliminate the need for water intensive washing to recover the catalyst [24]. Some examples of solid base catalysts are CaO, KF, and Eu_2O_3 on alumina or silica and ferric ion doped hydrotalcite (HTC) precursors [24]. An important aspect of these solid bases is high surface area. Table 5 compares surface areas of various heterogeneous base catalysts.

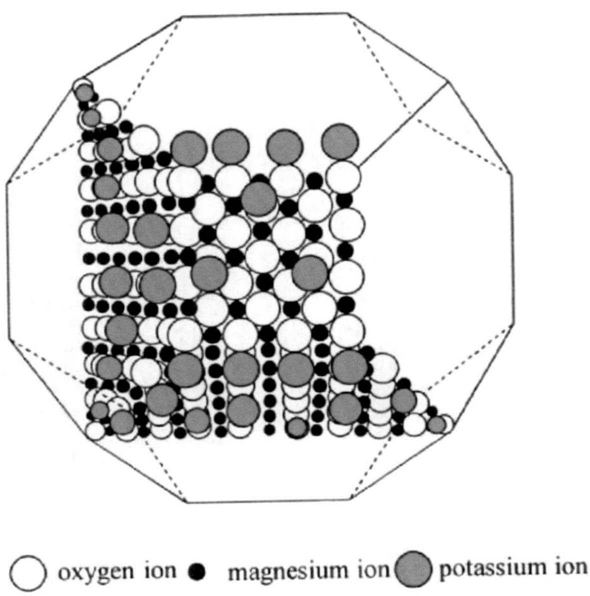

Figure 7. Schematic structure of K/AP-MgO. K+ are located at the edge/corner and adjacent sites. All particles are drawn to scale. [26]

Table 5. Surface area of the metal oxides

Catalyst	Surface Area ($m^2\ g^{-1}$)
PbO	0.55
MgO	157.4
MnO_2	50.55
BaO	0.76
CaO	61.39

Solid Super Base

According to Sun and Klabunde, "solid super bases are created when metal oxides are treated with alkali metals; for example, Na-Al_2O_3 or K-MgO." [26] The result of the treatment is highly active catalysts that are used for isomerization of alkenes or alkylation of alkenes at moderate conditions (room temperature) [26].

Solid Catalyst

Solid catalysts create a high surface area for reaction to occur upon. Since solids can be recovered easily, none of the water intensive washing needed in acid and base catalysis is required here. Washing takes long time periods to complete which makes up for the slower reaction time in solid catalysts. The glycerine product produced is also purer [1], which is important for the economics of the plant. An example of solid catalyst is mesoporous silica nano-particles [1]. There are also metal oxide based solid catalysts.

Enzymes

Enzymes are considered the environmentally friendly catalyst for this reaction. There is no by-product generation, which reduces waste. Lipase is the most commonly used enzyme catalyst [27-31], but enzymes have not been used commercially in biodiesel production to date [7]. The product is easily recovered and mild reaction conditions can be used (room temperature). Furthermore, the catalyst can easily be recycled. However, high conversions

Figure 8. Micrographs of a BSPs (a) before and (b) after cell immobilization [32].

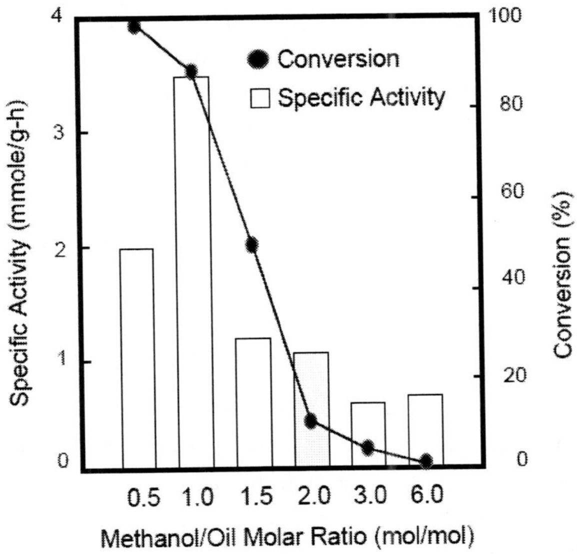

Figure 9. Effect of methanol content on methanolysis of soybean oils. The conversion was expressed as the amount of methanol consumed for the ester conversion of the oil (when the molar ratio of methanol/oil was less than 3), and as the ratio of methyl ester to the oil (more 3) [32].

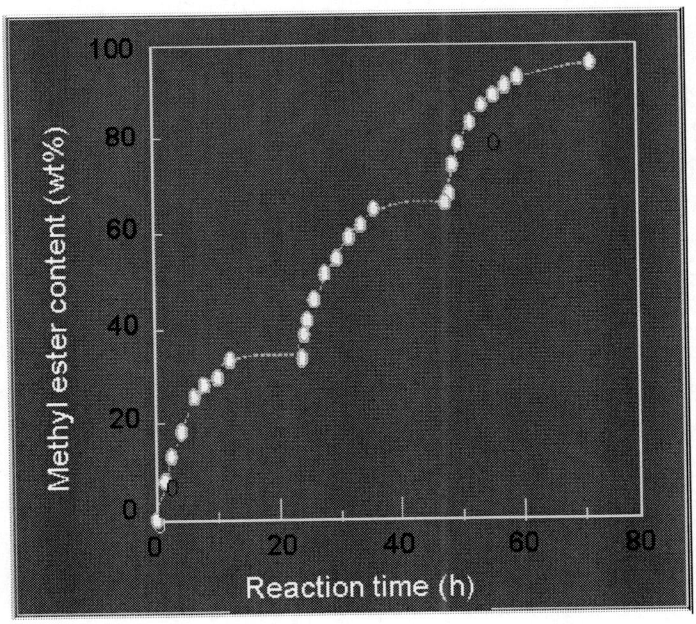

Figure 10. Time courses of ME content in a repeated methanolysis operation using immobilized-cells with and without glutaraldehyde treatment [32].

can only be achieved with organic solvents. There is also a high cost of production of enzymes and stringent controls which increases overall costs [27]. Enzymes can also be immobilized into a substrate, though would require replacement once yields decrease [19]. Lin et al (2005) demonstrated that *Rhizopus arrhizus* was employed as whole-cell catalyst and it can be immobilized in porous biomass support particles (BSPs) readily (see Figure 8). As the result presented, high concentration of methanol will inhibit the transesterification reaction and stepwise addition of methanol can avoid the cell denaturaliation. The methyl ester contain could reach about 90% within 72 hrs using cross-linked immobilizedcell with glutaraldehyde treatment as biocatalyst. The lipase activities of the cell could be maintained after 3 batch cycles. These results indicate that the use of immobilized- cell as biocatalyst shows a significant meaning and provides an alternative potential tool of biodiesel manufacture due to the simplicity of lipase preparation and long period stability of the biocatalyst [32].

3. Selection of Feedstock

The central reagent of the reaction is the actual triglyceride. As stated earlier, this can be any vegetable oil, animal fat, or waste cooking oil. While vegetable oil, or straight vegetable oil (SVO) has been used directly as fuel, it is not recommended due to the very high viscosity and boiling point. This has been shown to reduce engine lifespan and increase maintenance needs over the long term. [4, 33] These short term sources must be converted into biodiesel to avoid these issues. Currently, there are a wide range of available oils and fats that can and are used to produce biodiesel. The choice of which to use comes from scientific reasoning on their characteristics, but also on economic and social factors as well. In the United States, soybean oil is the most commonly used source, accounting for 80% of the biodiesel produced in the country. Canada uses mostly canola oil [7] and in Europe, 84% of their biodiesel feedstock is rapeseed [9]. Odor and color can also be factors in choosing feedstock, as this can affect public acceptance [19]. If biodiesel produced from tropical oils in warm climates are exported to colder climates, they would most likely require additional thinning agents [2]. There is a lot of concern over the overall sustainability of biodiesel especially in feedstock selection. Choosing vegetable based oil such as soybean and sunflower causes controversy due to the competition for food sources. Animal fats are very resource intensive to produce. Waste cooking oils (WCO) are thus receiving more positive opinions because this would otherwise be disposed of. However, WCO must be pretreated to remove non-oil substances which offset some of the reduced cost of purchase [7]. There is also not a large enough source of WCO to supply all potential biodiesel demand. Algae-oil is thus a highly regarded feedstock because it can be produced quickly to meet demand. There is also high energy content in algae compared to the other sources. Furthermore, it does not compete with farm land or water [2]. Algae are prevalent naturally in waste streams from dairy farms and food processing and sewage ponds [34]. Waste streams and sewage can themselves be used as feedstock. Angerbauer et al. propose that sewage sludge can be converted to lipids which are prime sources for biodiesel production [35]. Jatropha seed oil is another feedstock that shows promise at combating arguments against biodiesel.

Jatropha is one of many inedible vegetable oils and its use would therefore not compete with food uses [36]. Furthermore, Jatropha grows naturally in many areas that are considered "developing," including sub-Saharan Africa, India, South East Asia, and China [16]. Among its many pros are its ability to survive intertility, drought and is pest-resistant and high yield [37]. These characteristics allow it to thrive even in wastelands that exist in places like India which has 80-100 million hectares of wasteland [37]. Table 6 shows fatty acid composition of Jatropha seed oil.

3.1. Fatty Acid Content

The fatty acid content of the feedstock is one of the most important physical properties used in choosing. Different feedstock will have varying types of fatty acids that vary in terms of chain length and degree of saturation. Table 7 shows the dominant fatty acid contents of common vegetable oils. The effect of chain length and degree of saturation will be discussed later.

Additionally, when using conventional production methods, feedstocks with minimal (less than 1%) FFA content is necessary or else expensive pre-treatment must be included in process design. Free fatty acids are any fatty acids that are not part of a triglyceride. The following table shows the typical FFA content of the different types of biodiesel feedstock. As FFA content increases, cost of the feedstock decreases. For example, trap grease is very cheap, and often restaurants will pay to have them removed, but they require intensive pre-treatment.

Table 6. Fatty acid composition of crude Jatropha curcas oil

Fatty acid	Formula	Structure	Wt%
Myristic	$C_{14}H_{28}O_2$	14:0	0-0.1
Palmitic	$C_{16}H_{32}O_2$	16:0	14.1-15.3
Palmitoleic	$C_{16}H_{30}O_2$	16:1	0-1.3
Stearic	$C_{18}H_{36}O_2$	18:0	3.7-9.8
Oleic	$C_{18}H_{34}O_2$	18:1	34.3-45.8
Linoleic	$C_{18}H_{32}O_2$	18:2	29.0-44.2
Linolenic	$C_{18}H_{30}O_2$	18:3	0-0.3
Arachidic	$C_{20}H_{40}O_2$	20:0	0-0.3
Behenic	$C_{22}H_{44}O_2$	22:0	0-0.3

Table 7. Fatty acid content of various oils (wt%)

	Myristic (C14:0)	Palmitic (C16:0)	Stearic (C18:0)	Oleic (C18:1)	Linoleic (C18:2)
Safflower		6.4-7.0	2.4-29	9.7-13.8	75.3-80.5
Canola		4-5	1-2	55-63	20-31
Coconut	13-18.5	7.5-10.5	1-3	5-8.2	1-2.6
Olive	1.3	7-18.3	1.4-3.3	55.5-84.5	4-19
Palm	0.6-2.4	32-46.3	4-6.3	37-53	6-12
Peanut	0.5	6-12.5	2.5-6	37-61	13-41
Rapeseed	1.5	1-4.7	1-3.5	13-38	9.5-22
Soybean		2.3-11	2.4-6	22-30.8	49-53
Sunflower		3.5-6.5	1.3-5.6	14-43	44-68.7
Tallow	3-6	25-37	14-29	26-50	1-2.5
Recycle oil	1-3	13-25	5-12	43-52	7-22

Table 8. Free fatty acid content in biodiesel feedstock

Biodiesel Feedstock	Common FFA Content
Refined vegetable oils	<0.05%
Crude vegetable oil	0.3 – 0.7%
Restaurant waste grease	2 – 7%
Animal fat	5 – 30%
Trap Grease	40 – 100%

3.2. Viscosity

Viscosity is the, "Measure of the internal friction or resistance of an oil to flow....The most common method for designation of viscosity is kinematic viscosity..." [39] Biodiesel is typically much more viscous than petrodiesel. Diesel has a kinematic viscosity of 1.3-4.1 at 40°C while biodiesel is 4.0 to 6.0. Viscosity is an important chemical property to observe because too high a viscosity can damage the engine over long periods of time, while too low can result in power loss due to leakage [4]. Biodiesel should never reach the minimum viscosity of 1.9, so the latter should not be an issue [4].

3.3. Flash Point

Flash point is, "the lowest temperature at which a liquid will generate sufficient vapor to flash (ignite) when exposed to a source of ignition." [39] For biodiesel, the flash point standard is set well above that for petro-diesel. This is for fire safety reasons. Since methanol is used in the manufacturing process, and can reduce the flash point of biodiesel significantly if even trace residue is left, it is recommended to have a flash point of 150°C to ensure that all the methanol is burned off. Biodiesel's flash point ranges from 100 to 170°C. The minimum flash point for petrodiesel is only 70°C [4].

3.4. Cold Flow Properties: Cloud Point and Pour Point

One of the downsides of biodiesel is its high cloud point and pour point. This means that in cold temperature conditions, biodiesel will begin to solidify or gel at warmer temperatures than diesel. Therefore, the lower the cloud point, the better. Cloud point is, "the temperature at which small solid crystals are first visually observed as the fuel is cooled." Fuels can often continue to be used below the cloud point, but some can quickly reach the pour point [4]. Pour point is, "the temperature at which the fuel contains so many agglomerated crystals it is essentially a gel and will no longer flow." [4] Table 9 shows the different cloud and pour points of biodiesel derived from different feedstocks. Soybean derived biodiesel, the most common in the US has the highest cloud point of those shown.

Table 9. Physical characterisations of biodiesel from different feedstocks

Feedstock	Iodine value	Cetane number	Heat value (kJ kg^{-1})	Viscosity mm^2 s^{-1}	Cloud point(°C)	Pour point(°C)
Diesel	6-12	47	45343	2.7	-15.0	-33.0
Tallow	35-48	-	40054	51.15	-	-
Sunflower	110-143	37.1	39575	37.1	7.2	-15.0
Soybean	117-143	37.9	39623	32.6	-3.9	-12.2
Safflower	126-152	41.3	38519	31.3	18.3	-6.7
Rapeseed	94-120	37.6	39709	37.0	-3.9	-31.7
Peanut	80-106	41.8	39782	39.6	12.8	-6.7
Palm	35-61	42	-	-	-	-
Cotton	90-119	41.8	39468	33.5	1.7	-15.0
Recycle oil	-	46.5	41132	38.5	12	2

3.5. Chain Length and Degree of Saturation

While diesel fuel typically contains hundreds of compounds, biodiesel, regardless of the feedstock, has similar chemistry. As discussed earlier, the feedstock oil or fat consists of triglycerides. The triglyceride has a glycerin backbone of three carbons. Attached to each carbon is a long chain fatty acid. The variation comes in the length and saturation of the long chain. In the common types of fatty acids in the various types of oils and fats used for biodiesel, the chains range from 12 to 22 carbons in length. The degree of saturation refers to the amount of hydrogens bonded to each carbon along the chain. A saturated chain has the maximum number of hydrogens on the carbons. Therefore, the amount of unsaturation is shown by less than maximum number of hydrogen which results in double bonds. One double bond is considered monounsaturated and two or more double bonds on one chain are polyunsaturated. Fatty acids are often characterized by the number of carbon to double bonds. For instance, 16:1 indicates a carbon chain length of 16 and 1 double bond (monounsaturated). Different feedstocks are made of differing combinations of saturated, monounsaturated, and polyunsaturated fatty acids. According to the Department of Energy, the "'perfect' biodiesel would be made only from monounsaturated fatty acids." [1] Figure 11 shows the compositions of common feedstocks.

Table 10. Fuel properties as a function of fuel composition in diesel engines [4]

	Saturated	Monounsaturated	Polyunsaturated
Fatty acid	12:0, 14:0, 16:0, 18:0, 20:0, 22:0	16:1, 18:1, 20:1, 22:1	18:2, 18:3
Cetane Number	High	Medium	Low
Cloud Point	High	Medium	Low
Stability	High	Medium	Low

As shown by Table 10, while it would be ideal to have a moderate fuel composed of monounsaturated fatty acids, the actual composition of oils is not this simple. It is therefore important to observe the climate and conditions of where the biodiesel would be used. Colder regions would have to sacrifice higher cetane numbers in order to achieve lower cloud points and vice versa [4]. Coconut oil derived biodiesel may be fine in regions near the equator, but would do very poorly in Canada. Soybean oil, commonly used in the United States, is mostly polyunsaturated.

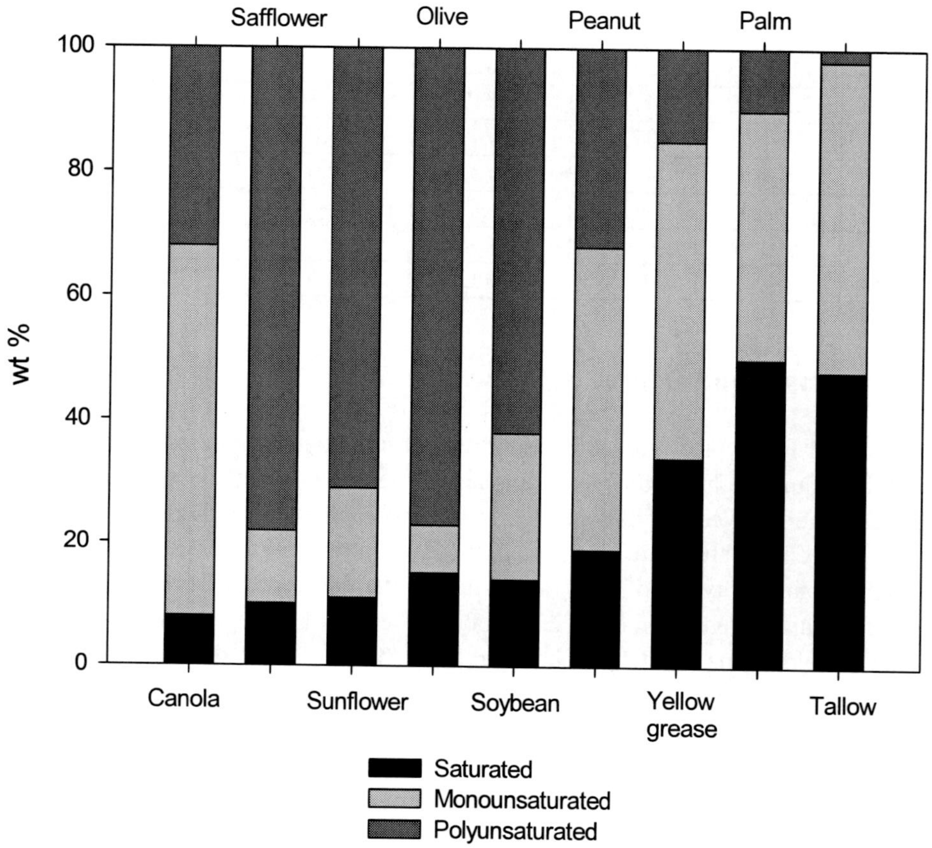

Figure 11. Composition of various biodiesel feedstocks.

Stability of the fuel is also affected by degree of saturation. Stability decreases by a factor of ten at each increase of un-saturation. This is because reactions with oxygen can occur at the double bond sites, which forms peroxides. Peroxides further break down into acids, sediments, and gums which are harmful to the fuel containers [4]. As alluded to in environmental considerations, longer chain lengths showed decreased NOx emissions, while increasing unsaturation resulted in higher NOx emissions compared to diesel fuel (see Figure 12).

Work by Soriano, et al. show the various implications of chain length and un-saturation on the physical properties discussed earlier. Their explanation is that the double bonds disrupt the attractive forces in the hydrocarbon chain, causing the physical properties to decrease [40].

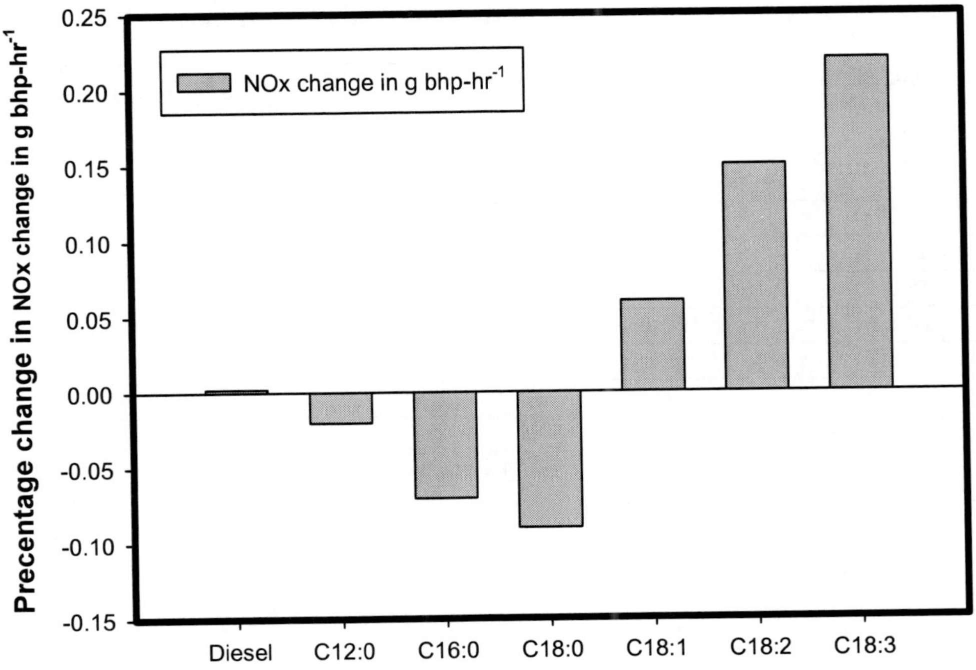

Figure 12. NOx emissions of B100 made from single types of fatty acids. [4]

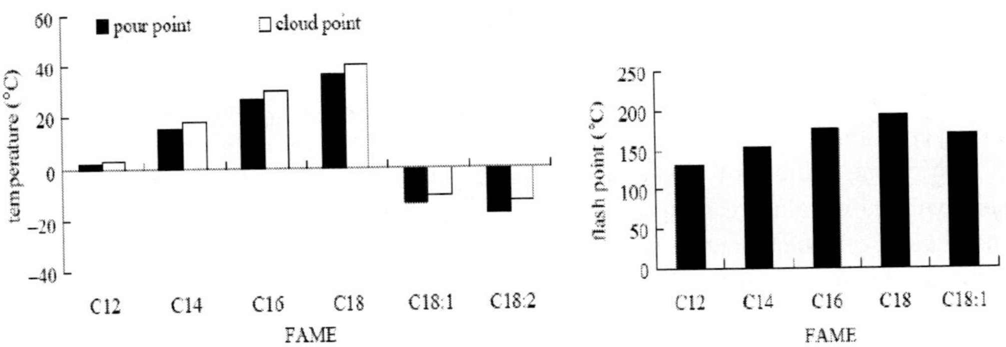

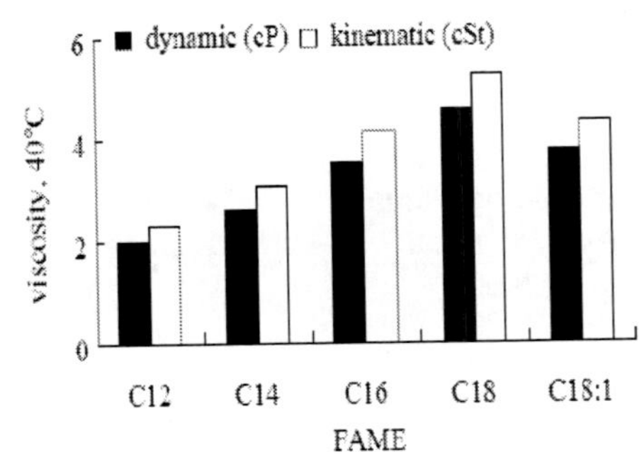

Properties	Equation of the line*	r^2
Kinematic viscosity	y = 0.50x – 3.87	0.9913
Dynamic viscosity	y = 0.43x – 3.31	0.9913
Flash point	y = 10.65 – 2.50	0.9920
Pour point	y = 5.7x – 65.5	0.9936
Cloud point	y = 6x – 67	0.9945

*Regression for saturated FAME only (C12:0-C18:0)

Figure 13. Effect of chain length and unsaturation on pour point, cloud point, flash point, and viscosity of FAME [40].

Table 11. Reported values of the cetane number for biodiesel

Soybean Methyl Ester	Rapeseed Methyl Ester	Palm Methyl Ester	Tallow Mthyl Ester
45.0(1)	51.9(9)	54.0(4)	58.0(17)
46.2(2)	48.0(10)	54.0(8)	62.9(13)
54.7(3)	54.4(11,12)		
67.0(4)	49.9(13)		
45.0(5)	54.5(14)		
54.8(6)	54-65(15)		
60.0(7)	61.2(6)		
51.9(8)	61.8(16)		
48.6(8)			

(1)Reed, 1993 [48]; (2)Wagner et al., 1984 [49]; (3)McDonald et al., 1995[50]; (4)Pischinger et al., 1982[51]; (5)Peterson et al., 1994[52]; (6)Sharp, 1994[53]; (7)Midwest Biofuels, 1993 [54]; (8)Marchetti et al., (in press)[55]; (9)Rakopoulos et al., 2006[56]; (10)McCormick, 2001[57]; (11)Van Gerpen et al., 2004[58]; (12)Wang et al., 2000[59]; (13)Kinast, 2003[60]; (14)Gragg, 1994[61]; (15)Peterson et al., 1993[62]; (16). Mittlebach et al., 1985[63]; (17) Schwab et al., 2000[64]

3.6. Cetane Number

Cetane number is defined as, "Number equal to the percentage by volume of cetane added to basic diesel fuel to achieve specific ignition performance characteristics." [24]. The cetane number is one of the most commonly cited indicators of diesel fuel quality. It measures the readiness of the fuel to autoignite when injected into the engine. It is generally dependent on the composition of the fuel and can impact the engine's startability, noise level, and exhaust emissions.

The cetane number of biodiesel is generally observed to be quite high. Data presented below will show values varying between 45 and 67. In the United States, No. 2 diesel fuel usually has a cetane number between 40 and 45. Table 11 shows the range of reported values for the cetane number of four different types of biodiesel. The range of values for SME varies from 45.0 to 67.0. The numbers in the brackets are the references from which the cetane numbers were taken.

Table 12. Energy content in diesel and biodiesel

Average Density and Heating Value of Biodiesel and Diesel Fuel

Fuel	Density, g/cm3	Net Heating Value Avg., Btu/gal.	% Difference vs. No. 2 Diesel Avg.
No. 2 Diesel	0.850	129,500	
Biodiesel (B100)	0.880	118,296	8.65 %
B20 Blend (B20)	0.856*	127,259*	1.73 %*
B2 Blend (B2)	0.851*	129,276*	0.17 %*

* Calculated Values from those of No. 2 Diesel and Biodiesel (B100)

Figure 14 shows a comparison of cetane numbers of various FAMES to diesel. While they vary slightly for different FAMEs, they are all considerably higher than diesel's cetane number.

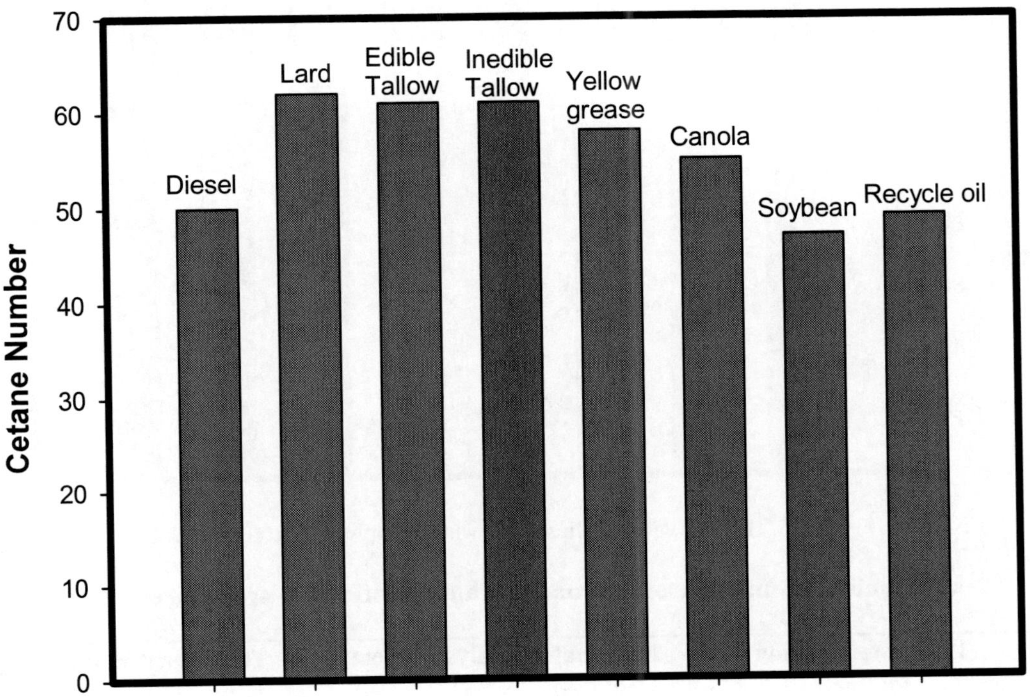

Figure 14. Cetane number of FAMEs, petroleum diesel and various biodiesel fuels.

3.7. Energy Content

The energy content (also referred to as heating value) of diesel fuel is its heat of combustion; the heat released when a known quantity of fuel is burned under specific conditions. Biodiesel has less energy density than diesel. Biodiesel is denser than diesel as well [4]. This relates to slightly lower fuel economy. A common blending fuel is B20, using 20% biodiesel mixed in 80% of diesel. At this mixing, the differences are only 1-2%, and continue to decrease with lower mixes [4].

As shown in Table 12 and Figure 15, while biodiesel does have lower energy content or heating value, regardless of the feedstock, they are very similar. This allows for better customer assurance of standards and less need for variation of production parameters to achieve ASTM specifications.

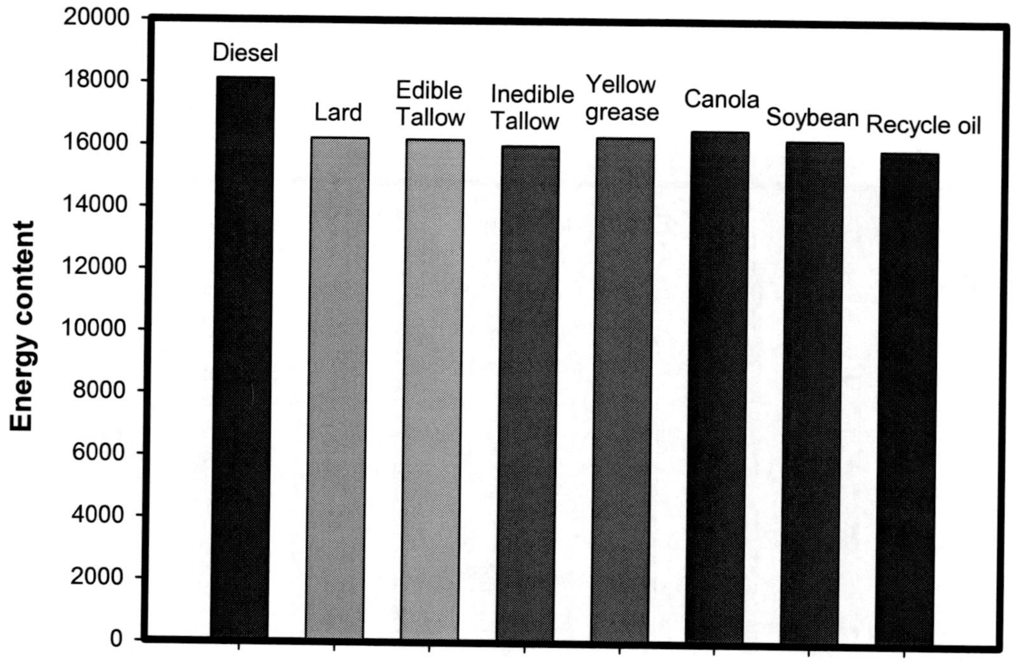

Figure 15. Heating value of diesel and varied biodiesel (B100) fuels.

Table 13. Summary of alternative transesterification procedures

Transesterification Methods	Maximal yield (%)	Catalyst (%)	Temp. (°C)	Time (minute)	Alcohol:oil ratio
Solid super base catalyst	93	1.5	70	150	9:1
In situ (without oil extraction step)	87	1.0	60	60	-
Supercritical alcohols	99	0	200-250	40	50:1
Enzymatically in the presence of supercritical CO_2	70	0	45	480	5:1

4. Manufacturing Art

There are many methods of producing biodiesel. In this section, a few of these methods are described in detail. Other methods are often combinations or slight deviations from the methods described. As an example, Table 13 includes a summary of a variety of methods to produce biodiesel. Most biodiesel production facilities utilize the conventional methods of a batch or continuous reactor process. These processes utilize the classic ingredients discussed above. Typically, the proportions used are:

- Reactants: Fat or Oil (e.g. 100 kg soybean oil)
- Primary Alcohol (e.g. 10 kg methanol)
- Catalyst: Mineral Base (e.g. 0.3 kg sodium hydroxide)
- Neutralizer: Mineral Acid (e.g. 0.25 kg sulfuric acid)

Pre-treatment is necessary to reduce free fatty acid content in feedstock [20]. This is because FFAs react with the base catalyst to produce soaps and water [20]. The typical method for pretreatment is esterification of the FFA with methanol in the presence of sulfuric acid [20].

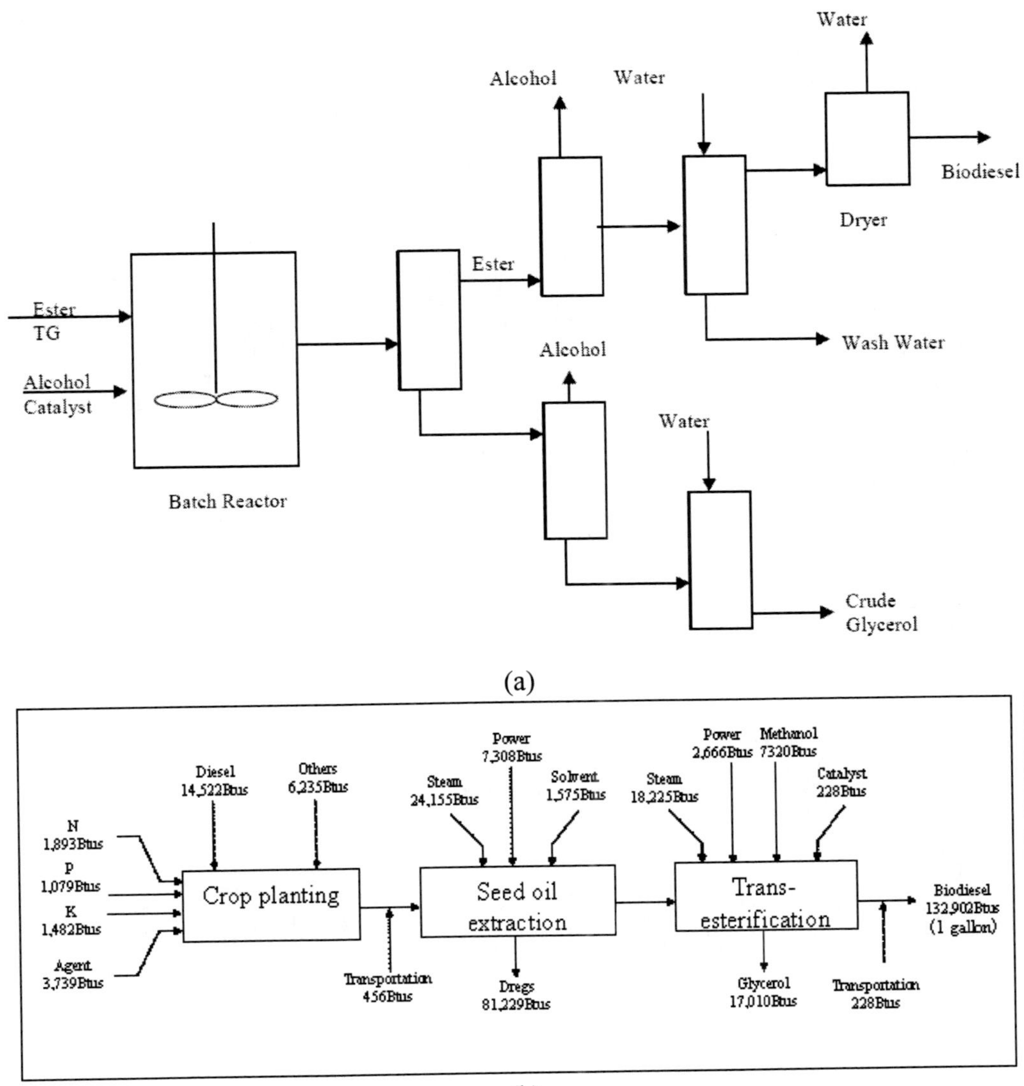

(a)

(b)

Figure 16. Batch process design for biodiesel production.

Batch reactor systems are beneficial for small scale production. This can be scaled up however by running multiple batch reactors in a cycle that simulates continuous processing to the downstream steps. Mixing is very important in the batch reactor process. Early mixing is important to enhance reaction between the immiscible reagents, and can be decreased as the reaction progresses. Figure 16(a) displays the typical batch reaction process The triglyceride is charged first; then alcohol and catalyst are added with thorough mixing during the reaction as detailed in Table 14. After the reaction, the mixture can be allowed to settle to separate some of the esters from the glycerol, or can be separated using another unit such as a settling vessel or centrifuge. Both the glycerol and ester streams must have the alcohol removed in an evaporator or flash unit. The ester stream is then neutralized and washed to take out residual methanol and salts before going through a dryer to remove the water. The remainder is sold as biodiesel. The glycerol stream is also washed and neutralized then sent to another refining section to improve the quality of glycerol able to be sold. If a triglyceride high in FFA is used, pretreatment equipment must be added to the process prior to the batch reactor. Figure 16(b) indicated a positive impact upon the overall energy consumed/produced ratio for whole processing.

Table 14. Typical specifications for a batch reaction process

Condition	Reported Range	Most common
Alcohol:triglyceride mole ratio	4:1 to 20:1	6:1
Operating temperature	25 to 85°C	65°C
Catalyst	NaOH, KOH	NaOH
Catalyst Loading	0.3% to 1.5%	
Reaction Extent	85% to 94%, or 95%+ with 2 step	
Reaction Time	20min to >1hr	

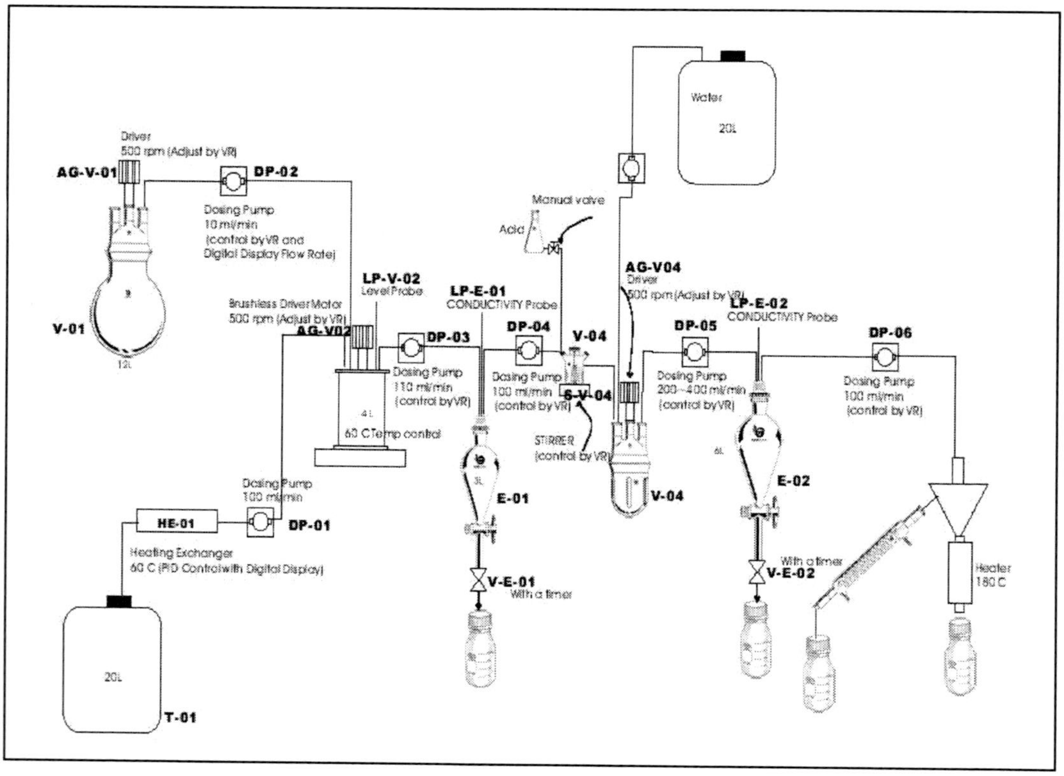

Figure 17. Continuous stirred tank reactors system [32].

For continuous reactor process design, the batch reactor design has been updated to become a continuous process, which allows for better heat economization, enhanced product purity, better recovery of excess methanol, less labour requirements for adjusting plant parameters, and lower per unit capital costs [9]. Most large scale commercial plants use continuous processing [11]. Continuous processes can be in two forms, using continuous stirred tank reactors (CSTRs) in series, or using a plug-flow reactor (PFR) as shown in Figure 17. Important in the CSTR process is ensuring that inputs into each CSTR is well mixed. PFR processes apply intense mixing inside the reactor to achieve a faster reaction time [7].

Table 15. Product information from continuous process [9]

Biodiesel Yield*	>98%
Untreated Crude Glycerine	~130kg of 75-80% glycerine/metric ton product
Fatty matter	<10 kg
Methanol stream	99.9% methanol (recycled)
Water stream (0.2% methanol)	nil

For high Free Fatty Acid Systems, an alternative to the acid catalyzed pre-treatment in the conventional method is required. A caustic is added to the feedstock that creates soaps which are removed using a centrifuge. This is called "caustic stripping." [7]

The Biox Process was nearing commercialization in 2004 [7]. It utilises tetrahydrofuran (THF) as a cosolvent to improve the miscibility of methanol in the triglyceride. Since THF has a boiling point near methanol, both can be recovered in the same step [7]. A recently developed process for producing biodiesel is the *McGyan Process*. The inventors propose that this process eliminates many of the downsides of the typical batch reactor process [42]. This process combines the alcohol and feedstock at high pressure and temperature into a continuous fixed bed reactor. The reactor is filled with solid metal oxide catalyst. Both transesterfication and esterification occur simultaneously in this reactor, so feedstock with any percentage of FFA can be used [43]. After reaction, the product stream is cooled in a heat exchanger that takes the heat to increase the temperature in the feed. The product stream is then put through a polishing unit called the Easy Fatty Acid Removal (E-FAR) system [18]. The E-FAR contains aluminum oxide particles that remove residual FFAs. Nitrogen gas is passed through to combust any FFAs in the particles, and this gas is recovered in order to recycle the residue collected [18]. The catalyst was tested to be able to run for 4000 minutes (67 hours) of continuous operation before conversion began to decrease [18]. The main equipment required for this process is a specially designed heat exchanger, a fixed-bed reactor, an E-FAR, rotary evaporator to recover alcohol, and distillation columns for separation.

The supercritical methanol method has similar benefits to the Mcgyan Process, but does not require a catalyst. The table below shows a comparison of the supercritical method to conventional methods. Methanol (with propane co-solvent) at supercritical conditions is mixed with the feedstock in a reactor. The supercritical methanol method can be performed as one step (Saka Process) or two-steps (Saka and Dadan Process) [44]. The comparison to conventional methods is depicted in Table 17. The downside of this is that the severe reaction conditions cause iso-merization of the FAME to their trans-form, which has poor cold-flow properties [44]. The remedy of this is the two step process which first hydrolyzes the triglycerides in sub-critical water, then esterifies the fatty acids in supercritical methanol. Other solvents can be used, and their conditions are listed in Table 18. The main equipment required for this process are heat exchangers to heat up the feed while cooling down the product, a tubular reactor, a distillation column and flash evaporator to recover the methanol and propane, and finally a settler to separate the glycerol and biodiesel.

Table 16. Typical specifications for Mcgyan process [18]

Condition	Value	Range
Alcohol:TAG mole ratio	32.7:1	
Reaction temperature	360°C	
Catalyst	Porous metal oxides	
Reaction Extent	87.5%	86-95%
Reaction time	56.9 s	5.4-59.0 s
Feedstock	Soybean oil	Any type of feedstock

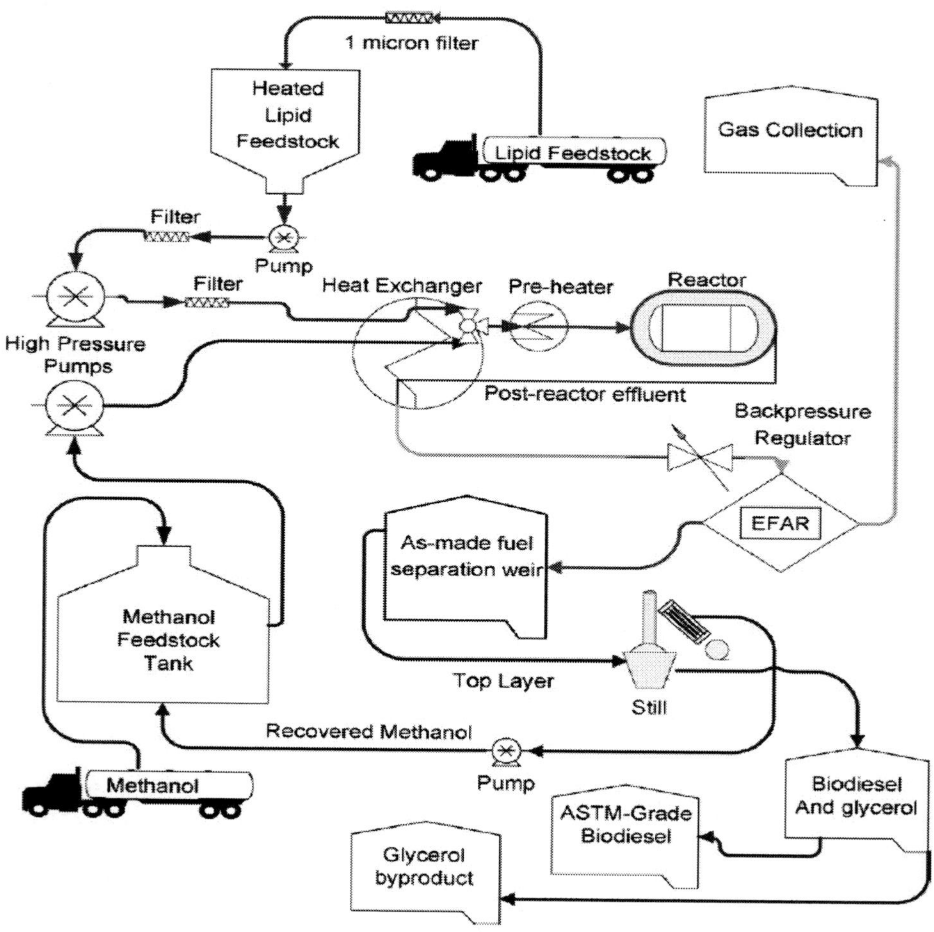

Figure 18. Process flow diagram of a biodiesel plant based on the Mcgyan process [18].

Table 17. Properties of supercritical and conventional transesterification

Properties	Supercritical	Conventional
Catalyst required	NO	YES
Reaction time	Second to minutes	Minutes to hours
Reaction Temp. (°C)	200-300	50-80
Pressure (bar)	100-200	1
Free fatty acid sensitive	NO	YES
Water sensitive	NO	YES
Pretreatment	NO	YES
Catalyst removal	NO	YES
Soap removal	NO	YES

Table 18. Critical state of various solvents studied and their pressure in batch-type reaction system for designated temperatures

Solvent	Critical temperature(°C)	Critical pressure (MPa)	Pressure		
			270°C	300°C	350°C
Methanol	239	8.09	43	20	17
Ethanol	243	6.38	25	15	-
1-Butanol	290	4.41	23	10	-
Methyl formate	214	6.00	31	-	5
Diiosprophy ether	227	2.88	12	-	5

Post-Reaction Processing

A common issue with biodiesel is its stability in oxygen environments. In the presence of oxygen, it can hydrolyze to alcohol and acid [37]. When alcohol and acid are formed, this reduces the flash point and increases acid number of the fuel [37]. A solution is to include antioxidants in the biodiesel. The two most common types are phenolic-types and aminic-types, whose mechanisms are shown in Figure 19. They basically work by interrupting the chain oxidation process [37]. The stability problem is especially pertinent to Jatropha oil based biodiesel [36, 37].

Biodiesel Quality Control

Biodiesel processing and quality are closely related. The processes used to refine the feedstock and convert it to biodiesel determine whether the fuel will meet the applicable specifications. The primary criterion for biodiesel quality is adherence to the appropriate standard. In the United States, this standard is ASTM D 6751-02 “Standard Specification for Biodiesel Fuel (B100) Blend Stock for Distillate Fuels” and EN14214 standard for European union. Generally, the fuel quality of biodiesel can be influenced by several factors:

- The quality of the feedstock.
- The fatty acid composition of the parent vegetable oil or animal fat.

- The production process and the other materials used in this process.
- Post-production parameters.

All biodiesel production facilities should be equipped with a laboratory so that the quality of the final biodiesel product can be monitored. It is also important to monitor the quality of the feedstocks. One strategy used by many producers is to draw a sample of the oil (or alcohol) from each delivery and use that sample to produce biodiesel in the laboratory. This test can be fairly rapid (1 or 2 hours) and can indicate whether serious problems are likely in the plant. Measuring feedstock quality can usually be limited to acid value and water content. To monitor the completeness of the reaction according to the total glycerol level specified in ASTM D 6751 requires the use of a gas chromatograph and a skilled operator. Large producers will find that having this equipment on-site is necessary. Commercial laboratories (i.e. Magellan Midstream Partners) are available that can analyze the samples but the cost is \$80-\$150/test and the time required may be several days. Smaller producers will need to use a more robust production process involving extra methanol and probably multiple reaction steps. Then the product quality can be monitored through periodic testing by an outside laboratory. To circumvent this, comparison to a reaction and product known to meet national standards is needed.

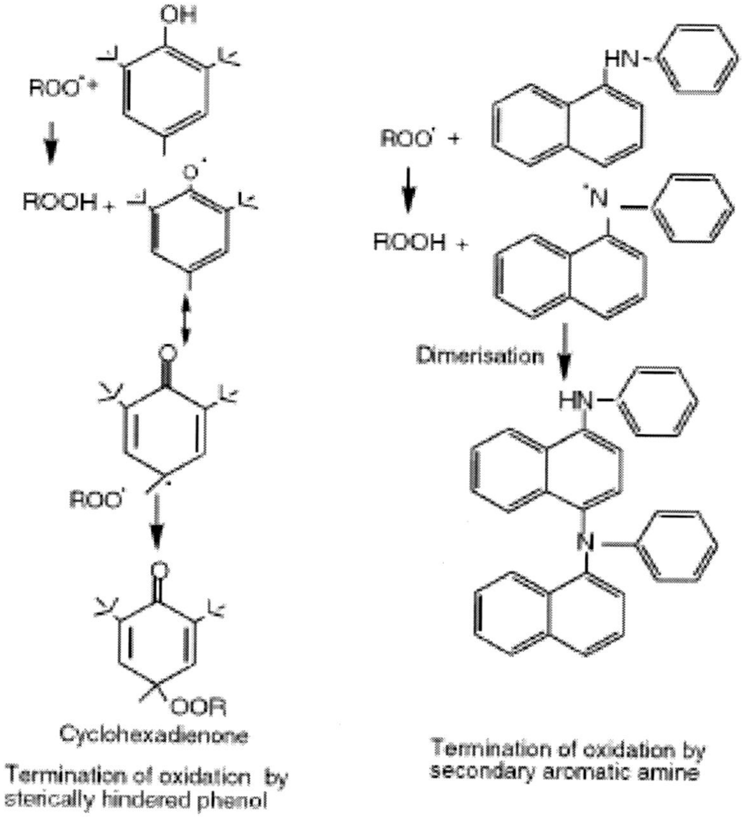

Figure 19. Mechanism of antioxidants reaction.

5. Economic Evaluation

Marchetti and Errazu compared four types of processes. They determined that a two step supercritical method was better technically and environmentally than more conventional processes, but not economically. It was proposed that a one step supercritical process with a propane co-solvent would be economically suitable. [45].

However, many studies indicated several issues must be highlighted:

- The cost structure of bio-diesel production makes it so that its competitiveness crucially depends on the price of the bio-mass feedstock and of the by-products obtained.
- Technological improvements may contribute to increase the market penetration of bio-diesel, but a mature bio-diesel sector is conditioned by a stable and cheap supply of feedstock.
- The competitiveness of bio-diesel also depends on the evolution of the fossil diesel prices.
- These three parameters are highly volatile and difficult to predict. Policies to ensure price stability should be endeavoured to favour a deeper penetration of bio-diesel crops.
- Bio-diesel is an energy carrier, whose manufacturing and production at industrial scale is totally determined by non-energy policies, i.e. agriculture policy (and, in particular, the set-aside schemes adopted), fiscal policy, etc.)
- Bio-diesel blends, i.e. mixing bio-diesel and fossil diesel, are viable way to foster a less carbonintensive automotive sector.
- However, in the long run and considering the agricultural yields, bio-diesel is not likely to supply a two-digit percentual share of the global road transport fuel needs.
- The set aside policy is a valuable instrument to introduce bio-fuels in transport market, but given the land area needed to cover the global demand, the competition with food-production crops will principally impeach their massive market penetration.

5.1. Economic Variables

Important factors for economic feasibility of a plant are raw material price, production capacity, glycerol price, and capital cost. [20, 46] These will be discussed below.

5.2. Raw Material Price

Feedstock cost is often the major factor in biodiesel production. Capital cost is only 7% of final product cost [11]. It is estimated that 88% of cost is feedstock [18]. Table 19 provides a summary detailing the percentage of importance of certain factors in manufacturing cost in the supercritical methanol method. Raw material cost is reduced by the ability to use feedstock with high percentage of FFAs as in waste cooking oils and animal tallow such as in the Mcgyan process and supercritical methanol method. At subcritical conditions, such as with the conventional method, if cheaper feedstock is used, an additional pretreatment processing step must be added, which reduces but does not completely offset the savings of cheaper feedstock.

5.3. Glycerol (Glycerine) Content

Glycerol by-product is important when looking at various processes because this can generate credit to supplement the plant's revenue [11]. Table 20 shows the percentage reduction of operating cost due to glycerol by-product sale in the supercritical methanol method; it clearly contributes significantly to the profitability of the plant. It is also important to purify glycerol as much as possible. As biodiesel production volumes increase in the future it is expected that the concomitant increase in glycerol supplies will reduce its market value. The impact of changes in the glycerol credit price on the production cost of biodiesel also was examined by Michael et al (2006) [47]. Decreases in the value of glycerol are linearly correlated with an increase in biodiesel production costs, with each US$0.01 reduction in glycerol value causing an approximately $0.008 rise in production cost. Since the amount of glycerol produced from a fixed amount of biodiesel feedstock, as well as the cost of glycerol production, purification, storage and etc. are constant irrespective of its selling price, the market value of glycerol would be expected to impact the net biodiesel production price solely in the context of a financial return at sale [47].

5.4. Energy Requirements

Mcgyan process claims to be comparable to the conventional process, requires 10 cents/gallon (at 5.5 cents kWh^{-1} energy cost) [18]. This is balanced because, while the Mcgyan process requires more strenuous operating conditions, other steps, such as water washing, are removed.

5.5. Capital Costs

Mcgyan process would require less equipment due to elimination of process chemicals and large batch reactors and water washing tanks [18]. However for a typical biodiesel production process from soybean oil at a scale of 37,000 tonnes a year, the estimated total capital cost was approximately US$11.3 million. One third of this was for actual hardware, and two thirds was based on our assumption of a construction cost roughly double the equipment costs. Of the equipment costs, nearly one third is for feedstock and product storage tanks. These were modeled at a 25 day working supply capacity. Substantial savings would accrue from reducing storage capacity, as in the case of collocating a facility at an oil production site, arranging for timely removal of product by rail, or accepting smaller inventory holding capabilities[47].

Table 19. Summary of sensitive key factors as percentage of manufacturing cost

tonnes/year	Netherland	USA	Taiwan
125,000	Waste oil (80%)	Waste oil (73%), capital(15%)	-
80,000	Waste oil (77%)	Waste oil (68%), capital(16%)	-
8,000	Waste oil(49%), capital (18%), human labour (16%)	Waste oil(35%), capital (20%), human labour (21%)	Waste oil(32%), capital (28%), human labour (18%)

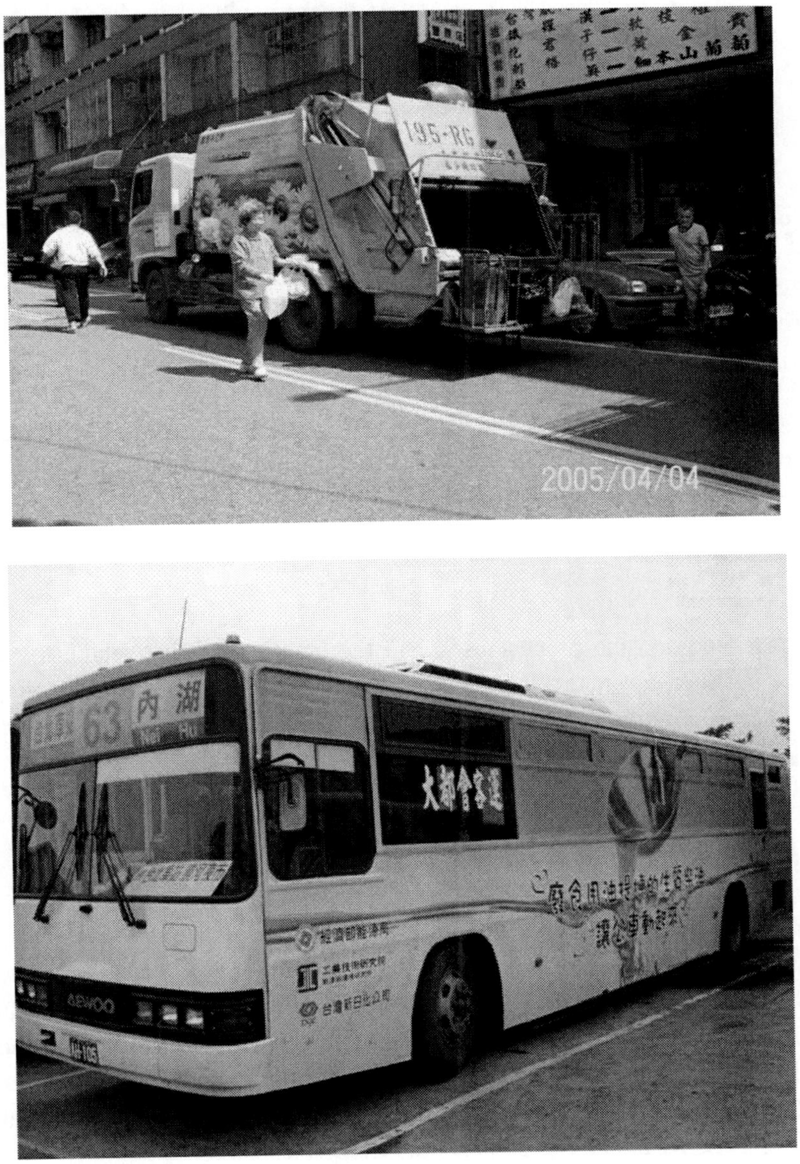

Figure 20. Biodiesel demonstration programme in Taiwan (2005-2007).

Figure 21. Effect of biodiesel fuel on pollutant emissions from diesel engines [32].

Table 20. Percentage reduction of operating cost by by-product sale [20]

Tonnes/year	Netherlands	USA
125,000	Glycerol (22%)	Glycerol (36%)
80,000	Glycerol (13%)	Glycerol (20%)
8,000	Glycerol (12%)	Glycerol (16%)

5.6. Selling Cost

Sell cost depends on the process used and the plant's capacity. Using the supercritical method, price per liter can range from \$0.52/l to only \$0.17/l depending on capacity due to economy of scale. Therefore, it is more profitable to produce from a large scale plant, assuming the demand is available. Demand can be created, such as in Europe.

6. Environmental Protection

Scientists believe carbon dioxide is one of the main greenhouse gases contributing to global warming. Biodiesel is the first and only alternative fuel to have a complete evaluation of emission results and potential health effects submitted to the U.S.Environmental Protection Agency (EPA) under the Clean Air Act Section 211(b). Among those countries have completed the examination of exhaust emissions or demonstration, most results indicated that neat Biodiesel (100% Biodiesel) reduces carbon dioxide emissions by more than 75% over petroleum Diesel. Using a blend of 20% Biodiesel reduces carbon dioxide emissions by 15%. The government in Taiwan became the first country in far-east Asia to launch the "Biodiesel Demonstration Programme" during 2005 to 2007 (see Figure 20) and subsequently initiated the "National Biodiesel Promotion Programmes" by selling B1 at gas stations in 2008. In the mean time, an examination of emission from diesel engine by using different ratio of blending biodiesel was carried out under the support of Environmental Protection Administration, Taiwan (Figure 21). It showed that biodiesel also produces fewer particulate matter, carbon monoxide, and sulphur dioxide emissions.

In addition, when the whole life cycle analysis (LCA) of biodiesel from production of the biomass via conversion to use as an energy source illustrated a significant reduction on green house gas emission (see Figure 22). In summary, LCA is an important tool for describing environmental (dis)advantages and therefore serves in decision-making. It can be complemented by a socioeconomic assessment to ensure complete and sustainable long-term production of biodiesel.

7. Conclusion

Biodiesel (fatty acid alkyl ester) is a cleaner-burning diesel replacement fuel made from natural, renewable and biodegradable sources. It is a stable diesel which performs reliably in all diesel engines without any modification, is mixable with petroleum diesel fuel, easy to make and safe to handle. Of the several methods available for producing biodiesel, transesterification of natural oils and fats is currently the method of choice. Biodiesel has become more attractive recently because of its environmental benefits and the fact that it is made from renewable resources. The remaining challenges are its cost and limited availability of fat and oil resources. There are two aspects of the cost of biodiesel, the costs of raw material (fats and oils) and the cost of processing. The cost of raw materials accounts for 60 to 75% of the total cost of biodiesel fuel. The use of other materials such as waste cooking oil

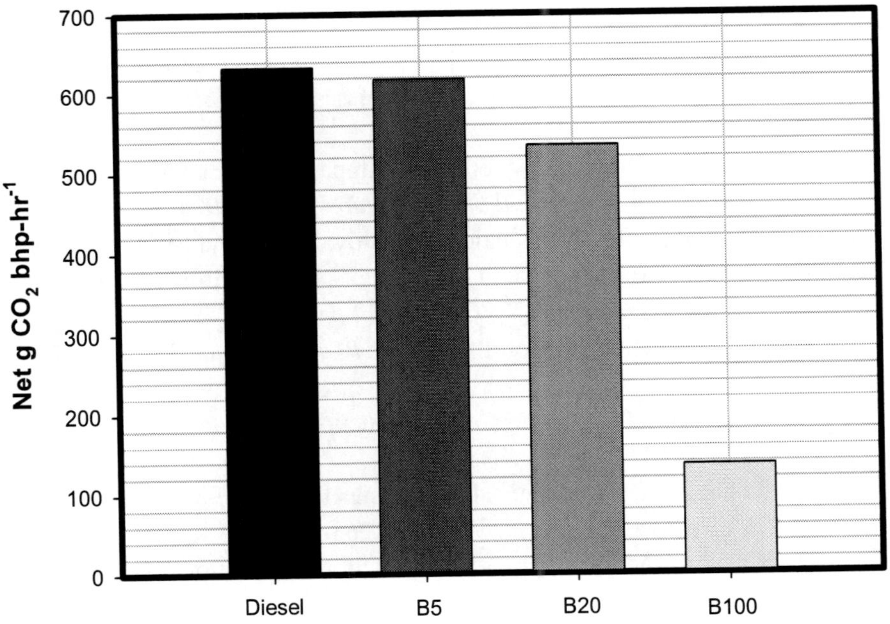

Figure 22. Comparison of Net CO_2 Life Cycle Emissions for Petroleum Diesel and Biodiesel Blends.

can lower the cost significantly. However, the quality of produced biodiesel shall be taken in concern. With the increase in global human population, more land may be needed to produce food for human consumption (indirectly via animal feed). The problem already exists in Asia. Vegetable oil prices are relatively high there. The same trend will eventually happen in the rest of the world. This is the potential challenge to biodiesel. From this point of view, biodiesel can be used most effectively as a supplement to other energy forms, therefore, is a very promising alternative fuel that can lead to a cleaner environment.

Reference

[1] Canter, N. "Catalytic approach to biodiesel." *Tribology and Lubrication Technology, January* 2008, Pages 12-13.

[2] Johnston, M., Holloway, T., "A Global Comparison of National Biodiesel Production Potentials." *Environmental Science and Technology,* Vol. 41, No. 23, 2007

[3] Marchetti, J.M., Errazu, A.F. "Technoeconomic study of supercritical biodiesel production plant." *Energy Conversion and Management* (2008), doi: 10.1016/j.enconman.2008.02.002

[4] "Biodiesel: Handling and Use Guidelines." US Department of Energy, *Energy Efficiency and Renewable Energy,* Third Edition, September 2006.

[5] Johnston, M., Holloway, T. "A global comparison of national biodiesel production potentials." *Environmental Science & Technology.* Vol. 41, No. 23, 2007. Pages 7967-7973

[6] Yang H. H., Chien S. M., Lo M. Y., Lan C. W., Lu W. C. and Ku Y. C., Effect of biodiesel on emissions of regulated air pollutants and polycyclic aromatic hydrocarbons under engine durability testing, *Atmospheric Environment,* Vol. 41, 2007. Pages 7232-7240

[7] Zhang, Y., Dube, M.A., McLean, D.D., Kates, M. "Biodiesel production from waste cooking oil: 1. Process design and technological assessment." *Bioresource Technology* **89**, 2003. Page 1-16

[8] Nunez T, "Biodiesel in Oregon:Environmental Impacts and Economic Feasibility on the Columbia Plateau." *Environmental Studies Senior Thesis,* 2004.

[9] Anderson, D., Masterson, D., McDonald, B., Sullivan, L., "Industrial Biodiesel Plant Design and Engineering: Practical Experience." Crown Iron Works Company. *International Palm Oil Conference* (PIPOC), 24-28 August 2003.

[10] Bruice, P.Y., Organic Chemistry, Fourth Edition. *Pearson Education,* Inc. 2004, Page. 690-719

[11] Van Gerpen, J., Biodiesel processing and production. *Fuel Processing Technology,* 2005. 86: Page. 1097-1107

[12] Nourredini, H., Zhu, D., 1997. Kinetics of transesterification of soybean oil. *Journal of the American Oil Chemists' Society,* v. 77, Page 1263.

[13] Darnoko, D., Cheryan, M., Kinetics of Palm Oil Esterification in a batch reactor. *Journal of the American Oil Chemists' Society,* v. 74, Page 1263-1267

[14] Ma, H., Li, S., Wang, B., Wang, R., Tian, S. "Transesterification of Rapeseed Oil for Synthesizing Biodiesel by K/KOH/γ-Al2O3 as Heterogeneous Base Catalyst." *Journal of American Oil Chemists' Society.* 2008 85: Page 263-270

[15] Gryglewicz, S., "Rapeseed oil methyl esters preparation using heterogeneous catalysts." *Bioresource Technology* **70**: 1999. Page 249-253

[16] Tamalampudi, S., Talukder, M.R., Hama, S., Numata, T., Kondo, A., Fukuda, H. "Enzymatic production of biodiesel from Jatropha Oil: A comparative study of immobilized-whole cell and commercial lipases as a biocatalyst." *Biochemical Engineering Journal* **39**: 2008, Page 185-189

[17] Lang, X., Dalai, A.K., Bakshi, N.N., Reaney, M.J., Hertz, P.B. "Preparation and characterization of biodiesels from various bio-oils." *Bioresource Technology* **80**: 2001. Page 53-62

[18] McNeff, C.V., McNeff, L.C., Yan, B.W, Nowlan, D, Rasmussen, M., Gyberg, A., Krohn, B., Fedie, R, Hoye, T. "A continuous catalytic system for biodiesel production." A*pplied Catalysis A: General:* 22 March 2008

[19] Van Gerpen, J., Shanks, B., Pruszko, R., Clements, D., Knothe, G., "Biodiesel Production Technology." *National Renewable Energy Laboratory Subcontractor Report.* July 2004

[20] Van Kasteren, JMN., Nisworo, A.P., "A Process model to estimate the cost of indiustrial scale biodiesel production from waste cooking oil by supercritical transesterification." *Resources, Conservation, and Recycling* **50**: 2007. Page 442-458

[21] Demirbas, A., Comparison of transesterification methods for production of biodiesel from vegetable oils and fats. *Energy Conversion and Management,* 2007, 49: Page . 125-130

[22] Kulkarni, M.G., Dalai, A.K. 2006. Waste cooking oil – an economical source for biodiesel: a review. *Industrial and Engineering Chemistry Research* **45**, Page. 2901-2913

[23] Mustafa Canakci, J.V.G, A Pilot Plant to Produce Biodiesel from High Free Fatty Acid Feedstocks. *American Society of Agricultural and Biological Engineers,* 2001. 46: Page. 945-954

[24] Macala, G.S., Robertson, A.W., Johnson, C.L., Day, Z.B., Lewis, R.S., White, M.G., Iretskii, Alexei, Ford, P.C. "Tranesterification Catalysts from Iron Doped Hydrotalcite-like Precursors: Solid Bases for Biodiesel Production." *Catalysis Letters* **12**:2008. Page. 205-209

[25] Singh, A.K, Fernando, S. "Reaction Kinetics of Soybean Oil Transesterification Using Heterogeneous Metal Oxide Catalysts." *Chemical Engineering Technology* 2007, 30, No. 12, Page. 1716-1720

[26] Sun, N., Kladunde, K.J. "High Activity Solid Super Base Catalysts Employing Nanocrystals of Metal Oxides: Isomerization and Alkylation Catalysts, Including Conversion of Propylene-Ethylene Mixtures to Pentenes and Heptenes." *Journal of Catalysis* **185**: 1999. Page. 506-512

[27] Hama, S., et al. "Biodiesel-fuel production in a packed-bed reactor using lipase-producing Rhizopus oryze cells immobilized with biomass support particles." *Biochemical Engineering Journal,* **34**: 2007. Page. 273-278

[28] Kuduka, HKA, Noda, H. Biodiesel fuel production by transesterification of oils. *J Bio Sci Bioeng,* **92**(5): 2001 Page. 405-416

[29] Shimada, Y., et al. Enzymatic alcoholysis for biodiesel fuel production and application of the reaction to oil processing, *Journal of Molecular Catalysis B: Enzymatic,* Vol. 17, Issues 3-5, 7 June 2002, Page.133-142

[30] Nourreddini, H., Gao, Z., Philkana, RS, Immobilized Pseudomonas cepacia lipase for biodiesel fuel production from soybean oil, *Bioresource Technology,* Vol. 96, Issue 7, May 2005, Page. 769-777

[31] Soumanou, M.M, Bornscheuer, U.T. Improvement in lipase-catalyzed synthesis of fatty acid methyl esters from sunflower oil, *Enzyme Microbial Technology,* Vol 33, Issue 1, 16 July 2003, Page. 97-103.

[32] Lin Y. H., Lan C. W., Lu W. J., Liau P. R., Chan I. H., and Chen J. P., *Biodiesel Production utilising immobilised whole cell biocatalyst,* 5th APCSEET (Asia Pacific Conference on Sustainable Energy and Environmental Technologies), Wellington, New Zeland :2005

[33] "Straight Vegetable Oil as a Diesel Fuel?" Clean Cities Fact Sheet. US Department of Energy, *Energy Efficiency and Renewable Energy.* April 2006

[34] Kiong, Errol. *"NZ Firm makes bio-diesel from sewage in world first."* New Zealand Herald. 12 May 2006

[35] Angerbauer, C., Sienbenhofer, M., Mittelback, M., Guebitz, G.M. "Conversion of sewage sludge into lipids by Lipomyces starkeyi for biodiesel production." *Bioresource Technology* **99**: 2008. Page. 3051-3056

[36] Berchmans, H.J., Hirata, S., "Biodiesel production from crude Jatropha curcas L. seed oil with a high content of free fatty acids." *Bioresource Technology* **99**: 2008. Page. 1716-1721

[37] Sarin, R., Sharma, M., Sinhary, S., Malhotra, R.K. "Jatropha-Palm biodiesel blends: An optimum mix for Asia." *Fuel* **86***:* 2007. Page. 1365-1371

[38] Goodrum, J.W. "Volatility and boiling points of biodiesel from vegetable oils and tallow." *Biomass and Bioenergy* **22**: 2002. Page. 205-211

[39] Alken-Murray Corp. *"Glossary of Petroleum Industry Common Terms and Symbols."* Updated Dec 1999. Retrieved May 2008. http://www.alken-murray.com/fuel-glossary.htm

[40] Soriano Jr., N., Migo, V., Matsumura, M. "Ozonized vegetable oil as pour point depressant for neat biodiesel." *Fuel* **85**:2006. Page. 25-31

[41] Achten, W.M.J, Verchot, L., Franken, Y.J., Mathijs, E., Singh, V.P., Aerts, R., Muys, B. "Jatrohpa biodiesel production and use." *Biomass and Bioenergy:* 2008, doi: 10.1016/j.biombioe.2008.03.003.

[42] Kytonen, Rachel. *"Isanti Welcomes biodiesel facility."* ECM-Inc. 17 October, 2007.

[43] Green Car Congress. *"Researchers Develop New Continuous Transesterification Process for Biodiesel Production."* 8 March 2008.

[44] Isayama, Y., Saka, S., "Biodiesel production by supercritical process with crude bio-methanol prepared by wood gasification." *Bioresource Technology* **99**: 2008. Page. 4775-4779.

[45] Marchetti, J.M., Errazu, A.F. "Technoeconomic study of supercritical biodiesel production plant." *Energy Conversion and Management* (2008), doi: 10.1016/j.enconman.2008.02.002.

[46] Zhang, Y., Dube, M.A., McLean, D.D., Kates, M. "Biodiesel production from waste cooking oil: 2. Economic assessment and sensitivity analysis." *Bioresource Technology* **90**: 2003. Page. 229-240

[47] Michael J.H., Andrew J.M., Winnie C.Y., Thomas A. F., "A process model to estimate biodiesel production costs" *Bioresource Technology* **97**: 2006. Page.671-678

[48] Reed T.B., "An overview of the current status of biodiesel.Proc." 1st Biomass Conf. Americas: *Energy, Environment, Agri- culture, and Industry.* August 30 - September 2. Burlington, Vermont, 1993. USA

[49] Wagner L.E., Clark S.J., and Schrock M.D., *"Effects of Soybean Oil Esters on the Performance, Lubricating Oil, and Wear of Diesel Engines." SAE Paper* No. 841385., 1984

[50] McDonald J.F., Purcell D.L., McClure B.T., and Kittelson D.B., *"Emissions Characteristics of Soy Methyl Ester Fuels in a IDI Compression Ignition Engine." SAE Paper* No. 950400., 1995.

[51] Pischinger G.M., Falcon A.M., Siekmann R.W., and Fernandes F.R., "Methylesters of plant oils as diesels fuels, either straight or in blends vegetable oil fuels." *ASAE Paper* No. 4-82., 1982.

[52] Peterson C.L., Reece D.L., Hammond B.J., Thompson J., and Beck S.M., *"Processing, characterization and performance of eight fuels from lipids." ASAEPaper* No. 946531., 1994.

[53] Sharp C.A., *"Transient Emissions Testing of Biodiesel in a DDC 6V-92TA DDEC Engine. Final Report to the National Biodiesel Board."* SWRI Report No. 6602 & 6673 Southwest Research Institute. San Antonio, TX. USA, 1994.

[54] Midwest Biofuels, Inc. *"Biodiesel Cetane Number Engine Testing Comparison to Calculated Cetane Index Number."* Final Report to the National Biodiesel Board, New York, USA. 1993.

[55] Marchetti J.M, Miguel V.U., and Errazu A.F., "Possible methods for biodiesel production." *Renewable and Sustaina-Ble Energy Reviews,* FAO Press, Rome, Italy (in press), 2007.

[56] Rakopoulos C.D., Antonopoulos K.A., Rakopoulos D.C., Hountalas D.T., and Giakoumis E.G., "Comparative performance and emissions study of a direct injection Diesel engine using blends of diesel fuel with vegetable oils or bio-diesels of various origins." *Energy Conversion and Manag.*, 2006.

[57] McCormick R., Graboski M., Alleman T., Herring A., and Shainetyson K., "Impact of biodiesel source material and chemical structure on emissions of criteria pollutants from a heavy-duty engine." *Environ. Sci. Technol.,* **35**, 2001.Page: 1742-1747.

[58] Van Gerpen J., Shanks B., Pruszko R., Clements D., and Knothe G., *"Biodiesel Analytical Methods."* NREL/SR-510-36240. 2004.

[59] Wang W.G., Lyons D.W., Clark N.N., and Gautam G., and Norton P.M., "Emissions from nine heavy trucks fueled by diesel and biodiesel blend without engine modification." *Environ. Sci. Technol.,* **34**, 2000. Page:933-939.

[60] Kinast J.A., "Production of Biodiesels from Multiple Feedstocks and Properties of Biodiesels and Biodiesel/ Diesel Blends." Final Report. *National Renewable Energy Laboratory.* 1617 Cole Boulevard Golden, CO, USA. 2003.

[61] Gragg K., "Effects of Environmentally Classified Diesel Fuels, RME and Blends of Diesel Fuels and RME on the Exhaust Emissions." *Final Report., Swedish Environ. Prot. Agency.* Bilprovning Press, Stockholm, Sweden. 1994.

[62] Peterson C.L., Haines H., and Chase C., "Rapeseed and Safflower Oils as Diesel Fuels." Proc. 1st Biomass Conf., Americas: *Energy, Environment, Agriculture, and Industry,* August 30 - September 2, Burlington, Vermont, USA. 1993.

[63] Mittelbach M., Tritthart P., and Junek H., *"Diesel fuel derived from vegetable oils." Energy in Agric.*, **4,** 1985. Page:207-215.

[64] Schwab A.W., Bagby M.O., and Freedman B., Preparation and properties of diesel fuels from vegetable oils. *Fuel*, **66**, 1987. Page:1372-1378.

In: Biochemical Engineering
Editors: F.E. Dumont and J.A. Sacco, pp. 41-74

ISBN: 978-1-60741-257-1

Chapter 2

ENZYMATIC SYNTHESIS OF ACYL ASCORBATE AND ITS FUNCTION AS A FOOD ADDITIVE

***Yoshiyuki Watanabe*[1,*] *and Shuji Adachi*[2]**

[1]Department of Biotechnology and Chemistry, School of Engineering, Kinki University, 1 Umenobe, Takaya, Higashi-Hiroshima 739-2116, Japan

[2]Division of Food Science and Biotechnology, Graduate School of Agriculture, Kyoto University, Sakyo-ku, Kyoto 606-8502, Japan

Abstract

Acyl ascorbates were synthesized through the condensation of various fatty acids with L-ascorbic acid using immobilized lipase in a water-soluble organic solvent, and their properties as food additive were examined. The optimal conditions, which were the type of organic solvent, reaction temperature, the initial concentrations of substrates and the molar ratio of fatty acid to ascorbic acid, for the enzymatic synthesis in a batch reaction were determined. The continuous production of acyl ascorbate was carried out using a continuous stirred tank reactor (CSTR) and plug flow reactor (PFR) at 50°C, and each productivity was *ca.* 6.0 x 10 for CSTR and 1.9 x 10^3 g/(L-reactor·d) for PFR for at least 11 days, respectively. The temperature dependences of the solubility of acyl ascorbate in both soybean oil and water could be expressed by the van't Hoff equation, and the dissolution enthalpy, ΔH, values for the soybean oil and water were *ca.* 20 and 90 kJ/mol, respectively, irrespective of the acyl chain length. The decomposition kinetics of saturated acyl ascorbate in an aqueous solution and air was empirically expressed by the Weibull equation, and the rate constant, k, was estimated. The activation energy, E, for the rate constant for the decomposition in both systems depended on the acyl chain length. The surface tensions of acyl ascorbates in an aqueous solution were measured by the Wilhelmy method, and the critical micelle concentration (CMC) and the residual area per molecule were calculated. The CMC values were independent of temperature but dependent on the pH. The effect of pH of aqueous phase on the stability of O/W emulsion prepared using acyl ascorbate as an emulsifier was examined, and the high stability at pHs 5 and 6 was ascribed to the largely negative surface-charge of droplets in the emulsion. The addition of saturated acyl ascorbate, whose acyl chain length was from 8 to 16, lengthened the induction period for the oxidation of linoleic acid in a

[*] E-mail address: wysyk@hiro.kindai.ac.jp. telephone to: +81-82-434-7000; fax: +81-82-434-7890; (Author to whom correspondence should be addressed).

bulk and microcapsule with maltodextrin as a wall material. The oxidative stability in bulk system increased with increasing the acyl chain length, whereas that in the microcapsule was the highest at the acyl chain length of 10. The esterification of various polyunsaturated fatty acids, such as linoleic, α- and γ-linolenic, dihomo-γ-linolenic, arachidonic, eicosapentaenoic, docosahexaenoic and conjugated linoleic acids with ascorbic acid and subsequent microencapsulation significantly improved their oxidative stability.

Keywords: Acyl ascorbate; Immobilized lipase; Continuous production; Amphiphilic food additive; Antioxidative emulsifier

Introduction

Oxidation of lipids has received much attention due to its involvement in food spoilage and the relevance of lipid peroxidation *in vivo* to membrane damage, aging, heart disease and cancer [1]. Lipid oxidation is a complicated process including the steps of initiation, propagation and termination. Antioxidants may be used at very low concentrations to lengthen the oxidation induction period [2]. They are usually phenolic compounds such as BHA (butylated hydroxyanisole), BHT (butylated hydroxytoluene), propyl gallate and tocopherols (vitamin E), and are used to inhibit free radical chain reactions. L-Ascorbic acid, which is a water-soluble compound and called as vitamin C, is also used widely as an additive in foods and cosmetics due to its strong reducing ability. Lipophilic derivatives of ascorbic acid acylated with a long-chain fatty acid such as palmitic acid are also used as additives in foods rich in lipid [3]. These esters, which are mainly synthesized by a chemical method [4], would be edible surfactants because each substrate is eaten daily as a food component. It has been reported that 6-*O*-palmytoyl ascorbate has antitumor activity and metastasis-inhibitory effects [5, 6]. The 6-*O*- and 2-*O*-monoesters of ascorbic acid and phosphate or fatty acid have physiological activity equivalent to unmodified vitamin C [7]. It has been recognized that medium-chain fatty acids or their glycerides enhance the absorption of hydrophilic substances in the intestine [8]. Therefore, it is expected that an ester of ascorbic acid and medium-chain fatty acid itself could be easily absorbed in the intestine with the physiological activity of vitamin C or that the ester could facilitate the absorption of hydrophilic substances. In addition, not only medium-chain fatty acid but also polyunsaturated fatty acid may be used as a substrate for the synthesis of acyl ascorbate. Polyunsaturated fatty acid has important physiological functions such as antithrombotic, cholesterol depressant and antiallergenic properties [9, 10]. For example, arachidonic acid is one of n-6 polyunsaturated fatty acids and the precursor of prostaglandins, leukotrienes and related compounds, which play important roles in inflammation and in the regulation of immunity [11]. However, it is well known that polyunsaturated fatty acid, especially high unsaturated fatty acid, is extremely susceptible to the oxidation [12]. The oxidative stability of polyunsaturated fatty acid may be improved by esterification of the acid with ascorbic acid.

Acyl ascorbate consists of ascorbyl moiety as a hydrophilic group and acyl residue as a hydrophobic group, and therefore it seems to be promising emulsifiers with both surface-activity and reductivity. However, 6-*O*-palmitoyl and stearoyl ascorbates, which are commercially available, are practically so insoluble in water that they would be difficult to use as an emulsifier for preparation of O/W emulsions. Acyl ascorbates with shorter acyl chains would be more amphiphilic than palmitoyl and stearoyl ascorbates. The ascorbates with short acyl chains would be more water-soluble and have more potential as an emulsifier for preparing O/W emulsions than those with long acyl chains.

Enzymatic preparation of the esters would be more advantageous than a chemical method because of the simplicity of its reaction process and its high regioselectivity. Although the enzymatic esterification of ascorbic acid and lauric acid in an aqueous system has been reported, the yield was low because of the high water content in the system [13]. The synthesis of esters of sugar and fatty acids by lipase in organic solvents has been extensively studied [14-17]. 6-*O*-Palmitoyl ascorbate has been synthesized in an organic solvent using lipase through the reaction shown in Figure 1 [18-23]. We also synthesized various 6-*O*-acyl ascorbates through condensation of ascorbic acid with medium-chain and polyunsaturated fatty acids using lipases from various origins in a water-miscible organic solvent with low water content, and examined their function as a food additive, such as antioxidant and emulsifier [24-37]. Fatty acids, which we used for the synthesis of acyl ascorbate, are shown in Figure 2.

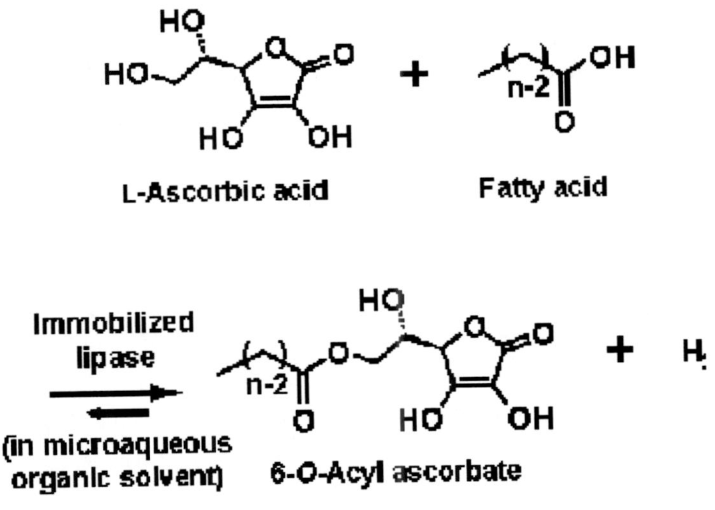

Figure 1. Scheme of condensation reaction of ascorbic acid with fatty acid using lipases from various origins in a microaqueous organic solvent with low water content.

In this chapter, our researches for enzymatic synthesis of 6-*O*-acyl ascorbates and their functions as food additives are described. At first, the optimal conditions for enzymatic synthesis of acyl ascorbate using an immobilized lipase in a batch reaction system are found in Section 1. Second, acyl ascorbate is produced continuously using a CSTR or PFR (Section 2). For the synthesized ascorbates, their solubility in water and oil and stabilities in an aqueous solution and air are measured under various conditions (Sections 3 and 4). Further, emulsifier property and antioxidative ability of acyl ascorbate against a lipid in a bulk system are evaluated (Sections 5 and 6). At last, acyl ascorbate is applied for microencapsulation of a lipid to demonstrate the example for the effective use of acyl ascorbate (Section 7).

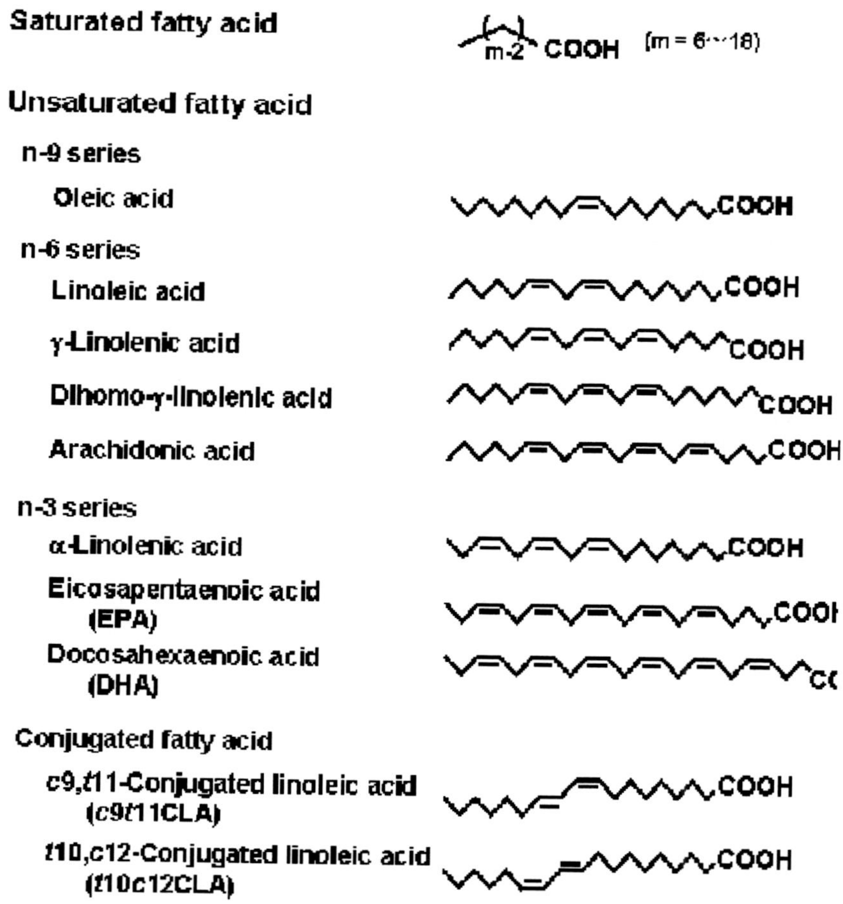

Figure 2. Molecular structures of fatty acids used for the synthesis of acyl ascorbate.

1. Optimal Conditions for Enzymatic Synthesis of Acyl Ascorbate in a Batch Reaction

The condensation of L-ascorbic acid and lauric acid was carried out in dehydrated acetonitrile using various lipases from different origins, such as Chirazyme®L-2 C2 from *Candida antarctica* (Roche Molecular Biochemicals, Germany), which was the same as Novozym®435 (Novo Nordisk Bioindustry Japan, Japan) and immobilized on a macroporous acrylic resin, lipases PS from *Pseudomonas cepacia*, A from *Aspergillus niger*, F from *Rhizopus oryzae* and M from *Mucor javanicus* (Amano Enzyme, Japan), lipases PL and QL from *Alcaligenes* sp., AL from *Achrobacter* sp., OF from *Candida cylindracea* and MY from *C. cylindracea nov.* sp. (Meito Sangyo, Japan), lipase type VII from *C. cylindracea* (Sigma Chemical, USA), to find the lipase which could catalyze the condensation in the medium [24]. Acetonitrile was dehydrated by adding molecular sieves 5A at a concentration of 0.2 g/mL-solvent. Twenty milliliters of the dehydrated solvent, 1 mmol of ascorbic acid and 5 mmol of a fatty acid were put into a glass vial with a screw-cap, then 400 mg of lipase was added. The vial was tightly sealed and immersed in a water-bath. The condensation reaction was carried out at 60°C with vigorous shaking. At appropriate intervals, a portion of the reaction mixture was sampled and used for HPLC analysis. Among the lipases tested, Chirazyme®L-2 C2 produced a sufficient amount of product and lipase PS made a trace amount of product during 24h-reaction. The other lipases did not produce any product. The two methods to synthesize acyl ascorbate are condensation and transfer reaction. For the transfer reaction, methyl laurate was used as the acyl donor, and the conversion was calculated based on the amount of ascorbic acid added, which was the limiting reactant. The conversion during condensation was much higher than that for the transfer reaction. For the synthesis of 6-*O*-palmitoyl ascorbate in 2-methyl-2-butanol by Chirazyme®L-2 C2, the transfer reaction was superior to condensation [18]. Although the reason for the difference is not clear, a possible one might be a difference in water content of medium between their studies and ours. The gradual decrease in conversion was observed at prolonged reaction times for both the condensation and the transfer reaction. This would be due to the degradation of the product to the compounds corresponding to dehydroascorbic acid and 2,3-diketo-gulonic acid [38]. Therefore, too long a reaction time would be undesirable for the production of acyl ascorbate. The condensation of ascorbic acid and lauric acid was carried out at various temperatures. The reaction rate was faster at higher temperature. The conversion leveled off at lower value at 70°C than at 55 and 60°C. This low conversion seemed ascribable to denaturation of the enzyme at that temperature. Therefore, condensation at 60°C was adopted.

The transient changes in the conversion of conjugated linoleoyl ascorbate in various organic solvents, such as acetone, acetonitrile, 2-methyl-2-propanol and 2-methyl-2-butanol, were measured [37]. The conversion was high in acetonitrile, acetone and 2-methyl-2-propanol, and that in 2-methyl-2-butanol followed. It seemed that the conversion was in proportion to the polarity of the solvent used. We studied the equilibrium constant based on the concentrations, K_C, for lipase-catalyzed condensation of mannose and lauric acid in water-miscible solvents [17, 39]. The K_C value correlated with the relative dielectric constant of the solvent, because the solubility of hydrophilic substrate such as ascorbic acid was higher in the solvent with the higher relative dielectric constant. Furthermore, the equilibrium constant for the formation of fatty acid butyl esters by the lipase-catalyzed condensation depended on the type of solvent polar group [40]. Both the polarity and polar group of the solvent would also affect the productivity for the lipase-catalyzed condensation of conjugated linoleic acid with ascorbic acid, because of the stronger interaction between conjugated linoleic acid and the solvent than that between conjugated linoleoyl ascorbate and the solvent.

The condensation of eicosapentaenoic acid and ascorbic acid was carried out at the various molar ratios in acetone dehydrated with molecular sieves 5A, the initial water content of which was 0.011% (v/v) [25]. The molar ratios were varied in two ways: 1) the amount of eicosapentaenoic acid was fixed at 0.124 mmol, and 2) the amount of ascorbic acid was fixed at 0.125 mmol. Figure 3(a) shows the transient changes in the conversion of the desired product. The experimental points were empirically connected to form a curve for each run, as shown by the solid curve in Figure 3(a). The initial reaction rate and the apparent maximum conversion were read from the curve, and are plotted against the molar ratio of eicosapentaenoic acid to ascorbic acid in Figure 3(b). The apparent maximum conversion was higher at the higher molar ratio. When the amount of ascorbic acid was fixed and that of eicosapentaenoic acid was changed (open circle), the initial reaction rate increased as the ratio increased. In another case where the amount of eicosapentaenoic acid was fixed and that of ascorbic acid was changed (open triangle), the reaction rate became slower at the higher ratios due to the decrease in the initial ascorbic acid concentration. The relationships between the maximum conversion and the molar ratio for the condensations of ascorbic acid and saturated fatty acids with various chain lengths from 6 to 12 were examined [24]. There was no significant difference in the relationship among the fatty acids tested. The maximum conversion was higher at the higher molar ratio. Since high conversion and a lower amount of unreacted fatty acid were favorable for the purification process, the molar ratio of 5 would be the most appropriate for production of acyl ascorbate in the present reaction system.

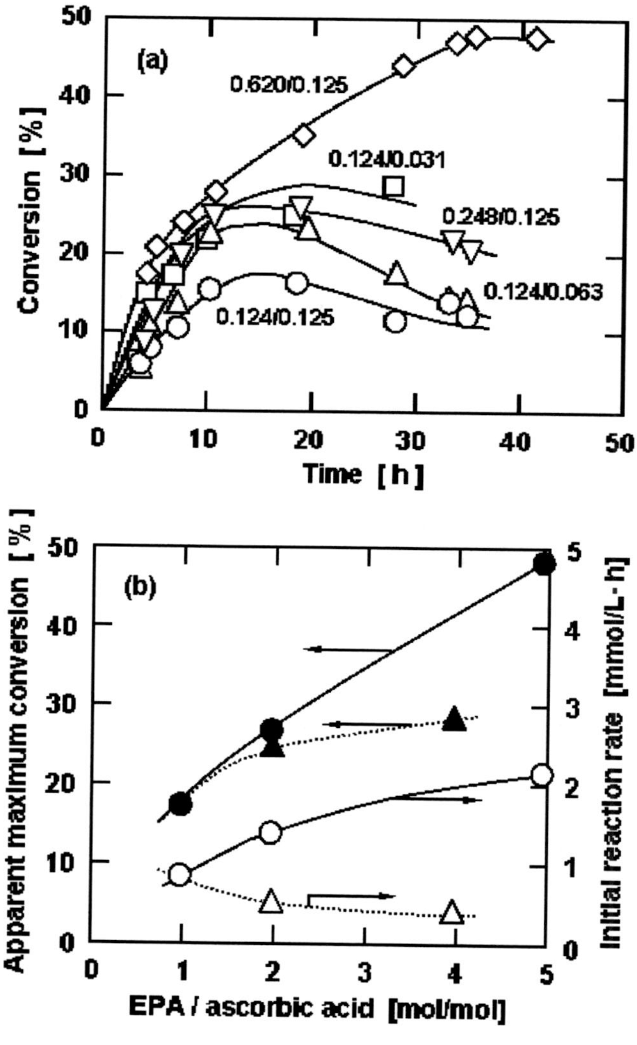

Figure 3. Lipase-catalyzed condensation of eicosapentaenoic acid (EPA) and ascorbic acid at their various molar ratios in acetone at 55°C. (a) Time courses for conversion of 6-*O*-eicosapentaenoyl ascorbate: the amounts of EPA/ascorbic acid (mmol/mmol) added in 2.5 ml acetone are shown in the figure. (b) Dependence of apparent maximum conversion (closed symbols) and initial reaction rate (open symbols) on the EPA/ascorbic acid molar ratio. Circles and triangles represent the results obtained at the fixed amounts of ascorbic acid (0.125 mmol) and EPA (0.124 mmol), respectively. The curves were empirically drawn.

The water content of reaction medium would be an important factor affecting the conversion because water is one of the condensation products. The synthesis of eicosapentaenoyl ascorbate was conducted in acetone having various initial water contents and at the fixed molar ratio of eicosapentaenoic acid : ascorbic acid = 0.129 mmol:0.125 mmol [25]. The transient changes in the conversion are shown in Figure 4(a). The condensation proceeded faster at the lower water contents. The apparent maximum conversion was also higher at the lower water contents as shown in Figure 4(b). The maximum conversion significantly decreased as the water content increased. This would be ascribed to the predominance of hydrolysis at higher water content. We attempted to evaluate the equilibrium constant for the formation of lauroyl ascorbate through condensation [24]. The equilibrium constant, K_C, is defined by

$$K_C = \frac{C_{P_e} C_{W_e}}{C_{A_e} C_{F_e}} \tag{1}$$

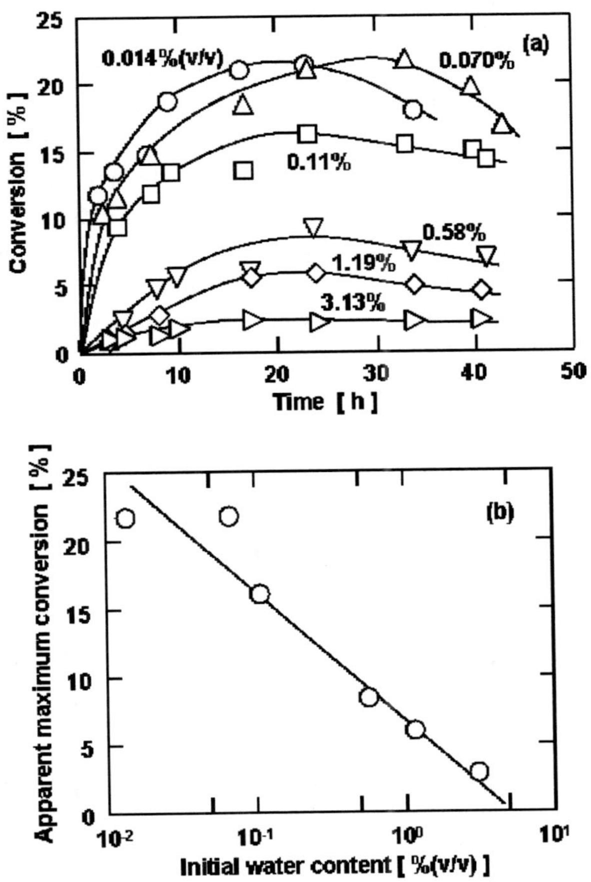

Figure 4. Effect of initial acetone water content on the synthesis of 6-*O*-eicosapentaenoyl ascorbate through lipase-catalyzed condensation. (a) Transient changes in the conversion at various water contents shown in the figure. The amounts of eicosapentaenoic acid and ascorbic acid added to 2.5 ml acetone were 0.129 mmol and 0.125 mmol, respectively. (b) Dependence of the apparent maximum conversion on the initial water content. The curves were empirically drawn.

where C is the concentration and the subscripts A, F, P, and W represent ascorbic acid, lauric acid, lauroyl ascorbate, and water, respectively. The subscript e indicates the equilibrium. The values of C_{Pe} and C_{We} were experimentally determined. As shown in Figures 3(a) and 4(a), the conversion increased with time, reached a plateau and then decreased gradually due to degradation of the product. Since the degradation rate was much slower than the formation rate, the maximum conversion was regarded as the equilibrium conversion. The C_{Ae} values were estimated from the solubility of ascorbic acid in used organic solvent. The C_{Fe} values were evaluated by subtracting C_{Pe} from the initial concentration of lauric acid C_{F0}. The plots of $C_{Pe}C_{We}$ versus $C_{Ae}C_{Fe}$ gave a nearly straight line passing through the origin with the correlation coefficient of 0.95, and the equilibrium constant K_C was evaluated to be about 1.6.

2. Continuous Production of Acyl Ascorbate Using a CSTR or PFR

A schematic diagram of a continuous stirred tank reactor (CSTR) system is shown in Figure 5(a) [29]. An immobilized lipase (3.0 g by dry weight), Chirazyme®L-2 C2, was packed into a stainless steel basket. The volume of the solvent in the reactor was 350 mL. At the beginning of the operation, 40 g of ascorbic acid was added to the reactor after removing the upper lid. A fatty acid (decanoic, lauric or myristic acid) was dissolved with acetone at the concentration of 200 mmol/L, and the mixture was fed to the reactor (85 mmϕ × 90 mm) at a specified flow rate. The reactor was immersed in a water-bath at 50°C and the reaction mixture was mixed by a magnetic stirrer. The flow rate was varied in the range of 1.5 - 9.9 mL/min for each fatty acid. A scheme of a plug flow reactor system (PFR) is shown in Figure 5(b) [31]. Ascorbic acid powders (*ca.* 40 g) were packed into a cylindrical glass column (10 mmϕ × 150 mm) and Chirazyme®L-2 C2 particles (*ca.* 1.5 g by dry weight) were packed into a stainless column (4.6 mmϕ × 150 mm). The columns were connected in series with a stainless steel tube. A fatty acid (decanoic, lauric, myristic, oleic, linoleic or arachidonic acid) was dissolved in acetone at a concentration of 25 - 250 mmol/L. The fatty acid solution was

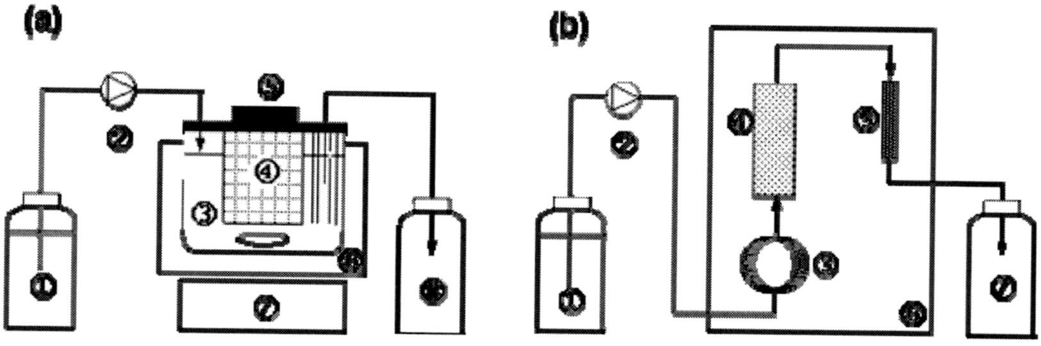

Figure 5. (a) Scheme of a CSTR for the continuous synthesis of saturated acyl ascorbates; 1: feed reservoir, 2: pump, 3: reactor, 4: basket packed with immobilized lipase, 5: lid, 6: water-bath, 7: magnetic stirrer, 8: effluent reservoir; (b) Scheme of a PFR; 1: feed reservoir, 2: pump, 3: pre-heating coil, 4: column packed with ascorbic acid, 5: column packed with immobilized lipase, 6: thermo-regulated chamber, 7: effluent reservoir.

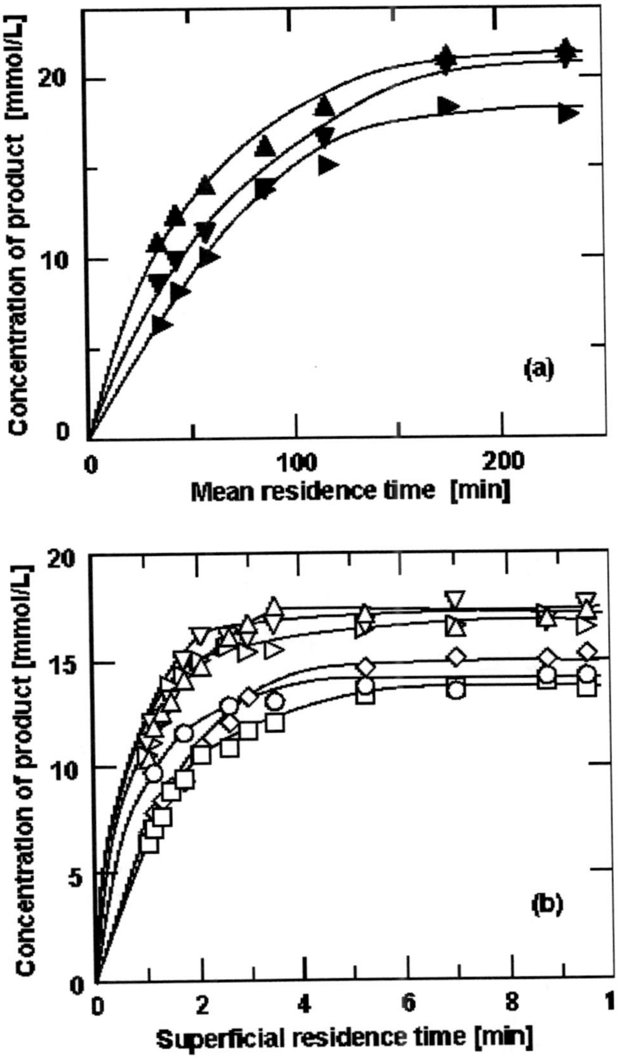

Figure 6. (a) Relationship between the mean residence time, τ, and the concentration of (▲) decanoyl, (▼) lauroyl or (▶) myristoyl ascorbate in the CSTR at 50°C. (b) Relationship between the superficial residence time, τ, and the concentration of (○) arachidonoyl, (□) oleoyl, (◇) linoleoyl, (△) decanoyl, (▽) lauroyl or (▷) myristoyl ascorbate using the PFR at 50°C. In both reactor systems, the fatty acid concentration in the feed was 200 mmol/L. The curves were empirically drawn.

fed to the column packed with ascorbic acid through the pre-heating coil (1.0 mmϕ × *ca.* 1.0 m) and then to the immobilized-enzyme column at a specified flow rate. The pre-heating coil and columns were installed in a thermo-regulated chamber at 50°C. In each reactor, the effluent was sampled after a steady-state was achieved. The product concentration in it was determined using an HPLC.

A fatty acid solution was fed to the CSTR with Chirazyme® L-2 C2 at various flow rates, and the concentration of acyl ascorbate in the effluent was observed after a steady-state was attained. The concentration of fatty acid in the feed solution was 200 mmol/L, since it gave the maximum concentration of the product. The relationships between the mean residence times in the reactor, τ, and the concentrations of the decanoyl, lauroyl and myristoyl ascorbates in the effluent are plotted in Figure 6(a). For each acyl ascorbate, the concentration of more than 18 mmol/L was attained at $\tau \geq 175$ min. Therefore, the reactor was operated at the residence time of 175 min for the long-term production of the acyl ascorbates. The long-term operational stability of the immobilized enzyme was examined for the synthesis of the decanoyl, lauroyl and myristoyl ascorbates at $\tau = 175$ min. As shown in Figure 7, the productivity of lauroyl ascorbate for 5 days was constant, and was determined to be *ca.* 57 g/L-reactor·day. The productivities of decanoyl and myristoyl ascorbate were calculated to be *ca.* 53 and 66 g/L-reactor·day, respectively. The decanoyl and myristoyl ascorbates were also produced at similar conversions. The immobilized enzyme was stable and the reactor was steadily operated for at least 11 days.

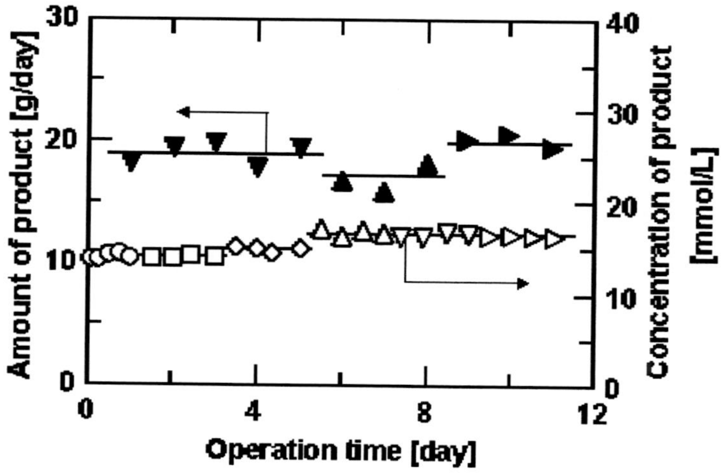

Figure 7. Continuous production of (▲, △) decanoyl, (▼, ▽) lauroyl or (▶, ▷) myristoyl, (○) arachidonoyl, (□) oleoyl, and (◇) linoleoyl ascorbate using the CSTR (closed symbols) or PFR (open symbols) with immobilized lipase, Chirazyme® L-2 C2, at 50°C. The flow rates were 2.0 mL/min for the CSTR and 0.5 mL/min for the PFR, respectively. The fatty acid concentration was 200 mmol/L. The curves were empirically drawn.

A decanoic, lauric, myristic, oleic, linoleic or arachidonic acid solution dissolved in acetone was fed to the PFR system at a flow rate of 0.5 mL/min, which corresponded to a surperficial residence time in the immobilized-enzyme column, τ, of 5.0 min. After a steady state was achieved, the concentration of the corresponding acyl ascorbate in the effluent was determined. For every fatty acid, the product concentration increased as the fatty acid concentration in the feed increased, and it reached the maximum value at the fatty acid concentrations higher 200 mmol/L. Because the solubility of ascorbic acid in acetone at 50°C was 25.3 mmol/L, the molar ratio of a fatty acid to ascorbic acid at the inlet of the immobilized-enzyme column would be *ca.* 8. The concentrations of oleoyl and linoleoyl ascorbates were lower than those of other ascorbates at any fatty acid concentration. This would be due to the lower purity of oleic and linoleic acids used (*ca.* 90%). In the subsequent experiments, the fatty acid concentration in the feed was fixed at 200 mmol/L. A fatty acid solution was fed through the ascorbic acid column to the column packed with the immobilized lipase at various flow rates, and the corresponding acyl ascorbate concentration in the effluent at a steady state was measured. Figure 6(b) shows the relationship between the superficial residence time in the immobilized-lipase column, τ, and the acyl ascorbate concentration in the effluent. For every acyl ascorbate, the maximum product concentration was given at $\tau = 5$ min or longer. Arachidonoyl ascorbate was synthesized for 1 day using the reactor system. The fatty acid to be fed was changed to oleic acid, and the synthesis of oleoyl ascorbate was continued for 2 days. Then, linoleic, decanoic, lauric and myristic acids were, in this order, fed to the system to produce the corresponding ascorbates every two days (Figure 7). For the synthesis of each acyl ascorbate, the τ value was fixed at 5 min. The system could be stably operated for 11 days, and no loss of the enzyme activity was observed. The product concentrations in the effluent were in the range of 14 to 17 mmol/L. The lower concentration of unsaturated acyl ascorbates than those of saturated acyl ascorbates would be ascribed to the lower purity of the unsaturated fatty acids. These product concentrations corresponded to the productivity of 1.6 to 1.9 kg/L-reactor·day, depending on the molecular mass of the product. The productivity was much higher than that for the synthesis of acyl erythritol and mannose using a packed-bed reactor [41, 42].

3. Solubility of Acyl Ascorbate in Water and Oil

The solubilities of the saturated acyl ascorbates in water or soybean oil were measured at various temperatures [29]. The purified acyl ascorbate (10 mg - 1 g) was weighed in an amber vial, and 1 - 4 mL of distilled water or soybean oil was added. After the vial was immersed in a water-bath at 50°C for 1 h, the vial was transferred to a chamber or water-bath kept at 5, 25, 37 or 50°C for 24 h. The solution was filtered using a membrane filter as quickly as possible, and the filtrate was analyzed using an HPLC.

Although ascorbic acid is insoluble in soybean oil [27] and highly soluble in water, the acylation of ascorbic acid significantly improved its solubility in soybean oil but decreased the solubility in water. The temperature dependence of the solubilities, *S*, of the decanoyl, lauroyl, myristoyl, palmitoyl and stearoyl ascorbates in water or soybean oil (Figure 8) could be expressed by the following van't Hoff equation:

$$\frac{dS}{d(1/T)} = -\frac{\Delta H}{R} \tag{2}$$

where ΔH is the dissolution enthalpy, R is the gas constant, and T is the absolute temperature. The HPLC peaks of the palmitoyl and stearoyl ascorbates in aqueous solution were not detected at every temperature because of their very low solubilities in water. The solubilities of the acyl ascorbates in both soybean oil and water were higher for those with a shorter acyl chain. The dependence of the solubility in water on the acyl chain length of the acyl ascorbate was much stronger than that of the solubility in soybean oil. The plots in Figure 8 produced a straight line for each acyl ascorbate. The dissolution enthalpy, ΔH, was evaluated from the slope. The relationship between the acyl chain length and ΔH for the solubilization of the saturated acyl ascorbates in soybean oil or water is elucidated. The ΔH values were about 20 kJ/mol for soybean oil and about 90 kJ/mol for water.

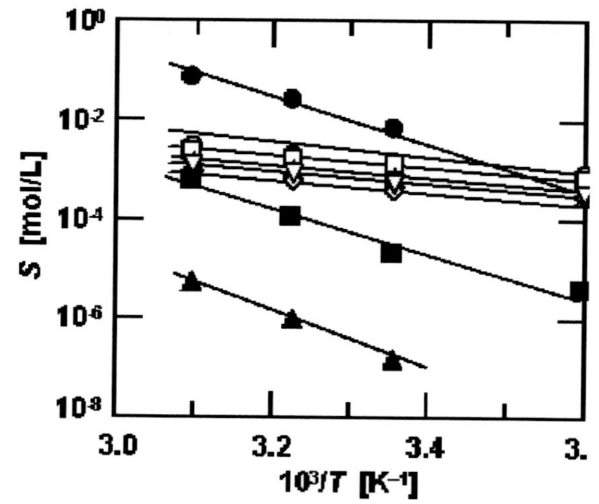

Figure 8. van't Hoff plots for the solubilities, *S*, of (○, ●) decanoyl, (□, ■) lauroyl, (△, ▲) myristoyl, (▽) palmitoyl and (◇) stearoyl ascorbates in soybean oil (open symbols) and water (closed symbols). The straight lines were empirically drawn.

4. Stability of Acyl Ascorbate in an Aqueous Solution and Air

The decomposition of saturated acyl ascorbate in an aqueous solution was measured at various pHs and temperatures [34]. An appropriate amount of an acyl ascorbate, which was determined to give a final concentration of 5 x 10^{-5} to 1 x 10^{-2} mol/L, was weighed into an amber vial with a screw-cap. Five milliliters of a buffer solution was added to the vial to dissolve the ascorbate. The buffer solutions used were 0.1 mol/L sodium citrate-HCl for pHs 2, 3 and 4, 0.1 mol/L sodium citrate-NaOH for pHs 5 and 6, and 0.1 mol/L Tris-HCl for pHs 7, 8 and 9. The vial was immersed in a water-bath controlled at a specified temperature of 30 to 60°C, and then vigorously shaken. At appropriate intervals, a portion of the solution (10 μL) was sampled to determine the remaining ascorbate by HPLC. The Weibull model is flexible and has a potential for describing many degradation kinetics [43]. Therefore, we used the model to describe the degradation of acyl ascorbate:

$$\frac{C}{C_0} = \exp\left[-(kt)^n\right] \tag{3}$$

where C is the acyl ascorbate concentration at time t, and C_0 is the initial ascorbate concentration. k is the rate constant, the reverse of which is called the scale constant, and n is the shape constant. The kinetic parameters, k and n, were evaluated by fitting the experimental results by non-linear regression using the Solver of Microsoft Excel® for Windows [44-46] or the Origin® Ver. 6.1 (Microcal Software, Inc., USA). The stability of acyl ascorbates was assessed at 30, 40, 50 and 60°C. Each ascorbate was dissolved in the sodium citrate-NaOH buffer, pH 5, at a concentration of 0.2 mmol/L. The parameters, k and n, were evaluated at the temperatures, and the temperature dependence of the rate constant, k, for each ascorbate was analyzed based on the Arrhenius equation:

$$k = A\exp\left(-\frac{E}{RT}\right) \tag{4}$$

where A is the frequency or pre-exponential factor, E is the activation energy.

The transient changes in decanoyl ascorbate concentration at 40°C and various pHs were measured. The initial ascorbate concentration was 2.0 x 10^{-4} mol/L at every pH. Decanoyl ascorbate was the most stable at pH 3 in the tested range. The decrease in the ascorbate concentration was faster at the higher pHs. The ascorbate also decreased faster at pH 2 than at pH 3. The decrease with time at pH 3 showed a clear sigmoidal pattern. This pattern was also recognized at pH 2, 4, 5 and 6, although it was weak. The pattern was observed in the decrease of ascorbic acid in orange juice, and the Weibull model was adopted to describe the decrease with time [47]. The rate constant k was the lowest at pH 3, and it increased almost exponentially with an increase in pH. On the other hand, the shape constant n was the largest at pH 3, and there was a tendency that the constant was lower at higher pH. The Weibull model has characteristics of a sigmoidal pattern that can be described when $n > 1$, that the model expresses the simple first-order kinetics at $n = 1$, and that C/C_0 steeply decreases during the early stage when $n < 1$. The largest n value at pH 3 reflected the strong sigmoidal pattern. Because the C/C_0 values at pHs 7 to 9 simply decreased with time, the n values at those pHs were almost 1 and the degradation of the ascorbate obeyed pseudo first-order kinetics.

Two mechanisms can be considered for the decrease in acyl ascorbate concentration: one is the decomposition of the ascorbyl moiety of the ascorbate, and another is the hydrolysis of the ester bond in the acyl ascorbate. In the latter case, free fatty acid is liberated. In order to know which mechanism causes the decrease in the ascorbate concentration, the increase in the decanoic acid concentration was measured at pHs 3, 6 and 9 as well as the decrease in the decanoyl ascorbate concentration. No decanoic acid was detected within 24 h at pH 3 although the half of the decanoyl ascorbate disappeared at that time. A slight amount of decanoic acid was liberated at pH 6. Therefore, the decomposition of the ascorbyl moiety was the predominant mechanism for its disappearance at a pH equal to or lower than 6. At pH 9, the decanoic acid concentration increased during the early stage and its increase continued even after decanoyl ascorbate completely disappeared. The increase in the latter stage would be ascribed to the hydrolysis of the oxidized product of acyl ascorbate such as acyl dehydroascorbate. The initial rate of the decrease in the decanoyl ascorbate concentration was higher than that of the increase in the decanoic acid concentration. Therefore, the decomposition of the ascorbyl moiety and the hydrolysis of the ester bond should simultaneously occur at pH 9. These results indicated that the hydrolysis became more significant at higher pHs although the decomposition of the ascorbyl moiety was the predominant mechanism for the decrease in the decanoyl ascorbate concentration.

It was reported that the initial concentration of ascorbic acid affected its degradation rate in aqueous solution under anaerobic conditions, and the degradation was analyzed based on the pseudo first-order kinetics [48]. The degradation of decanoyl ascorbate at pH 5 and 40°C was measured at different initial concentrations. The initial concentration was changed from 5.0 x 10^{-5} to 1.0 x 10^{-2} mol/L. The kinetic parameters, k and n, in the Weibull model were estimated at the respective initial concentrations, and were plotted versus the initial concentration. There were no significant dependencies of the shape constant, *n*, and the rate constant, *k*, on the initial concentration.

The degradation of acyl ascorbates was measured at pH 5 and different temperatures (Figure 9). The initial concentration of each ascorbate was fixed at 2.0 x 10^{-4} mol/L. The *k* value for each ascorbate is plotted versus the reciprocal of *T* in Figure 10(a). The plots for each ascorbate lay on a straight line, and the *E* and *A* values were estimated. Both the *E* and *A* values seemed to depend on the acyl chain length of ascorbate. As mentioned above, the hydrolysis of the ester bond did not significantly occur at pH 5, but only the oxidation of the ascorbyl moiety of the ascorbate was responsible for the decomposition of each acyl ascorbate. Therefore, it would be expected that the enthalpy-entropy compensation holds during the decomposition at that pH. Equation 5 is one of the expressions describing the compensation [49-51]:

$$E = R\, T_\beta \ln A + \gamma \tag{5}$$

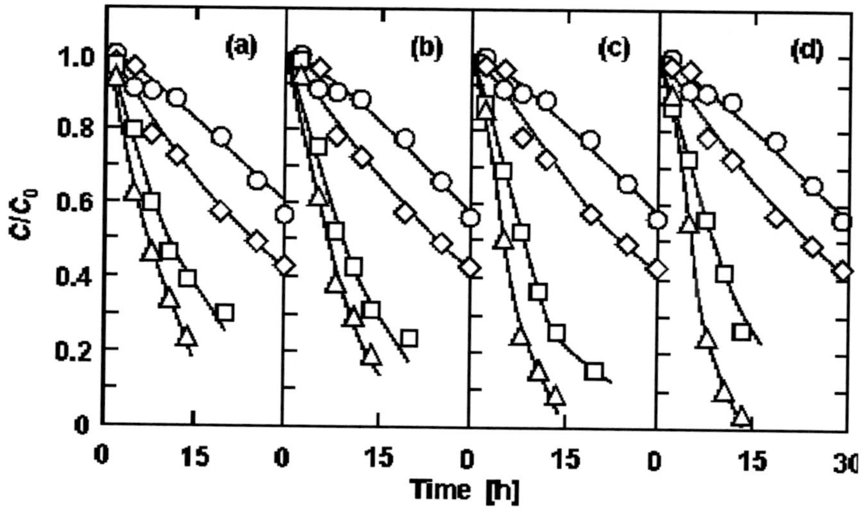

Figure 9. Decomposition of (a) hexanoyl, (b) octanoyl, (c) decanoyl and (d) lauroyl ascorbates at pH 5 and different temperatures. The temperatures were (○) 30, (◇) 40, (□) 50 and (△) 60°C. The initial concentration of each ascorbate was 2.0 x 10^{-4} mol/L. The *C* and C_0 represent the concentration of each ascorbate at any time and the initial one, respectively. The solid curves were calculated using the estimated kinetic parameters of the Weibull model.

where T_β is an isokinetic temperature and γ is a constant. The *E* values are plotted versus the logarithms of the *A* values in Figure 11. The plots lay on a straight line (R^2 = 0.994). This indicated that the decomposition of every acyl ascorbate essentially proceeded by the same mechanism. The T_β value was estimated to be 39.2°C from the slope.

The decomposition of saturated acyl ascorbate in air was also measured at various relative humidity and temperatures [36]. The transient changes in the fractions of the remaining octanoyl, decanoyl, lauroyl, myristoyl and palmitoyl ascorbates at 80°C and at 12, 44 and 75% relative humidity. The relative humidity significantly affected the decomposition of the acyl ascorbates. The decomposition rate of all ascorbates increased with increasing

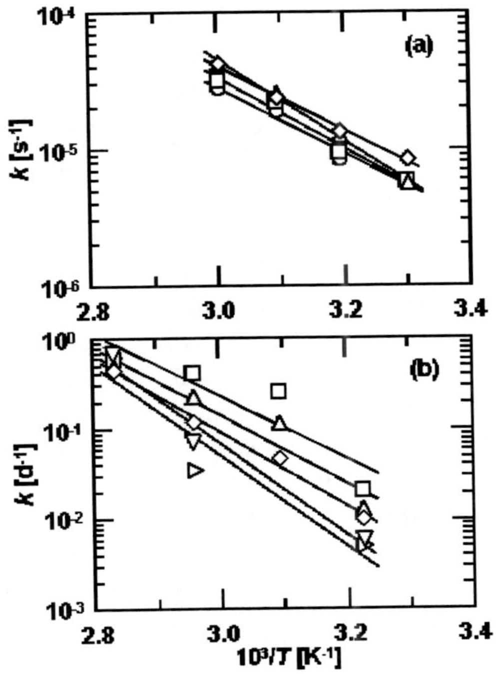

Figure 10. Temperature dependencies of the rate constants, k, of the Weibull model for the decomposition of (○) hexanoyl, (□) octanoyl, (△) decanoyl, (◇) lauroyl, (▽) myristoyl, and (▷) palmitoyl ascorbates (a) in aqueous solution at pH 5 and (b) in air at 75% relative humidity.

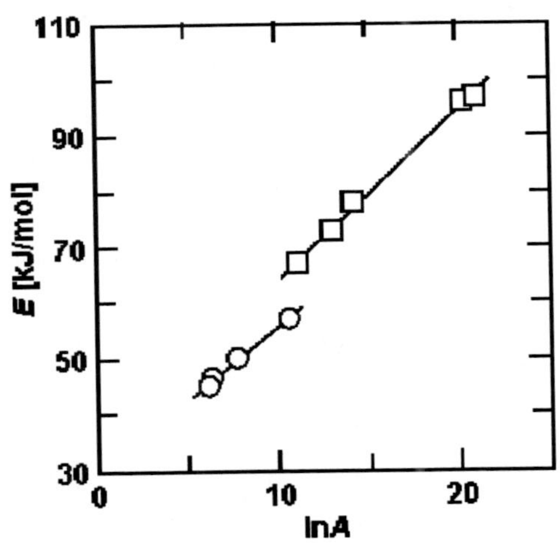

Figure 11. Relationship between the activation energy, E, and the logarithm of the frequency factor, A, for the rate constant, k. of the Weibull model for the decomposition of saturated acyl ascorbate in (○) aqueous solution at pH 5 and (□) air at 75% relative humidity. The A value is in units of s^{-1}.

relative humidity. At 12 and 44% relative humidity the rate of ascorbates increased as acyl chains decreased. The decomposition kinetics of the acyl ascorbates were empirically expressed by the Weibull equation (Eq. 3). The k values increased with increasing relative humidity for all acyl ascorbates at 80°C. The acyl chain length affected the k value at 44% relative humidity, and the k value increased as acyl chain length decreased. The n value for the decomposition of any acyl ascorbate at 75% relative humidity was nearly equal to or slightly less than unity, while the *n* values at 12% and 44% relative humidities were higher than unity. In particular, the *n* value at 44% relative humidity strongly depended on the acyl chain length of the ascorbate, and the *n* value for the decomposition of an ascorbate having a long acyl chain was high. This indicates that the decomposition of the ascorbate with a longer acyl chain progressed via a longer induction period. It is presumed that sorption of water onto the ascorbate with a longer acyl chain was weak at 44% relative humidity due to its high hydrophobicity, and that the induction period was consequently prolonged.

The decomposition processes of the acyl ascorbates at 37, 50 and 65°C and at 75% relative humidity were shown in Figure 12. All acyl ascorbates decomposed fast as temperature increased. The acyl chain length also affected the decomposition of the ascorbates, and ascorbates with shorter acyl chains decomposed faster. This tendency was significant at 65°C. The decomposition of all acyl ascorbates at all temperatures was analyzed based on the Weibull equation in order to estimate the kinetic parameters. The curves in the figure were drawn using the estimated *k* and *n* values. Figure 10(b) shows the temperature dependency of the *k* values for the decomposition of the acyl ascorbates at 75% relative humidity. Although the *k* value increased with increasing temperature, there was no strong temperature dependency for the *n* value and the values were almost 1.0 ± 0.4. The *k* values for the decomposition of ascorbates increased as acyl chains decreased. The Arrhenius plots in Figure 10(b), in each case, produced a straight line to evaluate the activation energy *E* and the frequency factor *A* from the slope and the intercept of the line, respectively. Both the *E* and *A* values for the ascorbate increased as acyl chain length decreased. The *E* values are plotted versus the natural logarithms of the *A* values in the inset of Figure 11. The plots lie on a straight line (R^2 = 0.997). The linear relationship indicated that the compensation held during the decomposition and that the decomposition of every acyl ascorbate essentially proceeded by the same mechanism. The T_β value was estimated to be 101°C from the slope. At this temperature, the rate constant *k* for the decomposition of all acyl ascorbates in air would be the same.

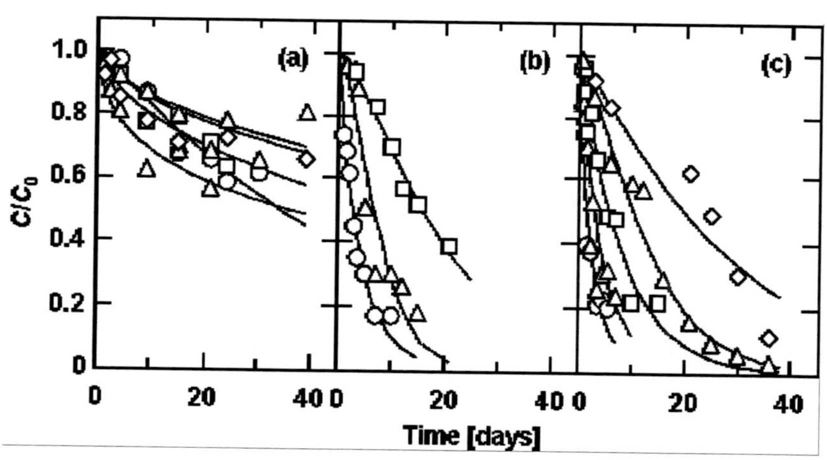

Figure 12. The decomposition processes of (○) octanoyl, (△) decanoyl, (□) lauroyl, (▽) myristoyl and (◇) palmitoyl ascorbate at 75% relative humidity and (a) 37°C, (b) 50°C and (c) 65°C. The solid curves were calculated using the estimated kinetic parameters of the Weibull model.

The transient changes in the fractions of the remaining lauroyl ascorbate and lauric acid liberated by the hydrolysis of lauroyl ascorbate at 65°C and 75% relative humidity were examined. During the early stage during the storage, the decomposition of lauroyl ascorbate rapidly proceeded, but free lauric acid was scarcely formed. Lauric acid gradually formed when the fraction of the remaining lauroyl ascorbate became less than 0.5. The amount of consumed lauroyl ascorbate was calculated by subtracting the fraction of the remaining lauroyl ascorbate, which was estimated using the parameters, k and n, by the Weibull equation, from unity. No lauric acid was liberated during the early stage of the decomposition of lauroyl ascorbate. This indicated that the oxidative degradation of the ascorbyl moiety occurred first, and then the hydrolysis of the ester bond followed to liberate the lauric acid; that is, the decomposition of acyl ascorbate appears to be a consecutive process.

Ascorbic acid and various polyunsaturated fatty acids, such as α- and γ-linolenic, dihomo-γ-linolenic, arachidonic, eicosapentaenoic, and docosahexaenoic acids, were condensed by the immobilized lipase, Chirazyme®L-2 C2, in acetone dehydrated with molecular sieves to produce polyunsaturated acyl ascorbates [25, 27]. Figure 13 shows the autoxidation processes of the unmodified polyunsaturated fatty acid and the polyunsaturated acyl moiety of the polyunsaturated acyl ascorbates at 65°C and nearly 0% relative humidity. The unmodified polyunsaturated fatty acids were almost completely autoxidized within 5 h, whereas all the polyunsaturated acyl ascorbates were significantly resistant to autoxidation. The polyunsaturated acyl moiety of 90% or more remained in the unoxidized state during the test period for every ascorbate. Esterification of polyunsaturated acid with ascorbic acid significantly improved its oxidative stability. The processes of the unmodified eicosapentaenoic acid mixed simply with equimolar ascorbic acid were also measured at 65°C and 0% relative humidity. The oxidation process of eicosapentaenoyl ascorbate was almost the same as that of eicosapentaenoic acid mixed with an equimolar amount of ascorbic acid. Acyl ascorbate was also synthesized using conjugated fatty acid as a substrate for condensation reaction. The oxidation processes of the unmodified conjugated linoleic acid isomers, which were *c*9,*t*11- and *t*10,*c*12-conjugated linoleic acid, and the linoleoyl moiety of the linoleoyl ascorbates at 65°C and 75% relative humidity were examined [37]. The unmodified conjugated linoleic acid isomers were almost completely oxidized within 4 h, whereas all the conjugated linoleoyl ascorbate were significantly resistant to oxidation. The oxidation kinetics was empirically expressed by the Weibull equation (Eq. 3). The k values for conjugated linoleoyl ascorbates and linoleoyl ascorbate were about 1/30 and 1/60, respectively, of that for the corresponding unmodified conjugated linoleic acid isomers. Although it is reported that *t*10,*c*12-conjugated linoleic acid was more easily oxidized than *c*9,*t*11-conjugated linoleic acid [52], there was no significant difference in the susceptibility to oxidation between the conjugated linoleic acid isomers esterified with ascorbic acid. The n values were greater than unity for all the unmodified conjugated linoleic acid isomers, whereas the values for conjugated linoleoyl ascorbates were smaller than unity. This fact indicated that there were induction periods, which reflected the slow progress of the oxidation during the early stage of storage, for unmodified conjugated linoleic acid isomers, whereas the oxidation for conjugated linoleoyl ascorbates progressed very slowly without an induction period. Thus, it was quantitatively demonstrated that the oxidation process for conjugated linoleoyl moiety of conjugated linoleoyl ascorbate was different from that for unmodified conjugated linoleic acid.

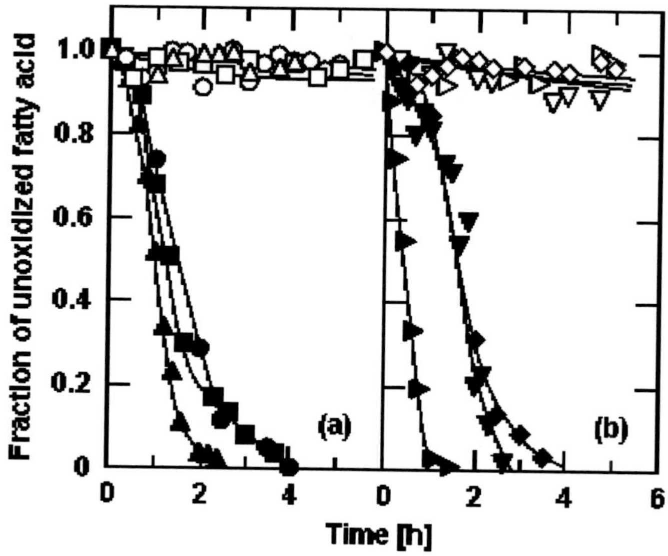

Figure 13. Oxidation processes of polyunsaturated fatty acids and their ascorbates at 65°C and nearly 0% relative humidity. (a) n-6 series: (●)γ-linolenic acid, (○)γ-linolenoyl ascorbate, (■) dihomo-γ-linolenic acid, (□) dihomo-γ-linolenoyl ascorbate, (▲) arachidonic acid and (△) arachidonoyl ascorbate; (b) n-3 series; (◆)α-linolenic acid, (◇)α-linolenoyl ascorbate, (▼) eicosapentaenoic acid, (▽) eicosapentaenoyl ascorbate, (▶) docosahexaenoic acid and (▷) docosahexaenoyl ascorbate.

5. Emulsifier Property of Acyl Ascorbate

The surface tensions of the aqueous solutions of acyl ascorbates were observed at the various concentrations and at 25°C by the Wilhelmy method [26]. The critical micelle concentration for each acyl ascorbate, CMC, was estimated from the results. The surface excess, Γ, was evaluated according to the following equation:

$$-\frac{d\gamma}{d\log C} = 2.30\Gamma RT \tag{6}$$

where γ is the surface tension, and *C* is the concentration of acyl ascorbate. The residual area per molecule, *a*, was calculated from the Γ value by:

$$a = \frac{1}{\Gamma N_{\mathrm{A}}} \tag{7}$$

where N_{A} is Avogadro's number. The CMCs of the ascorbates with longer acyl chains were lower, while the Γ values scarcely depended on the acyl chain length. The CMCs and *a* values of acyl ascorbates are plotted versus the carbon number of the acyl chains in Figure 14, together with the CMCs at 25°C for 6-*O*-acyl mannoses [42] and 1-alkyl $_{\beta}$-D-glucosides [53]. The change in the CMC as a function of the chain length *n* is expressed by [54]:

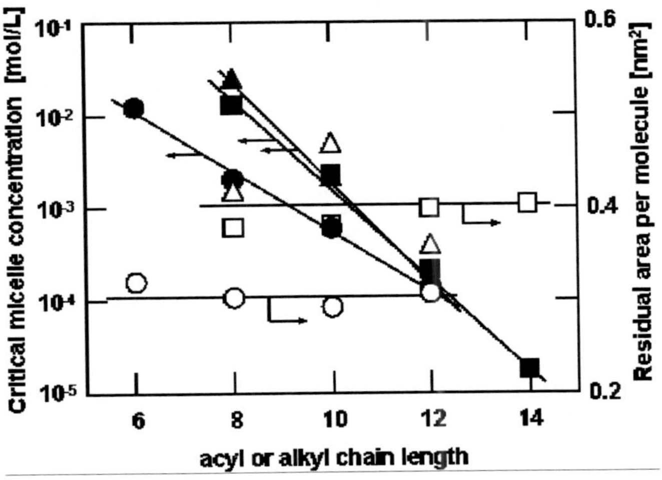

Figure 14. Dependences of the critical micelle concentration, CMC, (closed symbols) and the residual area per molecule, a, (open symbols) on the carbon number of the acyl or alkyl chain. The symbols (●, ○), (■, □) and (▲, △) indicate acyl ascorbates, acyl mannose and alkyl glucosides, respectively. The CMC and a values for the acyl mannoses and alkyl glucosides were cited from the literature [42] and [54], respectively.

$$\log\text{CMC} = -\frac{w}{k_{\text{B}}T}n + b \tag{8}$$

where w is the cohesive energy change per methylene group passing from the bulk of the solution to the micelle, k_{B} is Boltzmann's constant, and b is a constant. The w values were estimated to be 1.3×10^{-21}, 2.0×10^{-21} and 2.2×10^{-21} J for acyl ascorbates, acyl mannoses and alkyl glucosides, respectively. The w value for acyl ascorbates was much smaller than those for the acyl mannoses and alkyl glucosides. The a values of *ca.* 0.40 nm^2 were practically the same for the acyl mannoses and alkyl glucosides, while the a values of the acyl ascorbates were *ca.* 0.30 nm^2, which was much smaller than those of acyl mannoses and alkyl glucosides. Since acyl or alkyl derivatives would be oriented so as to stick their acyl or alkyl residues into the air, the a value seemed to be exclusively determined by the hydrophilic moiety. The molar volumes of mannose and glucose were 0.114 L/mol [55]. If the hexoses are assumed to be spherical, the cross-sectional area is 0.40 nm^2, which coincides with the a value evaluated from the surface tension measurement. The molar volume of ascorbic acid was estimated to be 0.106 L/mol, from which the cross-sectional area was calculated to be 0.38 nm^2 under the assumption that the molecule was a sphere. The a value of 0.30 nm^2 was much smaller than the cross-sectional area. Ascorbic acid has a γ-lactone ring, which is smaller than the pyranose one of mannose or glucose. The γ-lactone ring might act as the hydrophilic moiety of the acyl ascorbate. The surface tensions of decanoyl and lauroyl ascorbates dissolved in distilled water were measured at various temperatures and pHs to examine the temperature and pH dependences of the CMC values [31]. The temperature dependence of the CMC was very weak for both the ascorbates. The a value was also independent of the temperature, and was about 0.35 nm^2. The effect of pH on the surface tensions of decanoyl ascorbate solution at 30°C was examined. The CMC value was higher at the higher pH. The pK_{a} value of ascorbic acid is 3.77 [56]. If we assume the pK_{a} value of ascorbyl moiety is the same as that of ascorbic acid, the degree of dissociation of the ascorbyl moiety increases at the higher pH. The exponential increase in the CMC with an increase in pH would be explained by the increase in the degree of dissociation [57]. The *a* value was almost constant at any pH, and the value was the same as those obtained at different temperatures for decanoyl and lauroyl ascorbates.

The emulsifier properties of saturated acyl ascorbates with even carbon numbers of acyl chains from 8 to 14 for preparation of O/W emulsions using soybean oil as an oily phase were examined [30]. An appropriate amount of an acyl ascorbate, which was specified to give a final concentration of 5 x 10^{-7} to 5 x 10^{-3} mol/kg-solution (water plus soybean oil), was weighed into an amber glass vial, and 19 g of distilled water and 1 g of soybean oil were added into the vial. The buffer solutions used were 0.1 or 0.01 mol/L sodium citrate-HCl for pH 2, 3 and 4, 0.1 or 0.01 mol/L sodium citrate for pH 5, and 0.1 or 0.01 mol/L sodium citrate-NaOH for pH 5.5 and 6. The mixture was emulsified to produce an O/W emulsion with a rotor/stator homogenizer for 1 min at a power control setting of 70. The emulsion was further homogenized to produce a fine O/W emulsion with small oil droplets using a high-pressure homogenizer at a pressure of *ca.* 98 MPa. The particle size distribution of the oil droplets was measured using a laser diffraction particle size analyzer after appropriate dilution with distilled water. The stability of the emulsion was estimated by a turbidity method [58]. The emulsion was stored at 30°C with gentle stirring. At appropriate intervals, aliquots of the emulsion were removed and diluted 50 and 400 times for the emulsions with 0.1% (w/v) sodium dodecyl sulfate. The absorbance at 500 nm of the diluted emulsion was then measured using a spectrophotometer. Figure 15 shows the median diameters of oil droplets in the O/W emulsions prepared at various concentrations of acyl ascorbates with different acyl chain lengths. Except for palmitoyl ascorbate, the median diameter decreased with the increasing emulsifier concentration. The inner figure of Figure 15 shows the relationship between the median diameter, d_p, observed at an emulsifier concentration of 0.5 mmol/kg and the carbon number of the acyl chain of the emulsifier, *n*. The emulsifiers with acyl chains shorter than or equal to 14 gave fine emulsions with median diameters of *ca.* 1 μm. O/W emulsions consisting of distilled water and soybean oil were prepared at various concentrations of acyl ascorbates with different acyl chain lengths, and their stabilities were measured at 30°C by the turbidity method (Figure 16). The emulsions prepared using octanoyl, myristoyl and palmitoyl ascorbates as the emulsifiers were unstable at the concentrations tested. The stabilities of the emulsions prepared with decanoyl and lauroyl ascorbates depended on the emulsifier concentrations and increased with higher concentrations. The hydrophilic-lipophilic balance (HLB) numbers of octanoyl, decanoyl, lauroyl, myristoyl and palmitoyl ascorbates, which were evaluated according to the Davies' equation [59], are 11.8, 10.9, 9.90, 8.95 and 8.00, respectively. McClements (1999) mentioned that the maximum stability of O/W emulsions is obtained using surfactants with an HLB number around 10 to 12 [60]. This criterion was almost applicable for acyl ascorbates, and decanoyl and lauroyl ascorbates produced fine and stable O/W emulsions. The pH and ionic strength of the aqueous phase of O/W emulsion are factors affecting the stability. Their effects on the stability were examined for the emulsions prepared using lauroyl ascorbate at a concentration of 0.5 mmol/kg. When the buffer concentrations of 0.1 mol/L were high, and the emulsions were very unstable at any pH value. The instability would be ascribed to the short Debye length surrounding the oil droplets. When the buffer concentrations were 0.01

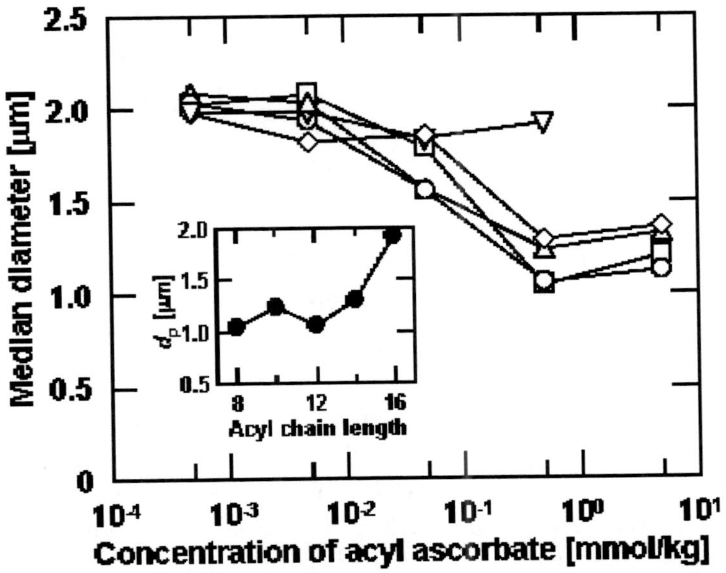

Figure 15. Median diameter of oil droplets in O/W emulsions prepared using acyl ascorbates with different acyl chains at various concentrations as emulsifiers. The emulsifiers used were (□) octanoyl, (△) decanoyl, (○) lauroyl, (◇) myristoyl and (▽) palmitoyl ascorbates. The inner figure shows the dependence of the median diameter, d_p, of oil droplets in the emulsions prepared at an emulsifier concentration of 0.5 mmol/kg on the carbon number of the acyl chain, n.

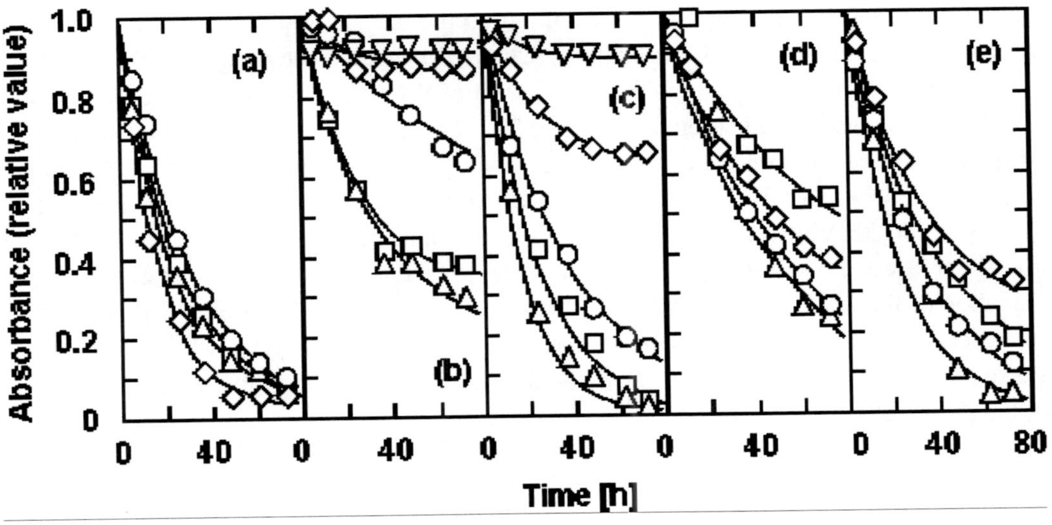

Figure 16. Stability at 37°C of O/W emulsions prepared using (a) octanoyl, (b) decanoyl, (c) lauroyl, (d) myristoyl and (e) palmitoyl ascorbates at various concentrations as emulsifiers. The concentrations of each acyl ascorbate were (□) 0.0005, (△) 0.005, (○) 0.05, (◇) 0.5, and (▽) 5.0 mmol/kg. The ordinate indicates the absorbance at 500 nm relative to the initial one.

mol/L, the stability of the O/W emulsions greatly depended on the pH value of the aqueous phase, and the emulsions were more stable at higher pH. Acyl ascorbates, which adsorb on the surface of oil droplets in O/W emulsions, also seem to have similar pK_a values. Therefore, the surface charge of the oil droplets would be affected by the pH of the aqueous phase. The ζ-potentials of oil droplets in the O/W emulsions prepared using decanoyl and lauroyl ascorbates at a concentration of 0.5 mmol/kg were measured at various pH values of the aqueous phase. For the emulsions with decanoyl and lauroyl ascorbates, the potential largely decreased at a pH higher than 4. These largely negative surface charges at pH 5 and 6 would be the reason for the high stability of the emulsions at these pH values.

6. Antioxidative Ability of Acyl Ascorbate against a Lipid in a Bulk System

The oxidation processes of linoleic acid in the presence of ascorbic acid or saturated acyl ascorbate were analyzed by applying a kinetic equation of the autocatalytic type to the processes [35]. Linoleic acid and ascorbic acid or octanoyl, lauroyl or palmitoyl ascorbate were mixed at the molar ratio of 0.025, 0.05, 0.075, 0.1 or 0.2 (ascorbate/ linoleic acid), and preserved in a plastic container to maintain the relative humidity at 12%. The container was stored in the dark at a given temperature (37, 50, 65 or 80°C). Samples were periodically taken, and the amount of unoxidized linoleic acid was measured by the GC analysis with an FID [61, 62]. The oxidative stabilities at 37°C and 12% relative humidity of linoleic acid mixed with ascorbic acid and octanoyl ascorbate at various molar ratios of the ascorbates to linoleic acid were measured. The higher the molar ratio was, the more strongly the oxidation of linoleic acid was suppressed in both cases. The induction period for the oxidation of linoleic acid was elongated by the addition of the ascorbates. The oxidation processes of linoleic acid mixed with lauroyl and palmitoyl ascorbates showed similar results. Octanoyl ascorbate had stronger antioxidative ability than ascorbic acid. The oxidation of linoleic acid mixed with octanoyl ascorbate at the molar ratio of 0.1 was suppressed for at least 130 h. Figure 17(a) shows the oxidation processes at 80°C and 12% relative humidity of linoleic acid with no additive and that mixed with ascorbic acid or octanoyl, lauroyl, or palmitoyl ascorbate at the molar ratio (additive/ linoleic acid) = 0.05. Linoleic acid with no additive was quickly oxidized. When ascorbic acid was added to linoleic acid, the oxidative stability of linoleic acid was slightly improved. Octanoyl, lauroyl, and palmitoyl ascorbates retarded the oxidation of linoleic acid more than ascorbic acid. There seemed to be little difference in the antioxidative ability among the three acyl ascorbates. We showed previously [63, 64] that the entire oxidation process of linoleic acid could be expressed by the following kinetic equation of the autocatalytic type:

$$\frac{dY}{dt} = -kY(1-Y) \tag{9}$$

where Y is the fraction of the unoxidized substrate, and k is the rate constant. Under the condition of $Y = Y_0$ at $t = 0$, the integration of Eq. 9 gives:

$$\ln\frac{1-Y}{Y} = kt + \ln\frac{1-Y_0}{Y_0} \tag{10}$$

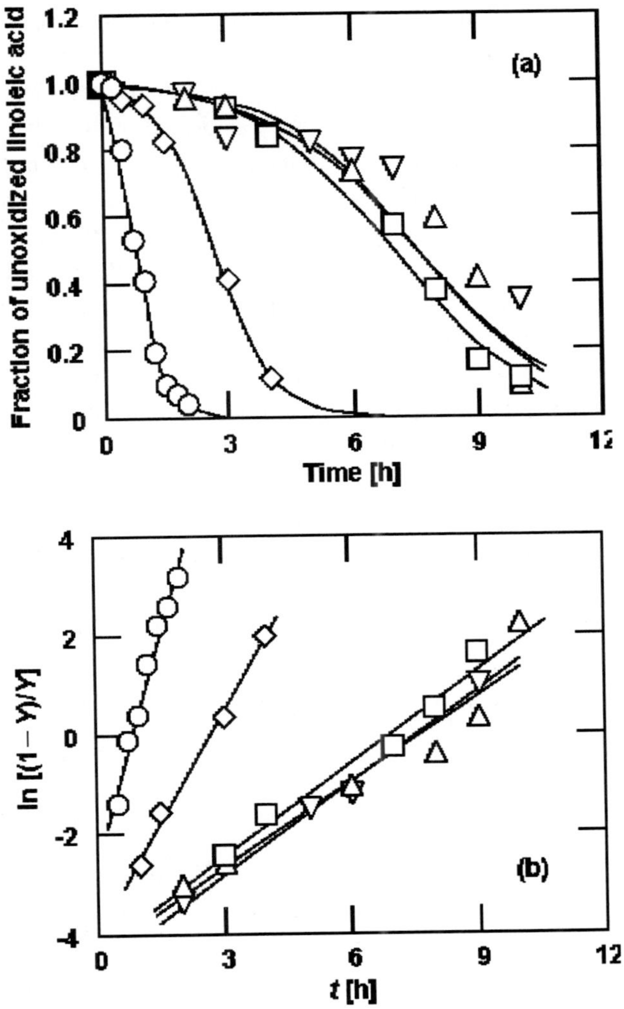

Figure 17. (a) Oxidation processes of (○) linoleic acid with no additive and that mixed with (◇) ascorbic acid, (□) octanoyl ascorbate, (△) lauroyl ascorbate or (▽) palmitoyl ascorbate at the molar ratio = 0.05 and at 80°C. The solid curves were drawn using the k and Y_0 values estimated in Figure 17(b); (b) Determination of the rate constant, k, in the rate expression of the autocatalytic type at 80°C and with the molar ratio = 0.05. The symbols are the same as those in Figure 17(a). Y denotes the fraction of unoxidized linoleic acid. The solid curves were drawn based on Eq. 10.

where Y_0 is the initial fraction of unoxidized substrate and determines the induction period due to the mathematical nature of the equation. The applicability of Eq. 9 to the oxidation processes of linoleic acid mixed with various ascorbates was examined. Linear plots of ln $[(1-Y)/Y]$ versus t for the oxidation process are shown in Figure 17(b). Based on a linear regression analysis, the k and Y_0 values were determined from the slope and the intercept, respectively. The k and Y_0 values for the oxidation processes of linoleic acid at 37, 50 and 65°C were also estimated in the same manner. Figure 18 shows the relationship between the acyl chain length of the ascorbates and the k value at various temperatures. At any temperature, the k value for linoleic acid with no additive was greater than that for LA with ascorbic acid or ascorbate. When octanoyl, lauroyl or palmitoyl ascorbate was added to linoleic acid, there was a tendency for the *k* value in the presence of ascorbate with a larger acyl chain length to be slightly smaller. The *k* value for linoleic acid mixed with ascorbic acid (the acyl chain length equal to zero) seemed to be greater than that for linoleic acid mixed with the acyl ascorbates at any temperature. As mentioned above, the Y_0 value reflects the initial state of the substrate, which affects the induction period. The large Y_0 value indicates the elongation of the induction period. The Y_0 values for the additives were greater than that for linoleic acid with no additive although there were some exceptions. This indicated that the addition of an additive delayed the induction period in the oxidation process of linoleic acid. The temperature dependence of the rate constant *k* was analyzed based on the Arrhenius equation (Eq. 4). The Arrhenius plots using the results shown in Figure 18, in each case, produced a straight line to evaluate the apparent activation energy, *E*, and the frequency factor, *A*, from the slope and the intercept of the line, respectively. The *E* values for linoleic acid mixed with every ascorbate were 50 to 70 kJ/mol and there was a tendency for *E* value to decrease with increasing acyl chain length. The *A* values for linoleic acid mixed with the ascorbates were smaller than that for linoleic acid with no additive or mixed with ascorbic acid. The *A* value also decreased with increasing acyl chain length. This would indicate that the presence of acyl ascorbate lowers both the height of the energy barrier for the oxidation of linoleic acid and the probability of the reaction resulting in the oxidation of linoleic acid.

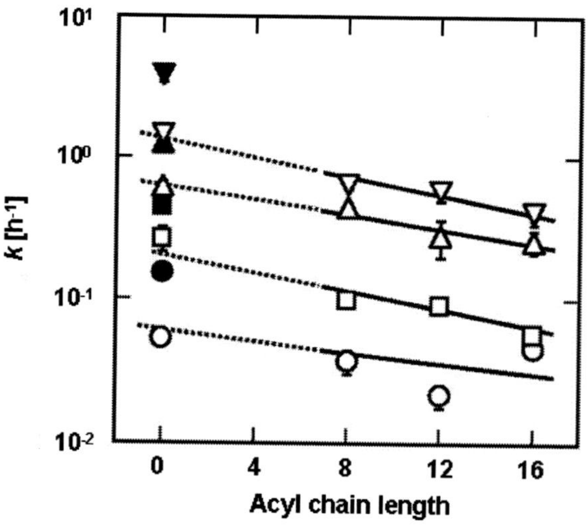

Figure 18. Relationship between acyl chain length of ascorbates and the rate constant *k* at (○) 37, (□) 50, (△) 65 and (▽) 80°C. Open symbols represent the rate constants (mean ± standard error) for the oxidation of linoleic acid mixed with various ascorbates, and the closed symbols that for no additive.

7. Application of Acyl Ascorbate for Microencapsulation of a Lipid

Microencapsulation of a lipid with a wall material is a promising technology in the food and other industries [65]. Microencapsulation of polyunsaturated fatty acid or its acylglycerol suppresses or retards its oxidation [62, 66-68]. Microencapsulation consists of emulsification of a core material such as a lipid with a dense solution of a wall material such as a polysaccharide, followed by drying of the emulsions. Spray-drying is commonly used to prepare the microcapsules [69]. Application of acyl ascorbate for microencapsulation of a lipid was examined to propose the example for the effective use of acyl ascorbate.

Linoleic acid mixed with various acyl ascorbates was microencapsulated with maltodextrin or gum arabic by spray-drying, and its oxidation process was evaluated to examine the antioxidative ability of the ascorbates toward the encapsulated linoleic acid [28]. The oxidation process of the encapsulated linoleic acid mixed with methyl palmitate or unmodified ascorbic acid and then encapsulated with maltodextrin was also measured for comparison. A specific amount of maltodextrin or gum arabic was dissolved in distilled water, and linoleic acid was mixed with the wall material solution. Methyl palmitate, ascorbic acid or acyl ascorbates, such as octanoyl, decanoyl, lauroyl and palmitoyl ascorbates, was then added to the solution. The mixture was emulsified with a rotor/stator homogenizer for 1 min at power control 8. The particle size distribution of the oil droplets in the emulsion was measured using a centrifugal particle size analyzer or a laser diffraction particle size analyzer. The emulsion was fed into a spray-dryer at a flow rate of 3.0 kg/h and was atomized by a centrifugal atomizer operated at *ca.* 3×10^4 rpm. The emulsion in the reservoir was gently magnetically-stirred to prevent flotation of the emulsion droplets. The temperatures of air at the inlet and the outlet were 200°C and 100 - 110°C, respectively. The flow rate of air was *ca.* 7.5 m^3/min. The microcapsules prepared were collected in a cyclone. Because spray-drying was completed within 3 or 6 min, no significant change in the size of oil droplets would occur during the drying. Spray-dried microcapsules were weighed into a flat-bottom glass cup, and the cups were placed in a plastic container in which a Petri dish filled with a saturated lithium chloride solution to regulate the relative humidity at 12%. The container was tightly closed and stored in the dark at 37°C. At appropriate intervals, a cup was removed from the container. The unoxidized linoleic acid in the microcapsules was extracted and determined by a GC analysis.

The oxidation processes of linoleic acid to which palmitoyl ascorbate was added at various molar ratios of the ascorbate to the acid and which was encapsulated with maltodextrin or gum arabic were measured. Because linoleic acid encapsulated without the addition of the ascorbate was oxidized relatively fast, there was a possibility that oxidation of the acid partially proceeded during emulsification and spray-drying although the peroxide values of the acid before and after microencapsulation were not measured. Therefore, the present results are overall estimate of the antioxidative ability of the ascorbate for all the steps of emulsification, spray-drying and storage. The oxidative stabilities of the linoleic acid in both the maltodextrin- and gum arabic-based microcapsules were higher at the higher molar ratios. The induction period for oxidation of the encapsulated linoleic acid was elongated by the addition of palmitoyl ascorbate at high molar ratios due to its radical scavenging ability. Oxidation of the encapsulated linoleic acid quickly proceeded after the induction period and then reached a level where further oxidation seemed to proceed very slowly or to cease. The level was higher for the microcapsules with the higher ratios. The molar ratio would affect the particle size of the oil droplets in the O/W emulsion because of the emulsification ability of palmitoyl ascorbate. Therefore, the distribution of the particle size was measured using the centrifugal particle size analyzer. The dependence of the median diameter of the oil droplet on the molar ratio was evaluated. As expected, the median diameter decreased with an increase in the ratio. We reported that the oxidative stability of linoleic acid in the microcapsule prepared from an emulsion with a smaller median diameter of oil droplets was higher because of high probability in interaction of the acid with the wall material within the microcapsule [62, 70]. The high oxidative stability of linoleic acid with the high molar ratio would be attributed to the synergistic effect of the small oil droplet in the emulsion and the antioxidative ability of palmitoyl ascorbate. It has been reported, in a non-encapsulated system, that the addition of a large amount of a saturated fatty acid to a polyunsaturated fatty acid decreased the polyunsaturated fatty acid fraction in the mixture and delayed the oxidation of the polyunsaturated fatty acid [64]. Thus, it is possible that the addition of palmitoyl ascorbate to linoleic acid resulted in improved oxidative stability of the encapsulated linoleic acid by decreasing the fraction of linoleic acid in the mixture. To examine this possibility, methyl palmitate was added to linoleic acid at various molar ratios, and the mixture was then microencapsulated with maltodextrin by spray-drying. The encapsulation efficiency of linoleic acid, which was defined as a ratio of the actual amount of the acid in the microcapsules to the one calculated from the composition of emulsion used for preparing the microcapsules, was 0.4 to 0.6. The efficiency was almost the same as that of the microcapsules prepared with no additive. At the ratio of methyl palmitate to linoleic acid of 0.1, the oxidation of linoleic acid was suppressed, but the extent was very small. Therefore, the suppressive effect of palmitoyl ascorbate on the oxidation of the encapsulated linoleic acid is not due to the decrease in the fraction of linoleic acid by the addition of the ascorbate. To compare the effects of unmodified ascorbic acid and acyl ascorbate on the oxidation of encapsulated linoleic acid, the oxidation processes of linoleic acid encapsulated with maltodextrin to which unmodified ascorbic acid was added to produce specified molar ratios of ascorbic acid to linoleic acid were examined. The addition of unmodified ascorbic acid at any ratio had no effect on the encapsulation efficiency of linoleic acid. Unmodified ascorbic acid slightly suppressed the oxidation of linoleic acid at a molar ratio of 1. Because this ratio was 10 times that for palmitoyl ascorbate, the antioxidative effect of unmodified ascorbic acid for encapsulated linoleic acid was very weak. Elongation of the induction period, which was observed for palmitoyl ascorbate, did not occur. It is expected that saturated acyl ascorbates with different acyl chain lengths also possess antioxidative ability for the encapsulated linoleic acid. Figure 19 shows the oxidation processes of linoleic acid mixed with various saturated acyl ascorbates at a molar ratio of 0.1 and encapsulated with maltodextrin. Mixing of any ascorbate significantly increased the encapsulation efficiency of linoleic acid and the efficiency was greater than 0.9. The median diameters of the oil droplets in the O/W emulsions before microencapsulation are also shown in Figure 20. Ascorbates with acyl chain lengths of 10 or more gave emulsions with small median diameters and significantly suppressed the oxidation of the encapsulated linoleic acid. Especially, decanoyl ascorbate was superior both in emulsification ability (small diameter of emulsion) and in suppression of the oxidation. The oil droplet size of the emulsion prepared with octanoyl ascorbate was large, and the ascorbate was less effective than other ascorbates for suppression of the oxidation of the encapsulated linoleic acid. The solubility of decanoyl ascorbate in water was almost the same as that in soybean oil. The saturated acyl ascorbates with acyl chain lengths of more than 10 are apt to dissolve in the oil, while the ascorbates with the lengths of less than 10 preferentially dissolve in water rather than in the oil. Therefore, when octanoyl ascorbate was used for microencapsulation of linoleic acid, most of the molecules would exist in the dehydrated wall material layer and could not scavenge the polyunsaturated fatty acid radicals generated in the oil phase. Because ascorbates which produced emulsions with small oil droplets were surface-active, they were apt to locate in the interface of linoleic acid and the wall material layer. Some linoleic acid molecules would be replaced with the saturated groups of the ascorbates and the replacement would increase the stability of linoleic acid toward the oxidation.

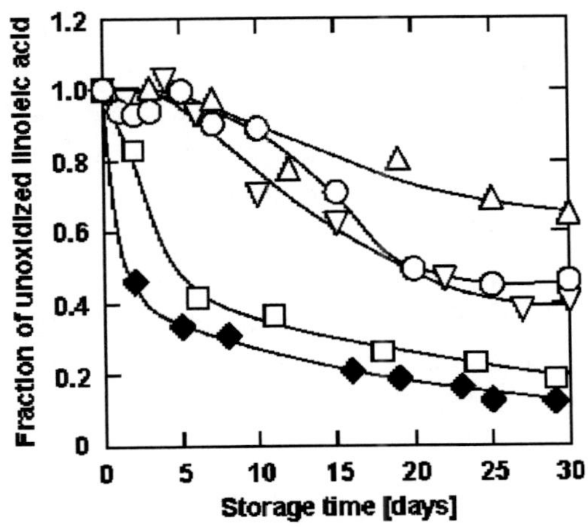

Figure 19. Effect of the addition of various saturated acyl ascorbates to linoleic acid on the oxidation of linoleic acid encapsulated with maltodextrin; (◇) no addition, (□) octanoyl, (△) decanoyl, (▽) lauroyl and (○) palmitoyl ascorbate. The molar ratio of the saturated acyl ascorbate to linoleic acid was 0.1. The oxidation was observed at 37°C and a relative humidity of 12%.

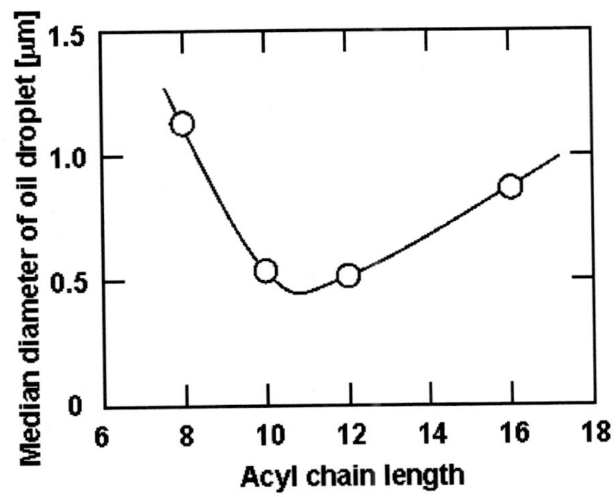

Figure 20. Median diameters of oil droplets in emulsions prepared with various saturated acyl ascorbates.

Arachidonoyl ascorbate, which was a polyunsaturated acyl ascorbate, was microencapsulated with various wall materials, which were maltodextrin, gum arabic and soluble soybean polysaccharide, by spray-drying [33]. Arachidonic acid was also microencapsulated with the wall materials in the presence or absence of ascorbic acid. The oxidative stabilities of the microencapsulated arachidonoyl ascorbate and arachidonic acid were then examined at 37°C and 12% relative humidity. For comparison, the oxidative stabilities of the nonencapsulated arachidonoyl ascorbate and arachidonic acid were also observed under the same conditions. The median diameters of the oil droplets in the O/W emulsions, the aqueous phase of which contained the polysaccharides, were measured. Because gum arabic and soluble soybean polysaccharide have an emulsifier ability [71, 72], the oil droplets in the emulsions prepared with these wall materials were small. Especially, the soluble soybean polysaccharide was the best wall material among the tested polysaccharides to prepare the emulsion with small oil droplets. The addition of ascorbic acid to the aqueous phase had no significant effect on the median diameter for every wall material. Because acyl ascorbates were surface active and acted as emulsifiers, the median diameters of the oil droplets in the emulsions, in which the mixture of arachidonoyl ascorbate and methyl oleate was used as the oil phase, were smaller than those of the oil droplets in the emulsions which included arachidonic acid as the oil phase. Figure 21(a) shows the oxidation of the nonencapsulated arachidonic acid and arachidonoyl ascorbate at 37°C and 12% relative humidity. The oxidation of arachidonic acid was observed in the presence and absence of ascorbic acid. Arachidonic acid was very rapidly oxidized in the absence of ascorbic acid. The addition of ascorbic acid to arachidonic acid retarded the oxidation of arachidonic acid. The arachidonoyl moiety of the ascorbate was very slowly oxidized during the long-term storage. We reported that the oxidation of an n-6 polyunsaturated fatty acid could be expressed by the kinetic equation of the autocatalytic type [63, 64]. Although the equation was applicable for the oxidation of arachidonic acid in the absence and presence of ascorbic acid, and for that of arachidonoyl ascorbate, it was not easy to apply the equation to the oxidation of the microencapsulated arachidonic acid and arachidonoyl ascorbate because of complicated phenomena, such as the diffusion of oxygen and interaction of the fatty acid with the wall material that occurred in the microcapsules [73]. Therefore, the oxidation kinetics was empirically expressed by the Weibull equation (Eq. 3). The k values for arachidonic acid in the presence of ascorbic acid and for arachidonoyl ascorbate were about 1/10 and 1/100, respectively, of that for arachidonic acid in the absence of ascorbic acid. The shape constant, n, was greater than unity for every substrate, reflecting the slow progress of the oxidation during the early stage of storage. Figures 21(b), (c) and (d) show the oxidation of arachidonic acid and the arachidonoyl moiety of the ascorbate encapsulated with maltodextrin, gum arabic and soluble soybean polysaccharide, respectively. The k values for arachidonic acid and arachidonoyl ascorbate encapsulated with the soluble soybean polysaccharide were very small, indicating that the polysaccharide had a high antioxidative ability to both arachidonic acid and arachidonoyl ascorbate. The oxidation of the microencapsulated unsaturated fatty acid often has two characteristics: one is the rapid oxidation progress during the early stage of storage and the other is the leveling off of the oxidation during the late stage [74]. The n values smaller than unity, which were obtained for some microcapsules, reflected the former characteristic. The oxidation of arachidonic acid, in both the presence and absence of ascorbic acid, and arachidonoyl ascorbate was slightly suppressed by the microencapsulation with maltodextrin. Gum arabic and soluble soybean polysaccharide were effective wall materials for the suppression of the oxidation of arachdonic acid in the absence of ascorbic acid. The addition of ascorbic acid to the water phase during the preparation of the O/W emulsions could suppress the oxidation of arachidonic acid encapsulated with any polysaccharide. The encapsulated arachidonoyl ascorbate exhibited the highest resistance to the oxidation for every wall material. The resistance would be ascribed to the antioxidative ability of the ascorbate moiety of the ester. Fatty acid molecules interact with the wall material within the microcapsule, and a molecule interacting with the material is oxidized more slowly than that not interacting with it [73]. The oxidative stability of linoleic acid in the microcapsule prepared from the O/W emulsion with a smaller median diameter of oil droplets was higher because of the higher probability of interaction of the acid with the wall material within the microcapsule [62]. The median diameter of the oil droplets in the emulsions prepared with gum arabic or the soluble soybean polysaccharide was small, especially for arachidonoyl ascorbate, which would be surface-active. This would also be a reason for the high oxidative stability of the arachidonoyl ascorbate encapsulated with the polysaccharide.

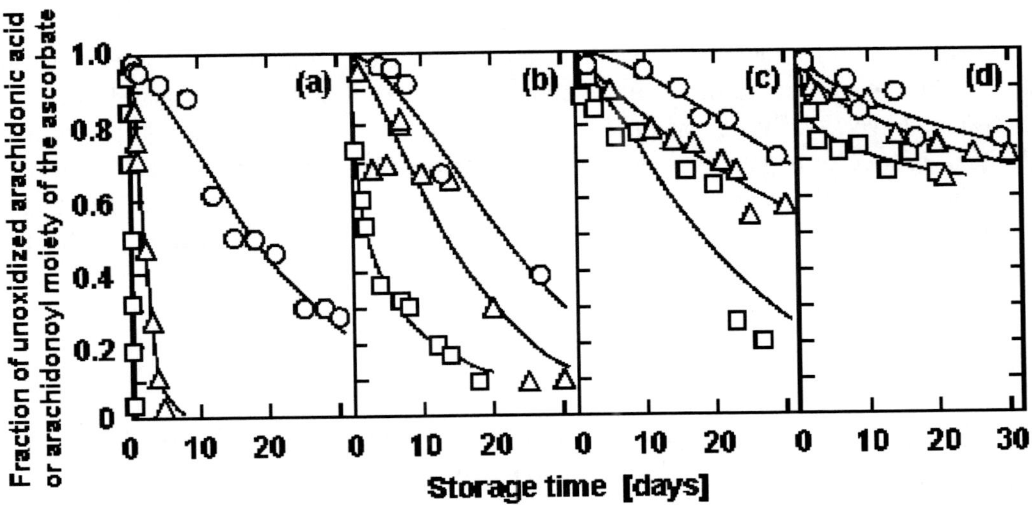

Figure 21. Oxidation processes at 37°C and 12% relative humidity of (□) arachidonic acid without ascorbic acid, (△) arachidonic acid in the presence of ascorbic acid and (○) arachidonoyl moiety of arachidonoyl ascorbate in (a) nonencapsulated system, and in microcapsules prepared with (b) maltodextrin, (b) gum arabic, and (d) a soluble soybean polysaccharide as the wall material. The solid curves were drawn based on the Weibull equation.

Conclusion

Acyl ascorbate is synthesized through lipase-catalyzed-condensation of ascorbic acid with fatty acid. Enzymatic synthesis of acyl ascorbate would be more advantageous than a chemical method due to the simplicity of its reaction process and its high regioselectivity. Further, it would be edible, because each substrate is eaten daily as a food component. Acyl ascorbate might be hydrolyzed in digestion or on intestinal membrane by lipase, and then be absorbed. Each substrate would separately function as an essential nutrient *in vivo*. Thus, it is very useful that the oxidative stability of polyunsaturated fatty acid, which has various physiological functions, is improved by esterification with ascorbic acid. Moreover, acyl ascorbate would be used as an emulsifier. The emulsifier property changes by altering acyl chain length. The application of acyl ascorbate for microencapsulation of a lipid as antioxidative emulsifier would be a useful technology for suppressing or retarding the oxidation of polyunsaturated fatty acid.

Acknowledgements

Most of this study was supported by the Program for the Promotion of Basic Research Activities for Innovative Biosciences (PROBRAIN), Japan, and it was partly supported by a Grant-in-Aid for Young Scientists (B) 19780106 from the Ministry of Education, Culture, Sports, Science and Technology (MEXT), Japan. The *c*9,*t*11- and *t*10,*c*12-conjugated linoleic acids were gifts from Dr. Yuji Shimada, the director of Osaka Municipal Research Institute.

The authors are very grateful to Dr. Ryuichi Matsuno, Professor Emeritus of Kyoto University and Professor of Ishikawa Prefectural University, for his valuable discussions and advises throughout this study, and sincerely appreciate Dr. Kazuhiro Nakanishi, Professor of Okayama University, for his patient and accurate guidance. The authors would like to thank all the members of the Laboratory of Biomolecular Engineering, Department of Biotechnology and Chemistry, School of Engineering, Kinki University, and that of Bioengineering, Division of Food Science and Biotechnology, Graduate School of Agriculture, Kyoto University.

References

[1] Cosgrove, J. P., Church, D. F. & Pryor, W. A. (1987). The kinetics of the autoxidation of polyunsaturated fatty acids. *Lipids*, **22**, 299-304.

[2] Gwo, Y. Y., Flick Jr., G. J. & Dupuy, H. P. (1985). Effect of ascorbyl palmitate on the quality of the frying fats for deep frying operations. *Journal of American Oil Chemists' Society*, **62**, 1666-1671.

[3] Enomoto, K., Miyamori, T., Sakimae, A. & Numazawa, R. (1991). Production of organic acid ester. Japanese patent, (H3-259089).

[4] Tanaka, H. & Yamamoto, R. (1966). Pharmaceutical studies on ascorbic acid derivatives. (1) Synthesis of esters of ascorbic acid and their physicochemical properties. *Yakugaku Zasshi*, **86**, 376-383 (in Japanese).

[5] Miwa, N. & Yamazaki, H. (1986). Potentiated susceptibility of ascites tumor to acyl derivatives of ascorbate caused by balanced hydrophobicity in the molecule. *Experimental Cell Biology*, **54**, 245-249.

[6] Nagao, N. Matsubara, H., Yamanaka, T., Etoh, T., Miwa, N., Iwagaki, H., Saiki, I., Yamaoka, S., Itoh, S. & Kokata, E. (1996). Functions of vitamin C for cells. (V) Mechanism for tumor infiltration- and metastasis-inhibitory effects of L-ascorbic acid. *Vitamins*, **70**, 199-200 (in Japanese).

[7] Inagaki, C. & Kawaguchi, K. (1966). Studies on the metabolism of fatty acid esters of L-ascorbic acid. (1) Excretion and accumulation of ascorbic acid after administration of mono- and dipalmitates of ascorbic acid. *Vitamins*, **34**, 293-297 (in Japanese).

[8] Shima, M., Yohdoh, K., Yamaguchi, M., Kimura, Y., Adachi, S. & Matsuno, R. (1997). Effects of medium-chain fatty acids and their acylglycerols on the transport of penicillin V across Caco-2 cell monolayers. *Bioscience Biotechnology and Biochemistry*, **61**, 1150-1155.

[9] Yoshii, H., Furuta, T., Itoh, T., Yoshimura, K., Misawa, Y., Hata, N., Kobayashi, H. & Linko, Y. Y. (1999). Kinetic analysis of the autoxidation of ethyl eicosapentaenoate at different oxygen levels. *Bioscience Biotechnology and Biochemistry*, **63**, 662-665.

[10] Linko, Y. Y. & Hayakawa, K. (1996). Docosahexaenoic acid: A valuable nutraceutical? *Trends in Food Science and Technology*, **7**, 59-63.

[11] Calder, P. C. (2001). Polyunsaturated fatty acids, inflammation, and immunity. *Lipids*, **36**, 1007-1024.

[12] Song, J. H., Inoue, Y. & Miyazawa, T. (1997). Oxidative stability of docosahexaenoic acid-containing oils in the form of phospholipids, triacylglycerols, and ethyl esters. *Bioscience Biotechnology and Biochemistry*, **61**, 2085-2088.

[13] Arakawa, N., Ebata, F., Hunag, L. & Inagaki, C. (1971). Studies on the metabolism of fatty acid esters of L-ascorbic acid. (4) The formation of fatty acid esters of ascorbic acid by enzyme preparations. *Vitamins*, **43**, 166-171 (in Japanese).

[14] Okamura, S., Iwai, M. & Tsujisaka, Y. (1979). Synthesis of various kinds of esters by four microbial lapases. *Biochimica et Biophysica Acta*, **575**, 156-165.

[15] Therisod, M. & Klibanov, A. M. (1987). Regioselective acylation of secondary hydroxyl groups in sugars catalyzed by lipases in organic solvents. *Journal of American Chemists' Society*, **109**, 3977-3981.

[16] Khaled, N., Montet, D., Farines, M., Pina, M. & Graille, J. (1992). Synthesis of sugar mono-esters by biocatalysis. *Oleagineux*, **47**, 181-190.

[17] Watanabe, Y., Miyawaki, Y., Adachi, S., Nakanishi, K. & Matsuno R. (2000). Synthesis of lauroyl saccharides through lipase-catalyzed condensation in microaqueous water-miscible solvents. *Journal of Molecular Catalysis B: Enzymatic*, **10**, 241-247.

[18] Humeau, C., Girardin, M., Coulon, D. & Miclo, A. (1995). Synthesis of 6-*O*-palmitoyl L-ascorbic acid catalyzed by *Candida antarctica* lipase. *Biotechnology Letters*, **17**, 1091-1094.

[19] Bradoo, S., Saxena, R. K. & Gupta, R. (1999). High yields of ascorbyl palmitate by thermostable lipase-mediated esterification. *Journal of American Oil Chemists' Society*, **76**, 1291-1295.

[20] Stamatis, H., Sereti, V. & Kolisis, F. N. (1999). Studies on the enzymatic synthesis of lipophilic derivatives of natural antioxidants. *Journal of American Oil Chemists' Society*, **76**, 1505-1510.

[21] Yan, Y., Bornscheuer, U. T. & Schmid, R. D. (1999). Lipase-catalyzed synthesis of vitamin C fatty acid esters. *Biotechnology Letters*, **21**, 1051-1054.

[22] Humeau, C., Rovel, B. & Girardin, M. (2000). Enzymatic esterification of bixin by L-ascorbic acid. *Biotechnology Letters*, **22**, 165-168.

[23] Tang, L., Zhang, H., Shehate M. M. & Sun, Y. (2000). A kinetic study of the synthesis of ascorbate fatty acid esters catalysed by immobilized lipase in organic media. *Biotechnology and Applied Biochemistry*, **32**, 35-39.

[24] Watanabe, Y., Adachi, S., Nakanishi, K. & Matsuno, R. (1999). Condensation of L-ascorbic acid and medium-chain fatty acids by immobilized lipase in acetonitrile with low water content. *Food Science and Technology Research*, **5**, 188-192.

[25] Watanabe, Y., Minemoto, Y., Adachi, S., Nakanishi, K., Shimada, Y. & Matsuno, R. (2000). Lipase-catalyzed synthesis of 6-*O*-eicosapentaenoyl L-ascorbate in acetone and its autoxidation. *Biotechnology Letters*, **22**, 637-640.

[26] Watanabe, Y., Adachi, S., Fujii, T., Nakanishi, K. & Matsuno, R. (2001). Surface activity of 6-*O*-hexanoyl, octanoyl, decanoyl and dodecanoyl ascorbates. *Japan Journal of Food Engineering*, **2**, 73-75.

[27] Watanabe, Y., Adachi, S., Nakanishi, K. & Matsuno, R. (2001). Lipase-catalyzed synthesis of unsaturated acyl L-ascorbates and their ability to suppress autoxidation of polyunsaturated fatty acid. *Journal of American Oil Chemists' Society*, **78**, 823-826.

[28] Watanabe, Y., Fang, X., Minemoto, Y., Adachi, S. & Matsuno, R. (2002). Suppressive effect of saturated acyl L-ascorbate on the oxidation of linoleic acid encapsulated with maltodextrin or gum arabic by spray-drying. *Journal of Agricultural and Food Chemistry*, **50**, 3984-3987.

[29] Watanabe, Y., Kuwabara, K., Adachi, S., Nakanishi, K. & Matsuno, R. (2003). Production of saturated acyl L-ascorbate by immobilized lipase using a continuous stirred tank reactor. *Journal of Agricultural and Food Chemistry*, **51**, 4628-4632.

[30] Kuwabara, K., Watanabe, Y., Adachi, S., Nakanishi, K. & Matsuno, R. (2003). Emulsifier property of saturated acyl L-ascorbate in O/W emulsion. *Food Chemistry*, **82**, 191-194.

[31] Kuwabara, K., Watanabe, Y., Adachi, S., Nakanishi, K. & Matsuno, R. (2003). Continuous production of acyl L-ascorbates using a packed-bed reactor with immobilized lipase. *Journal of American Oil Chemists' Society*, **80**, 895-899.

[32] Kuwabara, K., Watanabe, Y., Adachi, S., Nakanishi, K. & Matsuno, R. (2003). Synthesis of 6-*O*-unsaturated acyl L-ascorbates by immobilized lipase in acetone in the presence of molecular sieve. *Biochemical Engineering Journal*, **16**, 17-22.

[33] Watanabe, Y., Fang, X., Adachi, S., Fukami, H. & Matsuno, R. (2004). Oxidation of 6-*O*-arachidonoyl L-ascorbate microencapsulated with a polysaccharide by spray-drying. *Lebensmittel-Wissenschaft und –Technologie*, **37**, 395-400.

[34] Kuwabara, K., Watanabe, Y., Adachi, S. & Matsuno, R. (2005). Stability of saturated acyl L-ascorbates in aqueous solution. *Journal of Food Science*, **70**, 7-11.

[35] Watanabe, Y., Ishido, E., Fang, X., Adachi, S. & Matsuno, R. (2005). Oxidation kinetics of linoleic acid in the presence of saturated acyl L-ascorbate. *Journal of American Oil Chemists' Society*, **82**, 389-392.

[36] Watanabe, Y., Sawahara, Y., Asai, S. & Adachi, S. (2008). Decomposition kinetics of 6-*O*-monoacyl ascorbate in air. *Food Science and Technology Research*, **14**, 139-143.

[37] Watanabe, Y., Sawahara, Y., Nosaka, H., Yamanaka, K. & Adachi, S. (2008). Enzymatic synthesis of conjugated linoleoyl ascorbate in acetone. *Biochemical Engineering Journal*, **40**, 368-372.

[38] Kurata, T. (1976). Mechanism of degradation and discoloration reaction of L-ascorbic acid. *Nippon Nogeikagaku Kaishi*, **50**, 209-216 (in Japanese).

[39] Watanabe, Y., Miyawaki, Y., Adachi, S., Nakanishi, K. & Matsuno, R. (2001). Equilibrium constant for lipase-catalyzed condensation of mannose and lauric acid in water-miscible organic solvents. *Enzyme and Microbial Technology*, **29**, 494-498.

[40] Kobayashi, T., Furutani, W., Adachi, S. & Matsuno, R. (2003). Equilibrium constant for the lipase-catalyzed synthesis of fatty acid butyl ester in various organic solvents. *Journal of Molecular Catalysis B: Enzymatic*, **24-25**, 61–66.

[41] Adachi, S., Nagae, K. & Matsuno, R. (1999) Lipase-catalyzed condensation of erythritol and medium-chain fatty acids in acetonitrile with low water content. *Journal of Molecular Catalysis B: Enzymatic*, **6**, 21-27.

[42] Watanabe, Y., Miyawaki, Y., Adachi, S., Nakanishi, K. & Matsuno, R. (2001). Continuous production of acyl mannoses by immobilized lipase using a packed-bed reactor and their surfactant properties. *Biochemical Engineering Journal*, **8**, 213-216.

[43] Cunha, L. M., Oliverira, F. A. R. & Oliverira, J. C. (1998). Optimal experimental design for estimating the kinetic parameters of processes described by Weibull probability distribution function. *Journal of Food Engineering*, **37**, 175-191.

[44] Harris, D. C. (1998). Nonlinear least-square curve fitting with Microsoft Excel Solver. *Journal of Chemical Education*, **75**, 119-121.

[45] De Levie, R. (1999). Estimating parameter precision in nonlinear least squares with Excel's Solver. *Journal of Chemical Education*, **76**, 1594-1598.

[46] Carter, A. (2000). A simplified mathematical model for calculating the lattice energies of binary group I and II ionic compounds: using Solver to estimate the empirical constants. *Journal of Chemical Education*,, **77**, 1081-1084.

[47] Manso, M. C., Oliveira, F. A. R., Oliveira, J. C. & Frias, J. M. (2001). Modelling ascorbic acid thermal degradation and browning in orange juice under aerobic conditions. *International Journal of Food Science and Technology*, **36**, 303-312.

[48] Anmo, T., Washitake, M., Hayashi, H. & Takahashi, M. (1971). Studies on the decomposition of L-ascorbic acid in the coexistence of D-araboascorbic acid. Anaerobic decomposition of L-ascorbic acid in the coexistence of D-araboascorbic acid. *Yakugaku Zasshi*, **91**, 444-453.

[49] Leffer, J. E. (1955). The enthalpy-entropy relationship and its implications for organic chemistry. *Journal of Organic Chemistry*, **20**, 1202-1231.

[50] Exner, O. (1964). Concerning the isokinetic relationship. *Nature*, **201**, 488-490.

[51] Liu, L. & Guo, Q. X. (2001). Isokinetic relationship, isoequilibrium relationship, and enthalpy-entropy compensation. *Chemical Review*, **101**, 673-695.

[52] Minemoto, Y., Adachi, S., Shimada, Y., Iwata, T., Yamaguchi, Y., Yamamoto, T., Kometani, T. & Matsuno, R. (2003). Oxidation kinetics for cis-9,trans-11 and trans-10,cis-12 isomers of conjugated linoleic acid. *Journal of American Oil Chemists' Society*, **80**, 675-678.

[53] Zhang, L., Somasundaran, P. & Maltesh, C. (1996). Electrolyte effects on the surface tension and micellization of n-dodecyl β-D-maltoside solutions. *Langmuir*, **12**, 2371-2373.

[54] Shinoda, K., Yamaguchi, T. & Hori, R. (1961). The surface tension and the critical micelle concentration in aqueous solution of β-D-alkyl glucosides and their mixtures. *Bulletin of the Chemical Society of Japan*, **34**, 237-241.

[55] Adachi, S. & Matsuno, R. (1997). Effect of eluent composition on the distribution coefficient of saccharides onto a cation-exchange resin in sodium-ion form. *Bioscience Biotechnology and Biochemistry*, 61, 1296-1301.

[56] Kröger-Ohlsen, M. & Skibsted, L. H. (1997). Kinetics and mechanism of reduction of ferrylmyoglobin by ascorbate and D-isoascorbate, *Journal of Agricultural and Food Chemistry*, **45**, 668–676.

[57] Shinoda, K. & Becher, P. (1978). *Principles of solution and solubility*. New York, Marcel Dekker.

[58] Pearce, K. N. & Kinsella, J. E. (1978). Emulsifying properties of proteins: evaluation of a turbidimetric technique. *Journal of Agricultural and Food Chemistry*, **26**, 716-723.

[59] Davies, J. T. (1957). A quantitative kinetic theory of emulsion type, I. Physical chemistry of the emulsifying agent. *Proceedings of the Second International Congress of Surface Activity*. London, Butterworths Scientific Publications.

[60] McClements, D. J. (1999). *Food emulsions: principles, practice, and techniques*. Boca Raton, CRC Press.

[61] Hashimoto, N., Aoyama, T. & Shioiri, T. (1981). A simple efficient preparation of methyl easters with trimethylsilyldiazomethane ($TMSCHN_2$) and its application to gas chromatographic analysis of fatty acids. *Chemical and Pharmaceutical Bulletin*, **29**, 1475-1478.

[62] Minemoto, Y., Hakamata, K., Adachi, S. & Matsuno, R. (2002). Oxidation of linoleic acid encapsulated with gum arabic or maltodextrin by spray-drying. *Journal of Microencapsulation*, **19**, 181-189.

[63] Adachi, S., Ishiguro, T. & Matsuno, R. (1995). Autoxidation kinetics for fatty acids and their esters. *Journal of American Oil Chemists' Society*, **72**, 547-551.

[64] Ishido, E., Minemoto, Y., Adachi, S. & Matsuno, R. (2001). Autoxidation process of linoleic acid and methyl linoleate mixed with saturated fatty acid or its methyl ester, *Lebensmittel-Wissenschaft und –Technologie*, **34**, 234-238.

[65] Gibbs, B. F., Kermasha, S., Alli, I. & Mulligan, C. N. (1999). Encapsulation in food industry. *International Journal of Food Science and Nutrition*, **50**, 213-224.

[66] Maloney, J. F., Labuza, T. P., Wallace, D. H. & Karel, M. (1966). Autoxidation of methyl linoleate in freeze-dried model system I. Effect of water on the autocatalyzed oxidation. *Journal of Food Science,* **31**, 878-884.

[67] Lin, C. C., Lin, S. Y. & Hwang, L. S. (1995). Microencapsulation of squid oil with hydrophilic macromolecules for oxidative and thermal stabilization. *Journal of Food Science,* **60**, 36-39.

[68] Moreau, D. L. & Rosenberg, M. (1996). Oxidative stability of ahhydrous milkfat microencapsulated in whey proteins. *Journal of Food Science,* **61**, 39-43.

[69] Ré, M. I. (1998). Microencapsulation by spray-drying. *Drying Technology*, **16**, 1195-1236.

[70] Ishido, E., Hakamata, K., Minemoto, Y., Adachi, S. & Matsuno, R. (2002). Oxidation process of linoleic acid encapsulated with a polysaccharide by spray-drying. *Food Science and Technology Research*, **8**, 85-88.

[71] Randall, R. C., Phillips, G. O. & Williams, P. A. (1988). The role of the proteineous component on the emulsifying properties of gum arabic. *Food Hydrocolloids*, **2**, 131–140.

[72] Maeda, H. (2000). Soluble soybean polysaccharide. In: Phillips, G. O. and Williams, P. A. (Ed), *Handbook of Hydrocolloids*. Cambridge, CRC Press, Woodhead Publishing.

[73] Ishido, E., Minemoto, Y., Adachi, S. & Matsuno, R. (2003). Heterogeneity during autoxidation of linoleic acid encapsulated with a polysaccharide. *Journal of Food Engineering*, **4**, 19–23.

[74] Minemoto, Y., Adachi, S. & Matsuno, R. (1999). Autoxidation of linoleic acid encapsulated with polysaccharides of different weight ratio, *Bioscience, Biotechnology and Biochemistry*, **63**, 866–869.

In: Biochemical Engineering
Editors: F.E. Dumont and J.A. Sacco, pp. 75-115

ISBN: 978-1-60741-257-1

Chapter 3

APPLICATION OF A NATURAL BIOPOLYMER POLY (γ-GLUTAMIC ACID) AS A BIOFLOCCULANT AND ADSORBENT FOR CATIONIC DYES AND CHEMICAL MUTAGENS: AN OVERVIEW

***B. Stephen Inbaraj*† *and B.H. Chen* ***

Department of Food Science, Fu Jen University, Taipei 242, Taiwan

Abstract

Poly(γ-glutamic acid) (γ-PGA), a novel polyanionic and multifunctional macromolecule synthesized by *Bacillus* species, has attracted considerable attention because of its eco-friendly, biodegradable and biocompatible characteristics. Recently, its application in a wide range of fields such as food, agriculture, medicine, hygiene, cosmetics and environment has been explored. This book chapter reviews the literature reports on the application of γ-PGA as a flocculating agent, and adsorbent for cationic dyes and chemical mutagens, affected by several process parameters including pH, temperature, contact time, metal cations, concentration and molecular weight of γ-PGA.

1. Introduction

Biopolymers are molecules produced by biological cells including bacteria, fungi, plant and animal cells. Poly(γ-glutamic acid) (γ-PGA) is a natural and water-soluble biopolymer produced by a variety of *Bacillus* species through fermentation. The γ-PGA was first discovered as a major constituent in the capsule of *Bacillus anthracis*, which was subsequently released into the growth medium upon autoclaving or aging and autolysis of the cells [1,2]. In a later study, Fujji [3-5] reported fermented soybean (Natto), a traditional health food in Japan, also contained γ-PGA in the form of viscous sticky mucilage, with *B. subtilis*

* E-mail address: 002622@mail.fju.edu.tw. Tel.: 886-2-29053626, Fax: 886-2-29053415.
† E-mail address: sinbaraj@yahoo.com

natto being responsible for its synthesis. Ever since Bovarnick [6] reported the free secretion of γ-PGA in the growth medium of *B. subtilis*, several researchers have explored the extracellular synthesis of γ-PGA from a variety of *Bacillus* species [5]. Although most studies on γ-PGA was carried out between 1950s and 1970s [7-13], an increasing attention was drawn in recent years mainly because of its biodegradable and biocompatible properties.

The γ-PGA is primarily composed of repetitive glutamic acid monomer units, which are connected by γ-amide linkages between α-amino and γ-carboxyl groups and are thus synthesized in a ribosome-independent manner [5,14,15] (Figure 1). The naturally-produced γ-PGA contains nearly equal proportion of D- and L-glutamic acid units, yet the ratio of two optical isomers can be partially controlled by technological methods, yielding γ-PGA with different degree of stereoselectivity [5,14-16]. In addition, γ-PGA can be synthesized in different salt forms (Na, K, Ca, Mg and NH_4) and varying molecular weights (10,000 to 2 million Daltons) [5,14,15]. Owing to its non-toxic, polyanionic and multifunctional characteristics, γ-PGA finds potential application in a wide range of fields such as food, agriculture, cosmetics, medicine and environment [5,14,15]. Some specific applications of γ-PGA include its use as a health food, thickener, humectant, bitterness-relieving agent, osteoporosis-preventing agent, cryoprotectant, drug carrier, sustained release material, curable biological adhesive, biodegradable fibers, hydrogel (super water absorbent), moisturizer, tissue engineering material, biodegradable packing material, dispersant, flocculant, adsorbent of toxic cations and chemical mutagens, animal feed additive, enzyme immobilizing material, liquid display and conductive display materials [17-32]. This book chapter intends to review the reported studies on the application of γ-PGA as a bioflocculant, and an adsorbent for scavenging cationic dyes and chemical mutagens.

2. Biosynthesis and Physico-chemical Properties of γ-PGA

2.1. Biosynthesis

Several *Bacillus* species produce γ-PGA as a capsular component or an extracellular viscous material outside the cell body for subsequent release into the fermentation broth. To enhance the productivity of γ-PGA, researchers have investigated the nutrient requirements [5], which varied according to the type of *Bacillus* strain. Based on the nutrient requirements, γ-PGA-producing bacteria are divided into two groups: one requires the addition of L-glutamic acid (*de nova* method) to the medium to stimulate both cell growth and γ-PGA synthesis, while the other does not (salvage bioconversion method) [5,33-35]. Table 1 summarizes the nutrient requirements, cultivation condition, production yield, D/L glutamic acid ratio and molecular weight of γ-PGA synthesized from various *Bacillus* strains [7-9,36-40]. Besides carbon and nitrogen sources, factors such as ionic strength, aeration and medium pH, all of which can affect the productivity, molecular weight, stereochemical composition and quality of γ-PGA [5,41]. For instance, the molecular weight of γ-PGA synthesized from *B. licheniformis* ATCC 9945A increased by a factor of approximately 1.8 (1.2 to 2.2 million g/mol) for a concentration rise of NaCl from 0 to 4% [5,42]. The schematic pathway proposed for synthesis of γ-PGA from *B. subtilis* IFO 3335 is depicted in Figure 2 [42]. A more detailed description on the biosynthesis of γ-PGA has been well reviewed [5,8,14,15,41,43], and an elaborate discussion here is beyond the scope of this chapter.

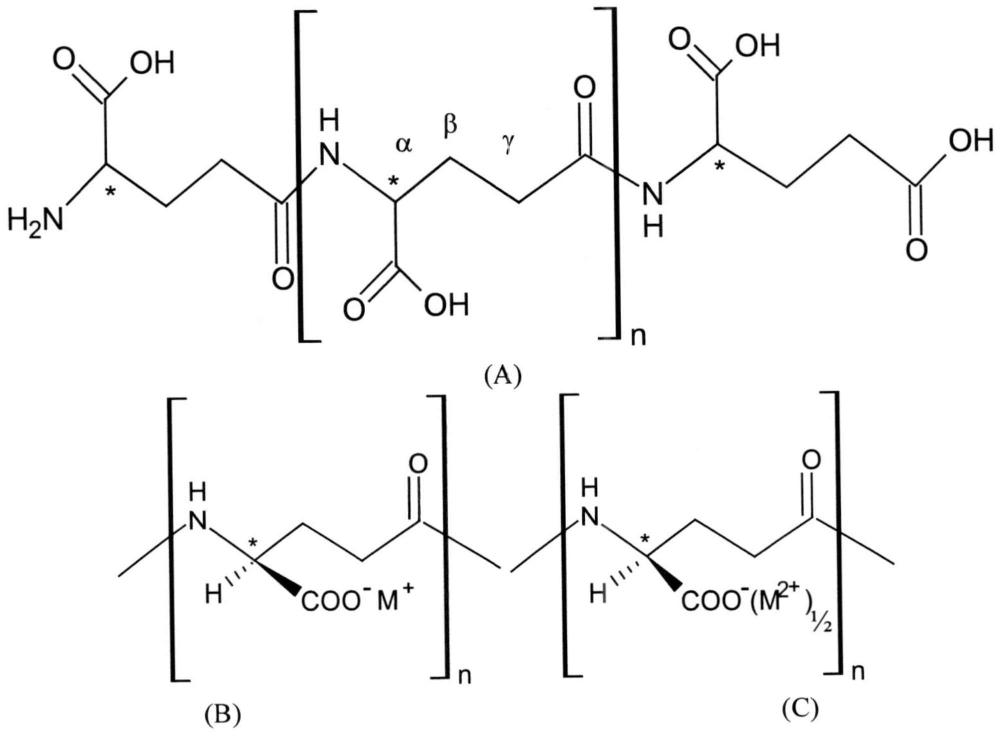

Figure 1. The chemical structure of poly(γ-D,L-glutamic acid) and poly(γ-glutamates). (A) H-form of γ-PGA; (B) monovalent metal salt form of γ-PGA; (C) divalent metal salt form of γ-PGA; $M^+ = Na^+$, K^+ or NH_4^+; $M^{2+} = Ca^{2+}$ or Mg^{2+}; Greek symbols (α, β, γ) denote carbon positions; Asterisk (*) symbol indicate the chiral carbon atom.

2.2. Physico-chemical Properties

The γ-PGA biosynthesized from *B. subtilis* (natto) was characterized for various physical and chemical properties by using several instrumental techniques including Fourier transform infrared spectrophotometry (FT-IR), ^{1}H- and ^{13}C-nuclear magnetic resonance spectroscopy (^{1}H- and ^{13}C-NMR), differential scanning calorimetry (DSC) and thermal gravimetric analysis (TGA), and the results are furnished in Table 2 [15]. Elemental analysis of purified H-form of γ-PGA at 1.23 x 10^6 kDa (C: 44.86%; H: 5.91%; N: 10.49%; S: 0%) was in close agreement with the calculated values (C: 46.51%; H: 5.43%; N: 10.85%; S: 0%) based on the formula composition [15]. The viscosity of γ-PGA was reported to be strongly dependent on pH, γ-PGA concentration, temperature and ionic strength [15,44], which should be due to the abrupt changes caused by these parameters on the conformation of γ-PGA. The viscosity of 4% Na-γ-PGA solution at 25°C increased following a rise in pH and γ-PGA dose, but decreased with temperature and ionic strength [15]. The γ-PGA remained unaltered on heating at 80°C for 60 min, however, it was rapidly hydrolyzed at 120°C because of random chain scission [33].

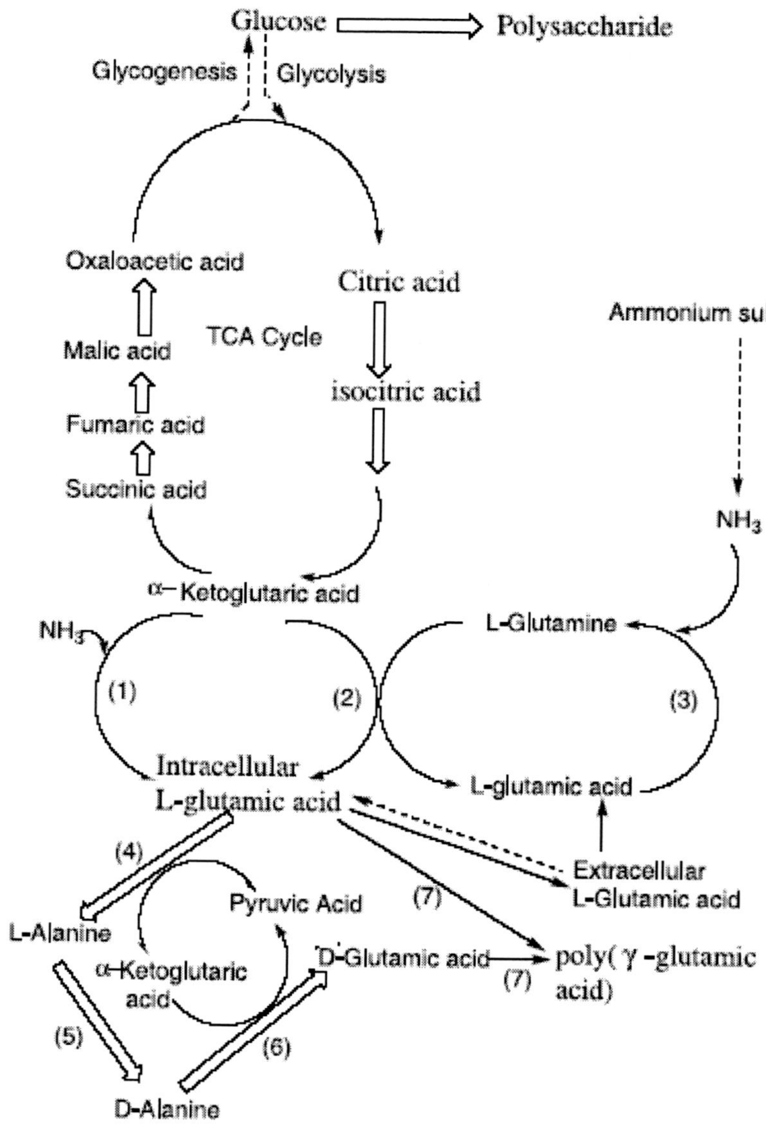

Figure 2. Proposed pathway of biosynthesis of poly(γ-glutamic acid) from *B. subtilis* IFO 3335. (1) Glutamate dehydronase; (2) Glutamate 2-oxoglutarate (α-ketoglutarate) aminotrasferase; (3) Glutamine synthetase; (4) Pyruvic acid aminotransferase (L-Glutamic acid); (5) Alanine racemase; (6) Pyruvic acid aminotransferase (D-Glutamic acid); (7) PGA synthetase (Source: Kunioka 1997).

Table 1. Synthesis of γ-PGA from various *Bacillus* strains

***Bacillus* strains**	**Nutrients**	**Cultivation conditions**	**Yield (g/L)**	**Molecular weight**	**D : L ratio**	**References**
B. licheniformis ATCC 9945	Glutamic acid, Glycerol,	37°C, 4 days	17-23	1.4×10^4 – 9.8×10^5	45-85 : 55-15	[8]
B. licheniformis ATCC 9945	Citric acid, NH_4Cl,	37°C, 4 days	17-23	1.4×10^4 – 9.8×10^5	45-85 : 55-15	[36]

Table 1. Continued

Bacillus strains	Nutrients	Cultivation conditions	Yield (g/L)	Molecular weight	D : L ratio	Refere-nces
B. subtilis IFO 3335	Glutamic acid (30 g/L), citric acid (20 g/L)	37°C, 2 days	10-20	1.0×10^5 – 2.0×10^6	17 : 83	[37]
B. subtilis TAM-4	Fructose (75 g/L), NH_4Cl (18 g/L)	30°C, 4 days	20	6.0×10^5 – 1.6×10^6	78 : 22	[38]
B. licheniformis A35	Glucose (75 g/L), NH_4Cl (18 g/L)	30°C, 3-5 days	8-12	3.0 – 5.0×10^5	59 : 41	[39]
B. subtilis F02-1	Glutamic acid (70 g/L), glucose (1 g/L), veal infusion broth (20 g/L)	30°C, 2-3 days	50	1.20×10^6	55 : 45	[34,35]
B. subtilis NRRL B2612	Wheat gluten (200 g/L)	33°C, 2-3 days	10-14	2.0×10^4	52 : 48	[7]
B. subtilis var. polyglutamicum	Glucose (50 g/L), urea (7.5 g/L)	30°C, 3-4 days	17-19	1.1×10^6	40 : 60	[9]
B. subtilis (natto) MR 141	Maltose (60 g/L), soy sauce (70 g/L), sodium glutamate (30 g/L)	40°C, 3-4 days	35	-	-	[40]
B. subtilis (natto)	Glucose (8%), sodium L-glutamate (10%)	37°C, 2 days	35	1.5 – 3.0×10^6	52 : 48	[15]

Table 2. Physical and chemical parameters of γ-PGA characterized by several instrumental techniques (Source: Ho et al. 2006)

Instrumental analysis	Form of poly(γ-glutamic acid)[a]					
	H^+	Na^+	K^+	NH_4^+	Ca^{2+}	Mg^{2+}
^{1}H-NMR (400 MHz, D_2O, 30°C)						
Chemical shift in ppm:						
α CH		3.98	4.00	3.68	4.18	4.08
β CH_2		1.98, 1.80	1.99, 1.80	1.68, 1.48	2.16, 1.93	2.05, 1.88
γ CH_2		2.19	2.19	1.93	2.38	2.31
^{13}C-NMR (67.9 MHz, D_2O, 30°C)						
Chemical shift in ppm:						
α CH		56.43	62.21		62.21	62.10
β CH_2		31.61	35.16		36.17	35.11

Table 2. Continued

Instrumental analysis	Form of poly(γ-glutamic acid)[a]					
	H^+	Na^+	K^+	NH_4^+	Ca^{2+}	Mg^{2+}
γ CH_2		34.01	39.74		39.68	39.60
CO		182.21	182.11		182.16	182.12
COO^-		182.69	185.46		185.82	185.16
FT-IR absorption (KBr), cm^{-1}						
C=O, stretch	1739					
Amide I, N-H bending		1643		1643	1622	1654
Amide II, stretch		1585				
C=O, symmetric stretch	1454	1402		1395	1412	1411
C-N, stretch	1162	1131		1139	1116	1089
N-H, oop bending	698	707		685	669	616
O-H, stretch	3449	3436		3443	3415	3402
Thermal analysis:						
Hydrated water (%)	0	10	42		20	40
Dehydration temp. (°C)		109	139		110	122
Melting point (°C)	206	160	193, 238	219		160
Decomposition temp. (°C)	209.8	340	341	223	335.7	331.8

[a] Molecular weight (kDa) of H-form=1.23 x 10^6, Na-form=1.23 x 10^6, K-form=0.98 x 10^6, NH_4-form=0.89 x 10^6, Ca-form=0.49 x 10^6, Mg-form=0.89 x 10^6.

2.3. Structural Characteristics

The monomer glutamic acid in γ-PGA possesses three chemically-active functional groups in the following order of reactivity: α-NH_2 > α-COOH > γ-COOH. In a chemical-catalyzed polymerization of glutamic acid, the condensation occurs between the α-COOH and α-NH_2 groups, resulting in the formation of poly(α-glutamic acid) through α-peptide linkages. However, in the submerged fermentation process, a portion of L-glutamic acid is enzymatically racemized to D-glutamic acid and eventually both D- and L- glutamic acids are co-polymerized through a novel γ-peptide bonding between a less reactive γ-COOH and α-NH_2 groups to form poly(γ-D,L-glutamic acid). An α-peptide bond, which is normally found in protein structures, can be easily decomposed by most of the protease enzymes, whereas γ-peptide bond can be hydrolyzed only by a rarely available γ-glutamyltranspeptidase and γ-PGA is thus considerably resistant to microbial attack [5,14,15]. Likewise, poly(γ-glutamic acid) contains four intramolecular hydrogen bonds

(3_{19}, 3_{17}, 3_{14} and 3_{12}), formed between the carbonyl group of one γ-peptide bond and the amino group of another γ-peptide bond within every 3 glutamic acid moieties [15,45], as opposed to only one hydrogen bond within an average of 3.6 amino acid residues in proteins found in nature. Optical rotatory dispersion (ORD) studies [15] have shown these strong hydrogen bonds in γ-PGA to be responsible for the compact α-helix conformation rendering a strong hydrophobic character (insoluble nature) at pH 2.0. But, as pH rises, the hydrogen bonding breaks down and the α-helix conformation is converted into a linear random-coil conformation resulting in ionization of α-COOH groups [5,14,15]. At pH = pK_a (4.09, determined by potentiometric titration), about 50% of α-COOH groups are ionized to α-COO^- anions (50% polyanionic random-coil transition). Beyond pH 6.0, γ-PGA mainly exists in the linear random-coil conformation and exhibits polyanionic characteristics, making it a novel multi-functional biopolymer for application in a wide range of fields.

3. γ-PGA as a Bioflocculant

Flocculation is a process of promoting the agglomeration of smaller particles into larger and more easily sedimentable flocs. Flocculating agents have been frequently used in wastewater treatment, food and fermentation industries, drinking water treatment and industrial downstream processing [46-50]. Among various flocculants, synthetic organic flocculants (polyacrylamide derivatives, polyacrylic acids and polyethylene imine) are widely applied due to their low cost and high efficiency [51]. Nevertheless, their use often gives rise to environmental and health problems as they are not readily biodegradable and some of their monomers such as acrylamide are neurotoxic and potential carcinogens [52-54]. Thus, the development of safe, biodegradable flocculants is imperative to reduce the health risks associated with synthetic flocculants. This chapter subsection unveils the literature reports on flocculation capability of γ-PGA as affected by pH, temperature, γ-PGA concentration, metal cations, cross-linking γ-PGA and molecular weight of γ-PGA.

3.1. Assay of Flocculation Activity and Flocculation Rate

Measurement of flocculation activity was based on decrease in turbidity through addition of γ-PGA in a reaction mixture containing test suspension and metal cations [30,31,54,55]. After the visible flocs were settled, the optical density (OD) of the supernatant of reaction mixture with or without γ-PGA (control) was measured at 550 or 660 nm using a spectrophotometer, and the flocculation activity and flocculation rate were calculated using the following equations:

$$\text{Flocculation activity}\,(1/\text{OD}) = \frac{1}{\text{OD}_{\text{sample}}} - \frac{1}{\text{OD}_{\text{control}}} \tag{1}$$

$$\text{Flocculation rate (\%)} = \frac{OD_{control} - OD_{sample}}{OD_{control}} \times 100 \tag{2}$$

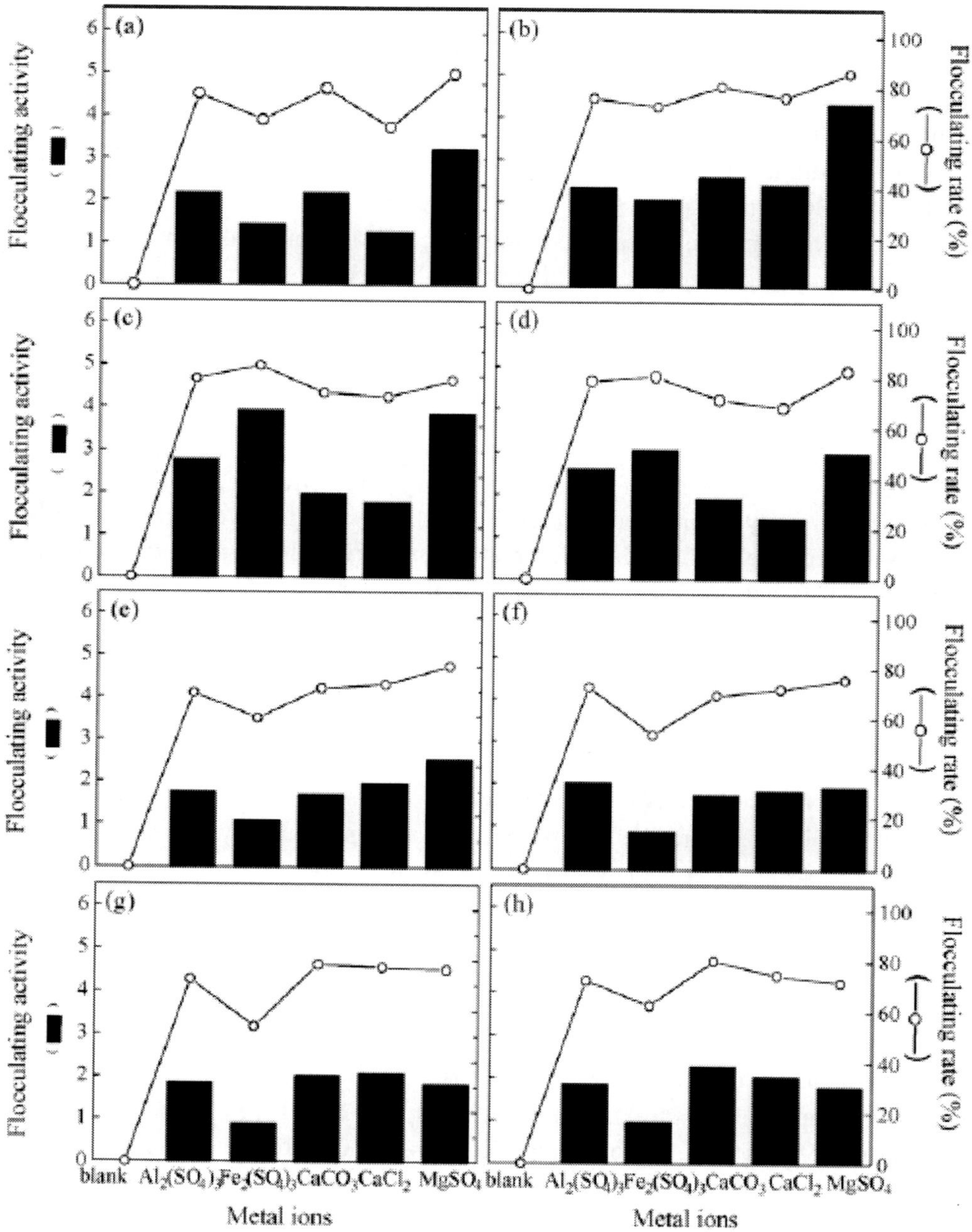

Figure 3. Effect of γ-PGA concentration on flocculation activity and flocculating rate at low metal concentration (5×10^{-3} mM) and pH 7.0. γ-PGA concentration (mg/L): (a) 10; (b) 20; (c) 30; (d) 40; (e) 50; (f) 60; (g) 70; (h) 80 (Source: Wu and Ye 2007).

3.2. Effect of γ-PGA Concentration and Molecular Weight

When evaluated at different concentrations (5-100 mg/L), the γ-PGA from PY-90 and IFO3335 strains of *B. subtilis* showed an optimum flocculation activity of 15 and 10.5 l/OD respectively, for a 20-mg/L kaolin suspension containing 4.5 and 10 mM $CaCl_2$ [54,55]. However, Shih et al. [30] found the optimum flocculation (8.5 l/OD) to occur at a γ-PGA dose of 3.7 mg/L in a 5 g/L kaolin suspension containing 9.0 mM $CaCl_2$. This variation in γ-PGA level may be attributed to different *Bacillus* species and culture conditions employed in the latter study [30], in which *B. licheniformis* (CCRC 12826) and a combination of glutamic acid, citric acid and glycerol was used as carbon source, as opposed to *B. subtilis* and only glutamic acid used in the former study [54,55]. Recently, Wu et al. [31] examined the flocculation activity at various concentrations of γ-PGA from DYU1 *B. subtilis* strain (10-80 mg/L) in a reaction mixture containing kaolin and 5 x 10^{-3} mM of metal cations (Ca^{2+}, Mg^{2+}, Fe^{2+}, Fe^{3+} or Al^{3+}) and a level ranging from 10-30 mg/L was found effective in attaining the optimum flocculation activity (Figure 3).

The molecular weight (MW) of γ-PGA also showed a significant impact on the flocculation activity. A high MW γ-PGA (2 x 10^6 kDa) from *B. licheniformis* (CCRC 12826) exhibited a higher flocculation activity compared to the purified low MW γ-PGA (~1 x 10^5 kDa) from Sigma [30]. Apparently, a high MW polymer contains a large number of free functional groups, which may act as bridges connecting numerous suspended particles to form flocs of large size during the flocculation reaction [49].

3.3. Effect of Added Metal Cations and pH

Incorporation of metal cations plays a key role in promoting flocculation, probably because of decline in charge density by cations leading to inter-particle bridging between suspended particles. Flocculation studies on bacterial γ-PGA isolate were unequivocal in reporting the synergistic effect of added metal cations, with the trivalent ions (Al^{3+} or Fe^{3+}) exerting the highest effect, followed by divalent (Ca^{2+}, Mg^{2+} or Fe^{2+}) and monovalent (Na^+ or K^+) ions [30,31,55]. The amount of metal ions required for maximum flocculating efficiency decreased following an increase in the valency of metal ions. Wu et al. [31] reported an optimum concentration of >10 mM for monovalent ions, but only 0.10-0.90 mM for divalent ions and <0.005 mM for trivalent ions in a reaction mixture of kaolin and γ-PGA from *B. subtilis* (DYU1) (Figure 4). Nonetheless, a high Ca^{2+} concentration of 2-8 mM and 13.5 mM was required to attain the maximum flocculation activity for γ-PGA produced from *B. subtilis* (PY-90) and *B. licheniformis* (CCRC 12826), respectively [30,54]. Unlike monovalent and divalent ions, a rise in concentration of trivalent ions (Al^{3+} or Fe^{3+}) sharply reduced the flocculation activity (Figure 4), which may be ascribed to the drop in solution pH caused by precipitation of these ions. However, the flocculation activity could be restored by adjusting pH of the reaction mixture to 7 as reported by Yokoi et al. [55], who proposed the hydroxide precipitates of trivalent ions and Fe^{2+} ions to be effective in stimulating the flocculation activity of γ-PGA at or near neutral pH. Likewise, for both Ca^{2+} and Mg^{2+} cations, the pH around 7 was shown to be optimum in achieving the largest flocculation activity and flocculating rate by γ-PGA from *B. subtilis* DYU1 (Figure 5) [30]. On the contrary, a slightly acidic pH around 4 was reported for Ca^{2+} ions to reach the maximum flocculating effect in the suspension mixture containing kaolin and γ-PGA from *B. subtilis* IFO 3335 [54,55]. The reason for this disparity is unclear.

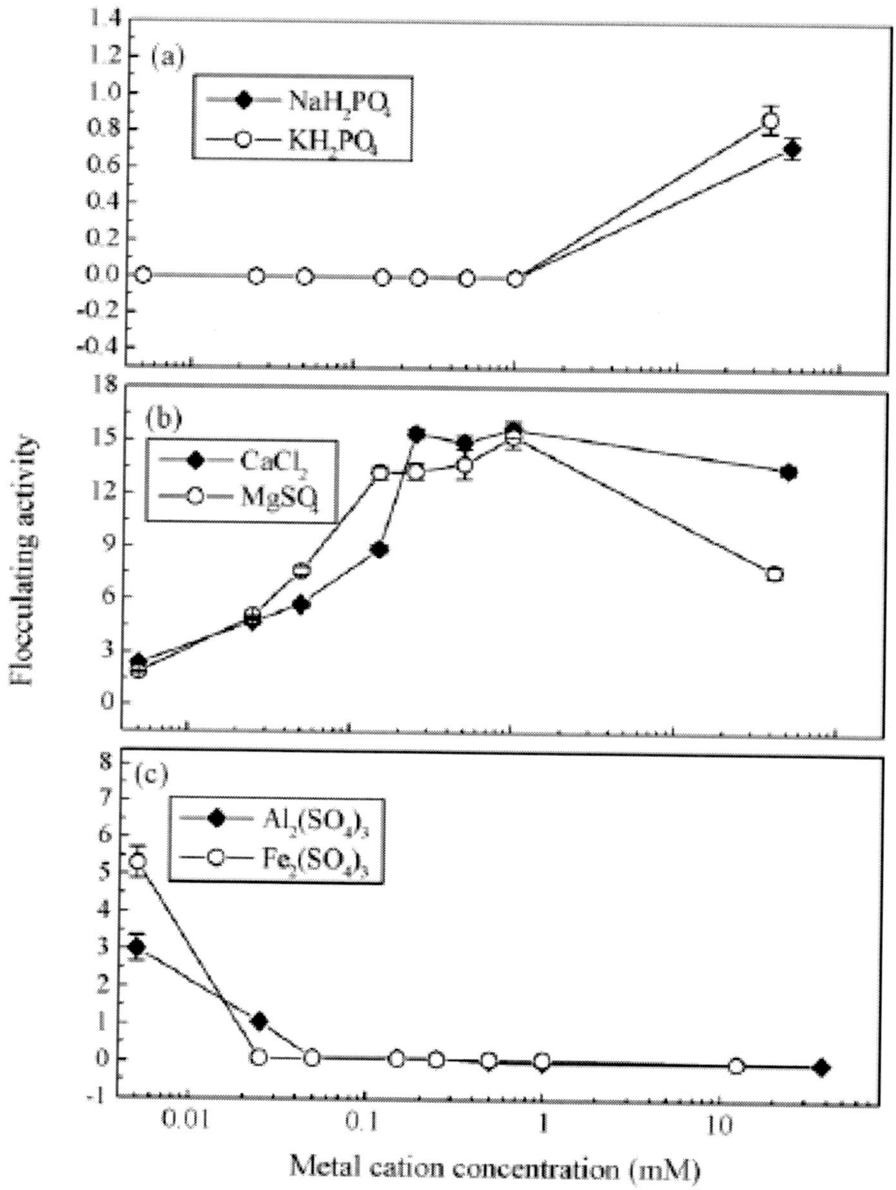

Source: Wsu and Ye 2007.

Figure 4. Effect of metal cation concentration on flocculation activity of γ-PGA from *B. subtilis* in kaolin suspension. (a) monovalent cations; (b) divalent cations; (c) trivalent cations.

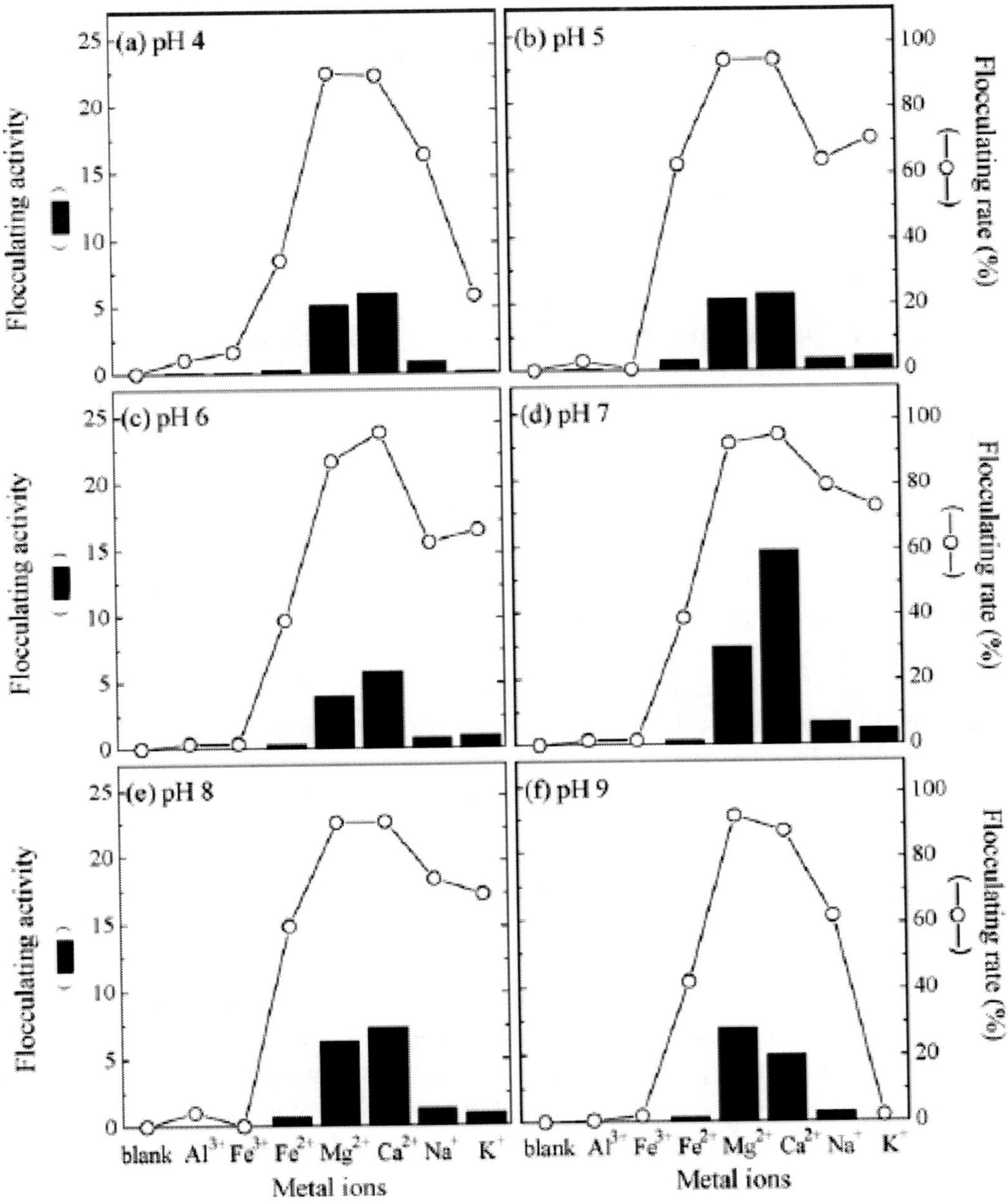

Source: Wu and Ye 2007.

Figure 5. Effect of metal ions and pH on flocculation activity and flocculating rate of kaolin suspension by γ-PGA from *B. subtilis* (DYU1). Metal concentration (mM): $Al_2(SO_4)_3$=37.5; $Fe_2(SO_4)_3$=12.5; $FeSO_4$=32.9; $CaCO_3$=49.9; NaH_2PO_4=32; KH_2PO_4= 36.7.

3.4. Effect of Temperature

The flocculation activity declined linearly with a rise in incubation temperature (30-120°C) of the reaction mixture containing kaolin suspension (5 g/L), metal cation (5 x 10^{-3} mM) and γ-PGA isolate from *B. subtilis* DYU1 (40 mg/L) [31]. The optimum flocculation activity (5.2 l/OD) and flocculation rate (90%) were attained at 30°C. But, the flocculation activity dropped by 50% after incubation at 60°C for 15 min, followed by a complete inactivation of γ-PGA at 120°C. Yokoi et al. [54] also reported a rapid decline in flocculation activity of γ-PGA from PY-90 *B. subtilis* strain at elevated temperatures, diminishing to zero on heating at 100°C (40 min). Obviously, the bioflocculants composed of protein or peptide backbone are susceptible to heat treatment, while those containing sugars and polysaccharides are thermally stable [56].

3.5. Effect of Various Inorganic and Organic Suspensions

The γ-PGA from different *Bacillus* species was not only effective in flocculating kaolin suspension, but also in organic suspensions and some other inorganic suspensions including active carbon, solid soil and acid clay (Table 3) [30,55]. The γ-PGA isolate from *B. subtilis* (IFO3335) at 20 mg/L exhibited a greater flocculating efficiency (l/OD) in acid clay suspension (10.9-34.7) compared to that in active carbon (2.8-3.3) and solid soil (1.0-1.9). In the absence of any added metal cations, γ-PGA flocculated suspensions of calcium and magnesium salts, with the highest flocculation activity (l/OD) being shown in $Ca(OH)_2$ suspension (21.3), followed by $MgCO_3$ (9.3), $Mg(OH)_2$ (8.1), Al_2O_3 (6.1), $[Ca(PO4)_2]_3{\cdot}Ca(OH)_2$ (5.7) and $CaCO_3$ (2.0) [30,55]. The γ-PGA from *B. licheniformis* (80 mg/L) showed a good flocculating effect in organic suspensions (5 g/L) as well, which equaled to 3.0, 2.6, 4.6 and 2.8 l/OD for cellulose powder, carboxymethylcellulose, *Saccharomyces cerevisiae* and *B. circulans*, respectively (Table 3) [30].

3.6. Effect of Crosslinking Poly(γ-glutamic Acid)

Crosslinking of γ-PGA (CL-γ-PGA) with an optimum γ-irradiation dose of 20 kGy was shown to generate the highest water absorption capacity (1005.6 mL/g) and viscosity (3.31 η) and possess twice the flocculation activity (5.48 l/OD) when compared to a non-crosslinked γ-PGA (2.74 l/OD) in kaolin suspension at 30°C and pH 5.0 [57]. However, a pretreatment with polyacrylamide (PAC) drastically raised the flocculation activity of CL-γ-PGA in different suspensions [57,58]. Table 4 summarizes the flocculation activity of γ-PGA as affected by pretreatment with different doses of PAC for various suspensions at 30°C and pH 5.0 [58]. Compared to the control (only CL-γ-PGA), the flocculation activity (l/OD) rose by 39.9, 32.8, 30.0, 23.5, 88.2 and 12.2 for kaolin, bentonite, diatomaceous earth, *E. coli*, *M. aeruginosa* and Mandai pond (real water sample) suspensions, respectively, for PAC and CL-γ-PGA at a dose (μg/mL) of 0.5 and 10, 2 and 5, 0.5 and 10, 0.5 and 10, 0.5 and 5, and 2 and 20 [58]. Apparently, the Al^{3+} ion in PAC could electrostatically interact with the negatively charged cell surface of *E. coli* and *M. aeruginosa* or with the silicate anion induced on the surface of kaolin, bentonite and diatomaceous earth suspensions to form small flocks, which may eventually become larger following the addition of CL-γ-PGA. Yet, the pretreatment with PAC did not show any effect on the real sample suspension from Yamato River (a dirtiest river in Japan). In contrast, a reversed trend was shown in crystal violet (CV) dye suspension, probably due to electrostatic interaction between the iminium cation in CV and

COO^- in CL-γ-PGA. Among various added metal cations, the trivalent ions (Al^{3+} and Fe^{3+}) demonstrated a marked effect on flocculation activity of CL-γ-PGA, whereas the divalent ions (except Mn^{2+}) and monovalent ions did not show any effect [57].

Under the optimum conditions of CL-γ-PGA and PAC, the flocculating rate expressed in "decrease in turbidity" increased linearly with incubation time for bentonite suspension, while it was rapid in the initial 5 min followed by a gradual decline for diatomaceous earth (Figure 6) [58]. This may be caused by the faster precipitating behavior of diatomaceous earth compared to bentonite. On the other hand, for *E. coli* suspension, no flocculation was observed until 5 min, but increased with time thereafter [58]. However, a sharp decline in turbidity was noticed within 5 min for *M. aeruginosa* and then reached a plateau (Figure 6).

Table 3. Flocculation activity of γ-PGA in various inorganic and organic suspensions[a]

Suspensions	Metal cations	Flocculation activity (l/OD)
Inorganic suspension		
Kaolin[b]	90 mM Ca^{2+}	8.5
	8 mM Ca^{2+}	3.3
Active carbon[c]	2 mM Mg^{2+}	2.9
	0.05 mM Fe^{2+}	2.8
	8 mM Ca^{2+}	1.1
Solid soil[c]	8 mM Mg^{2+}	1.0
	6 mM Fe^{2+}	1.9
	6 mM Ca^{2+}	10.9
Acid clay[c]	4 mM Mg^{2+}	14.0
	0.1 mM Fe^{2+}	34.7
$Ca(OH)_2$[c]	Nil	21.3
$CaCO_3$[c]	Nil	2.0
$[Ca(PO4)_2]_3 \cdot Ca(OH)_2$[c]	Nil	5.7
$Mg(OH)_2$[b]	Nil	8.1
$MgCO_3$[c]	Nil	9.3
Al_2O_3[b]	Nil	6.1
Organic suspension[b]		
Cellulose powder	90 mM Ca^{2+}	3.0
Carboxymethylcellulose	90 mM Ca^{2+}	2.6
Saccharomyces cerevisiae	90 mM Ca^{2+}	4.6
Bacillus circulans	90 mM Ca^{2+}	2.8

[a] suspension concentration: 5 g/L; [b] treated with γ-PGA from *B. licheniformis* (CCRC 12826) at 80 mg/L; [c] treated with γ-PGA from *B. subtilis* (IFO 3335) at 20, 10 and 10 mg/L for Ca^{2+}, Mg^{2+} and Fe^{2+}, respectively.

Source: Yokoi et al. 1996; Shih et al. 2001.

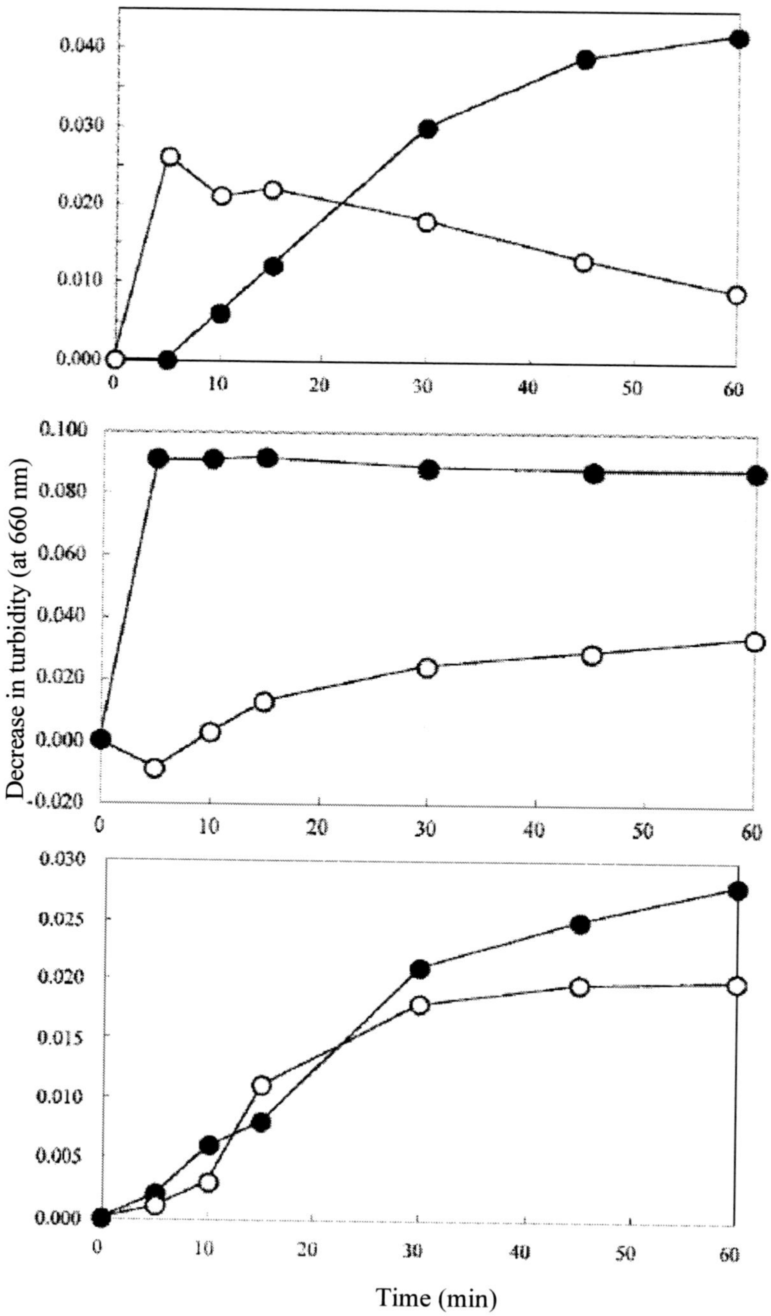

Source: Taniguchi et al. 2005.

Figure 6. Changes in the turbidity of various suspensions by treatment with PAC (incubation at 30°C and pH 5 for 10 min) followed by CL-γ-PGA. (A) *M. aeruginosa* M21 (closed circles) and *E. coli* IFO 3992 (open circles) pretreated with PAC (0.5 μg/mL) followed by CL-γ-PGA at 5 and 10 μg/mL, respectively. (B) Bentonite (closed circles) and diatomaceous earth (open circles) pretreated with 2 and 0.5 μg/mL PAC followed by CL-γ-PGA at 5 and 10 μg/mL, respectively. (C) Real sample suspensions, Mandai pond (closed circles) and Yamato river (open circles) pretreated with 2 and 0 μg/mL PAC, respectively, followed by CL-γ-PGA at 20 μg/mL.

Table 4. Flocculation activity of cross-linked γ-PGA (CL-γ-PGA) on various suspensions pretreated with different concentrations of polyacrylamide (PAC)

Suspension	CL-γ-PGA (μg/mL)	Flocculation activity (l/OD)[a] PAC (μg/mL) 0	0.1 (0.01)[b]	0.25 (0.025)[b]	0.5 (0.05)[b]	1 (0.1)[b]	2 (0.25)[b]	4 (0.5)[b]
Kaolin	10	–	–	–	14.54	39.94	41.56	–
Bentonite	5	0.6	–	–	-5.2	2.9	33.4	0.0
Diatomaceous earth	10	-0.2	–	23.3	30.0	29.6	33.4	–
Escherichia coli	10	2.4	-7.2	16.0	25.9	8.5	1.1	–
Mycrocystis aeruginosa	5	-0.3	6.6	9.5	29.3	41.0	71.7	88.2
Crystal violet	5	60.7	50.4	38.1	18.4	8.1	2.2	0.6
Mandai pond	20	10.8	7.4	7.9	6.1	12.6	23.0	–
Yamato river	20	28.2	25.1	28.7	26.6	26.7	26.1	–

Source: Taniguchi et al. 2005.

[a] incubation at 30°C and pH 5 for 30 min; [b] PAC dose for *Mycrocystis aeruginosa.*

3.7. Mechanism of Flocculation

Among various factors, the metal cations played a prominent role in affecting the flocculation activity of γ-PGA. In accordance with Schulze Hardy's law [59], although both suspended particles (like kaolin) and the γ-PGA-based-bioflocculants are anionic, their interaction to facilitate flocculation is possible only through charge neutralization and bridging by added metal cations. Based on this phenomenon, two flocculation mechanisms were suggested [31] as follows:

I. The metal cations may decrease the negative charge on suspended particles, which in turn result in a charge reversal from negative to positive. Finally, the negatively charged carboxyl group (COO^-) of γ-PGA may react with the positively charged site of the suspended cation-kaolin particles (Figure 7a).
II. The metal cations may neutralize and stabilize the residual negative charges of the γ-PGA first, followed by bridge formation with the negatively charged kaolin particles (Figure 7b).

Thus, metals cations could facilitate charge neutralization and bridging between γ-PGA and suspended particles, resulting in an increase in floc density, floc size and floc resistance to shear. This explains why both the trivalent and bivalent cations possess stronger synergistic effect than monovalent ions.

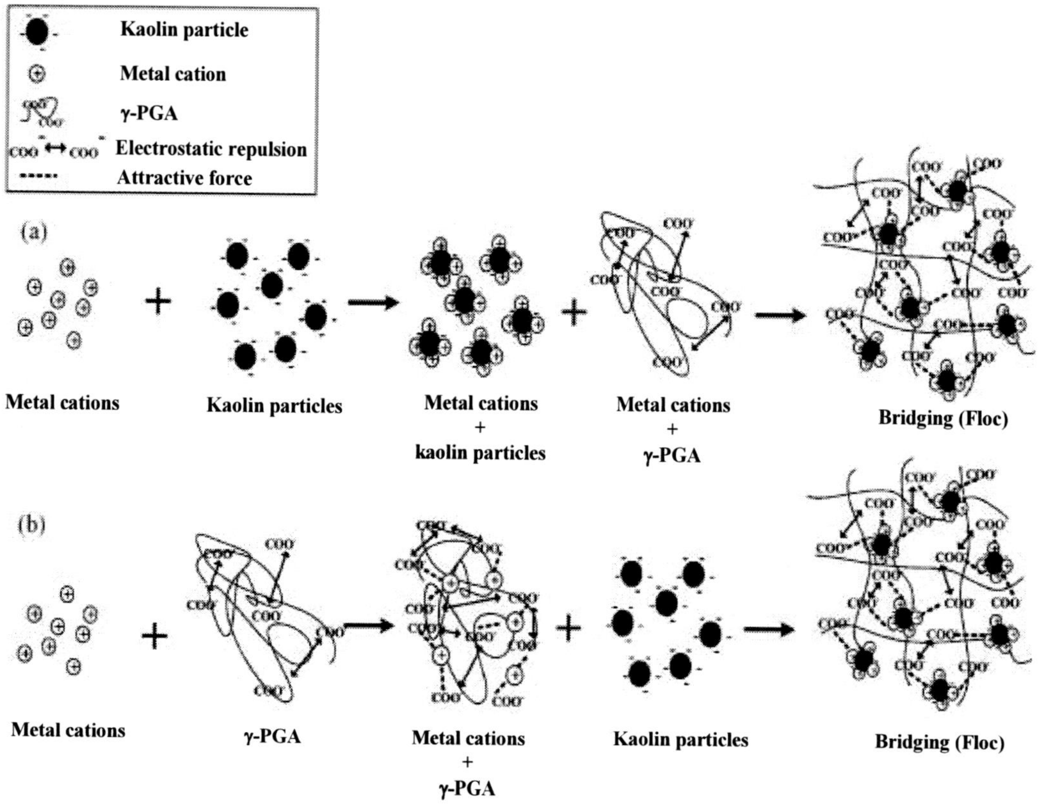

Source: Wu and Ye 2007.

Figure 7. Proposed mechanism of flocculation by γ-PGA.

4. γ-PGA as an Adsorbent of Cationic Dyes

Cationic dyes are the brightest class of soluble dyes used by the textile industry and the waste discharge accumulates heavy organic load in the environment causing a harmful impact on both aquatic organisms and human [60-64]. Because of the limitations prevailing in the conventional treatment methods [62], there has been a continuous search for a safe, eco-friendly and biodegradable adsorbent to treat dye-containing wastewaters. Lately, the bioremediation of dyes employing extracellular polymeric substances (EPS) produced by microorganisms is gaining attention [65-69], and this part of the chapter consolidates the published reports on evaluating γ-PGA as an adsorbent of cationic dyes. In all the three studies reported here [27,70,71], the γ-PGA (H-form; MW 990 kDa) produced from *B. subtilis* by salvage bioconversion pathway was employed. Figure 8 presents the molecular structure of cationic dyes studied along with their color index name, color index classification number, formula weight and wavelength of maximum absorption (λ_{max}).

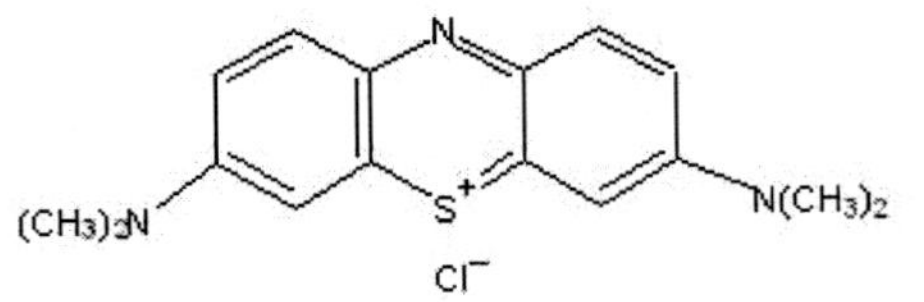

Methylene blue (MB)
(Basic blue 9, C.I. 52015, FW 373.9 g, λ_{max} 663 nm)

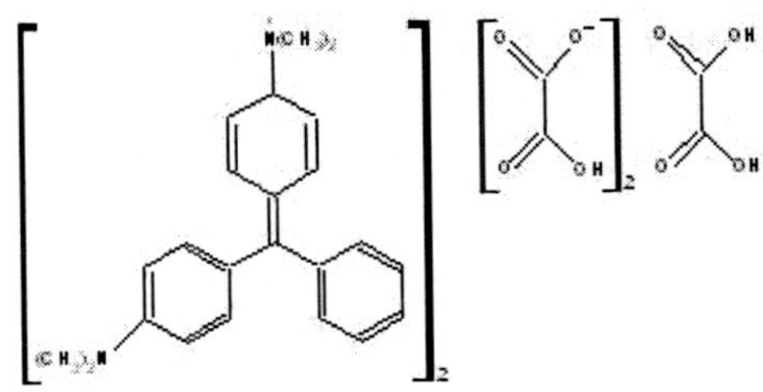

Malachite green (MG)
(Basic green, C.I. 42000, FW 929.0 g, λ_{max} 617 nm)

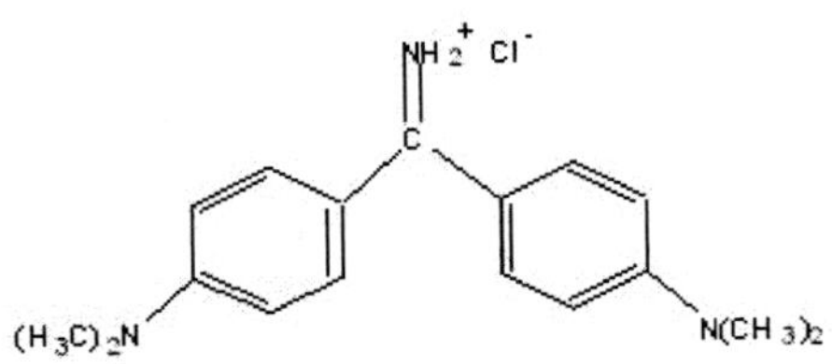

Auramine O (Au-O)
(Basic yellow 2, C.I. 41000, FW 303.8 g, λ_{max} 432 nm)

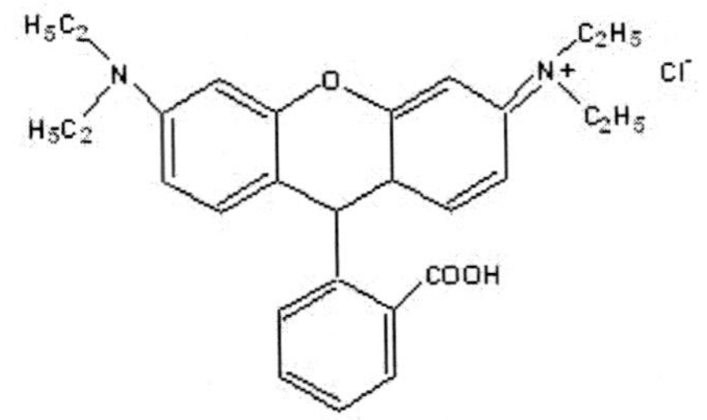

Rhodamine B (Rh-B)
(Basic violet 10, C.I. 45170, FW 479.0 g, λ_{max} 553 nm)

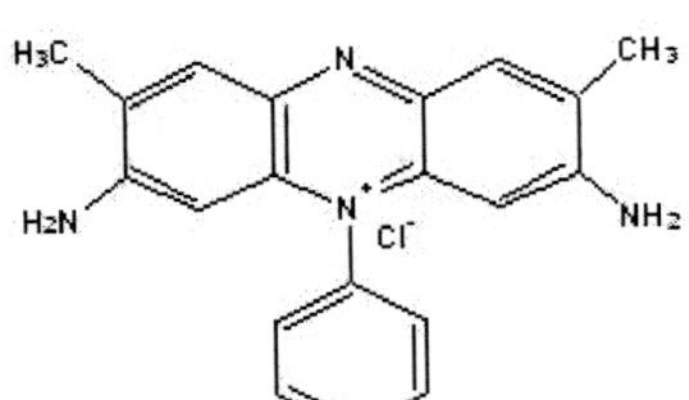

Safranin O (Sa-O)
(Basic red 2, C.I. 50240, FW 350.9 g, λ_{max} 518 nm)

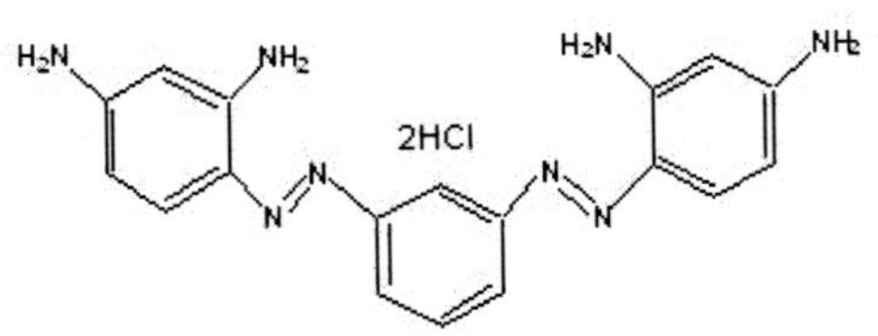

Bismarck brown Y (BB-Y)
(Basic brown 1, C.I. 21000, FW 419.3 g, λ_{max} 460 nm)

Figure 8. Molecular structure of cationic dyes. C.I.: Color Index classification number; FW: Formula weight.

4.1. Adsorption Kinetics

Process performance and ultimate cost of an adsorption system depend on the effectiveness of process design and efficiency of process operation, which requires an understanding of the kinetics of uptake or the time dependence of the concentration distribution of the solute in both bulk solution and solid adsorbent. The time (min)-uptake (mg/g) profile for adsorption of 100 or 200 mg/L of cationic dyes by γ-PGA was characterized by a fast dye uptake, attaining equilibrium within 30 or 60 min, respectively, for

Au-O, Rh-B and Sa-O dyes. However, a large portion (>85%) of the equilibrium uptake was accomplished within 12 or 20 min, indicating a rapid surface adsorption of cationic dyes did occur on γ-PGA [27]. The kinetic data were modeled with the following pseudo first order and pseudo second order equations [72-74] given by

$$q_t = q_e[1 - \exp(-k_1 t)] \tag{3}$$

$$q_t = \frac{t}{\frac{1}{k_2 q_e^2} + \frac{t}{q_e}} \tag{4}$$

where, q_t and q_e is the dye uptake (mg/g) at time t and equilibrium, respectively, k_1 (L/min) and k_2 (g/mg min) are first order and second order rate constants, and $k_2 q_e^2$ (mg/g min) represents initial adsorption rate (h). Fitted kinetic parameters obtained by non-linear regression (Marquart-Levenberg algorithm) are summarized in Table 5 [27,70,71]. Based on two error parameters, r^2 (coefficient of determination) and χ^2 (chi-square statistic), the pseudo second order model gave a more precise fit, suggesting the rate of dye uptake may be largely controlled by chemisorption mechanism involving valence forces through sharing or exchange of electrons [74]. Following a rise in dye concentration from 100-200 mg/L, the second order rate (g/mg min) of dye uptake dropped from $1.04 \times 10^{-2} - 4.75 \times 10^{-3}$ for Au-O, $6.95 \times 10^{-3} - 4.50 \times 10^{-3}$ for Rh-B and $2.18 \times 10^{-3} - 5.25 \times 10^{-4}$ for Sa-O. It is apparent that once the readily available active sites on γ-PGA are occupied, the rate of dye uptake decreases [27].

The adsorption kinetics of BB-Y dye (100 mg/L) conducted at three different temperatures (301, 318 and 333 K) showed a rapid uptake with time (92-98% within 20 min), but diminished on elevating the temperature [71]. The pseudo second order model fitted the kinetic data best, with the derived q_e values being declined by 59.58 mg/g for a temperature raised from 301 to 333 K, accompanied by the rate climbing from $1.50 \times 10^{-3} - 6.25 \times 10^{-3}$ g/mg min, probably because of acceleration in the mobility of dye cations at higher temperatures. In addition, the dominant surface adsorption through ion exchange and absence of any particle diffusion may account for the reduced dye uptake at an elevated temperature, as the particle diffusion is an endothermic process [74]. Furthermore, the higher the temperature, the lower the ion exchange capacity and the larger the physical adsorption capacity [75]. The activation energy (E_a, kJ/mol) was determined using the Arrhenius equation [71], which relates the rate constant (k) and temperature (T, Kelvin) according to the expression,

$$k = A\, e^{-\left(E_a/RT\right)} \tag{5}$$

where A (g/mg min) is the temperature-independent pre-exponential factor. The E_a value obtained was 37.21 kJ/mol, which was greater than the range (<30.00 kJ/mol) reported for a diffusion-controlled reaction [71], implying the adsorption of BB-Y dye by γ-PGA was only reaction-controlled.

Likewise, the kinetic data at different pH (2.00, 3.00, 3.41, 3.94 and 5.00) also fitted the pseudo second order model and the adsorption rate was characterized by a sharp rise from 5.49 x 10^{-4} – 2.48 x 10^{-3} g/mg min for a pH change from 2-3, accounting for a rapid decrease in competing hydrogen ions by 10 times [71]. However, a sudden drop in rate from 2.48 x 10^{-3} – 1.40 x 10^{-3} g/mg min occurred for a pH increase from 3-3.41. When the pH was raised from 2-3 and 3-3.94, the dye uptake increased proportionately by 97.72 and 101.14 mg/g for a 10-fold decrease in the hydrogen ion concentration, respectively [71].

4.1.1. Boyd's Ion Exchange Model

The kinetic data was also analyzed using an ion exchange model proposed by Boyd et al. [78], who postulated the adsorption kinetics for exchange of ions by organic zeolites to be a chemical phenomenon. The exchange reaction between two monovalent ions can be expressed according to the mass law as

$$A^+ + BR \rightleftharpoons B^+ + AR$$

If m_{A+} and m_{B+} denote the concentrations of the ions A^+ and B^+ in solutions, and n_{AR} and n_{BR} the moles of A^+ and B^+ in the adsorbent, respectively, then the net reaction rate can be written as follows:

$$\frac{dn_{AR}}{dt} = k_1(m_{A^+})(n_{BR}) - k_2(m_{B^+})(n_{AR}) = -n_{AR}(k_1 m_{A^+} + k_2 m_{B^+}) + k_1 m_{A^+} E \quad (6)$$

where k_1 and k_2 are the forward and reverse specific rate constants, respectively, and E is a constant defined by $E = n_{AR} + n_{BR}$. When both concentrations of A^+ and B^+ in solution are kept constant, then, on integration, Eq.(6) becomes

$$n_{AR} = \frac{k_1 m_{A^+} E}{k_1 m_{A^+} + k_2 m_{B^+}}(1 - e^{-St}) = q_t \quad (7)$$

where $S = k_1 m_{A^+} + k_2 m_{B^+}$, and Eq.(7) can be rewritten as

$$\log(1 - F) = -\left(\frac{S}{2.303}\right)t \quad (8)$$

where F is the fractional attainment of equilibrium at time t, calculated by the ratio between the amounts adsorbed (mg/g) at time t and at infinite time ($F=q_t/q\infty$), and S (L/min) is the ion exchange rate constant. The kinetic data fitted with Eq.8 gave linear curves (plots not shown) with high correlation (r^2=0.954-0.998) and the rate constant (S) values are summarized in Table 6 [27,70,71]. The difference in S values for different dye adsorption by γ-PGA may be due to variation in the structural features that affect their mass transfer rate from bulk solution onto γ-PGA.

4.1.2. Mass Transport Mechanism

An adsorption process involves three consecutive mass transport steps such as (i) film diffusion, (ii) intraparticle or pore diffusion and (iii) adsorption onto interior sites [70]. Among these three steps, the last step is considered negligible as it is assumed to be rapid and hence the rate of adsorption should be controlled by either film or particle diffusion. From the practical point of view, it is essential to determine the rate-determining step of an adsorption process. Therefore, the kinetic data were further analyzed using a particle-controlled diffusion expression proposed by Boyd et al. [78,79], which is given below

$$F = 1 - \frac{6}{\pi^2} \sum_{n=1}^{\infty} \frac{1}{n^2} \exp[-n^2 Bt] \tag{9}$$

$$B = \pi^2 D_i / r_o^2 \tag{10}$$

where r_o (cm) denotes the radius of the γ-PGA particle determined by sieve analysis, D_i (cm^2/s) is the effective diffusion coefficient and B is the time constant. The Bt values calculated for each value of F were plotted against time t and the plots obtained were shown to be linear (plots not shown), but did not pass through the origin, confirming that film diffusion or external mass transport mainly governs the uptake of cationic dyes by γ-PGA [70]. This tendency also suggested that intraparticle resistance is negligible during the initial period of adsorption. Assuming γ-PGA particles to be spherical in nature, the effective diffusion coefficient (D_i, cm^2/s) was determined using the Eq.10 to be 1.54×10^{-4} and 1.55×10^{-4} for MB and MG dye adsorption, respectively. Similar D_i values in the order of 10^{-4} were also reported for film-diffusion-controlled adsorption of MB dye by mango seed kernel powder [80]. It was further reported that the film diffusion controls adsorption process in a system with poor mixing, low solute concentration, small adsorbent size and high solute/adsorbent interaction, whereas the opposite is true for the particle diffusion-controlled adsorption process [80]. Apparently, the small-sized γ-PGA particles (1-150 μm) employed for dye adsorption with poor mixing (120 rpm), and the fast dye uptake rate implicating high dye/γ-PGA interaction ascertain that the ion exchange reaction through film diffusion is the rate-controlling process for cationic dye adsorption by γ-PGA.

4.2. Adsorption Isotherms

Adsorption isotherms are the basic requirements for designing an adsorption system as an isotherm expresses the relation between the liquid phase solute concentration and the mass of solute adsorbed at constant temperature per unit mass of adsorbent. Isotherm curves obtained at three different temperatures (301, 318 and 333 k) and at natural pH (without adjusting pH) were characterized by an increase in dye adsorption following a raise in dye concentration from 10-200 mg/L [27,70,71]. Conversely, the amount of dye adsorbed declined with a rise in temperature from 301-333 K, implying an exothermic nature of the adsorption process. The shape of isotherm curves resembled either L- or H-type according to classification by Giles et al. [81]. These isotherm types, commonly referred as Langmuir type, signified a high degree of adsorption at very low dye concentration, and suggested the uptake of dyes by γ-PGA to be associated with chemical forces rather than physical interaction [82,83].

An accurate mathematical description of equilibrium adsorption data is crucial for reliable prediction of adsorption parameters and quantitative comparison of adsorption behavior. The process variables and the underlying thermodynamic assumptions of the equilibrium models often provide insight into the surface property and affinity of the adsorbent as well as the adsorption mechanism. The equilibrium adsorption curves of Au-O, Rh-B, Sa-O and BB-Y by γ-PGA were fitted with three most commonly used isotherm models, namely, Freundlich [84], Langmuir [83] and Redlich-Peterson [85], represented as

$$q_e = K_F C_e^{1/n} \tag{11}$$

$$q_e = \frac{q_m K_e C_e}{1 + K_e C_e} \tag{12}$$

$$q_e = \frac{A C_e}{1 + K_R C_e^{\gamma}} \tag{13}$$

where C_e is the dye concentration in solution at equilibrium, K_F (mg/g) and n are Freundlich constants denoting adsorption capacity and intensity of adsorption, respectively, whereas q_m (mg/g) and K_e (L/mg) represent the maximum adsorption capacity and energy of adsorption. Non-linear regression analysis along with comparison of error parameters (r^2 and χ^2) revealed that the Redlich-Peterson model could describe the equilibrium data more closely than the other two models (Table 7) [27,71]. The maximum adsorption capacity q_m derived from the Langmuir model was higher or comparable to those reported for nonconventional adsorbents (Table 8) [86-111], demonstrating γ-PGA could be an effective adsorbent in scavenging cationic dyes. The main characteristics of the Langmuir isotherm can be expressed by a dimensionless separation factor or equilibrium parameter R_L, which is defined as $R_L=1/(1+K_eC_o)$, where K_e (L/mg) is the Langmuir constant and C_o (mg/L) is the initial dye concentration. Hall et al. [112] proposed four idealized types of equilibrium behavior, namely, irreversible ($R_L=0$), unfavorable ($R_L>1$), linear ($R_L=1$) and favorable ($0 < R_L < 1$). The R_L values for the dye concentration range 10-200 mg/L in Table 9 were 0-1, indicating a favorable adsorption of cationic dyes by γ-PGA [27,70,71].

Table 5. Kinetic model parameters for adsorption of cationic dyes by γ-PGA at different experimental conditions

Experimental conditions	Pseudo first order model				Pseudo second order model			
	q_e (mg/g)	k_1 (L/min)	r^2	$\chi^{2\,a}$	q_e (mg/g)	k_2 (g/mg min)	r^2	$\chi^{2\,a}$
Au-O at 100 mg/L [b]	164.54	0.9917	0.914	2.13	172.03	1.04×10^{-2}	0.996	0.09
Au-O at 200 mg/L [b]	249.99	0.7618	0.941	3.41	261.88	4.75×10^{-3}	0.992	0.66
Rh-B at 100 mg/L [b]	184.95	0.8590	0.853	5.32	195.94	6.95×10^{-3}	0.981	0.67
Rh-B at 200 mg/L [b]	295.69	0.5988	0.810	21.23	315.89	4.50×10^{-3}	0.962	4.39
Sa-O at 100 mg/L [b]	226.22	0.3851	0.931	15.42	249.14	2.18×10^{-3}	0.985	2.64
Sa-O at 200 mg/L [b]	414.97	0.1593	0.964	55.38	453.77	5.25×10^{-4}	0.993	5.64
MB at 100 mg/L [c]	226.17	0.8995	0.848	6.45	237.22	6.37×10^{-3}	0.981	0.80
MG at 100 mg/L [c]	185.45	0.9227	0.882	3.68	195.25	7.91×10^{-3}	0.992	0.25
BB-Y at 301 K [d]	267.83	0.2902	0.882	57.49	289.74	1.50×10^{-3}	0.961	15.17
BB-Y at 318 K [d]	233.57	0.5493	0.789	24.91	248.17	3.26×10^{-3}	0.950	6.19
BB-Y at 333 K [d]	221.27	0.8312	0.838	8.66	230.16	6.25×10^{-3}	0.968	1.62
BB-Y at pH 2.00 [e]	87.38	0.0539	0.994	0.74	104.54	5.49×10^{-4}	0.994	1.87
BB-Y at pH 3.00 [e]	177.41	0.3308	0.942	10.26	192.00	2.48×10^{-3}	0.993	1.10
BB-Y at pH 3.41 [e]	248.08	0.2642	0.923	31.20	270.19	1.40×10^{-3}	0.987	4.51
BB-Y at pH 3.94 [e]	267.83	0.2902	0.882	57.49	289.74	1.50×10^{-3}	0.961	15.17
BB-Y at pH 5.00 [e]	285.30	0.2854	0.872	63.63	309.52	1.36×10^{-3}	0.960	16.67

Source: Inbaraj et al. 2006a; 2006c; 2008.

[a] $\chi^2 = \Sigma((q_{ex}-q_{th})^2/q_{th})$, where q_{ex} and q_{th} are experimental and theoretical amounts of dye adsorbed (mg/g) at time t; [b]γ-PGA dose: 0.4 g/L, Temperature: 301 K, pH: natural pH; [c]γ-PGA dose: 0.4 g/L, Temperature: 301 K, pH: natural pH; [d]BB-Y concentration: 100 mg/L, γ-PGA dose: 0.3 g/L, pH: natural pH; [e]BB-Y concentration: 100 mg/L, γ-PGA dose: 0.3 g/L, Temperature: 301 K, pH: natural pH.

Table 6. Ion exchange rate constants (S) for cationic dye adsorption by γ-PGA at different experimental conditions

Experimental conditions	Boyd's ion exchange model	
	S (L/min)	r^2
Au-O at 100 mg/L[a]	0.2538	0.958
Au-O at 200 mg/L[a]	0.2294	0.934
Rh-B at 100 mg/L[a]	0.1087	0.965
Rh-B at 200 mg/L[a]	0.0864	0.976
Sa-O at 100 mg/L[a]	0.1331	0.967
Sa-O at 200 mg/L[a]	0.0769	0.974
MB at 100 mg/L[b]	0.1918	0.997
MG at 100 mg/L[b]	0.1799	0.988
BB-Y at 301 K[c]	0.1025	0.998
BB-Y at 318 K[c]	0.1082	0.992
BB-Y at 333 K[c]	0.2199	0.987
BB-Y at pH 2.00[d]	0.0548	0.998
BB-Y at pH 3.00[d]	0.1027	0.975
BB-Y at pH 3.41[d]	0.0979	0.985
BB-Y at pH 3.94[d]	0.1025	0.998
BB-Y at pH 5.00[d]	0.0910	0.998

Source: Inbaraj et al. 2006a; 2006c; 2008.

[a] γ-PGA dose: 0.4 g/L, Temperature: 301 K, pH: natural pH; [b] γ-PGA dose: 0.4 g/L, Temperature: 301 K, pH: natural pH; [c] BB-Y concentration: 100 mg/L, γ-PGA dose: 0.3 g/L, pH: natural pH; [d] BB-Y concentration: 100 mg/L, γ-PGA dose: 0.3 g/L, Temperature: 301 K, pH: natural pH.

Table 7. Isotherm model parameters for adsorption of cationic dyes by γ-PGA at different temperatures[a]

Isotherm model	Fitted isotherm parameters											
	Au-O/γ-PGA			Rh-B/γ-PGA			Sa-O/γ-PGA			BB-Y/γ-PGA		
	301 K	318 K	333 K	301 K	318 K	333 K	301 K	318 K	333 K	301 K	318 K	333 K
Freundlich												
K_F ($L^n\,mg^{1-n}/g$)	44.24	27.85	24.30	31.89	15.68	11.02	129.60	95.31	76.84	112.59	68.08	54.42
N	2.54	2.33	2.44	1.79	1.54	1.52	2.33	2.44	2.56	2.82	2.52	2.53
r^2	0.994	0.997	0.999	0.975	0.987	0.978	0.954	0.984	0.975	0.959	0.968	0.973
χ^2	6.33	3.43	0.52	30.68	18.21	20.55	77.84	32.61	33.69	45.58	28.61	18.77
Langmuir												
q_m (mg/g)	277.29	203.60	197.50	390.25	378.00	305.00	502.83	432.26	369.29	469.36	411.57	357.48
K_e (L/mg)	0.07	0.05	0.04	0.04	0.02	0.01	0.30	0.19	0.14	0.17	0.08	0.06
r^2	0.982	0.979	0.967	0.998	0.998	0.995	0.996	0.979	0.988	0.984	0.987	0.991
χ^2	38.79	21.07	44.34	2.84	4.26	6.34	7.04	47.18	25.02	16.23	10.28	5.28
Redlich-Peterson												
A (L/g)	67.32	41.23	259.32	15.61	7.56	7.56	169.51	190.86	88.98	111.41	43.16	29.22
K_R $(L/mg)^{\gamma}$	1.00	1.00	10.18	0.04	0.02	0.02	0.42	1.18	0.52	0.40	0.20	0.16
γ	0.70	0.65	0.60	1.00	1.00	1.00	0.93	0.73	0.81	0.87	0.86	0.86
r^2	0.999	0.999	0.999	0.998	0.998	0.995	0.997	0.996	0.997	0.992	0.992	0.996
χ^2	0.58	0.44	0.36	2.84	4.26	6.34	8.12	11.40	5.93	11.87	8.91	3.97

Source: Inbaraj et al. 2006a; 2008.

[a] Dye concentration range: 10-200 mg/L; γ-PGA dose: 0.4 g/L for Au-O, Rh-B, Sa-O and 0.3 g/L for BB-Y.

Table 8. Maximum adsorption capacities (q_m) of poly(γ-glutamic acid) compared with those reported for cationic dye removal by some non-conventional adsorbents

Adsorbent	Cationic dye	q_m (mg/g)	Reference
Bark	Safranin O	1119.00	[86]
Bark	Methylene blue	914.58	[86]
Cotton waste	Safranin O	875.00	[86]
Rice husk	Safranin O	838.00	[86]
Corn cob	Safranin O	790.00	[87]
Poly(γ-glutamic acid)[a]	Bismarck brown Y[a]	667.09[a]	[71]
Pinewood	Methylene blue	556.00	[88]
Sugar industry mud	Basic red 22	519.00	[89]
Rice husk carbon	Malachite green	511.00	[90]
Poly(γ-glutamic acid)	Safranin O	502.83	[27]
Poly(γ-glutamic acid)[b]	Bismarck brown Y[b]	469.36[b]	[71]
Tree fern	Basic red 13	408.00	[91]
Peat	Basic violet 14	400.00	[92]
Poly(γ-glutamic acid)	Rhodamine B	390.25	[27]
Waste newspaper	Methylene blue	390.00	[93]
Poly(γ-glutamic acid)	Methylene blue	352.76	[70]
Treated peat	Malachite green	350.00	[92]
Coal	Methylene blue	323.68	[86]
Palm-fruit bunch	Basic yellow	320.00	[94]
Rice husk	Methylene blue	312.00	[86]
Clay	Methylene blue	300.00	[95]
Poly-γ-glutamic acid	Malachite green	293.32	[70]
Poly(γ-glutamic acid)	Auramine O	277.29	[27]
Cotton waste	Methylene blue	277.78	[86]
Activated sludge biomass	Methylene blue	256.41	[96]
Palm-fruit bunch	Basic red 18	242.00	[97]
TriSyl silicas	Malachite green	208.00	[98]
Coir pith	Rhodamine B	203.00	[99]
Activated furniture	Methylene blue	200.00	[100]
Jackfruit peel carbon	Malachite green	166.37	[101]
Hair	Methylene blue	158.23	[86]
Mango seed kernel powder	Methylene blue	153.85	[80]
Spirodela polyrrhiza biomass	Methylene blue	144.93	[102]
Activated tyres	Methylene blue	130.00	[100]
Activated sewage char	Methylene blue	120.00	[100]
Pithophora species	Malachite green	117.65	[103]

Table 8. Continued

Adsorbent	Cationic dye	q_m (mg/g)	Reference
Jackfruit peel carbon	Rhodamine B	104.17	[60]
Pearl millet husk carbon	Methylene blue	82.37	[104]
Raw date pits	Methylene blue	80.30	[105]
Pyrolysed furniture	Methylene blue	80.00	[100]
Treated sawdust	Malachite green	74.50	[106]
Modified cyclodextrin	Methylene blue	56.50	[107]
Banana peel	Methylene blue	20.80	[108]
Iron humate	Malachite green	19.20	[109]
Orange peel	Methylene blue	18.60	[108]
Activated date pits (900℃)	Methylene blue	17.30	[105]
Sugarcane dust	Malachite green	4.88	[110]
Neem sawdust	Malachite green	3.42	[111]

[a] At natural pH 4.06 (average pH of concentration range 20-200 mg/L); [b] At pH 5.00.

Table 9. Separation factor (R_L) for removal of cationic dyes by γ-PGA at 301 K

Initial dye concentration (mg/L)	Separation faction (R_L)					
	MB / γ-PGA	MG / γ-PGA	Au-O / γ-PGA	Rh-B / γ-PGA	Sa-O / γ-PGA	BB-Y / γ-PGA
10	0.1354	0.4838	0.5865	0.7415	0.2488	0.2241
20	0.0726	0.3191	0.4149	0.5892	0.1421	0.1615
30	0.0496	0.2380	0.3210	0.4888	0.0994	0.1262
40	0.0377	0.1898	0.2617	0.4177	0.0765	0.1036
50	0.0304	0.1579	0.2210	0.3646	0.0621	0.0878
60	0.0254	0.1351	0.1912	0.3235	0.0523	0.0674
80	0.0192	0.1049	0.1506	0.2640	0.0398	0.0546
100	0.0154	0.0857	0.1242	0.2229	0.0321	0.0459
120	0.0129	0.0724	0.1057	0.1930	0.0269	0.0396
140	0.0111	0.0627	0.0920	0.1701	0.0231	0.0349
160	0.0097	0.0553	0.0814	0.1520	0.0203	0.0281
200	0.0078	0.0448	0.0662	0.1255	0.0163	0.2241

Source: Inbaraj et al. 2006a; 2006c; 2008.

[a] γ-PGA dose: 0.4 g/L for Au-O, Rh-B, Sa-O and 0.3 g/L for BB-Y; Temperature: 301 K.

4.3. Effect of Temperature

By using the equilibrium data obtained for a range of dye concentration at different temperatures (301, 318 and 333 K), several thermodynamic parameters such as change in free energy (ΔG°), enthalpy (ΔH°) and entropy (ΔS°) could be determined using the following equations and are summarized in Table 10 [27,71].

$$\Delta G^\circ = -RT \ln K_e \tag{14}$$

$$\ln K_e = -\frac{\Delta H^\circ}{RT} + \frac{\Delta S^\circ}{RT} \tag{15}$$

Table 10. Thermodynamic parameters for adsorption of cationic dyes by γ-PGA[a]

Temperature	K_e (L/mg)	ΔG° (kJ/mol)	ΔH° (kJ/mol)	ΔS° (J/mol K)
		Au-O/γ-PGA		
301	0.071	-7.69	-15.17	-25.13
318	0.046	-6.97		
333	0.040	-6.92		
		Rh-B/γ-PGA		
301	0.035	-7.06	-28.24	-70.85
318	0.016	-5.39		
333	0.012	-4.84		
		Sa-O/γ-PGA		
301	0.302	-11.67	-20.23	-28.46
318	0.194	-11.16		
333	0.139	-10.76		
		BB-Y/γ-PGA		
301	0.173	-10.72	-27.26	-55.45
318	0.079	-9.25		
333	0.062	-9.01		

Source: Inbaraj et al. 2006a; 2008.

[a] Dye conc. range: 10-200 mg/L; γ-PGA dose: 0.4 g/L for Au-O, Rh-B, Sa-O and 0.3 g/L for BB-Y.

where K_e is the equilibrium constant (Langmuir constant), R is the universal gas constant (8.314 J/mol K) and T is the absolute temperature (K). The negative ΔG° values indicated the spontaneity of dye adsorption, while the negative ΔH° value confirmed the exothermic nature of adsorption process [27,71]. Likewise, the negative ΔS° revealed the decreased randomness at the solid/solution interface and no structural modification occurred in γ-PGA [27,71]. Thus, the negative ΔH°, ΔS° and ΔG° values affirmed that the adsorption of cationic dyes Au-O, Rh-B, Sa-O and BB-Y on γ-PGA was favored at lower temperatures.

4.4. γ-PGA Dose-activity Relationship

On increasing the γ-PGA dose from 0.04-6.00 g/L for a 200 mg/L dye solution, the removal of Au-O, Rh-B and Sa-O dyes rose by 79.1, 81.8 and 67.9%, respectively, while the amount of dye adsorbed declined by 889.9, 800.0 and 1531.7 mg/g [27]. This contrasting behavior may be attributed to a less proportionate increase in dye adsorption following a rise in γ-PGA dose, and reduction in the vicinity of active sites because of aggregation of γ-PGA particles, which in turn may result in removal of some weakly bound dyes from the γ-PGA surface [60,113]. The relationship between γ-PGA dose (m, g/L) and percentage removal (R) fitted the mathematical equations (16)-(18) with r2 being 0.990, 0.988 and 0.978 for Au-O, Rh-B and Sa-O dyes, respectively [27].

$$R = \frac{m}{(3.39\times10^{-3}) + (9.57\times10^{-3}m)} \tag{16}$$

$$R = \frac{m}{(2.37\times10^{-3}) + (9.30\times10^{-3}m)} \tag{17}$$

$$R = \frac{m}{(1.13\times10^{-3}) + (9.05\times10^{-3}m)} \tag{18}$$

4.5. Effect of pH

Solution pH is a critical parameter which dictates the uptake and release of solutes in any adsorption process involving ion exchange mechanism. Following an increment in pH from 1-5 in a 100 mg/L dye solution, the removal of MB, MG, Au-O, Rh-B, Sa-O and BB-Y dyes rose from 6.7-94.7, 4.0-91.8, 7.6-69.1, 14.5-79.0, 23.0-95.6 and 15.0-93.5%, respectively, and remained constant thereafter [27,70,71]. The reduced uptake at low pH may be due to the excessive hydrogen ions in solution competing for active sites in γ-PGA. However, at pH ≥ pK_a of γ-PGA (4.09), the carboxyl groups would be largely deprotonated, facilitating maximum exchange of dye cations. A significant drop in solution pH substantiated the concomitant release of hydrogen ions during dye adsorption. For example, the pH of Au-O, Rh-B and Sa-O solutions declined to an average value of 3.86, 3.75 and 3.68, respectively, when the dye solutions at pH 7, 6, 5 and 4 were equilibrated with γ-PGA [27].

4.6. Effect of Electrolytes

Incorporation of different electrolytes (NaCl, KCl, $CaCl_2$ or $MgCl_2$) at varying concentrations (0.001-1 M) showed a declining trend on Au-O, Rh-B and Sa-O dyes adsorption by γ-PGA (Figure 9) [27]. The metal salts added may screen the electrostatic interaction between anionic γ-PGA and cationic dye, and an increase in salt concentration could reduce the amount of dye adsorbed on γ-PGA [114,115]. The degree of screening by added metal salts followed the order: $Ca^{2+} > Mg^{2+} > K^+ > Na^+$, which may be accounted for by the size and valence of added metal ions [114,115]. Accordingly, the divalent ions Ca^{2+} and Mg^{2+} showed a more pronounced effect than the monovalent ions Na^+ and K^+, with Ca^{2+} ions exerting a larger effect than Mg^{2+}, and K^+ ions than Na^+. Apparently, the larger the ions, the lower the degree of hydration and the less hydrated ions are drawn closer to the charged surface due to their smaller radii.

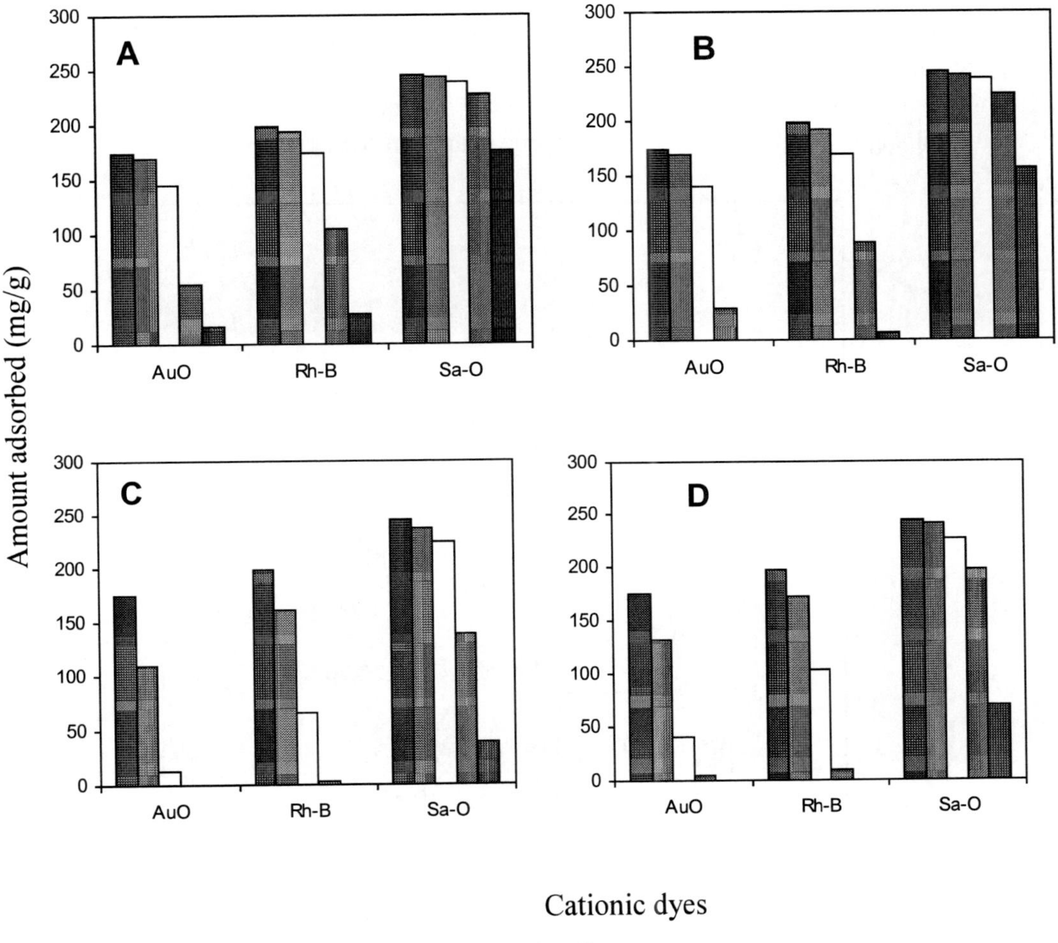

Source: Inbaraj et al. 2006a; Table data converted into bar diagram.

Figure 9. Effect of electrolytes at different concentrations on removal of cationic dyes by γ-PGA. Electrolyte type: (A) NaCl; (B) KCl; (C) $CaCl_2$; (D) $MgCl_2$; Electrolyte concentration: 0 M (control); 0.001 M; 0.01 M; 0.1 M; 1 M

4.7. Recovery of Adsorbed Dyes

Maximum recovery (>98%) of cationic dyes (MB, MG, Au-O, Rh-B, Sa-O and BB-Y) from dye-treated-γ-PGA could be achieved at pH 1 or 1.5, but tended to decline on raising the solution pH [27,70,71]. This desorption trend was opposite to that observed for dye adsorption, suggesting an ion exchange mechanism may be predominantly involved in adsorption of cationic dyes on γ-PGA. This outcome enables the reuse of γ-PGA by subjecting the dye-treated-γ-PGA to dye desorption and adopting proper regeneration procedures without losing its physico-chemical characteristics.

4.8. Infra-Red (IR) Spectra of Untreated and Dye-treated γ-PGA

The IR spectra of both untreated and dye-treated γ-PGA were almost identical (Figure 10) [71]. Nevertheless, by critical comparison the presence of two new peaks at ~1600 and 1405.72 cm^{-1} in dye-treated γ-PGA could be observed, which are characteristic of asymmetric and symmetric stretching vibrations of carboxylate anion (COO^-), respectively [71]. Thus, the adsorption of cationic dyes on γ-PGA may occur by interaction with carboxylate anions through exchange of hydrogen ions from the side chain carboxyl groups.

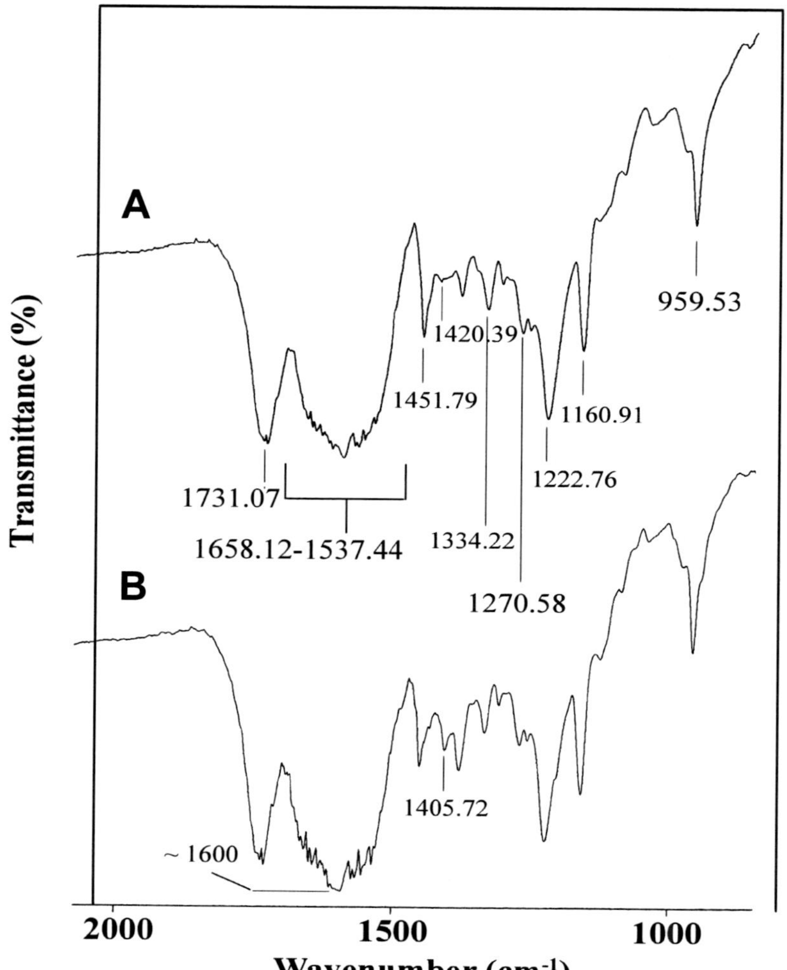

Figure 10. Infra-Red spectra of untreated γ-PGA (A) and BB-Y dye-treated γ-PGA (B).

5. γ-PGA as an Adsorbent of Chemical Mutagens

Many polysaccharides and dietary fibers from vegetables, fruits and cereals are known to be antimutagenic [116-121]. Nevertheless, the antimutagenic activity of extracellular polymeric substances derived from bacterial sources remains less explored. Recently, the adsorption and antimutagenic behavior of γ-PGA for several chemical mutagens, listed in Figure 11, have been investigated [28,32] and are discussed herein.

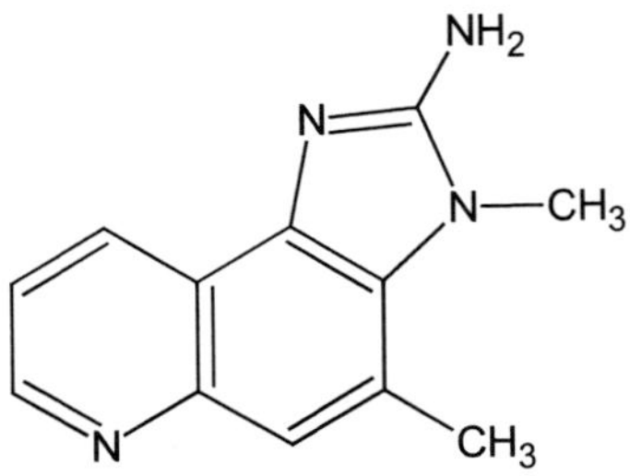

2-amino-3,4-dimethylimidazo[4,5-f] quinoline (MeIQ)

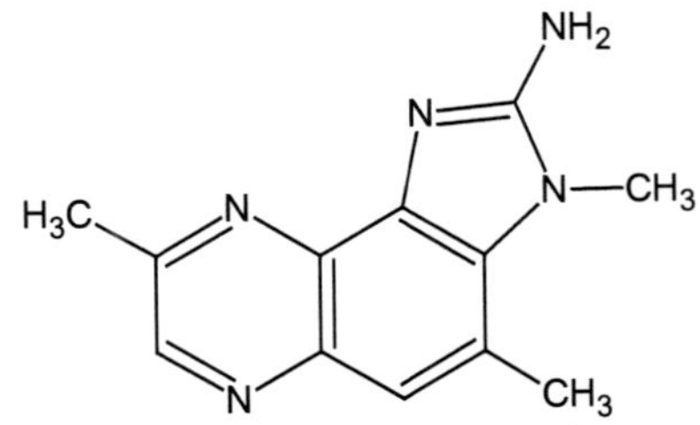

2-amino-3,4,8-trimethylimidazo[4,5-f] quinoxaline (4,8-DiMeIQx)

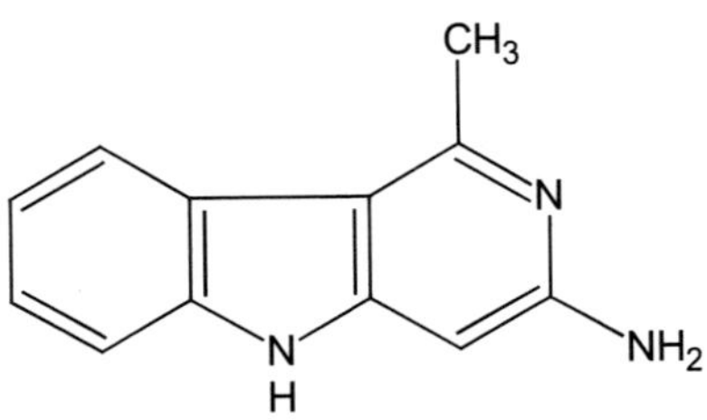

3-amino-1-methyl-5H-pyrido[4,3-b] indole (Trp-p-2)

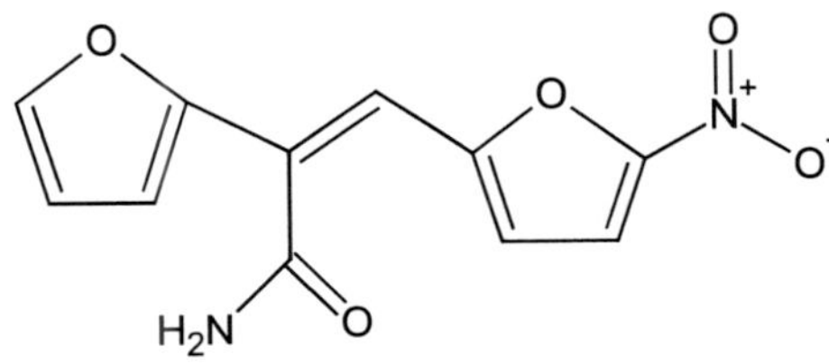

Furylframide (AF-2)

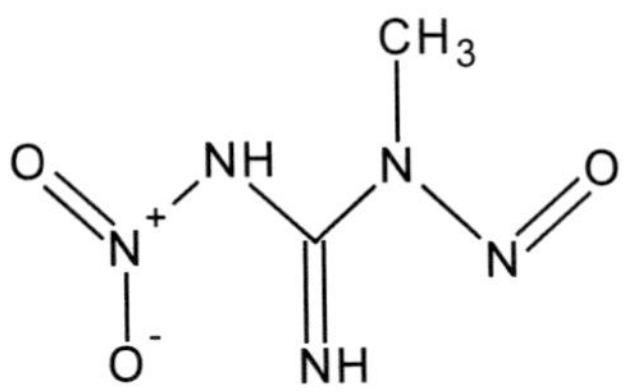

N-methyl-N′-nitro-N-nitrosoguanidine (MNNG)

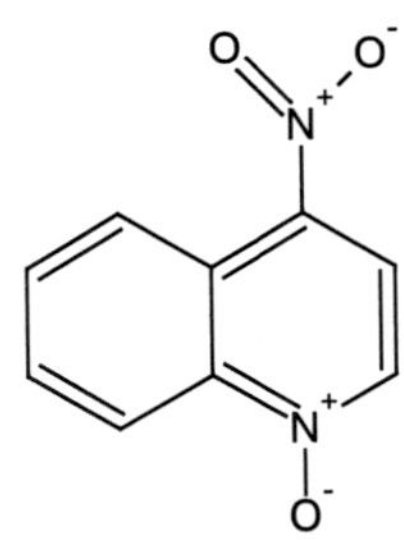

4-nitroquinoline 1-oxide (4NQO)

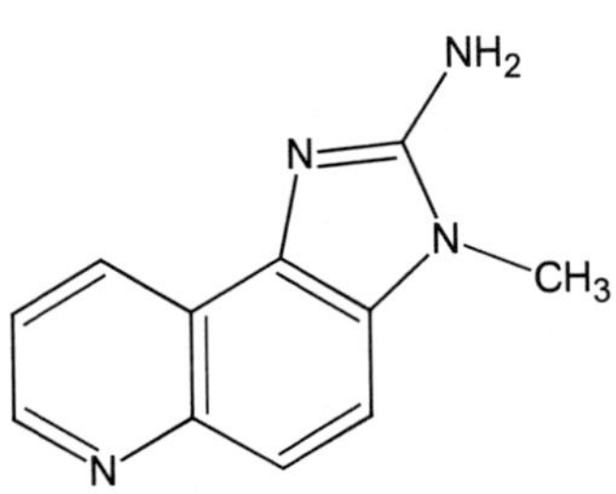

2-amino-3-methylimidazo[4,5-f] quinoline (IQ)

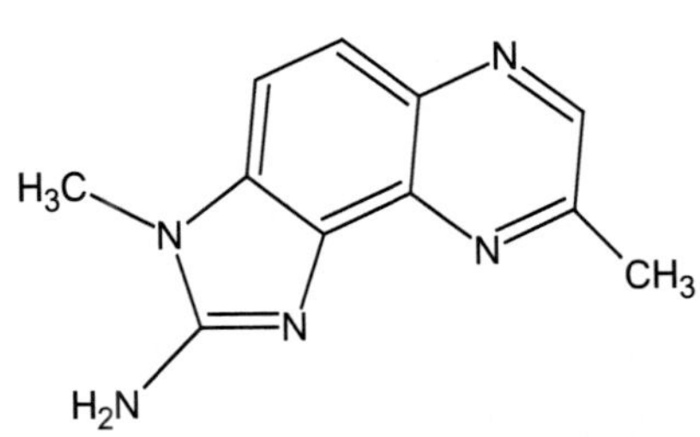

2-amino-3,8-dimethylimidazo[4,5-f] quinoxaline (MeIQx)

Figure 11. Molecular structure of chemical mutagens.

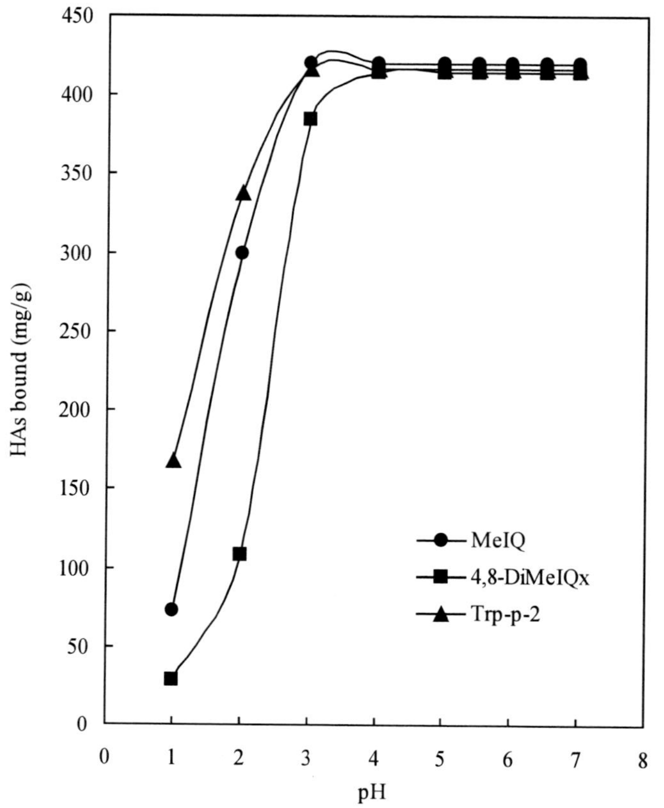

Source: Inbaraj et al. 2006b.

Figure 12. Effect of pH on binding of heterocyclic amines (HA) by γ-PGA. HA concentration: 500 mg/L; γ-PGA dose: 5 mg/mL.

5.1. Binding of Mutagenic Heterocyclic Amines at Gastrointestinal pH

Inbaraj et al. [28] investigated the adsorption characteristics of three mutagenic heterocyclic amines (HA), namely, MeIQ, 4,8-DiMeIQx and Trp-p-2, by γ-PGA from *B. subtilis* (H-form; MW 990 kDa) at gastrointestinal pH conditions. Heterocyclic amines were monitored by a HPLC method developed by Chen and Yang [122], who employed a binary solvent system of acetonitrile and 0.05 ammonium acetate solution (pH 3.6) with the following gradient elution: 9% A at 0 min, 15% A at 8 min, 27% at 18 min, 55% A at 28 min and 100% A at 30 min. A batch-mode equilibrium study at different solution pH (1-7) showed appreciable binding of all HAs over the entire gastrointestinal pH range typically being 2.5 (stomach), 7.5 (small intestine), 5.5 (cecum/ascending colon) and 6.8 (distal colon) (Figure 12). Binding curves developed at pH 2.5 and 5.5 exhibited different isotherm shapes belonging to S and L types, respectively (plots not shown), according to classification by Giles et al. [81]. The S-type curve at pH 2.5 was characterized by an initial concave portion signifying less or no adsorption, which may be attributed to clustering or agglomeration of HA molecules caused by repulsive force between protonated primary amino group of HA and secondary amino group of γ-PGA. In contrast, at high HA concentration, the interaction between HA and γ-PGA became stronger due to multiple physical interaction resulting in a convex shape. The maximum adsorption capacity (q_m, mg/g) derived by fitting the isotherm data at pH 2.5 with a linear form of the classical Langmuir equation ($1/q_e = 1/q_m K_e C_e + 1/q_m$) [83] followed the order: Trp-p-2 (1428.57) > MeIQ (1250.00) > 4,8-DiMeIQx (666.67) (Table 11). The difference in q_m values between HAs may be due to variation in their hydrophobic character as reported by Ferguson and Harris [122], who demonstrated the adsorption of nine HAs on dietary fiber α-cellulose to increase with hydrophobicity. Also, the number of bulky methyl groups in HA may be responsible for this effect. Accordingly, 4,8-DiMeIQx containing three methyl groups showed the lowest q_m value (666.67 mg/g), followed by MeIQ (1250.00 mg/g) and Trp-p-2 (1428.57 mg/g). Thus, the interaction between HAs and γ-PGA at pH 2.5 may largely depend on physical forces.

Table 11. Langmuir parameters for heterocyclic amines binding by γ-PGA at pH 2.5[a]

HA	Langmuir parameters		
	q_m (mg/g)	b (l/mg)	r^2
MeIQ	1250.00	0.0020	0.948
4,8-DiMeIQx	666.67	0.0004	0.933
Trp-p-2	1428.57	0.0005	0.952

Source: Inbaraj et al. 2006b.

[a] Experimental conditions: HA concentration range: 10-2000 mg/L; γ-PGA dose: 1 mg/mL.

On the other hand, the isotherm curves developed for HA concentrations 100-2000 mg/L at pH 5.5 were characterized by two distinct portions (curve I and curve II) with a curve shift appearing between the same C_o values (600 and 800 mg/L) for all the three HAs (plots not shown). This phenomenon suggested the presence of more than one adsorption site with varied affinities and binding energies for HA molecules. Fitting curve I and II individually with the Langmuir equation ($C_e/q_e = C_e/q_m + 1/K_e q_m$) provided good correlation ($r^2 > 0.98$), suggesting saturation of two binding sites with varying binding energies. Linear fitting of curve I yielded the isotherm parameters for initial monolayer saturation, while curve II gave relavant parameters for overall HA binding by γ-PGA (Table 12). Moreover, fitting with the Scatchard equation ($q_e/C_e = q_m K_e - q_e K_e$) [123] showed a significant deviation from linearity, but a good correlation ($r^2 > 0.91$) was observed when applied separately to curves I and II, confirming the presence of at least two different types of binding (Table 12). The multisite HA adsorption at pH 5.5 was also interpreted on the basis of a multimolecular adsorption by using the BET equation [124] as given below:

$$\frac{C_e}{(C_s - C_e)} = \frac{1}{aq_m} + \left(\frac{a-1}{aq_m}\right)\left(\frac{C_e}{C_s}\right) \tag{19}$$

where C_S (mg/l) is the solute concentration at the saturation of all layers, q_m (mg/g) is the amount of HA required to form a unimolecular layer and a is a BET constant representing average heat of adsorption in the first layer. Linear plots of $C_e/(C_S - C_e)$ versus C_e/C_S gave a high correlation for 4,8-DiMeIQx and Trp-p-2 ($r^2 > 0.96$), but not for MeIQ ($r^2 > 0.87$). From the fitted parameters, the q_m value for the unimolecular layer was found to be ~2.5 times higher than that obtained by the Langmuir equation for saturation of all layers (Table 13), implying the multimolecular HA adsorption did occur for at least two layers on γ-PGA. According to the generalized polarization theory by DeBoer and Zwicker [125], the adsorption of non-polar molecules on ionic adsorbents like γ-PGA may induce dipoles in the first layer of adsorbed molecules, which in turn may stimulate dipoles on the second layer favoring a multilayer adsorption. Based on the hydrophobic character of HAs and ionic nature of γ-PGA, the adsorption on first layer may be formed by hydrogen-bonding, hydrophobic and ionic interactions, while the subsequent layer(s) may result from weak dipole-dipole and hydrophobic interactions.

Table 12. Langmuir and Scatchard plot parameters for heterocyclic amines binding by γ-PGA at pH 5.5[a]

HA	Curve I			Curve II		
	q_m (mg/g)	b (l/mg)	r^2	q_m (mg/g)	b (l/mg)	r^2
			Langmuir model parameters			
MeIQ	500.00	0.2632	0.989	1428.57	0.0374	0.999
4,8-DiMeIQx	370.37	0.2030	0.991	1000.00	0.0152	0.999
Trp-p-2	500.00	0.6897	0.982	1666.67	0.0211	0.999
			Scatchard plot parameters			
MeIQ	499.16	0.1805	0.918	1465.09	0.0432	0.983
4,8-DiMeIQx	397.79	0.0706	0.994	944.68	0.0169	0.991
Trp-p-2	454.30	1.5903	0.913	1619.74	0.0228	0.995

Source: Inbaraj et al. 2006b.

[a] Experimental conditions: HA concentration range: 10-2000 mg/L; γ-PGA dose: 1 mg/mL.

Table 13. BET isotherm parameters for heterocyclic amines binding by γ-PGA at pH 5.5[a]

HA	BET model parameters		
	q_m (mg/g)	a	r^2
MeIQ	653.60	51.00	0.867
4,8-DiMeIQx	370.37	451.00	0.961
Trp-p-2	623.44	401.00	0.989

Source: Inbaraj et al. 2006b.

[a] Experimental conditions: HA concentration range: 10-2000 mg/L; γ-PGA dose: 1 mg/mL.

5.2. Suppressive Effect on SOS Response of *Salmonella typhimurium* Induced by Chemical Mutagens

Sato et al. [32] recently reported the suppressive effect of γ-PGA (Na form) from *B. subtilis* on SOS response (a post replication DNA repair system that allow DNA replication to bypass lesions) of *Salmonella typhimurium* (TA1535/pSK1002) induced by direct and indirect forms of chemical mutagens (AF-2, MNNG, 4NQO, Trp-p-2, IQ and MeIQx). The antimutagenicity was determined by a slightly modified *umu* test [126], which was based on the ability of γ-PGA to suppress the *umu* operon expression induced by DNA-damaging chemical mutagens. A plasmid (pSK 1002) carrying fusion gene *umu*C-*lac*Z was introduced into *S. typhimurium* TA 1535 and the percentage of suppression was calculated in terms of the β-galactosidase activity of cells resulting from the expression and suppression of the *umu* operon according to the equation: (1-A/B) x 100, where A and B are the β-galactosidase activity with and without γ-PGA, respectively. The concentration of chemical mutagens used to induce the SOS response in *S. typhimurium* was 0.3, 6, 3, 0.3, 0.3 or 3 μg/mL of AF-2, MNNG, 4NQO, IQ, MeIQx or Trp-p-2, respectively, which was chosen based on induction of β-galactosidase activity without affecting the cell growth.

The γ-PGA (Na form; 4000 kDa) tested at three different concentrations (1, 2 and 3%) showed a substantial suppressive effect on the SOS response induced by all the mutagens, with >80% suppression being achieved by a 3% solution (Figure 13). Irrespective of the type of chemical mutagen, the antimutagenic activity tested for γ-PGA (3%) with different molecular masses (50-8000 kDa) exhibited a lower effect at 50 kDa and >6000 kDa when compared to others, with maximum activity being shown for γ-PGA at 4000 kDa (Figure 3). The low water-holding capacity of γ-PGA at 50 kDa and the three dimensional structure or the rheological property of γ-PGA at >6000 kDa may be responsible for the less-pronounced effect in entrapping the mutagens [32]. However, the variation in salt form of γ-PGA (Na, K or Ca) did not show any remarkable suppressive effect on the SOS response induced by indirect mutagens (IQ, MeIQx and Trp-p-2), implying the difference in electronegativity or valency of metal ions poses no significant impact. The antimutagenic activity of γ-PGA (3% of 4000 kDa) was also compared with its monomer units, D- and L-glutamic acids (Na form), and some other reference compounds like carboxymethylcellulose (CMC) and xanthan gum (XG) [32]. The percentage of suppression by γ-PGA was the highest for all chemical mutagens, followed by XG, CMC, and D- and L-glutamates (Table 14). Although XG and CMC are similar to γ-PGA in possessing carboxyl groups, the difference in antimutagenic activity was large, probably because of variation in ion exchange capacity (meq/g) between γ-PGA (8.7) and XG (2.0) or CMC (0.68). Besides, the γ-PGA also exhibited significant antimutagenic effect on the SOS response of *S. typhimurium* induced by the active forms (nitrenium ions) of three indirect mutagens (Trp-p-2, IQ and MeIQx) [32].

Table 14. Suppressive effect of γ-PGA[a] and reference compounds on SOS response of *S. typhimurium* induced by various chemical mutagens

Mutagen (Final concentration)	Suppression (%)				
	CMC[b]	XG[c]	LG[d]	DG[e]	γ-PGA
AF-2 (0.3 μg/mL)	25	40	15	12	90
MNNG (6 μg/mL)	30	72	20	18	80
4NQO (3 μg/mL)	40	75	20	22	81
IQ (0.3 μg/mL)	20	48	38	43	90
MeIQx (0.3 μg/mL)	35	70	17	25	90
Trp-p-2 (3 μg/mL)	30	43	27	25	82

Source: Sato et al. 2008.

[a] 3% Na-form γ-PGA at 4000 kDa; [b]carboxymethylcellulose; [c]xanthan gum; [d]L-glutamic acid; [e]D-glutamic acid.

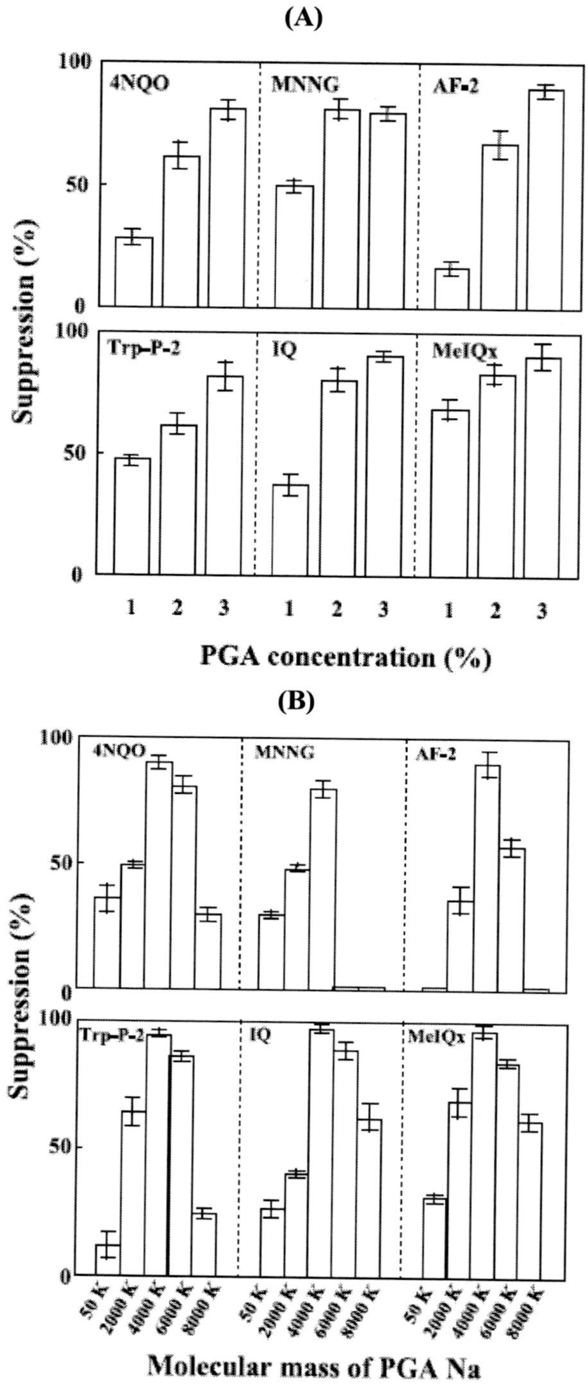

Source: Sato et al. 2008.

Figure 13. Effect of γ-PGA concentration (A) and molecular mass (B) on suppression of SOS response of *S. typhimurium* induced by various chemical mutagens. Mutagen concentration: AF-2=0.3 μg/mL; MNNG=6 μg/mL; 4NQO=3 μg/mL; Trp-p-2=3 μg/mL; IQ=0.3 μg/mL; MeIQx=0.3 μg/mL, γ-PGA: Na-form at 4000 kDa. The vertical error bars represent standard deviation for a mean of triplicate experiments.

6. Conclusion

The foregoing review has demonstrated the eco-friendly and biocompatible biopolymer poly(γ-glutamic acid) can be a potential material for flocculating turbid waters, and scavenging cationic dyes and chemical mutagens. Various physical and chemical factors affecting the flocculation as well as adsorption of cationic dyes and chemical mutagens were elaborated. Though the synthetic alternatives may be effective and inexpensive, their application could be restricted from the environmental and health point of view. On the contrary, biopolymers are generally non-toxic and biodegradable, but the major pitfall limiting the practical application is their high cost. Nevertheless, the commercial viability may be enhanced by synthesizing extracellular biopolymers through large-scale fermentation by employing simple and cost-effective downstream processing methods. Future research on these topics should focus on a more realistic system involving column operations for cationic dye removal and *in vivo* study for scavenging chemical mutagens.

References

[1] Ivánovics, G.; Bruckner, V. *Z. Immunitatsforsch.* 1937, 90, 304-318.

[2] Ivánovics, G.; Erdös, L. *Z. Immunitatsforsch.* 1937, 90, 5-19.

[3] Fujii, H. *Nippon Nogeikagaku Kaishi*, 1963, 37, 407-411.

[4] Fujii, H. *Nippon Nogeikagaku Kaishi*, 1963, 37, 474.

[5] Shih, I. L.; Van, Y. T. *Bioresour. Technol.* 2001, 79, 207-225

[6] Bovarnick, M. *J. Biol. Chem.* 1942, 145, 415-424.

[7] Ward, R. M.; Anderson, R. F.; Dean, F. K. *Biotechnol. Bioeng.* 1963, 5, 41-48.

[8] Troy, F. A. *J. Biol. Chem.* 1973, 248, 305-316.

[9] Murao, S. *Koubunshi* 1969, 16, 1204-1212.

[10] Thorne, C. B.; Gómez, C. G.; Blind, G. R.; Housewright, R. D. *J. Bacteriol.* 1953, 65, 472-478.

[11] Thorne, C. B.; Gómez, C. G.; Noyes, H. E. Housewright, R. D. *J. Bacteriol.* 1954, 68, 307-315.

[12] Housewright, R. D. In *The Bacteria: A Treatise on Structure and Function*; Gunsalus, I. C.; Stanier, R. Y.; Eds.; Academic Press: New York, 1962; pp 389-412.

[13] Nitccki, D. E.; Goodman, J. W. In *Chemistry and Biochemistry of Amino Acids, Peptides and Proteins*; Weinstein, B.; Ed.; Marcel Dekker: New York, 1971, pp 87-92.

[14] Sung, M. H. ; Park, C.; Kim, C. J.; Poo, H.; Soda, K.; Ashiuchi, M. *Chemical Record* 2005, 5, 352-366.

[15] Ho, G. H.; Ho, T. I.; Hsieh, K. H.; Su, Y. C.; Lin, P. Y.; Yang, J.; Yang, K. H.; Yang, S. C. *J. Chin. Chem. Soc.* 2006, 53, 1363-1384.

[16] Xu, H.; Jiang, M.; Li, H.; Lu, D.; Quyang, P. *Process Biochem.* 2005, 40, 519-523.

[17] Shih, I. L.; Van, Y. T.; Shen, M. H. *Mini-Rev. Med. Chem.* 2004, 4, 179.

[18] Matsusaki, M.; Serizawa, T.; Kishida, A.; Endo, T.; Akashi, M. *Bioconjugate Chem.* 2002, 13, 23.

[19] Mark, S. S.; Crusberg, T. C.; DaCunha, C. M.; Di Iorio, A. A. *Biotechnol. Prog.* 2006, 22, 523.

[20] Radu, J. E. F.; Novak, I.; Hartmann, J. F.; Borbely, *J. Polym. Prep.* 2006, 47, 420.

[21] Kang, H. S.; Park, S. H.; Lee, Y. G.; Son, T. I. *J. Appl. Polym. Sci.* 2007, 103, 386.

[22] Shimokuri, T.; Kaneko, T.; Serizawa, T.; Akashi, M. *Macromol. Biosci.* 2004, 4, 407.

[23] Perez-Camero, G.; Garcia-Alvarez, M.; Mrtinez de Iiarduya, A.; Fernandez, C.; Campos, L.; Munoz-Guerra, S. *Biomacromolecules* 2004, 5, 144.

[24] Matsusaki, M.; Serizawa, T.; Kishida, A.; Akashi, M. *Biomacromolecules* 2005, 6, 400.

[25] Chiu, Y. T.; Chiu, C. P.; Chien, J. T.; Ho, G. H.; Yang, J.; Chen, B. H. *J. Agric. Food Chem.* 2007, 55, 5123-5130.

[26] Kao, T. H.; Chen, B. H. *Molecules*, 2007, 12, 917-931.

[27] Inbaraj, B. S.; Chien, J. T.; Ho, G. H.; Yang, J.; Chen, B. H. *Biochem. Eng. J.* 2006a, 31, 204-215.

[28] Inbaraj, B. S.; Chiu, C. P.; Chiu, Y. T.; Ho, G. H.; Yang, J.; Chen, B. H. *J. Agric. Food Chem.* 2006b, 54, 6452-6459.

[29] Inbaraj, B. S.; Wang, J. S.; Lu, J. F.; Siao, F. Y.; Chen, B. H. *Bioresour. Technol.* 2009, 100, 200-207.

[30] Shih, I. L.; Van, Y. T.; Yeh, L. C.; Lin, H .G.; Chang, Y. N. *Bioresour. Technol.* 2001, 78, 267-272.

[31] Wu, J. Y.; Ye, H. F. *Process Biochem.* 2007, 42, 1114-1123.

[32] Sato, M.; Kanie, K.; Soda, K.; Yokoigawa, K. *J. Biosci. Bioeng.* 2008, 105, 690-693.

[33] Goto, A.; Kunioka, M. *Biosci. Biotechnol. Biochem.* 1992, 56, 1031-1035.

[34] Kubota, H.; Nambu, Y.; Endo, T. *J. Polym. Sci. Part A: Polym. Chem.* 1993, 31, 2877-2878.

[35] Kubota, H.; Matsunobu, T.; Uotani, K.; Takebe, H.; Satoh, A.; Tanaka, T.; Tanguchi, M. *Biosci. Biotechnol. Biochem.* 1993, 57, 1212-1213.

[36] Cromwick, A. M.; Gross, R. A. *Biotechnol. Bioeng.* 1996, 50, 222-227.

[37] Kunioka, M.; Goto, A. *Appl. Microbiol. Biotechnol.* 1994, 40, 867-872.

[38] Ito, Y.; Tanaka, T.; Ohmachi, T.; Asada, Y. *Biosci. Biotechnol. Biochem.* 1996, 60, 1239-1242.

[39] Cheng, C.; Asada, Y.; Aaida, T. *Agric. Biol. Chem.* 1989, 53, 2369-2375.

[40] Ogawa, Y.; Yamaguchi, F.; Yuasa, K.; Tahara, Y. *Biosci. Biotechnol. Biochem.* 1997, 61, 1684-1687.

[41] Kunioka, M. *Appl. Microbiol. Biotechnol.* 1997, 47, 469-475.

[42] Birrer, G. A.; Cromwick, A. M.; Gross, R. A. *Int. J. Biol. Macromol.* 1994, 16, 265-275.

[43] Troy, F. A. In *Peptide Antibiotics: Biosynthesis and Functions*; Kleinkauf, H.; Von Döhren, H.; Eds.; Walter de Gruyter: New York, 1982; pp 49-83.

[44] Crescenzi, V.; Dentini, M.; Mattei, B. *ACS Sump. Ser.* 1996, 627, 233 (Chapter 19).

[45] Zanuy, D.; Aleman, C.; Munoz-Guerra, S. *Intl. J. Biol. Macromol.* 1998, 23,175.

[46] Tong, Z.; Zhe, L.; Huai-Lan, Z. *J. Environ. Sci.* 1999, 11, 1-12.

[47] Kurane, R.; Takeda, K.; Suzuki, T. *Agric. Biol. Chem.* 1986, 50, 2301-2307.

[48] Salehizadeh, H.; Vossoughi, M.; Alemzadeh, I. *Biochem. Eng. J.* 2000, 5, 39-44.

[49] Gutcho, S. *Waste Treatment with Polyelectrolytes and other Flocculants*; Noyes Data Corp.: Park Ridge, NJ, 1977; pp 1-37.

[50] Nakamura, J.; Miyashiro, S.; Hirose, Y. *Agric. Biol. Chem.* 1976, 40, 1341-1347.

[51] Salehizadeh, H.; Shokaosadati, S. A. *Biotechnol. Adv.* 2001, 19, 371-385.

[52] Dearfield, K. L.; Abermathy, C. O. *Mutant. Res.* 1988, 195, 45-77.

[53] Vanhorick, M. Moens, W. *Carcinogenesis* 1983, 4, 1459-1463.

[54] Yokoi, H.; Natsuda, O.; Hirose, J.; Hayashi, S.; Takasaki, Y. *J. Ferment. Bioeng.* 1995, 79, 378-380.
[55] Yokoi, H.; Arima, T.; Hirose, J.; Hayashi, S.; Takasaki, Y. *J. Ferment. Bioeng.* 1996, 82, 84-87.
[56] Takagi, H.; Kadowaki, K. *Chitin Nat. Technol.* 1985, 3, 121-128.
[57] Taniguchi, M.; Kato, K.; Shimauchi, A.; Ping, X.; Fujita, K. I.; Tanaka, T.; Tarui, Y.; Hirasawa, E. *J. Biosci. Bioeng.* 2005, 99, 130-135.
[58] Taniguchi, M.; Kato, K.; Shimauchi, A.; Ping, X.; Nakayama, H.; Fujita, K. I.; Tanaka, T.; Tarui, Y.; Hirasawa, E. *J. Biosci. Bioeng.* 2005, 99, 245-251.
[59] Klute, R.; Neis, U. In *Proceedings of the International Conference, Colloid Interface Science*; 50th Edition; Ker, M.; Ed.; 1976; Vol.4, pp 113.
[60] Inbaraj, B. S.; Sulochana, N. *Indian J. Chem. Technol.* 2006, 13, 17-23.
[61] Pearce, C. I.; Lloyd, J .R.; Guthrie, J. T. *Dyes Pigments* 2003, 58, 179-196.
[62] Blackburn, R. S. *Environ. Sci. Technol.* 2004, 38, 4905-4909.
[63] Choy, K. K. H.; Porter, J. F.; McKay, G. *Langmuir* 2004, 20, 9646-9656.
[64] Crini, G. *Bioresour. Technol.* 2006, 97, 1061-1085.
[65] Wingender, J.; Neu, T. R.; Flemming, H. C. *Microbial Extracellular Polymeric Substances: Characterization, Structures and Function*; Springer-Verlag: Heidelberg, 1999.
[66] Liu, H.; Fang, H. H. P. *Biotechnol. Bioeng.* 2002, 80, 806-811.
[67] Gutnick, D. L.; Bach, H. *Appl. Microbiol. Biotechnol.* 2000, 54, 451-460.
[68] Kazy, S. K.; Sar, P.; Singh, S. P.; Sen, A. K; D'Souza, S. F. *World J. Microbiol. Biotechnol.* 2002, 18, 583-58.
[69] Kim, S. Y.; Kim, J. H.; Kim, C. J.; Oh, D. K. *Biotechnol. Lett.* 1996, 18, 1161-1164.
[70] Inbaraj, B. S.; Chiu, C. P.; Ho, G. H.; Yang, J.; Chen, B. H. *J Hazard. Mater.* 2006c, *B*137, 226-234.
[71] Inbaraj, B. S.; Chiu, C. P.; Ho, G. H.; Yang, J.; Chen, B. H. *Bioresour. Technol.* 2008, 99, 1026-1035.
[72] Lagergren, K. *Sven. Ventenskapsakad. Handl.* 1898, 24, 1-39.
[73] Blanchard, G.; Maunaye, M.; Martin, G. *Water Res.* 1984, 18 1501-1507.
[74] Ho, Y. S.; McKay, G. *Process Biochem.* 2003, 38, 1047-1061.
[75] Vlasov, A. V.; Prokof'ev, K. V.; Sangalov, Y. A.; Kotov, S. V. *Chem. Technol. Fuels Oils*, 1988, 24, 52-55.
[76] Ho, Y. S.; Ng, J. C. Y.; McKay, G. *Separ. Purif. Meth.* 2000, 29, 189-232.
[77] Lazaridis, N. K.; Asouhidou, D. D. *Water Res.* 2003, 37, 2875-2882.
[78] Boyd, G. E.; Adamson, A. W.; Myers Jr. L. S. *J. Am. Chem. Soc.* 1947, 69, 2836-2848.
[79] Reichenberg, D. *J. Am. Chem. Soc.* 1953, 75, 589-597.
[80] Kumar, K. V.; Kumaran, A. *Biochem. Eng. J.* 2005, 27, 83-93.
[81] Giles, C. H.; MacEwan, T. H.; Nakhwa, S. N.; Smith, D. *J. Chem. Soc.* 1960, 111, 3973-3993.
[82] Sumner, M. E. *Handbook of Soil Science*; CRC Press: Boca Raton, FL, 2000.
[83] Langmuir, I. *J. Am. Chem. Soc.* 1918, 40, 1361-1368.
[84] Freundlich, H. M. F. *Z. Phys. Chem.* 1906, 57, 384-470.
[85] Redlich, O.; Peterson, D. L. *J. Phys. Chem.* 1959, 63, 1024-1026.
[86] McKay, G.; Porter, J. F.; Prasad, G. R. *Water Air Soil Pollut.* 1999, 114, 423-438.

[87] Juang, R. S.; Wu, F. C.; Tseng, R. L. *Colloid Surf. A: Physiochem. Eng. Aspect* 2002, 201, 191-199.
[88] Tseng, R. L.; Wu, F. C.; Juang, R. S. *Carbon* 2003, 41, 487-495.
[89] Magdy, Y. H.; Daifullah, A. A. M. *Waste Manage.* 1998, 18, 219-226.
[90] Guo, Y.; Yang, S.; Fu, W.; Qi, J.; Li, R.; Wang Z.; Xu, H. *Dyes Pigments* 2003, 56, 219-229.
[91] Ho, Y. S.; Chiang, T. H.; Hsueh, Y. M. *Process Biochem.* 2005, 40, 119-124.
[92] Sun, Q.; Yang, L. *Water Res.* 2003, 37, 1535-1544.
[93] Okada, K.; Yamamoto, N.; Kameshima, Y.; Yasumori, A. *J. Colloid Interf. Sci.* 2003, 262, 194-199.
[94] Nassar, M. M. *Adsorpt. Sci. Technol.* 1997, 15, 609-617.
[95] Bagane, M.; Guiza, S. *Ann. Chim. Sci. Mater.* 2000, 25, 615-626.
[96] Gulnaz, O.; Kaya, A.; Matyar, F.; Arikan, B. *J. Hazard. Mater.* 2004, 108, 183-188.
[97] Nassar, M. M.; Hamoda, M. F.; Radwan, G. H. *Water Sci. Technol.* 1995, 32, 27-32.
[98] Karadağ, E.; Saraydin, D.; Aydin, F. *Turkish J. Chem.* 1998, 22, 227-236.
[99] Namasivayam, C.; Kumar, M. D.; Selvi, K.; Begum, R. A.; Vanathi, T.; Yamuna, R. T. *Biomass Bioenergy* 2001, 21, 477-483.
[100] Sainz-Diaz, C. I.; Griffiths, A. J. *Fuel* 2000, 79, 1863-1871.
[101] Inbaraj, B. S.; Sulochana, N. *Indian J. Chem. Technol.* 2002, 9, 201-208.
[102] Waranusantigul, P.; Pokethitiyook, P.; Kruatrachue, M.; Upatham, E. S. *Environ. Pollut.* 2003, 125, 385-392.
[103] Kumar, K. V.; Sivanesan, S.; Ramamurthi, V. *Process Biochem.* 2005, 40, 2865-2872.
[104] Inbaraj, B. S.; Selvarani, K.; Sulochana, N. J. Sci. Ind. Res. 2002, 61, 971-978.
[105] Banat, F.; Al-Asheh, S.; Al-Makhadmeh, L. *Process Biochem.* 2003, 39, 193-202.
[106] Garg, V. K.; Gupta, R.; Yadav, A. B.; Kumar, R. *Bioresour. Technol.* 2003, 89, 121-124.
[107] Crini, G.; Peindy, H. N. *Dyes Pigments* 2006, 70, 204-211.
[108] Annadurai, G.; Juang, R. S.; Lee, D. J. *J. Hazard. Mater.* 2002, 92, 263-274.
[109] Janoš, P. *Environ. Sci. Technol.* 2003, 37, 5792-5798.
[110] Ho, Y. S.; Chiu, W. T.; Wang, C. C. *Bioresour. Technol.* 2005, 96, 1285-1291.
[111] Khattri, S. D.; Singh, M. K. *Water Air Soil Pollut.* 2000, 120, 283-294.
[112] Hall, K. R.; Eagleton, L. C.; Acrivos, A.; Vermeulen, T. *Ind. Eng. Chem. Fundam.* 1966, 5, 212-219.
[113] Manohar, D. M.; Anoop Krishnan, K.; Anirudhan, T. S. *Water Res.* 2002, 36, 1609-1619.
[114] International Atomic Energy Agency, A*pplication of Ion Exchange Processes for the Treatment of Radioactive Waste and Management of Spent Ion Exchangers*; International Atomic Energy Agency: Vienna, Austria, 2002, Technical Report No. 408.
[115] Vermöhlen, K.; Lewandowski, H.; Narres, H. D.; Schwuger, M. J. *Colloids and Surfaces A: Physicochem. Eng. Aspects* 2000, 163, 45-53.
[116] Vikse, R.; Mjelva, B. B.; Klungsøyr, L. *Food Chem. Toxicol.* 1992, 30, 239-246.
[117] Ferguson, L. R.; Roberton, A. M.; Watson, M. E.; Triggs, C. M.; Harris, P. J. *Chem. Biol. Interact.* 1995, 95, 245-255.
[118] Harris, P. J.; Triggs, C. M,; Roberton, A. M.; Watson, M. E.; Ferguson, L. R. *Chem. Biol. Interact.* 1996, 100, 13-25.

[119] Ryden, P.; Roberton, J. A. *Mutat. Res.* 1996, 351, 45-52.
[120] Isobe, Y.; Yokoigawa, K.; Kawai, H. *Food Sci. Technol. Int.* 1998, 4, 77-79.
[121] Barnes, W. S.; Maiello, J.; Weisburger, J. H. *J. Natl. Cancer Inst.* 1988, 70, 757-760.
[122] Chen, B. H.; Yang, D. J. *Chromatographia* 1998, 48, 223-230.
[123] Ferguson, L. R.; Harris, P. J. *Mutat. Res.* 1996, 350, 173-184.
[124] Scatchard, G. *Ann. NY Acad. Sci.* 1949, 51, 660-672.
[125] Brunauer, S.; Emmett, P. H.; Teller, E. *J. Am. Chem. Soc.* 1938, 60, 309-319.
[126] DeBoer, J. H.; Zwicker, C. Z. *Physik. Chem.* 1929, *B*3, 407-418.
[127] Oda, Y.; Nakamura, S.; Oki, I.; Kato, T.; Shinagawa, H. *Mutat. Res.* 1985, 147, 219-229.

In: Biochemical Engineering
Editors: F.E. Dumont and J.A. Sacco, pp. 117-140

ISBN: 978-1-60741-257-1

Chapter 4

Molecular Imprinted Polymers in Biomacromolecules Recognition

***Jie Hu*[1,*], *Zhen Tao*[2], *Shunsheng Cao*[1] and *Xinhua Yuan*[1]**

[1]School of Materials Science and Engineering, Jiangsu University, Zhenjiang 212013, China

[2]The Affiliated Hospital of Jiangsu University, Zhenjiang 212001, China

Abstract

Molecular imprinting has proved to be an effective technique for generating specific recognition sites in synthetic polymers. These sites are tailor-made in situ by copolymerization of functional monomers and cross-linked around the template molecules. The print molecules are subsequently extracted from the polymer, leaving accessible complementary binding sites in the polymer network. Despite significant growth within the field, the majority of template molecules studied thus far are low molecular weight compounds and generally insoluble in aqueous systems. In biological systems, molecular recognition occurs in aqueous media. So, in order to create molecular imprinted polymers capable of mimicking biological processes, it is necessary to synthesize artificial receptors which can selectively recognize the target biological macromolecules such as peptides and proteins in aqueous media. Actually, the synthesis of molecular imprinted polymers specific for biomacromolecules has been a focus for many scientists working in the area of molecular recognition, since the creation of synthetic polymers that can specifically recognize biomacromolecules is a very challenging but potentially extremely rewarding work. The resulting molecular imprinted polymers with specificity for biological macromolecules have considerable potential for applications in the areas of solid phase extraction, catalysis, medicine, clinical analysis, drug delivery, environmental monitoring, and sensors. In this review, the authors discuss the challenges associated with the imprinting of peptides and proteins, and provide an overview of the significant progress achieved within this field. This review offers a comparative analysis of different approaches developed, focusing on their relative advantages and disadvantages, highlighting trends and possible future directions.

Keywords: molecular imprinted polymers; biomacromolecules; recognition; synthetic receptor

* E-mail address: hujiechem@hotmail.com. Corresponding author: Jie Hu.

1. Introduction

The selective recognition of biological molecules governs many essential biological interactions, which has attracted much attention in recent years. Mimicking these processes with synthetic receptors is still a major challenge in chemistry, physics and biology. Among the variety of approaches for the preparation of tailor-made artificial receptors that can selectively bind predetermined target molecules, molecular imprinting has been widely recognized as the most promising technique, which offers a number of distinct advantages [1, 2]. Molecular imprinting technology allows the creation of synthetic receptors with binding constants comparable to natural receptors, but capable of withstanding much harsher conditions, such as acids, bases, metal ions, organic solvents, extreme pH, high pressure and temperature. These artificial materials can be prepared in large quantities with good reproducibility and low cost [3].

The principle of molecular imprinting has been widely illustrated in some reviews [4-7]. Generally, the imprinting relies on the formation of a host-guest complex between a template molecule and one or more functional monomers in an appropriate solvent. After the addition of a cross-linking agent and subsequent polymerization, a highly cross-linked polymer matrix is formed which encapsulates the template molecule. Then, the template molecules are extracted from the polymer, leaving cavities with suitable size, shape, and functional group orientation which are complementary to the target molecule. From the 1990s, molecular imprinting technology has been extensively exploited and many kinds of molecules have been successfully imprinted for a variety of applications, including affinity chromatography supports, solid phase extraction membranes, biomimetic sensors, environmental monitoring and drug delivery [8-14].

However, despite significant growth within the field, the majority of template molecules studied thus far have been characterized by their low molecular weight and insolubility in aqueous systems. In this field, imprinting of biomacromolecules is one of the most challenging tasks. More physical and chemical factors need to be considered on the preparation of biomacromolecular receptors in comparison to the molecular receptors for small targets. The first report on protein imprinting was proposed by Glad T. et al. in 1985 [15]. In the following years, many contributions in the molecular imprinting of peptides and proteins were conducted by Mosbach K [16-19]. The molecular imprinted polymers synthesized in these early studies were tested for their ability to selectively recognize their respective target molecules in organic solvents. However, most molecules of biological importance are water soluble, and many natural recognition events such as antigen-antibody binding occur in aqueous media. Moreover, proteins have a flexible structure and conformation, which can be easily affected by changes in temperature or in the environment. According to the calculation reported by Nicholls [20], it is difficult, from thermodynamical viewpoint, to develop successful imprints for such molecules. So, the main challenge during the synthesis of molecular imprinted polymers selective to biomacromolecules is the fact that they need to be imprinted at conditions close to their natural environment ensuring conformational integrity. Furthermore, these materials need to provide access to the binding pockets, which is different from conventional molecular imprinted polymers expecting diffusion of the template molecules into the matrix during rebinding.

Considering the large number of functional groups on the peptides and proteins, which are available for the interaction with functional monomers, there are a number of different strategies for creating molecular imprinted polymers targeting these biomacromolecules. Functional groups forming strong template interactions with the target molecules are commonly used, including electrostatic and metal-chelating groups [21, 22], which is however limited by the structure and chemical properties of the selected template molecules. A more generally applicable approach seems to design molecular imprinted polymers utilizing the shape complementarity along with multi-point weak interactions, such as hydrophobic and hydrogen bonding, for providing molecularly selective binding sites.

This review presents the recent advances and developments of molecular imprinted polymers in biomacromolecules recognition, with an emphasis on the comparative analysis of different approaches developed, underlying the advantages and disadvantages.

2. Molecular Imprinting in Aqueous Media

There is strong evidence to suggest that the cooperative interaction between hydrogen bond and hydrophobic interaction is the most dominant form of template-receptor complexation. Hydrogen bonds are high directional, and many synthesized polymers rely on the high directionality of the hydrogen bonds and shape specificity of target molecules to prepare receptors with strong interaction and molecular recognition [23]. So, the majority of published papers on molecular imprinted polymers are based on polymers synthesized in organic solvents, because the hydrogen bond between template molecules and functional monomers considered among the dominant forces in imprinting is largely weakened in aqueous system [24]. It was observed that the separation factors of the molecular imprinted polymers were significantly lower in acetonitrile/water mixtures than in pure acetonitrile [25]. In addition, it was shown that water could significantly lower the effectiveness of hydrogen bonds between the template and the receptor. However, the recognition of biomolecules is most relevant in aqueous solution to avoid denaturing or degradation of the template molecules [26]. Hence, molecular imprinting in aqueous media has to take advantage of other types of interactions, such as π-π stacking, ionic interactions, and hydrophobic interactions to facilitating molecular recognition [27-29].

Recently, many researches have demonstrated that using chitosan in the polymer network can increase the affinity and selectivity to target proteins due to its high content of amino and hydroxyl functional groups. Xia et al. prepared a semi-interpenetrating molecular imprinted hydrogel for hemoglobin based on polyacrylamide/chitosan in aqueous solution [30]. The –NH_3^+ and the hydroxyl groups of chitosan will interact with the target protein by electrostatic forces rather than hydrogen bonding, which are suspected as the main recognition mechanism at high water content. Further research indicated that the amount of hemoglobin adsorbed increased with increase of temperature, and the calculated apparent activation energy for the adsorption was 24.6 kJ/mol, indicating that the adsorption has a low potential barrier [31]. Guo et al. [32-34] prepared hemoglobin imprinted polymers by physically entrapping or chemically grafting of selective polyacrylamide gels onto porous chitosan beads, and Fu et al. [35] reported the synthesis of bovine serum albumin imprinted polymers gels based on chitosan and polyacrylamide via homogeneous graft copolymerization of acrylamide on chitosan dissolved in acidic aqueous. Their results indicate that the incorporation of chitosan into traditional polyacrylamide gels may produce homogeneous molecular imprinted polymers for proteins with a fine affinity, selectivity and rebinding capacity. Using randomly distributed carboxylic groups on the backbone of the polymer chain as recognition sites, Zhao et al. [36] prepared molecular imprinted polymer with cloned bacterial protein template to enrich authentic target in cell extract. It seems that, in this case, the electrostatic force and cavity shape were the dominate factors for the selective recognition.

It has been suggested that binding in aqueous solution can be enhanced by exploiting cooperative interactions and selecting proper hydrophobic microenvironments to create better receptors. In one of most interesting example of molecular imprinting by the cooperative interactions of hydrophobic and electrostatic interaction, acrylated β-cyclodextrin was copolymerized with an electrostatic functional monomer 2-acryoylamido-2,2′ -dimethylpropane sulfonic acid. β-cyclodextrin has a hydrophobic cavity which can bind aromatic groups on the chain of biomacromolecules. So the hydrophobic effect is the dominant factor in imprinting water rich system, whereas electrostatic interaction dominate recognition in aqueous poor mixtures [37, 38]. Similarly, Qin et al. [39] pointed out that bonded β-cyclodextrin and acrylamide cooperated together could improve the selective recognition ability of the molecular imprinted polymers for the template tryptophan. Figure 1 showed the formation of template-receptor complexation. The high performance liquid chromatography column packed with this kind of molecular imprinted polymer could even separate the template from its enantiomer in aqueous mobile phase. Using toluene diisocyanate as the cross-linking agent, Egawa et al. [40] and Asanuma et al. [41] prepared molecular imprinting cyclodextrin polymers in the present of cholesterol template. Although the imprinting occurred in a dimethylsulfoxide emulsion, the recognition could be achieved in the solvents containing water since binding of the guest molecules was based on inclusion complex formation with cyclodextrins. Thus, these polymeric receptors are not only applicable to selective recognition of biologically important molecules, but also removal of toxic molecules from aqueous media.

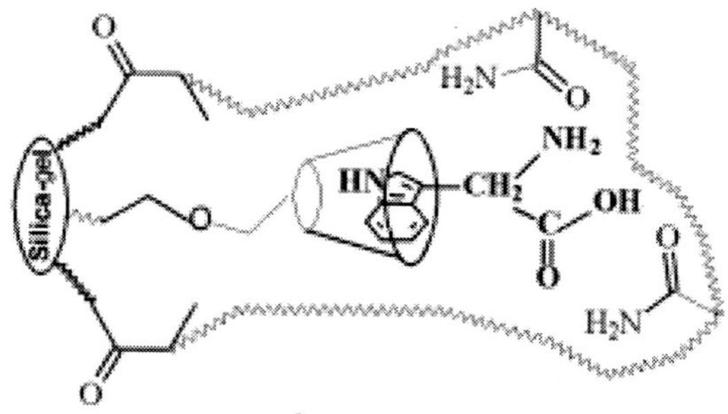

Figure 1. Schematic representation of tryptophan template imprinting.

In contrast to weak interactions like hydrogen bonding and hydrophobic interactions in water, metal complexation [22, 42] and ionic interaction [43] would provide much stronger interactions for molecular recognition. Janotta et al. [44] demonstrated that ionic interaction can play an important role during the selective recognition of 4-nitrophenol by molecular imprinted polymers in aqueous solution. Interestingly, polymers imprinted with 2-nitrophenol and 3-nitrophenol show less pronounced imprinting effects comparing the 4-nitrophenol imprinted polymer. The decrease of the imprinting effect may be result from the fact that the

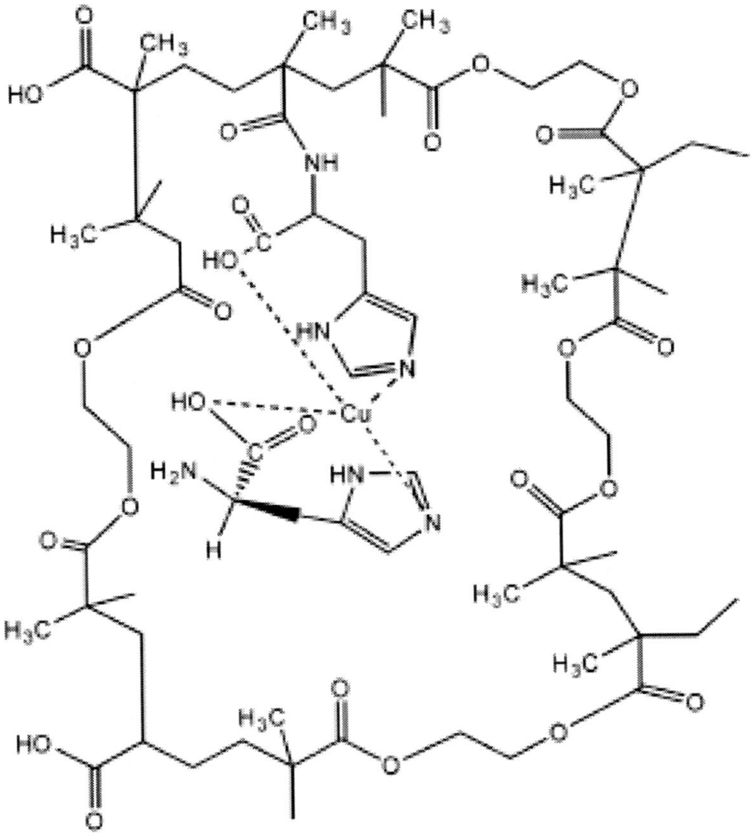

Figure 2. Schematic representation of L-histidine template imprinting.

functional noieties of 2- and 3-nitrophenol are in closer proximity, which presents the formation of the selective binding sites. Selective and reversible recognition of lysozyme by accordingly synthesized molecular imprinted polymers was observed in aqueous solution [45]. In this case, *N*-methacryloyl- (L)-histidine-copper(II) was chosen as the metal-chelating monomer via metal coordination interactions to form complex with lysozyme template and the lysozyme-imprinted cryogel was produced by free radical polymerization. It is found that this kind of imprinted cryogel could be used many times without decreasing the adsorption capacity significantly. Also using *N*-methacryloyl-(L)-histidinemethylester as the metal-coordinating monomer, Arzu Ersoz et al. produced histidine imprinted beads by suspension polymerization for the selective separation of cytochrome c [46]. Schematic representation of L-histidine template formation is shown in Figure 2. For the purification of lysozyme from aqueous solutions and egg white, choosing *N*-methacryloyl-(L)-histidinemethylester as the metal-ccmplexing monomer, Odabaşi et al. [47] prepared lysozyme imprinted particles by UV-initiated polymerization. The purity of the desorbed lysozyme was about 89% with recovery about 84%. For another example, macromolecular receptors for peptides were developed based on strong interactions between N-terminal histidine of peptides, and the Ni(II)-polymerizable NTA ligand, which was incorporated in water-soluble polyacrylamides. These molecular imprinted polymers show adequate water solubility, while providing strong histidine binding sites attracting the dipeptides to the polymer surface. It is pointed out that in terms of strength, specificity, and directionality, the metal coordination interaction is more like a covalent interaction than hydrogen bonding or electrostatic interactions in water. These features make metal coordination a promising binding for the preparation of highly specific templated polymers for the recognition of proteins, via the arrangements of metal coordinating ligands on their surface.

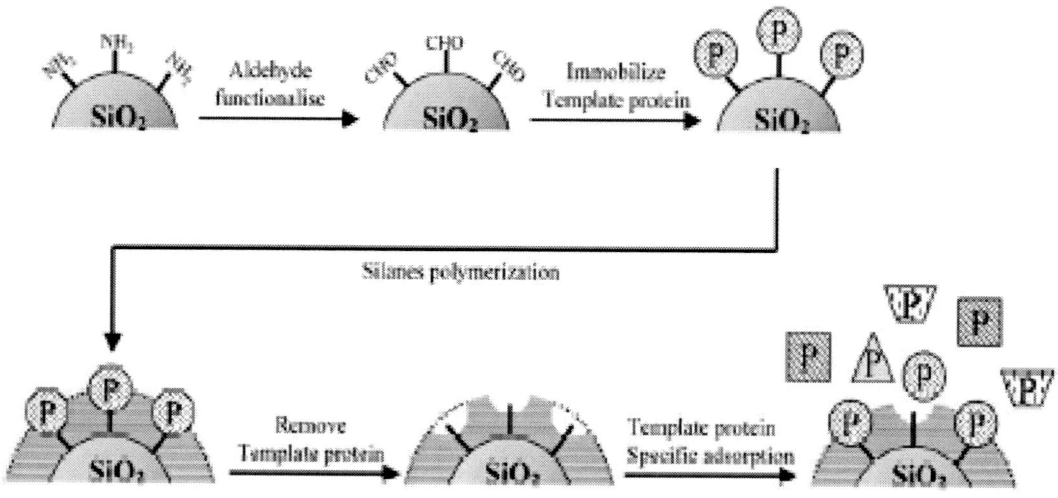

Figure 3. Schematic representation of protein imprinting using immobilized templates.

Because of the effects of the temperature, pH, and the ionic concentration on the status of biomacromolecules and the difficulty of the removal and rebinding of template molecules, covalent immobilization of template proteins has not been widely considered in molecular imprinting. However, there are several successful applications of covalent bond in molecular imprinting. The immobilization of the protein template on a supporting substrate provides a number of advantages, including the imprinting of templates independently of their solubility in the polymerization system, minimizing protein aggregation and creating more homogeneous binding sites. Using silanes based on template covalent immobilization, Shiomi et al. [48] first produced surface-imprinted microspheres for hemoglobin. Covalent immobilization was achieved by forming imine bonds between amino groups on the protein surface and anchored aldehyde groups on silica, these bonds being readily broken by oxalic acid [49] (Figure 3). The recognition appeared to not only depend on the molecular weight and the electric charge, but also on shape and hydrophobicity of the pores created on silica. Bonini et al. [50] modified the approach to fabricate silica-based imprinted beads for hunan serum albumin using polyaminophenylboronic acid instead of silanes. The prepared molecular imprinted polymer beads show recognition mainly due to shape interactions. Using a two-stage core-shell miniemulsion polymerization inspired by the above work of Bonini et al., Chau Jin Tan et al. [51] prepared bovine serum albumin imprinted particles with the immobilization of the template protein molecules on polymeric (instead of silica) support beads. In the competitive adsorption environment, it can be seen that the molecular imprinted polymer prepared using this strategy can display highly specific recognition of bovine serum albumin with no adsorption of the non-template proteins. Devanathan et al. reported [52], for the first time, the formation of a biomimetic covalently imprinted polymeric sensor for a target ligand, the δ-opioid G-protein coupled receptor agonist DPDPE, which reproducibly exhibits subpicomolar binding affinity in an aqueous environment. This imprinted polymer showed a broad structure-activity relationship profile, not unlike that found for protein receptors. Such sensitivity and robustness of molecular imprinted polymers suggest potential applications ranging from biowarfare agent detection to pharmaceutical screening.

Recently, considering the importance of the template shape and conformation in a pre-polymerization mixture for protein imprinting, possible sources causing protein template denaturation in the application of miniemulsion polymerization for molecular imprinting was studied [53, 54]. It was suggested that to avoid high temperature or high-energy irradiation, redox initiation would be a suitable alternative for protein imprinting. In addition, it was found that a certain degree of interaction between the template protein and the micelles was required to maintain the proteins at the particle surface. However, such an interaction should not be too extensive to cause a significant conformational change to ensure the recognition of proteins in their native states. Using poly(vinyl alcohol) as a co-surfactant had proved to be effective in preserving the template protein structural integrity. On top of that, the high-shear homogenization had been shown to cause negligible disruption to the template protein conformation. Employed fluorescein isothiocyanate derivative-albumin as the protein template molecule in an aqueous phase molecular imprinted polymer, for the first time, Hawkins et al. [55] reported the use of a fluorescently labeled template. The selective recognition material had been imaged using confocal microscopy. The resulting imformation would contribute to the understanding of aqueous phase molecular imprinting protocols, with the successful in situ imaging of real-time protein denaturation events being of particular worth.

3. Surface Imprinting

Molecular imprinting has been considered one of the most promising techniques for the preparation of synthetic receptors. In spite of the ease of the conventional imprinting methodology and its associated success with the imprinting of small molecules, the approach has its limititation for the imprinting of protein macromolecules. The major problem during templating of large molecules is the poor efficiency of specific binding due to the multitude of functional sites at these molecules, as well as the limited diffusion of the template molecules into and out of an imprinted polymer network and the incompatibility of the fragile protein template with the imprinting conditions. Confining the recognition sites of molecular imprinted polymers onto the surface of matrix is a technique which has been developed to circumvent the problems associated with the imprinting of biological macromolecules in aqueous solution [56].

Today, surface imprinting has been widely studied as one method to improve the performance of molecular imprinting polymers, for it can resolve the problems of limited mass transfer and template molecule removal often associated with the traditional molecular imprinting techniques, especially when imprinting biological macromolecules. Zhao et al. [57, 58] simultaneously prepared calcium phosphate/alginate hybrid polymer microspheres with bovine serum albumin embedded and coating on the surface. It is found that the surface imprinted beads exhibited better imprinting efficiency and higher rebinding capacity.

An interesting strategy for protein recognition at the quartz crystal microbalance chip surface by surface imprinting was reported by Ratner et al. [23]. Firstly, a layer of template protein was absorbed onto a mica substrate. Disaccharides were then deposited onto the protein molecule as a protective layer against denaturation and degradation. Subsequently, radio-frequency glow discharge plasma was used to deposit hexafluoropropylene onto the protein–disaccharide complexes. The outer surface of that film was attached to a glass substrate by adhesion facilitating detachment of the mica substrate. Finally, the mica substrate and template protein were removed from the hemispherically shaped binding pockets which exhibited highly selective recognition for a variety of template proteins, including albumin, immunoglobulin G, lysozyme, ribonuclease and streptavidin. The specificity of these surfaces prepared with this kind of imprinting strategy is predominantly based on the shape selectivity and hydrogen bonding interaction. This promising strategy certainly has potential for further optimization exploiting different recognition chemistries for functional mapping of protein surfaces. In addition, this strategy assumes that the conformation of the deposited protein used as the template is similar to the free one in solution. Hayden et al. [59] produced polymer receptors for the online monitoring of bioanalytes formed directly onto quartz crystal microbalances using surface imprinting technique. The molded materials were capable of enriching whole cells, viruses and enzymes on the sensor layer surface. Enzyme imprinted polymer layers were also effective as nucleation site for the induction of protein crystallization.

In another example of molecular imprinted polymers in which the recognition sites have been confined to the surfaces, Tan et al. [51] successfully synthesized bovine serum albumin surface-imprinted particles with a two-stage core-shell miniemulsion polymerization. The imprinting strategy was based on the surface immobilization of template molecules with a series of surface modification of the support beads. Actually, many surface-imprinted beads have been successfully prepared with the core-shell polymerization method [60-62]. In one notable work [63], cholesterol surface imprinted submicron beads were successfully synthesized. The significance of this work is the application of a carbonate ester sacrificial spacer which had been utilized in a molecular imprinting system through the conventional bulk imprinting approach [64]. In this technique, the template cholesterol is covalently linked to the functional monomer. After the removal of the template, rebinding of the template molecule to the molecular imprinted polymer would only require noncovalent interactions. So, this semicovalent strategy not only allows the creation of homogeneous binding sites with covalent imprinting, but also takes advantage of the fast rebinding rate of noncovalent adsorption. Also using core-shell emulsion polymerization, perez et al. [65] prepared surface imprinted beads in the presence of polymerizable surfactant, pyridinium 12-(4-vinylbenzyloxycarbonyl)dodecanesulfate and pyridinium 12-(cholesteryloxycarbon-yloxy)dodeca nesulfate. Using the special surfactant, the template molecules could be effectively constrained to the W/O phase boundary during the emulsion polymerization, thus successfully creating hydrophobic binding sites for cholesterol on the imprinted bead surfaces. Furthermore, the selectivity of the surface-imprinted beads was enhanced. In another example, immobilizing the protein template at the surface of the beads, Bonini et al. [50] reported the preparation of imprinted beads for the recognition of human serum albumin. The recognition and binding of the imprinted beads was tested with a complex sample, human serum and targeted removal of human serum albumin without a loss of the other protein components was demonstrated.

Whitcombe and coworkers [66] described a promising approach to prepare homogeneous molecular imprints where a combination of covalent imprinting and non-covalent rebinding techniques was developed. The template was prepared by covalently linking the target peptide with a functional molecule. Upon polymerization and chemical cleavage, the functional group was left in the cavities with precise spatial arrangement, and the essential void space facilitating removal and rebinding of the target peptide.

The creation of thin molecular films on suitable substrates, such as gold, mica and carbon, is a commonly used approach to surface imprinting. Various ways were used to produce such molecular imprinted films. For example, organic/inorganic hybrid thin films for protein recognition were prepared by the liquid-phase deposition (LPD) where pepsin was used as a template protein and titanium oxide was deposited on gold substrates in the presence of pepsin-poly-L-lysine complexes, just as shown in Figure 4 [67]. Surface plasmon resonance (SPR) signals on the deposited films were measured to examine the binding behaviors toward proteins. Matsunaga et al. [68] also prepared molecular imprinted polymer films which were covalently conjugated on gold-coated SPR sensor chips. It was found that the presence of salt in the pre-polymerization mixtures derived higher selectivity and the optimization of salt concentrations in the re-binding buffers could make the selective binding prominent due to the decrease of non-specific binding. Turner et al. [69] studied the crucial role that monolayer constituency and transfer play in the formation and preservation of template protein on multi-component lipid monolayers with both protein attracting and protein repelling elements. All the results indicated that the controllable formation of imprints was aided by a balance between promotion of the protein-monolayer interaction through electrostatics and hindrance through steric effects. Upon transfer to a hydrophobic substrate, these imprinted lipids were immobilized, thus "locking in" their structure (Figure 5).

For the thin film imprinting technique above mentioned, one major issue is controlling the molecular imprinted polymer film thickness, as it should be sufficiently thin for rapid mass transfer. This is especially important in sensor applications, where high sensitivity and short response time are desired. Using atom transfer radical polymerization, Wei et al. [70] reported the successful preparation of ultra thin (<10 nm), surface-confined, molecular imprinted polymer films on model gold substrates. It is found that the template removal from the imprinted films appeared to be 100% efficient. In addition, because the controllable nature of atom transfer radical polymerization allows the growth of uniform molecular imprinted films with adjustable thicknesses, it is possible to tailor the resulting materials to have high capacities by growing thicker films or high binding efficiencies by growing thinner films. Using single- and dual-site Langmuir adsorption models, further researching [71] on the adsorption kinetics of the molecular imprinted film indicated that the adsorption could be well described by any of the four isotherm models: Langmuir, dual-site Langmuir, Freundlich, or Langmuir-Freundlich. The relatively high heterogeneity index values regressed using the Freundlich and Langmuir-Freundlich isotherms suggested the formation of fairly homogeneous molecular imprinted polymer films, although Scatchard analysis revealed that binding site heterogeneity did exist.

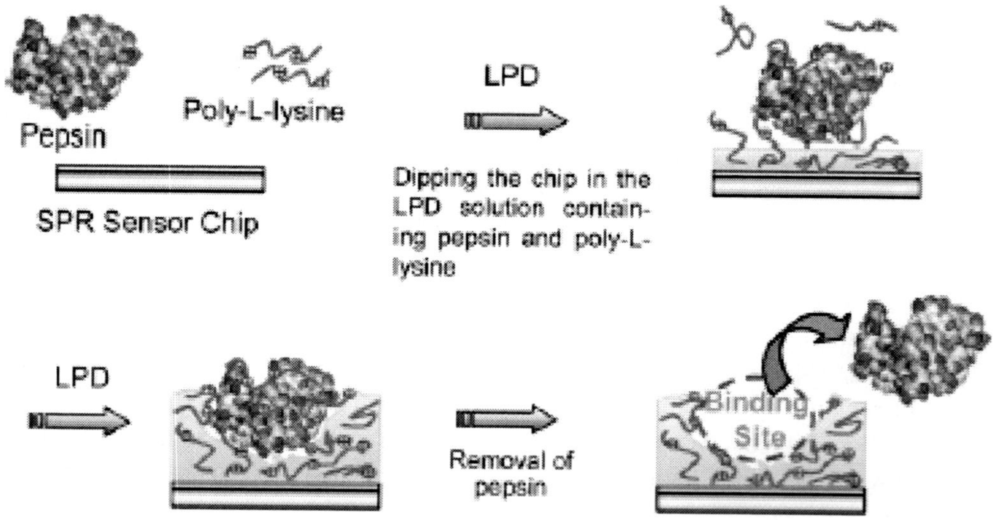

Figure 4. Schematic illustration of protein-templated LPD.

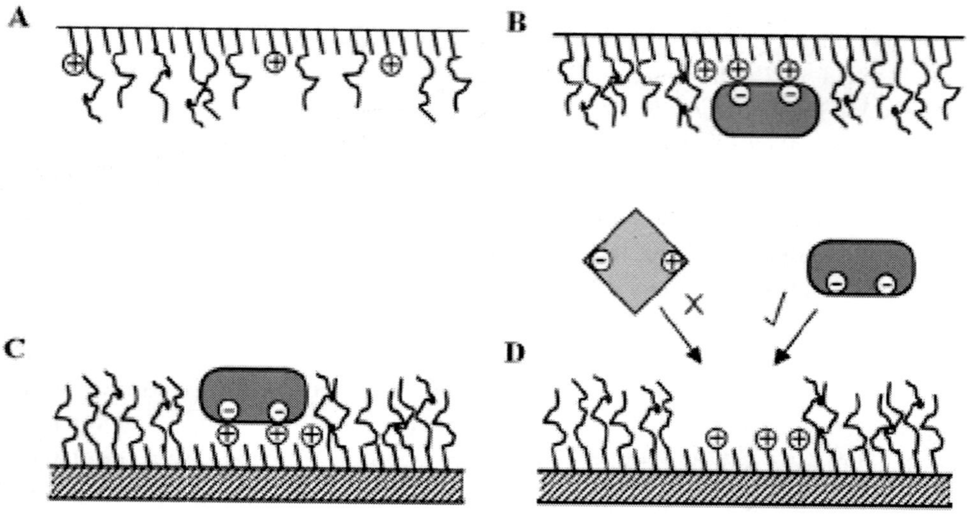

Figure 5. Schematic of protein imprinting in Langmuir monolayers. The native monolayer at the air/water interface (A) interacts with the protein to form imprints (B) that are complementary in shape and charge to the protein. The monolayer is locked into place when transferred to a hydrophobic support (C), followed by removal of the template protein and rebinding studies (D).

Recently, a promising way of printing molecular information on a planar film, so-called molecular mirrors, was presented for the first time [72]. Firstly, template surfaces were prepared by physical adsorption of the proteins onto freshly cleaved ruby mica. Then, cassettes for casting the metal surfaces were prepared from a borosilicate glass slide and a mica slide. Pure molten gallium (T=37□) was poured into the cassette and left to solidify at room temperature for 30 min in a humid controlled chamber. Finally the metal surface was peeled off the template. The metallic prints not only replicate the shape and size of the proteins used as template. They also show specific binding for the template species. Using a microcontact method and low melting point metal, this imprinting strategy is different significantly from the traditional approaches where organic or inorganic materials are generally used. It may open new avenues in optical and electronic sensor design resulting from the combination of the conductive property and the molecular selectivity.

4. Epitope Approach: A Small Structural Element for the Whole Molecule Recognition

In the development of the protein imprinting materials, many challenges, such as the accessibility of binding sites, non-specific binding, and flexible conformation of the macromolecules, remain in producing materials that selectively bind the target molecules. In addition, proteins are inherently capable of assuming a large number of conformations depending on a number of factors, including temperature, pH, and ionic strength. Steric factors also make molecular recognition of proteins difficult, as it is extremely difficult for their large structures to move freely through highly cross-linked polymer networks. It is known that well-defined recognition sites are more easily formed by the imprinting of relatively small molecules with rigid structure. So, to reduce the complications associated with the imprinting of proteins and peptides, Rachkov and coworkers [73] developed a new strategy for the synthesis of biomacromolecule recognition polymers, which were termed the "epitope approach". Generally, an epitope refers to the small active site located within the larger protein structure on an antigen, which combines with the antigen-binding site on an antibody or lymphocyte receptor. In this technique, instead of the whole proteins, a short peptide sequence, often exposed at the protein surface, was used as a template for molecular imprinting polymers preparation. Once the matrix has been polymerized the resultant imprinted material should be able to recognize and bind the whole protein. That is to say, when the original whole protein is exposed to the imprinted material, the region containing the peptide used as template can recognize its spatial and chemical mimic in the polymer.

Rachkov et al. [73] prepared an imprinting polymer specific for the nona-peptide oxytocin, a neurohypophyseal hormone. The polymer was made of methacrylic acid and ethylene glycol dimethacrylate. A small oxytocin sequence of three amino acids proved to be enough for recognition of the whole sequence by the molecular imprinting polymer. In this study, synthesis of the molecular imprinting polymer was performed in an organic environment, but subsequent rebinding experiments were performed using chromatographic methods in both aqueous-rich and aqueous-poor mobile phases. In the aqueous-poor mobile phase, hydrogen bonds and ionic interactions are the dominating factor in creating selective recognition sites. In the aqueous-rich phase, ionic and hydrophobic interactions provide the dominant binding interaction. Few years later, they exploited the epitope approach for the imprinting of the octapeptide hormone angiotensin II in aqueous media [29]. HPLC evaluation of these polymers with a water-based mobile phase showed their selectivity for the peptide, [Sar1,Ala8]angiotensin II, that had been used as the template, but not for its parent peptide angiotensin II. The binding capacity and selectivity of the molecular imprinted polymers depended on the ratio of template to functional monomer in the polymerization mixture, as well as on the ionic strength and pH of the chromatographic mobile phase.

For the first time, using a double stranded DNA (34 base pairs) identical to the sequence of the verotoxin gene as the template molecule, Slinchenko et al. reported the synthesis of DNA-imprinted polymers which could specifically recognize double-stranded DNA without the need to destroy the native double-stranded structure of the DNA [74]. It was believed that the hydrogen bond formation between NH_2 groups of the functional monomer, 2-vinyl-4,6-diamino-1,3,5-triazine, hydrogen bond acceptors of A-T base pair which are exposed in the major groove of double-stranded DNA, was the dominate factor for the recognition.

Nishino et al. prepared a molecular imprinted polymer film imprinted with the peptide sequence from the C-terminus of cytochrome c, alcohol dehydrogenase, and bovine serum albumen, which was able to selectively capture the target molecules from a mixture solution of different proteins. They attributed the binding specificity of the epitope-imprinted polymers, at least in this particular case, to collective hydrogen bonding. When the target protein contacts a complementary domain, multiple hydrogen bonds are formed between the oriented groups within the polymer domain and the target protein's C-terminus sequences. Their results indicated that such a method allowed for the recognition of target proteins based only on genomic information [75]. Using a 16-residue peptide (lysozyme C, 1.8kDa) as the template, Brown et al. [76] produced imprinting porous silica scaffolds that could lead to preferential binding of the whole protein (lysozyme, 14kDa). For comparison, the whole lysozyme was also imprinted into porous silica scaffolds using sol-gel processing. Both imprinting methods are presented in Figure 6. After removing template, both kinds of scaffolds were exposed to lysozyme and/or RNase A, which was used as a competitor molecule of comparable size. When comparing whole protein- to peptide-imprinted scaffolds, similar amounts of lysozyme and RNase were bound from single protein solutions. However, while whole lysozyme-imprinted scaffolds showed about 4:1 preferential binding of lysozyme to RNase, peptide-imprinted scaffolds failed to show statistical significance, even though a slight preferential binding trend was present. It is suggested that there is potential for further development of the epitope approach to improve selectivity.

Actually, concerning protein recognition, it is difficult to evaluate the true efficiency of the epitope approach. In most experiments reported by far, the epitope exploited were up to few amino acid long and the template they were intended to mimic were just small peptides, only 8–10 amino acid long, thus non-three-dimensionally structured. Concerning the exploitation of the epitope approach for the recognition a protein (sequences above 200 amino acids), attention should be paid to three-dimensional structure of the selected epitopes, as it would play a significant role in recognition.

5. Mechanism of the Molecular Imprinting and Recognition

It is well known now that molecular imprinting generally involves arranging polymerizable functional monomers around a template, followed by polymerization and template removal. The quality and the performance of the molecular imprinted polymers depend on carefully tailoring of the physical and chemical nature of the monomers and the interactions between them, the polymerization and its affect on the porous structure, and the rebinding ability of the imprinted cavities. Ensuring that the functional groups of the template molecule are interacting with complementary functionality of polymer-forming compounds is the major factor in this technique.

In most general cases, the binding process may be represented as the follow reaction [77]:

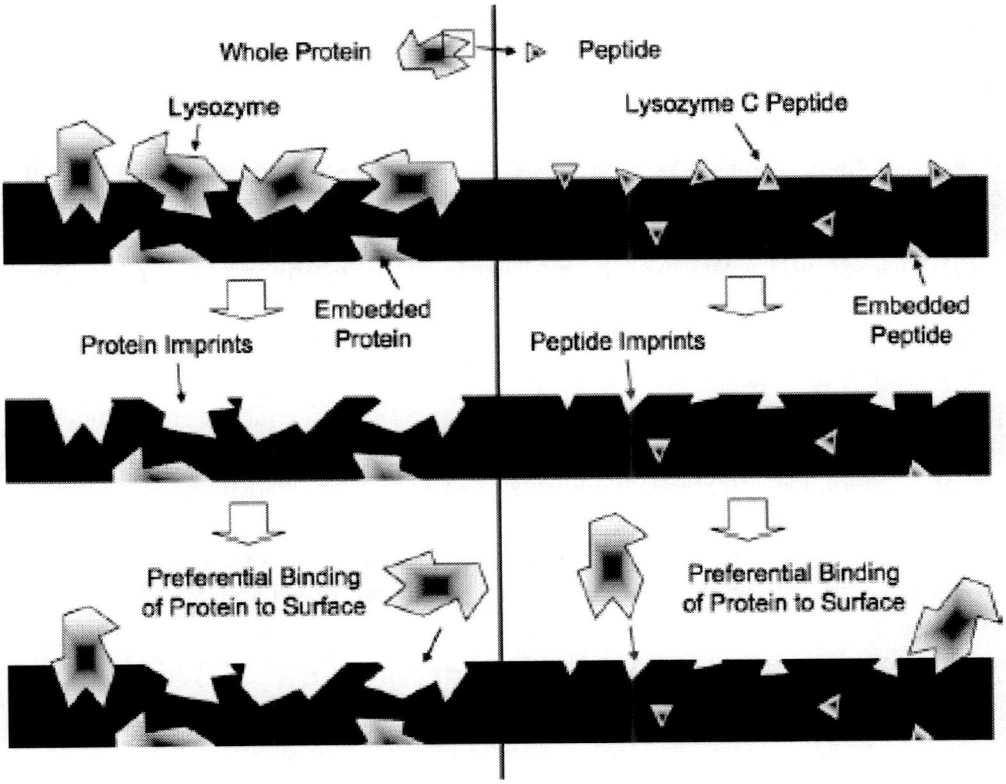

Figure 6. Presentation of whole protein imprinting (traditional approach) and peptide imprinting (epitope approach). On the left, the entire protein molecule creates large imprints with high surface availability. On the right, a small polypeptide sequence from the protein creates small imprints that bind the template molecule at only a specific location.

$$P + \sum_{i=1}^{n} T_i \rightarrow P - T_k + \sum_{i \neq k} T_i$$

Here, the polymer P specifically recognizes the target T_k from a mixture of molecules T_i (i=1, n) and forms a polymer/target complex.

If the compact molecules are larger than the characteristic size of the polymers, the targets are effectively two-dimensional surfaces, and the recognition becomes a process where the polymers selectively bind to specific heterogeneous surfaces. The common property of these targets is that they have a distribution of functional groups. It is the difference in the distribution of these functional groups on the target surfaces or along the linear chains that make the targets recognizable.

Concerning the interactions between the template molecule and the functional monomer, molecular imprinting technology can be categorized into covalent approach and non-covalent approach.

The covalent imprinting method was developed by Wulff [78]. In his studies, template species with covalently attached polymerizable functionality were copolymerized with cross-linking agents in porogenic solvents. These template-monomer complexes became integrated into a macroporous network. Subsequently the template molecules were removed result from the chemical cleavage of the supporting covalent bonds, leaving cavities with complementary size, shape and functional group arrangement to the template molecule. Covalently imprinted polymers are characterized by high yields of stable binding sites, thermodynamically rather homogeneous binding nature, and reduced non-specific adsorption [79]. However, the slow rebinding kinetics resulting from the high steric requirements imposed by the well defined binding site geometry and the difficulty of the cleavability of the bonds under relatively mild conditions limit the application of the covalently imprinting approach.

Mosbach [80, 81] established the most common strategy to circumvent disadvantages associated with the covalent imprinting by exploiting non-covalent interaction, such as Van der Waals forces, dipole-dipole bonds, hydrophobic interaction, ion pair interaction, and hydrogen bonds, between the functional monomers and the template to position the monomers in a particular orientation with respect to the template molecule prior to polymerization. Various polymerizable functional monomer has been use for non-covalent molecular imprinting [82, 83]. Following polymerization and template removal, the functional groups within the imprinted polymers can subsequently recognize and bind the target analyte using the same non-covalent interactions. Nevertheless, non-covalent imprinting is not particularly successful for template molecules that do not possess appropriate functional groups. Another limitation of the non-covalent imprinting is the relatively poor yield of high-affinity binding sites. Typically, less than 15% of the loaded template produces functional binding sites, indicating that most of the imprinted cavities are irreversibly lost, by shrinking, on template elution. Despite these disadvantages, non-covalent imprinting is considered to be very flexible in terms of the possible functional monomers and target molecules.

In order to combine the superiority of both covalently and non-covalently imprinting approach, hybrid techniques have been developed, which also be named semi-covalent and sacrificial-spacer methods [84, 85]. These approaches exploit covalent template-monomer complexes in the imprinting step, but entirely non-covalent interactions for rebinding. For example, Whitcombe et al. prepared a cholesterol imprinted polymers using a carbonate derivative as sacrificial linker motif [64]. So, this semicovalent strategy not only allows the creation of homogeneous binding sites with covalent imprinting, but also takes advantage of the fast rebinding rate of noncovalent adsorption.

Besides various interactions which are possible between the template and the molecular imprinted polymer receptor, the microenvironment surrounding the binding site can have a important role in determining how effective the molecular imprinted polymer will be in recognizing the target molecule. Hart et al. studied the effect of ionic strength on the template-receptor recognition where ionic interaction was the dominant force [22]. In this researching, a number of buffer solutions were utilized to observe the effect of ionic strength and buffer composition on the binding capacity of the molecular imprinted polymers. It was found that all buffers used resulted in a decrease in binding capacity of the receptor. Moreover, a 60mM KCl solution instead of a buffer showed the largest inhibition for the binding capacity. It was therefore concluded that the ionic strength of the binding solution, not the buffer composition itself, plays a important role in determining the effectiveness of the molecular imprinted polymers. It was pointed out that the inhibition of binding capacity in this case is most likely resulted from the change of macromolecule conformation or the hindrance of template-receptor interaction caused by the high salt concentration. It is possible that the variation of buffer composition and ionic strength will have a large effect on the binding systems, even when ionic interaction is not the dominant recognition factor. Just because proteins can adopt various conformations relying on the environment, it is extremely difficult to make a quantitative comparison between molecular imprinted polymers that have been prepared and tested under different conditions.

In addition, the different structure and composition of molecular imprinted polymers, which are usually conduced by different initial concentration of template and/or functional monomer in the prepared process, can have a important effect on their ability to recognize the target molecule [86]. This is most likely related to the fact that high template/functional monomer concentration during preparation will result in a larger number of available binding sites. However, it is not an advisable choice to increase the amount of template in order to gain larger adsorption capacity, because the template protein may be deeply buried in the inner of the molecular imprinted polymers. It is suggested that there exists some optimum distance between functional groups on the synthetic receptor which will correspond to a maximal binding capacity. Study of lysozyme imprinted silica beads indicated that the amount of lysozyme adsorbed onto the beads depended on the composition of functional monomers used during synthesis of the particles [87]. The determination of the zeta-potential of lysozyme molecules and the imprinted silica beads showed a coincidence between the zeta-potential of the template and that of the imprinted polymer beads at the value where the maximal amount of specifically adsorbed lysozyme was observed. It seemed that zeta potential matching may be an important factor in the design of synthetic receptors.

Type and concentration of the cross-linking agent is also a critical factor in creating synthetic receptors with high affinity for their target molecules. When synthesizing molecularly imprinted polymers, ethylene glycol dimethacrylate has been the most commonly used cross linking agent. The effect of other cross linking agents on the recognition by molecular imprinted polymers is still foggy. It is notable that ethylene glycol dimethacrylate is structurally related to polyethylene glycol, which has long been used to provide resistance to non-specific protein binding. Lin et al. [88] investigated the importance of the choice of cross-linking agents in molecular imprinted polymers. In addition, the isothermal titration of monomers to protein templates, including lysozyme, ribonuclease A and myoglobin, was studied to screen the functional monomer for molecular imprinted polymers. Four different cross-linking monomers, ethylene glycol dimethacrylate, tetraethylene glycol dimethacrylate, polyethylene glycol 400 dimethacrylate, and polyethylene glycol 600 dimethacrylate, were first used to synthesize protein imprinted polymers. The chemical structures of the cross-linking monomers suggest that the size of the protein template may be correlated with the optimal number of ethylene glycol-repeat units in the cross-linking monomers. Comparing the poly(ethylene glycol dimethacrylate) or poly(tetraethylene glycol dimethacrylate) protein-imprinted polymers, it is found that the non-specific hydrophobic binding appears to be the cause of the high binding to short polyethylene glycol-repeating units, especially since non-imprinted films bind lysozyme equally well, if not better than, "imprinted" films. It was interesting to note that more effective protein binding and imprinting was obtained with a cross-linking agent containing more ethylene glycol repeat units—even though poly(ethylene glycol) is often used to prevent protein binding. It may be that the greater flexibility of the longer cross linking agent is important in allowing the polymeric matrix to conform to the protein molecule. Moreover, microcalorimetry of the interaction between monomers and protein stamps indicated that styrene monomer was a good candidate as the functional monomer with tetraethylene glycol dimethacrylate to form poly(styrene-*co*-tetraethylene glycol dimethacrylate) lysozyme-imprinted polymers. The poly(styrene-*co*-polyethylene glycol 400 dimethacrylate) RNase A-imprinted polymers also showed good potential to achieve higher rebinding ability with RNase A. On the other hand, for myoglobin, the highest affinity functional monomer is methyl methacrylate.

Although generally neglected, the removal of target molecules from molecular imprinted polymers is a key factor in the rebinding capability of the receptor. The removal of template proteins is complicated by the fact that the high molecular weight of proteins makes it difficult for them to navigate the highly cross-linked matrix. The system of polyacrylamide incorporated with methacrylic acid and 2-(dimethylamino)ethyl methacrylate was studied for the possibility of imprinting of lysozyme [89]. The results showed that approximately 27% (w/w) of the lysozyme template was not able to be extracted from the molecularly imprinted acrylamide polymers. It was found that the amount of template could be extracted was lower with molecular imprinted polymers at higher concentration of cross-linking agent and functional monomer. It may be due to the fact that a more compact gel was formed at a higher preparation concentration. Therefore, it is getting more difficult for the diffusion and removal of templates. For the first time, Hawkins et al. [90] explored in detail a variety of template removal methods, including the use of sodium dodecylsulphate:acetic acid and trypsin digest, for removing template bovine haemoglobin from a polyacrylamide imprinted hydrogel. The optimum ratio of sodium dodecylsulphate:acetic acid was found to be a 10% (w/v):10% (v/v) for the most effective template removal. This resulted in >90% (imprinting efficiency) of re-loaded template molecule being selectively bound within the MIP. At 15%:15% of sodium dodecylsulphate:acetic acid, although there was even more initial template removal, subsequent re-binding studies showed a decrease in imprinting efficiency (67.9%). Trypsin solutions were also used as a method of template removal. Up to 87.4% of template was reproducibly removed initially; however, the imprinting efficiency was only 20.4%. In another example [91], horseradish peroxidase imprinted polymers was treated either with alkaline, neutral, or acidic solutions in order to select the most effective washing procedure. It is found that no template release was observed for alkaline washing, while washing with acid in the presence of a neutral detergent gave a logarithmic decrease of the template contents in the polymer. After five to six cycles, there was only minute amounts (<10%) of the template protein remained in the molecular imprinted polymer. Further experiments indicated that the quantity of template remaining trapped in the polymer, after the proteolytic enzyme trypsin treatment for 18 hours, did not differ significantly from the results obtained simply by acidic washing. It was pointed out that the stabilizing function of the support and spatial orientation of the polymer chains and template functional groups are the major factors affecting the imprint formation and template recognition.

Recently, the utilization of molecular modeling procedures to the field of molecular imprinting to understand more deeply the fundamental molecular mechanisms has attracted much attention [6, 92, 93]. Given the current state-of-the-art in rationally designing molecular imprinted polymers, such simulations may enable selecting combinations of template, functional monomers, cross-linkers, and solvents providing the most stable complex in the pre-polymerization solution, which will then be selected for molecular dynamics simulations investigating the interaction and conformation of the pre-polymerization complex. Enoki et al. studied the effects of frustration on molecular imprinting involving copolymerization of functional ligands (that interact with the imprint molecule) and backbone monomers. It was shown that post-crosslinking in the presence of the target molecule minimizes frustration and thus allowed the imprinted copolymer gel to approach its free energy minimum and that the affinity toward the target molecule increased exponentially with increasing concentrations of post-crosslinking agent. Using mean-field theory, Srebnik et al. [94] investigated the induced porosity of polymer networks obtained through various imprinting approaches, concentrating on the induced porous structure of the final imprinted gel under various preparation conditions. It was found that cross-linking induced microphase separation, and when accompanied by polycondensation, the pore size distribution was affected in an unexpected manner. The model also predicted the extent of microphase separation of the chemically different monomers as a function of the chemical constitution, the degree of cross-linking, and the size of the leaving group. A few years later, Srebnik further introduced a measure of imprinting efficiency (a measure of the degree of complexation) and researched the effect of size and functionality of the imprinting agent on the imprinting efficiency using a lattice model, concentrating on the effect of solvent-template-monomer interactions on the equilibrium population of monomer-template complexes. The extent of template complexation at equilibrium is governed by the change in Gibbs free energy of formation for each mode of template-monomer interactions, generally resulting in a population of sites with various affinities for the imprinted molecule. Neglecting template-template association, it was found that stronger template-monomer interactions resulted in higher functional complexes and thus higher imprinting efficiencies. However, the efficiency strongly depended on the preparation conditions.

In addition, using molecular dynamics simulations through a topological analysis of the imprinted network configuration before and after removal of the templates, the imprinting quality of cross-linked polymer networks was investigated [95]. Low qualities of the imprinted polymers were attributed to the aggregation of the templates in the pre-polymerization solution, aggregation of the imprinting-induced sites with small pores inherent to the polymer, and deformation of the binding sites due to relaxation of the gel after removal of the templates. The formation of distinct individual cavities that retained the size and shape of the template was enhanced by high degrees of cross-linking and low template concentrations. Henthorm et al. [96] developed an all-atom kinetic gelation simulation technique to study the recognitive polymeric network formation which incorporates both intramolecular as well as intermolecular interactions and the subsequent effects they have on the end network structure.

6. Conclusion

Molecular imprinting is a valuable technique for the preparation of synthetic materials for the selective recognition of biologically relevant molecules such as amino acids, peptides, and proteins. The key factors for the advancement in molecular imprinted polymer research for biomacromolecular targets lie in the preparation of molecular imprinted polymers providing high affinity to the macromolecular template molecule without disrupting the bioreactivit or biofunctionality of the target analyte. However, despite the recent advancements made in the synthesis and characterization of molecular imprinted polymers, there is still an enormous amount of work that needs to be conducted in the field. The exact recognition mechanism, and the way in which the interactions cooperate to recognize target molecules in many molecular imprinted polymer systems is not clearly understood. In addition, many fundamental properties of the molecular imprinted polymers prepared, such as the dependence of the affinity on cross-linking agent concentration, functional monomer concentration, buffer composition, ionic strength, pH, and temperature, have not yet to be systematically studied. It is evident that the development of the molecular modeling and analytical techniques will enable the better understanding of the effective molecular imprinted polymer-based receptors.

It is believed that combined with the increasing interest in creating materials which are capable of mimicking biological processes, the study of molecular imprinted polymers capable of the selective recognition of peptides, proteins, and other biological macromolecules will continuously receive considerable attention in the next decades.

Acknowledgement

This work is supported by National Natural Science Foundation of China (Approval No. 20704019).

References

[1] O'Mahony, J; Molinelli, A; Nolan, K; Smyth, MR; Mizaikoff, B. Towards the rational development of molecularly imprinted polymers: H-1 NMR studies on hydrophobicity and ion-pair interactions as driving forces for selectivity. *Biosensors and Bioelectrinics*, 2005, 20, 1884-1893.

[2] Molinelli, A; O'Mahony, J; Nolan, K; Smyth, MR; Jakusch, M; Mizaikoff, B. Analyzing the mechanisms of selectivity in biomimetic self-assemblies via IR and NMR spectroscopy of prepolymerization solutions and molecular dynamics simulations. *Analytical Chemistry*, 2005, 77, 5196-5204.

[3] Nicholls, IA; Andersson, LI; Mosbach, K; Ekberg, B. Recognition and enantioselection of drugs and biochemicals using molecularly imprinted polymer technology. *Trends in Biotechnology*, 1995, 13, 47–51.

[4] Whitcombe, MJ; Vulfson, EN. Imprinted Polymers. *Advanced Materials*, 2001, 13, 467–478.

[5] Quaglia, M; Sellergren, B; Lorenzi, ED. Approaches to imprinted stationary phases for affinity capillary electrochromatography. *Journal of Chromatography A*, 2004, 1044, 53–66.

[6] Wei, S; Jakusch, M; Mizaikoff, B. Capturing molecules with templated materials—Analysis and rational design of molecularly imprinted polymers. *Analytica Chimica Acta*, 2006, 578, 50–58.

[7] Bossi, A; Bonini, F; Turner, APF; Piletsky, SA. Molecularly imprinted polymers for the recognition of proteins: The state of the art, *Biosensors and Bioelectronics*, 2007, 22, 1131–1137.

[8] Greene, NT; Shimizu, KD. Colorimetric Molecularly Imprinted Polymer Sensor Array using Dye Displacement. *Journal of the American Chemical Society*, 2005, 127, 5695-5700.

[9] Haginaka, J; Sanbe, H; J. Uniformly sized molecularly imprinted polymer for (S)-naproxen: Retention and molecular recognition properties in aqueous mobile phase. *Journal of Chromatography A*, 2001, 913, 141-146.

[10] Watabe, Y; Hosoya, K; Tanaka, N; Kubo, T; Kondo, T; Morita, M. Novel surface modified molecularly imprinted polymer focused on the removal of interference in environmental water samples for chromatographic determination. *Journal of Chromatography A*, 2005, 1073, 363-370.

[11] Caro, E; Marce, RM; Cormack, PAG; Sherrington, DC; Borrull, F. On-line solid-phase extraction with molecularly imprinted polymers to selectively extract substituted 4-chlorophenols and 4-nitrophenol from water. *Journal of Chromatography A*, 2003, 995, 233-238.

[12] Perez-Moral, N; Mayes, AG. Direct rapid synthesis of MIP beads in SPE cartridges. *Biosensors and Bioelectrinics*, 2006, 21,1798-1803.

[13] Zhu, X; Yang, J; Su, Q; Cai, J; Gao, Y. Selective solid-phase extraction using molecularly imprinted polymer for the analysis of polar organophosphorus pesticides in water and soil samples. *Journal of Chromatography A*, 2005, 1092, 161-169.

[14] Wei, S; Molinelli, A; Mizaikoff, B. Molecularly imprinted micro and nanospheres for the selective recognition of 17β-estradiol. *Biosensors and Bioelectrinics*, 2006, 21, 1943-1951.

[15] Glad, T; Glad, M; Norrlow, S; Sellergren, B; Siegbahn, N; Mosbach, K. Use of silane monomers for molecular imprinting and enzyme entrapment in polysiloxane-coated porous silica. *Journal of Chromatography A*, 1985, 347, 11–23.

[16] Sellergren, B; Lepisto, M; Mosbach, K. Highly enantioselective and substrate-selective polymers obtained by molecular imprinting utilizing noncovalent interactions. NMR and chromatographic studies on the nature of recognition. *Journal of the American Chemical Society*, 1988, 110, 5853–5860; .

[17] Andersson, LI; O'Shannessy, DJ; Mosbach, K. Molecular recognition in synthetic polymers: preparation of chiral stationary phases by molecular imprinting of amino acid amides. *Journal of Chromatography A*, 1990, 513, 167–179.

[18] Andersson, LI; Miyabayashi, A; O'Shannessy, DJ; Mosbach, K. Enantiomeric resolution of amino acid derivatives on molecularly imprinted polymers as monitored by potentiometric measurements. *Journal of Chromatography A*, 1990, 516, 323–331.

[19] Kempe, M; Mosbach, K. Separation of amino acids, peptides and proteins on molecularly imprinted stationary phases. *Journal of Chromatography A*, 1995, 691, 317–323.

[20] Nicholls, IA. Thermodynamic Considerations for the Design of and Ligand Recognition by Molecularly Imprinted Polymers. *Chemistry letters*, 1995, 24, 1035–1036.

[21] Matsui, J; Nicholls, IA; Takeuchi, T; Mosbach, K; Karube, I. Metal ion mediated recognition in molecularly imprinted polymers. *Analytica Chimica Acta*, 1996, 335, 71–77.

[22] Hart, BR; Shea, KJ. Molecular Imprinting for the Recognition of N-Terminal Histidine Peptides in Aqueous Solution. *Macromolecules*, 2002, 35, 6192–6201.

[23] Shi, H; Tsai, W-B; Garrison, MD; Ferrari, S; Ratner, BD. Template-imprinted nanostructured surfaces for protein recognition. *Nature*, 1999, 398, 593–597.

[24] Rachkov, A; Minoura, N. Recognition of oxytocin and oxytocin-related peptides in aqueous media using a molecularly imprinted polymer synthesized by the epitope approach. *Journal of Chromatography A*, 2000, 889, 111–118.

[25] Nicholls, IA; Ramström, O; Mosbach, K. Insights into the role of the hydrogen bond and hydrophobic effect on recognition in molecularly imprinted polymer synthetic peptide receptor mimics. *Journal of Chromatography A*, 1995, 691, 349–353.

[26] Daniel, S; Kofinas, JP.r Molecular imprinting of peptides and proteins in aqueous media. *Analytical and Bioanalytical Chemistry*, 2007, 389, 399–404.

[27] Marx, S; Zaltsman, A. Molecular Imprinting of Sol Gel Polymers for the Detection of Paraoxon in Water. *International Journal of Environmental Analytical Chemistry*, 2003, 83, 671-680.

[28] Araki, K; Goto, M; Furusaki, S. Enantioselective polymer prepared by surface imprinting technique using a bifunctional molecule. *Analytica Chimica Acta*, 2002, 469, 173-181.

[29] Rachkov, A; Hu, M; Bulgarevich, E; Matsumoto, T; Minoura, N. Molecularly imprinted polymers prepared in aqueous solution selective for [Sar1,Ala8]angiotensin II. *Analytica Chimica Acta*, 2004, 504, 191-197.

[30] Xia, YQ; Guo, TY; Song, MD; Zhang, BH; Zhang, BL; Hemoglobin Recognition by Imprinting in Semi-Interpenetrating Polymer Network Hydrogel Based on Polyacrylamide and Chitosan. Biomacromolecules, 2005, 6, 2601-2606.

[31] Xia, YQ; Guo, TY; Song, MD; Zhang, BH; Zhang, BL. Adsorption dynamics and thermodynamics of Hb on the Hb-imprinted polymer beads. Reactive and Functional Polymers, 2008, 68, 63–69.

[32] Guo, TY; Xia, YQ; Hao, GJ; Song, MD; Zhang, BH. Adsorptive separation of hemoglobin by molecularly imprinted chitosan beads. *Biomaterials*, 2004,25, 5905-5912.

[33] Guo, TY; Xia, YQ; Wang, J; Song, MD; Zhang, BH. Chitosan beads as molecularly imprinted polymer matrix for selective separation of proteins. *Biomaterials*, 2005, 26,5737-5745.

[34] Guo, TY; Xia, YQ; Hao, GJ; Zhang, BH; Fu, GQ; Yuan, Z; He, BL; Kennedy, JF. Chemically modified chitosan beads as matrices for adsorptive separation of proteins by molecularly imprinted polymer. *Carbohydrate Polymers*, 2005, 62, 214–221.

[35] Fu, G; Zhao, J; Yu, H; Liu, L; He, B. Bovine serum albumin-imprinted polymer gels prepared by graft copolymerization of acrylamide on chitosan. *Reactive and Functional Polymers*, 2007, 67, 442–450.

[36] Zhao, Z; Wang, C; Guo, M; Shi, L; Fan, Y; Long, Y; Mi, H. Molecular imprinted polymer with cloned bacterial protein template enriches authentic target in cell extract. *FEBS Letters*, 2006, 580, 2750–2754.

[37] Piletsky, SA; Andersson, HS; Nicholls, IA. Combined Hydrophobic and Electrostatic Interaction-Based Recognition in Molecularly Imprinted Polymers. *Macromolecules*, 1999, 32, 633–636.

[38] Piletsky, SA; Andersson, HS; Nicholls, IA. The rational use of hydrophobic effect-based recognition in molecularly imprinted polymers. *Journal of Molecular Recognition*, 1998, 11, 94–97.

[39] Qin, L; He, XW; Li, WY; Zhang, YK. Molecularly imprinted polymer prepared with bonded β-cyclodextrin and acrylamide on functionalized silica gel for selective recognition of tryptophan in aqueous media. *Journal of Chromatography A*, 2008, 1187, 94–102.

[40] Egawa, Y; Shimura, Y; Nowatari, Y; Aiba D, Juni,K. Preparation of molecularly imprinted cyclodextrin microspheres, *International Journal of Pharmaceutics*, 2005, 293, 165–170.

[41] Asanuma, H; Hishiya, T; Komiyama M. Efficient separation of hydrophobic molecules by molecularly imprinted cyclodextrin polymers, *Journal of Inclusion Phenomena and Macrocyclic Chemistry*, 2004, 50, 51–55.

[42] Striegler, S. Carbohydrate Recognition in Cross-Linked Sugar-Templated Poly(acrylates). *Macromolecules*, 2003, 36, 1310-1317.

[43] Kubo, T; Hosoya, K; Nomachi,M; Tanaka, N; Kaya, K. Preparation of a novel molecularly imprinted polymer using a water-soluble crosslinking agent. *Analytical and Bioanalytical Chemistry*, 2005, 382, 1698-1701.

[44] Janotta, M; Weiss, R; Mizaikoff, B; Bruggemann, O; Ye, L; Mosbach, K. Molecularly Imprinted Polymers for Nitrophenols - An Advanced Separation Material for Environmental Analysis. *International Journal of Environmental Analytical Chemistry*, 2001, 80, 75-86.

[45] Bereli , N; Andac, M; Baydemir, G; Say, R; Galaev, IY; Denizli, A. Protein recognition via ion-coordinated molecularly imprinted supermacroporous cryogels, *Journal of Chromatography A*, 2008, 1190, 18–26.

[46] Ozcan, AAt; Say, R; Denizli, A; Erso1z, A. L-Histidine Imprinted Synthetic Receptor for Biochromatography Applications. *Analytical Chemistry*, 2006, 78,7253-7258.

[47] Odabaşi, M; Say, R; Denizli, A. Molecular imprinted particles for lysozyme purification. *Materials Science and Engineering C*, 2007, 27, 90–99.

[48] Shiomi, T; Matsui, M; Mizukami, F; Sakaguchi, K. A method for the molecular imprinting of hemoglobin on silica surfaces using silanes. *Biomaterials*, 2005, 26, 5564–5571.

[49] Corma, A; Fornes, V; Jorda, JL; Rey, F; Fernandez-Lafuente R; Guisan, JM; Mateo, C. Electrostatic and covalent immobilization of enzymes on ITQ-6 delaminated zeolitic materials. *Chemical Communications*, 2001, 5, 419–420.

[50] Bonini, F; Piletsky, S; Turner, APF; Speghini, A; Bossi, A. Surface imprinted beads for the recognition of human serum albumin. *Biosensors and Bioelectronics*, 2007, 22, 2322–2328.

[51] Tan, CJ; Chua, HG; Ker, KH; Tong, YW. Preparation of bovine serum albumin surface-imprinted submicrometer particles with magnetic susceptibility through core-shell miniemulsion polymerization. *Analytical Chemistry*, 2008, 80, 683-692.

[52] Devanathan, S; Salamon, Z; Nagar, A; Narang, S; Schleich, D; Darman, P; Hruby, V; Tollin, G. Subpicomolar sensing of α-ppioid receptor ligands by molecular-imprinted polymers using plasmon-waveguide resonance spectroscopy. *Analytical Chemistry*, 2005, 77, 2569-2574.

[53] Tan, CJ; Tong, YW. The effect of protein structural conformation on nanoparticle molecular imprinting of ribonuclease a using miniemulsion polymerization, *Langmuir*, 2007, 23, 2722-2730.

[54] Chau Jin Tan, Shalom Wangrangsimakul, Renbi Bai, and Yen Wah Tong, Defining the interactions between proteins and surfactants for nanoparticle surface imprinting through miniemulsion polymerization. *Chemistry of Materials*, 2008, 20, 118–127.

[55] Hawkins, DM; Trache, A; Ellis, E. A; Stevenson, D; Holzenburg, A; Meininger, GA; Reddy, SM. Quantification and confocal imaging of protein specific molecularly imprinted polymers. *Biomacromolecules*, 2006, 7, 2560-2564.

[56] Tan, CJ; Tong, YW. Molecularly imprinted beads by surface imprinting, *Analytical and Bioanalytical Chemistry*. 2007, 389, 369–376.

[57] Zhao, K; Cheng, G; Huang, J; Ying, X. Rebinding and recognition properties of protein-macromolecularly imprinted calcium phosphate/ alginate hybrid polymer microspheres. *Reactive and Functional Polymers*, 2008, 68, 732–741.

[58] Zhao, KY; Kan, BH; Wei, JF; Cheng,GX; Chen, L. Bovine serum albumin imprinted calcium phosphate/polyacrylate/alginate multi-hybrid polymer microspheres in inverse-phase suspension. *e-Polymers*, 2008, no. 100.

[59] Hayden, O; Bindeus, R; Haderspock, C; Mann, KJ; Wirl, B; Dickert, FL. Mass-sensitive detection of cells, viruses and enzymes with artificial receptors. *Sensors and Actuators B*, 2003, 91, 316–319.

[60] Carter, S; Lu, S; Rimmer, S. Core-shell Molecular Imprinted Polymer Colloids. *Supramolecular Chemistry*, 2003, 15, 213–220.

[61] Pérez-Moral, N; Mayes, AG. Comparative study of imprinted polymer particles prepared by different polymerisation methods. *Analytica Chimica Acta*, 2004, 504, 15–21.

[62] Carter, SR; Rimmer, S. Molecular recognition of caffeine by shell molecular imprinted core-shell polymer particles in aqueous media. *Advanced Materials*, 2002, 14, 667–670.

[63] Pérez, N; Whitcombe, MJ; Vulfson, EN. Molecularly imprinted nanoparticles prepared by core-shell emulsion polymerization. *Journal of Applied Polymer Science*, 2000, 77, 1851–1859.

[64] Whitcombe, MJ; Rodriguez, ME; Villar, P; Vulfson, EN. A new method for the introduction of recognition site functionality into polymers prepared by molecular imprinting: synthesis and characterization of polymeric receptors for cholesterol. *Journal of the American Chemical Society*, 1995, 117, 7105–7111.

[65] Pérez, N; Whitcombe, MJ; Vulfson, EN. Surface imprinting of cholesterol on submicrometer core-shell emulsion particles. *Macromolecules*, 2001, 34, 830–836.

[66] Klein, JU; Whitcombe, MJ; Mulholland, F; Vulfson, EN. Template-Mediated Synthesis of a Polymeric Receptor Specific to Amino Acid Sequences. *Angewandte Chemie International Edition*, 1999, 38, 2057-2060.

[67] Tatemichi, M; Sakamoto, M; Mizuhata, M; Deki, S; Takeuchi, T. Protein-templated organic/inorganic hybrid materials prepared by liquid-phase deposition. *Journal of the American Chemical Society*, 2007, 129, 10906-10910.

[68] Matsunaga, T; Hishiya, T; Takeuchi, T. Surface plasmon resonance sensor for lysozyme based on molecularly imprinted thin films. *Analytica Chimica Acta*, 2007, 591, 63–67.

[69] Nicholas, W; Turner, BE; Wright, VH; David, WB. Formation of protein molecular imprints within Langmuir monolayers: a quartz crystal microbalance study. *Journal of Colloid and Interface Science*, 2007, 308, 71–80.

[70] Wei, X; Li, X; Husson, SM. Surface molecular imprinting by atom transfer radical polymerization. *Biomacromolecules*, 2005, 6, 1113-1121.

[71] Li, X; Husson, SM. Adsorption of dansylated amino acids on molecularly imprinted surfaces: a surface plasmon resonance study. *Biosensors and Bioelectronics*, 2006, 22, 336–348.

[72] Bossi, A; Rivetti, C; Mangiarotti, L; Whitcombe, MJ; Turner, APF; Piletsky, SA. Patterned gallium surfaces as molecular mirrors. *Biosensors and Bioelectronics*, 2007, 23, 290–294.

[73] Rachkov, A; Minoura, N. Recognition of oxytocin and oxytocin-related peptides in aqueous media using a molecularly imprinted polymer synthesized by the epitope approach. *Journal of Chromatography A*, 2000, 889, 111–118 .

[74] Slinchenko, O; Rachkov, A; Miyachi, H; Ogiso, M; Minoura, N. Imprinted polymer layer for recognizing double-stranded DNA, *Biosensors and Bioelectronics*, 2004, 20, 1091–1097.

[75] Nishino, H; Huang, CS; Shea, KJ. Selective protein capture by epitope imprinting. *Angewandte Chemie International Edition*, 2006, 45, 2392–2396.

[76] Brown, ME; Puleo, DA. Protein binding to peptide-imprinted porous silica scaffolds. *Chemical Engineering Journal*, 2008, 137, 97–101.

[77] Papisov IM; Litmanovich, AA. Molecular 'Recognition' in interpolymer interactions and matrix polymerization. *Advances in Polymer Science*, 1989, 90, 139–179.

[78] Wulff, G. Molecular imprinting in cross-linked materials with the aid of molecular templates - a way towards artificial antibodies. *Angewandte Chemie International Edition in English*, 1995, 34, 1812-1832.

[79] Wulff, G. Enzyme-like Catalysis by Molecularly Imprinted Polymers. *Chemical Reviews*, 2002, 102, 1-28.

[80] Andersson, L; Sellergren, B; Mosbach, K. Imprinting of amino acid derivatives in macroporous polymers. *Tetrahedron Letters*, 1984, 25, 52115214.

[81] Mosbach, K. Molecular imprinting. *Trends in Biochemical Sciences*, 1994, 19, 9-14.

[82] Yan, M; Ramstrom O. (Eds.), Molecularly Imprinted Materials: Science and Technology, Marcel Dekker, New York, NY, 2005.

[83] Holthoff, EL; Bright, FV. Molecularly templated materials in chemical sensing. *Analytica Chimica Acta*, 2007, 594, 147–161.

[84] Mayes, AG; Whitcombe, MJ. Synthetic strategies for the generation of molecularly imprinted organic polymers. *Advanced Drug Delivery Reviews*, 2005, 57, 1742-1778.

[85] Maier, NM; Lindner, W. Chiral recognition applications of molecularly imprinted polymers: a critical review. *Analytical and Bioanalytical Chemistry*, 2007, 389, 377–397.

[86] Pang, X; Cheng, G; Zhang, Y; Lu, S. Soft-wet polyacrylamide gel beads with the imprinting of bovine serum albumin. *Reactive and Functional Polymers*, 2006,66, 1182–1188.

[87] Hirayama, K; Sakai, Y; Kameoka, K. Synthesis of polymer particles with specific lysozyme recognition sites by a molecular imprinting technique. *Journal of Applied Polymer Science*, 2001, 81, 3378–3387.

[88] Lin, HY; Hsu, CY; Thomas, JL; Wang, SE; Chen, HC; Chou, TC. The microcontact imprinting of proteins: The effect of cross-linking monomers for lysozyme, ribonuclease A and myoglobin. *Biosensors and Bioelectronics*, 2006, 22, 534–543.

[89] Ou, SH; Wu, MC; Chou, TC; Liu, CC. Polyacrylamide gels with electrostatic functional groups for the molecular imprinting of lysozyme. *Analytica Chimica Acta*, 2004, 504, 163–166.

[90] Hawkins, DM; Stevenson, D; Reddy, SM. Investigation of protein imprinting in hydrogel-based molecularly imprinted polymers (HydroMIPs). *Analytica Chimica Acta*, 2005, 542, 61–65.

[91] Bossi, A; Piletsky, SA; Piletska, EV; Righetti, PG; Turner, APF. Surface-grafted molecularly imprinted polymers for protein recognition, *Analytical Chemistry*, 2001, 73, 5281-5286.

[92] Pavel, D; Lagowski, J. Computationally designed monomers and polymers for molecular imprinting of theophylline—part II. *Polymer*, 2005, 46, 7543-7556.

[93] Pavel, D; Lagowski, J. Computationally designed monomers and polymers for molecular imprinting of theophylline and its derivatives. Part I. *Polymer*, 2005, 46, 7528-7542.

[94] Srebnik, S; Lev, O; Avnir, D. Pore size distribution induced by microphase separation: effect of the leaving group during polycondensation, *Chemistry of Materials*, 2001, 13, 811-816.

[95] Yungerman, I; Srebnik, S. Factors Contributing to binding-site imperfections in imprinted polymers, *Chemistry of Materials*, 2006, 18, 657-663.

[96] Henthorn, DB; Peppas, NA. Molecular simulations of recognitive behavior of molecularly imprinted intelligent polymeric networks, *Industrial and Engineering Chemistry Research*, 2007, 46, 6084 -6091.

In: Biochemical Engineering
Editors: F.E. Dumont and J.A. Sacco, pp. 141-164
ISBN: 978-1-60741-257-1

Chapter 5

Uncoupled Energy Metabolism for Sludge Reduction in the Activated Sludge Process

Bo Jiang[1], Yu Liu[1], Guanghao Chen[2] and Etienne Paul[3]

[1]School of Civil and Environmental Engineering, Nanyang Technological University, 50 Nanyang Avenue, Singapore 639798

[2]Department of Civil and Environmental Engineering, Hong Kong University of Science and Technology, Hong Kong

[3]Université de Toulouse; INSA, UPS, INP; LISBP, 135 Avenue de Rangueil, F-31077 Toulouse, France; INRA, UMR792 Ingénierie des Systèmes Biologiques et des Procédés, F-31400 Toulouse, France; CNRS, UMR5504, F-31400 Toulouse, France

Abstract

The activated sludge process is a mature and widely-adopted biotechnology for treating both municipal and industrial wastewater for more than one century. However, a large quantity of excess sludge is inevitably generated as a byproduct of biological conversion of organic matters during the process. Treatment and disposal of this byproduct usually accounts for up to 60% of the total capital and operation cost; thus it poses a great challenge in the field of environmental biotechnology. In order to solve this problem, some strategies for minimizing sludge production have been explored and developed, e.g. lysis-cryptic growth, bacteriovoric metabolism, maintenance metabolism and uncoupled energy metabolism-associated sludge reduction, etc. Lysis-cryptic growth technique is using either physical or chemical forces (e.g. heat treatment, ozonation, chlorination, etc.) to disintegrate and mineralize sludge. However, this method is difficult to control, expensive to implement and have a low efficiency. Such drawbacks weaken its capability in practice. Bacteriovoric metabolism method highly depends on the properties of predators and requires strict control of growth conditions to promote specific predator to bloom. Uncoupled energy metabolism of activated sludge is an alternative to reduce excess sludge generation in the activated sludge process. Microbial metabolism is basically includes interrelated catabolic and anabolic reactions. Under normal conditions, catabolism of microbes is tightly coupled with anabolism in the light of energy requirements. However, energy uncoupling can be triggered when some abnormal conditions are present, such as excess carbon source and nutrients limitation; high temperature; alternative aerobic-anaerobic cycle; and presence of metabolic inhibitors. Under such conditions, energy generation from catabolizing substrate is in excess with respect to the anabolism requirement, resulting in dissipation of part of energy through futile cycles (e.g. energy becomes heat). As a result, the biomass yield would be reduced significantly. This chapter aims to offer an overview of different excess sludge reduction methods with special focus on uncoupled energy metabolism-associated sludge reduction and particularly on addressing the mechanisms behind it. Meanwhile, potential interactions and genetic behaviors of microbes under uncoupled growth conditions, and some advanced microbiological tools are also discussed.

Introduction

The conventional activated sludge (CAS) process is in common use for the treatment of both municipal and industrial wastewater in the world involving the utilization of a mixed culture of microorganisms to degrade the contaminants aerobically with easy operation and high quality effluent. In this process, dissolved and suspended organic substrates are converted to new biomass and respiratory products (CO_2, CH_4, H_2O, N_2, SO_2, etc.). The new biomass then can be separated from the treated water by sedimentation and disposed of in a concentrated form usually called excess sludge [1]. The sludge yield coefficient in CAS process is typically 0.6 [2], indicating that 1 kg of chemical oxygen demand (COD) removed will generate 0.6 kg of excess sludge in dry weight. According to Metcalf and Eddy [2], some factors may significantly affect the biomass yield in CAS process, including (i) the status of the substrate and nutrients (oxidation state); (ii) the degree of polymerization and hydrolysis of substrate; (iii) metabolic pathways; (iv) biomass growth rates and (v) environmental physical parameters (such as pH, temperature, etc.). The large quantity of sludge produced is a secondary solid waste eventually containing many highly toxic species, such as heavy metals, pathogenic organisms, polychlorinated biphenyls (PCBs), polychlorinated dibenzo-p-dioxins/dibenzofurans (PCDD/Fs) or polycyclic aromatic hydrocarbons (PAHs) due to diffused discharges from industrial sites [3], thus need to be disposed of safely and cost-effectively. So far, handling, treatment and disposal of excess sludge usually account for about 50%, sometimes even up to 60% of the total operation cost of wastewater treatment, which poses a rising challenge for wastewater treatment plants (WWTPs) [4,5].

In the last decades, the main alternative solutions in EU for the disposal of excess sludge were agricultural landspreading, landfill and incineration that accounted for nearly 90% of total sludge production in EU [3,5]. Currently, sludge disposal has entered an epoch of distinct change due to the increased environmental and legislative constraints. Due to the more and more stringent regulations of food safety, sludge disposal in most countries are no longer allowed for application in agriculture or dumping into the sea [6]. The further increase of excess sludge caused by the expansion of population and rapid proliferation of new industry, accompanied by the stringent legislations, will cause a dramatic increase of operation costs in the near future. In the sense of environmental engineering, it is much more attractive and preferable reduce biomass production at the site of WWTPs rather than post-treatment of the sludge produced in the CAS process. As a result, more and more attention for operation of the CAS process has been turned to the reduction of biomass production rather than the volume of wastewater turnover and maximum contaminants removal [7].

This chapter synthetically overviews the most promising methods applied for sludge minimization in recent research with special focus on the uncoupled energy metabolism for sludge reduction in CAS process. It is expected that this chapter can provide a comprehensive picture for researchers and engineers to better understand the mechanism of uncoupled energy metabolism as well as to develop efficient strategies for sludge reduction in practice.

2. Strategies for Minimizing Biomass Production

Nowadays, various technical means have been developed for minimizing biomass production, such as sludge alkaline thermal treatment [8], ozonation [9,10], chlorination [11], advanced oxidation process [12], enzyme-induced biological hydrolysis [13], biodegradation of sludge in membrane-assisted reactors [14,15], predation on bacteria [16,17], oxic-settling-anaerobic (OSA) process [18,19], chemical uncoupler-associated activated sludge process [20,21], etc. All these strategies could be roughly classified into four major categories as specified in the following sections.

2.1. Lysis-cryptic Growth

This strategy is to use physical, chemical or biological forces to induce cell lysis and release cell contents (e.g. organelles, proteins, DNA, RNA, lipids, etc.) into the medium as an internal food source for wastewater microorganisms, thus leading to the increase of system organic loading (Figure 1). This amount of organic loading is usually defined as autochthonous substrate, and will be reutilized in microbial metabolism. Part of autochthonous substrate will be assimilated into biomass to build up cell matrix, while part of them will be released as respiratory products, resulting in a reduction of biomass production. The biomass growth by absorbing the autochthonous substrate and by taking the original organic substrate always happen simultaneously and cannot be distinguished one from the other, and is therefore termed as cryptic growth [22]. There are two linked processes in lysis-cryptic growth, including lysis and biodegradation. Between these two steps, lysis of particulates to soluble substrate is the rate-limiting stage, because the cells usually have a protein-lipid bi-layer plasma membrane, forming a barrier separating cell contents from the surroundings, which is resistant to lysis. Therefore, an increased cell lysis efficiency can contribute to the overall decrease of sludge yield [5,23].

It is known that moderate osmosis through simply lowering the ionic strength of the medium is usually sufficient enough to lyze most cells which induces them to swell, burst and finally release their contents into the solution [24]. So far, several physical, chemical or biological methods have come into practice to enhance the complete dissociation of the cells and their cellular components to reduce sludge burden, including (i) physical disintegration using sonication, mechanical mills, mechanical agitation with glass or ceramic beads, freeze/thaw cycles and nitrogen bomb/nitrogen burst cell disruption method (Table 1); (ii) chemical disintegration applying alkaline thermal treatment, ozonation, chlorination or advanced oxidation processes with Photo-Fenton reagent and (iii) biological method utilizing enzyme-induced biological hydrolysis, etc [5].

Ozone is a very strong chemical oxidant with a remarkably short life span that has potential to reduce excess sludge in CAS process. During ozonation, part of recycled sludge is contacted with ozone to be mineralized to carbon dioxide, water and biodegradable organics which will be further decomposed in the subsequent biological treatment unit (Figure 2) [1].

It was shown that the application of ozonation treatment on waste sludge could induce substantial reduction in the volume of sludge and the release of intracellular and extracellular materials [26]. With the increase of ozone dose to above 0.04 g O_3/g MLSS [27,28], the structure of sludge flocs could be disintegrated completely and the characteristics of sludge (e.g. the settleability and dewaterability) would also be improved remarkably. Nagare et al. [29], employing sludge ozonation on wastewater treatment processes, demonstrated that ozonation would reduce excess sludge production by 93%. In this process, the ratio of return sludge to inflow rate was controlled at 1.2% and a solubilization degree of 30% was achieved. The solubilization degree per ozone consumption for general sludge ranged from 2.4 to 5.8 g SS/g O_3 or from 4.1 to 7.7 g COD/g O_3. Sludge ozonation would effectively reduce sludge production and has been successfully applied in practice. However, several possible drawbacks exist, such as the release of biorefractory organic compounds in the effluent that could pose problems to the receiving bodies [3], the increase of effluent total organic carbon and nutrient concentrations, high capital and operation cost [28], etc.

As chlorination is known to be cheaper than ozonation, Saby et al. [11] applied the chlorination method to treat excess sludge, and found that at the dose of 0.066 g Cl_2/g MLSS, a 65% reduction of excess sludge would be achievable. It was found in this process that chlorination only weakened the cell membrane, but did not lyze the cells, i.e. cells through chlorination could be hydrolyzed easily in the following steps. It should be pointed out that the chlorination treatment would result in poor sludge settleability, potential formation of trihalomethanes (THMs), volatile organic compounds (VOCs) and hazardous air pollutant (HAP), as well as significant increase of soluble COD in the effluent, thus posing serious challenges for the application of chlorination-based sludge minimization technique in CAS process.

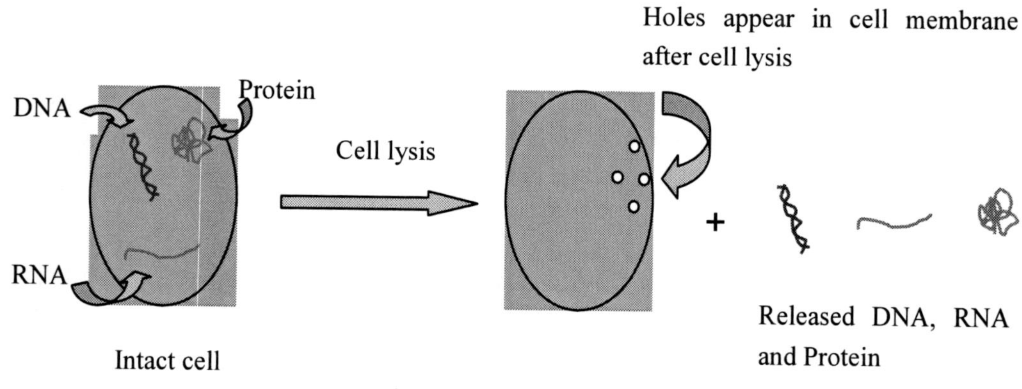

Figure 1. The sketch map of cell lysis process. After lysis, holes appear in the cell membrane, leading to the release of cell contents, such as DNA, RNA, protein, etc.

It has been demonstrated that sludge production increased at low temperature operation [30]. Combination of thermal and alkaline treatment is a very efficient cell lysis technique to reduce excess sludge. Saiki et al. [8] reported that self-digestion of excess sludge occurred by anaerobic incubation at 60°C after the addition of NaOH at a final concentration of 0.01 N. This method induced approximately 40% solubilization of the organic matter in the excess sludge. However, using thermal and alkaline treatment will cause corrosion problem, implying that corrosion-resistant materials are needed. Furthermore, the chemical/energy input and equipment maintenance in the operation will lead to dramatic increase in the investment.

Table 1. Some physical techniques for cell lysis [25]

Lysis method	Apparatus	Method description
Sonication	Sonicator/homogenizer	Shear cells via cavitation and impaction by pulsed, high frequency sound waves (typically 20-50 KHz)
Agitation with beads	Electric motor	Disrupt cells through vigorous agitation provided by electric motor
Freeze/thaw cycle	Freezer or dry ice/ethanol	Repeat cycles of freezing and thawing to disrupt cells through ice crystal formation
nitrogen bomb/nitrogen burst	High pressure nitrogen	Rapid release of high pressure nitrogen gas (usually 25,000 psi) as bubbles inducing cell lysis

The Photo-Fenton reaction, one of the advanced oxidation processes (AOPs), has been exploited for minimization of excess sludge. In this process, the cell wall of a microbe is destroyed and becomes easily decomposable. The results by Tokumura et al. [12] showed that the chemical oxidation of excess sludge by Photo-Fenton reaction contained two phases: (i) oxidative decomposition of the cell wall leading to the increased dissolved COD due to the discharge of organic substances into the medium; (ii) the maximum COD release and followed by gradual decrease. Solubilization of suspended solids and mineralization of dissolved organic substances by the Photo-Fenton reaction might be dominant in these two interrelated phases.

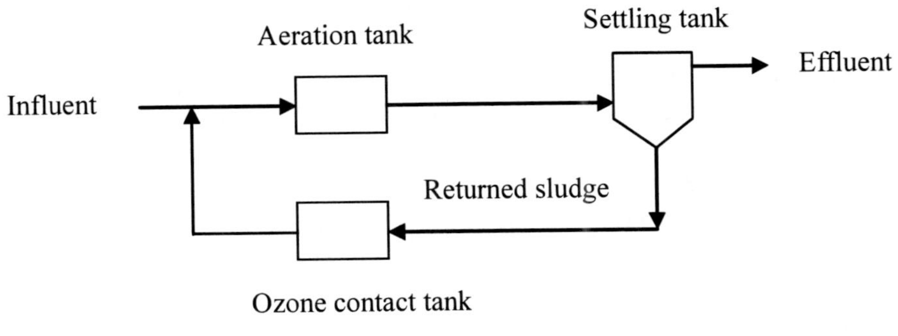

Figure 2. The flow sheet of ozonation process. The returned sludge is contacted with ozone and then re-circulated to the aeration tank for biodegradation.

The lysis-cryptic growth techniques sound reasonable in practice to achieve effective minimization of excess sludge production in CAS process; however, there are some constraints in those approaches, i.e. two-step correlative reactions (solubilization and oxidation of autochthonous substrate) may limit the efficiency of excess sludge reduction [23]. According to previous research, these techniques on the basis of lysis-cryptic growth also incur additional costs because cumbersome equipments are always needed, and other shortcomings, such as difficulty in control, not suitable for large scale application, make their capabilities unacceptable in the environmental engineering sense.

2.2. Bacteriovoric Metabolism

Appropriate practice to maximize, encourage and optimize the growth of higher organisms in the food chain which predate on bacteria would be one possible way to prevent excess sludge accumulation. During energy transfer from a low trophic level (bacteria) to a high level (protozoa or metazoa), energy is reduced due to respiration, growth, reproduction, defecation and nonpredatory death (organisms that die but are not consumed by consumers). Typically, on average, about 10% of net energy production at one trophic level is passed on to the next level. Such inefficient biomass conversion induces energy lost and subsequently sludge reduction in the process of bacteriovoric metabolism [15,31]. Promotion of macroinvertebrates can minimize the amount of assimilated energy in the system [32]. Bacteriovoric metabolism in excess sludge reduction process can be achieved via bacteriovory by higher organisms, such as protozoa and metazoa (e.g. nematodes, rotifers and oligochaete worms), which play an important role in CAS process because they maintain the density of bacterial populations by consuming dispersed bacteria and contribute to the flocculation process, keeping the effluent clear and with high quality [33]. Predation by phagotrophic protists, the protozoa, are considered to be the most common predators and a major mortality factor of bacteria, constituting around 5% of the total dry weight of a wastewater biomass [34,35]. Bacteriovoric metazoa are also widely distributed in various aquatic and terrestrial habitats and play a part in enhancement of excess sludge reduction. It has been demonstrated that increased protozoa/metazoa grazing on bacteria results in shifts in bacterial communities from edible to grazing-resistant species or ecotypes [36]. In aerobic biological wastewater treatment processes, floc-/film-growing bacteria or granules are more resistant from the predators. In order to achieve an efficient excess sludge reduction, it is necessary to fully understand the specificity and population dynamics of prey-predator relationships and identify predators capable of grazing on the present biomass.

Investigations of bacterial predation by protozoa and metazoa showed that grazing efficiency was limited by their physical/mechanistic properties, such as the size of the mouth and pharynx of protozoa and metazoa [37], thus different grade of sludge yield reduction would be possible when different kinds of protozoa and metazoa were predominant in the system. Worms are almost the largest organisms observed during the microscopic investigation of activated sludge including *Aeolosomatidae*, *Naididae*, *Pristina*, *Dero* and *Tubificidae* [38], and may have more potential on sludge reduction in practical application due to their bigger sizes. *T. tubifex*, which has the largest size among oligochaete worms, was chosen to be a sludge predator by Huang et al. [39], demonstrating that sludge minimization could be triggered by *T. tubifex*'s predation on sludge in the recycled sludge reactor where *T. tubifex* was introduced and grew. The sludge reduction rate of *T. tubifex* R was from 0.18 to 0.81 mg VSS mg^{-1} *Tubifex* d^{-1}. The sludge reduction capacity of the recycled sludge reactor E was from 650 to 1080 mg VSS L^{-1} d^{-1}. The optimum density of *T. tubifex* and the optimum sludge recycled ratio were 2500 mg L^{-1} and 1, respectively. Although the total amount of sludge can be reduced substantially by the introduction of oligochaete worms, the practical application is still uncontrollable because of unpredictable factors affecting worm growth making them unstable, e.g. how to stimulate the initiation of worm bloom and maintain a stable worm density under controlled full-scale conditions are still not well understood [40].

Lapinski et al. [41] reported that cultured bdelloid rotifers can significantly improve municipal wastewater clarification through two-fold mechanism on suspended particles: (i) consumption of small, dissociative flocs due to feeding activity and (ii) improved settling due to the function of triggering flocs aggregation to settle out of suspension. In their study, particle size was in the range of 0.2-10 μm, corresponding to the size range of particles consumed by rotifers. As a result, rotifers might be exploited as biotechnological tools for reduction of bio-solids from wastewater by grazing on dispersed bacteria.

Lee et al. [17] investigated the possibility of reducing sludge production in aerobic treatment using a continuous-flow two-stage laboratory reactor system with two different carbon sources (acetic acid and methanol): the first stage working as a completely mixed, aerobic reactor at a short solid retention time (SRT) without biomass retention to prevent the growth of predators and to favor the growth of fast-growing, dispersed bacteria that consume the readily biodegradable organic matter in the wastewater, and the second stage being designed to be a biofilm reactor as a predator stage, in order to provide a long sludge age and optimal conditions for the growth of the protozoa and metazoa feeding on the bacteria. Their results showed that the sludge yield decreased considerably both with acetic acid (61% reduction) and methanol (80% reduction). Generally, for the two-stage process, a large volume of bioreactor is needed in the first stage because of long hydraulic retention time (HRT), resulting in increasing the capital investment to install such a chamber compared to CAS process. Moreover, the increased mineralization of organic matter in the process leading to a decreased sludge yield will involve higher operation costs in the form of oxygen requirements. It was reported that the oxygen consumption of the process was almost doubled due to the increase in oxidation of organic matter [42]. Furthermore, with these systems, significant amount of ammonium, nitrate and phosphate as consequences of biomass mineralization released in the predator stage is another limitation of the two-stage principle [15,17].

So far, the relationship between operation conditions and predators growth is still not very clear. It is challenging to create favorable niches for their maturation and successful sexual reproduction [43]. Further research needs to be focused on the connection between food chain (prey-predator) and the sludge flocs formation-disintegration for bacteriovoric metabolism-associated sludge reduction [5].

2.3. Maintenance Metabolism

It is known that microorganisms satisfy their maintenance energy requirements in preference to producing additional biomass [44]. The utilization of carbon source by cells to provide energy requirements for non-growth activities (in particular maintenance functions and survival functions) would divert the organic substrates from assimilation for biosynthesis of new biomass. Exothermic maintenance functions include the turnover of enzyme, osmotic work to maintain concentration gradients, cell viability, nutrient storage and cell motility [34,45]. Selection of microorganisms present in the wastewater treatment process with high maintenance energy, i.e., substrate is utilized for cellular maintenance and endogenous metabolism rather than for biomass synthesis could reduce the excess sludge production. This selection can be achieved by lowering the substrate to biomass ratio [42,46].

In order to reduce biomass production during wastewater treatment with respect to maintenance metabolism, Low et al. [44] looked into how microorganisms would allocate organic substrates between maintenance and biosynthesis, and additional utilization of the carbon source for energy generation was found, with more carbon dioxide being evolved per unit substrate removed and less carbon being incorporated into biomass with continuously feeding substrate at low dilution rates to a culture. In the study by Sun et al. [47], a pilot scale submerged membrane bioreactor (MBR) was used to treat high strength industrial wastewater with a higher sludge concentration of 14,500 mg L^{-1} at prolonged SRT. The results showed that the sludge yield and endogenous decay rate were 0.115 g VSS g^{-1} COD and 0.024 d^{-1}, respectively, which was almost half the lower value reported for CAS process. The specific oxygen utilization rate (SOUR) was also lowered due to the less energy required by microorganisms during the endogenous phase. It was demonstrated that the energy usage by microorganisms during endogenous phase was much small and most of the incoming energy (oxygen and carbon sources) could be used for cell maintenance instead of cell synthesis. For the purpose of reducing excess sludge production in the wastewater treatment process, an MBR system has also been introduced by Canales et al. [48] to study the maintenance and cryptic growth phenomena of *Pseudomonas fluorescens* culture, where a thermal treatment was added in line on a recirculation loop to increase the death and the lysis of the biomass. The results of this study showed that the biomass with high sludge age had a low viability level and COD removal associated with requirements for maintenance was higher than for growth. A maintenance coefficient of 0.035 g COD g^{-1} biomass h^{-1} was obtained.

MBR can work with high biomass concentration that may eventually lower the ratio of the food/microorganism concentrations, and subsequently amplify the maintenance phenomena. However, effluent quality from the MBR system is still dependent on the biodegradability directly. Moreover, membrane fouling remains an important issue in MBR operation.

2.4. Uncoupled Energy Metabolism

Senez [49] suggested that a separation between the catabolic and the anabolic chain of reactions exists in the case of organisms and culture conditions where the source of energy is distinct from the materials utilized for cell biosynthesis. In order to consume this part of free energy, several possibilities have been considered, including the dissipation of energy as heat or futile cycles by adenosine triphosphatase systems, reducing the efficiency of ATP generation by deletion of oxidative phosphorylation sites, the activation of alternative metabolic pathways by-passing free energy conserving reactions, disposing of intracellular energy by dissipation of membrane potential, the accumulation of polymerized products in storage form or as secreted waste, branching of the respiratory chain and ATP hydrolysis [34]. In uncoupling process, energy generation from catabolism-associated substrate consumption is beyond that required for anabolism, including growth and maintenance. The portion of excess energy will be wasted as heat [50]. The following sections will especially focus on the mechanism of uncoupled energy metabolism and their applications in excess sludge reduction.

3. Uncoupled Energy Metabolism

Bacteria have a complete set of metabolic maps forming the metabolic network which are striking features of metabolism and important to their life (Figure 3). The catabolic pathways are those in which complicated molecules are taken in and broken down into simple ones and free energy is taken out, gathered and stored in energy carriers (ATP). Anabolic pathways are the construct biochemical processes involving the use of free energy to increase complexity of molecules from simple ones and build up the structures required by the cell [42]. To reduce the production of excess sludge, wastewater treatment processes must be well adjusted to divert from biosynthesis via substrate assimilation to energy requiring functions associated with non-growth activities, e.g. maintenance, mobility, and so on, which is called uncoupling, spillage or overflow metabolism [44]. Russel et al. [51] defined "uncoupling" as being the inability of chemosmotic oxidative phosphorylation to generate the maximum theoretical quantity of metabolic energy in the form of ATP, which was also redefined by Low et al. [34] as "uncoupled oxidative phosphorylation" to differentiate it from other mechanisms of uncoupling metabolism.

3.1. Mechanism of Energy Coupling

For better understanding uncoupled energy metabolism, attention is given to oxidative phosphorylation first. Cells generate ATP during cellular respiration through oxidative steps in the degradation of carbohydrates, fats, and proteins. In eukaryotes, it occurs in mitochondria, involving the reduction of O_2 to H_2O using electrons donated by NADH and $FADH_2$. The light-dependent photophosphorylation occurs in chloroplasts involving the conversion of H_2O to O_2, using $NADP^+$ as ultimate electron acceptor. Bacteria are a simple life form whose cells typically consist of a single compartment surrounded by a plasma membrane that separates the cell from its environment. In such a system, the energy production by electron transport and oxidative phosporylation through electron flow from NADH and $FADH_2$ to O_2 is carried out at (and across) the plasma membrane. Correspondingly, there are three types of membranes taking part in coupling the electron transport and phosphorylation reactions including mitochondrial membrane in eukaryotes, chloroplast membrane in photosynthetic organisms and plasma membrane in prokaryotes [52]. Though they capture different kinds of energy, all these three types of membranes allow proton move through them from specific protein channels down proton concentration gradient, and then provides the free energy for ATP synthesis. So far, three different mechanisms of energy coupling have been proposed for energy transfer between electron transport and ATP synthesis as specified in the following sections [53].

1.The Chemical Coupling Hypothesis

According to this hypothesis, electron transport is coupled to ATP synthesis by a sequence of consecutive reactions and free energy is converted to the assumed nonphosphorylated high energy intermediate and is used finally for binding ADP and inorganic phosphate [54]. However, this hypothesis has been argued for decades because the proposed nonphosphorylated high energy intermediate has never been identified. On the other hand, several experimental findings are against this hypothesis, for example, it could not account for the uncoupling of electron transfer system and phosphorylation system in the presence of uncoupling agents.

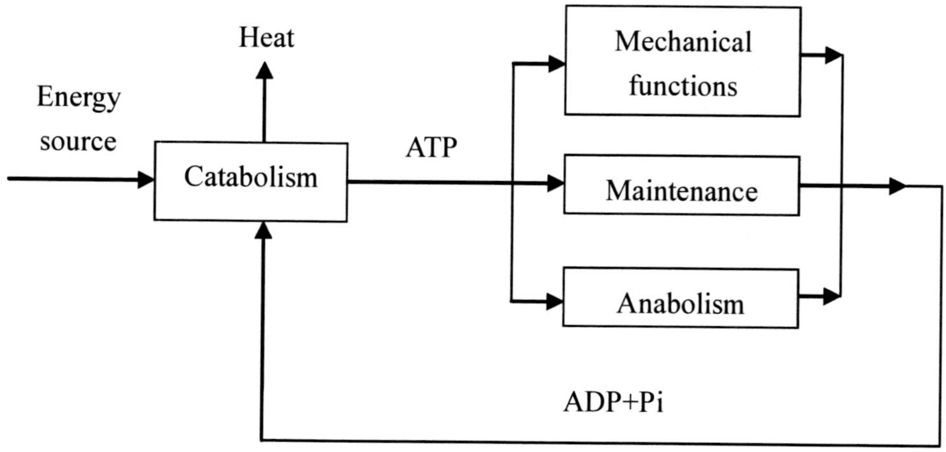

Figure 3. The role of the ATP-ADP cycle in cell metabolic system. During catabolism, energy transformation occurs from energy source. Some portion of the produced energy (in the form of ATP) will be utilized in mechanical functions, maintenance and anabolism, while some portion of energy will be lost in the form of heat.

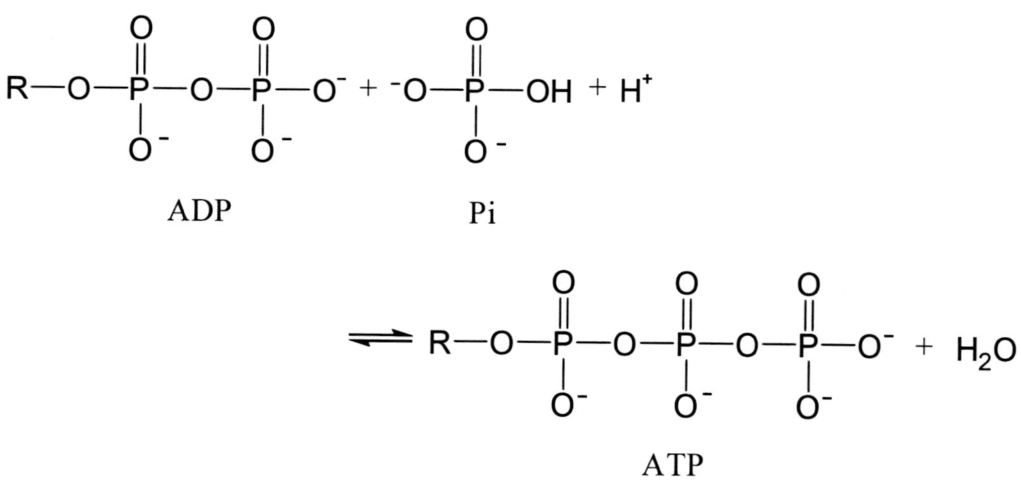

Figure 4. The scheme of ATP synthesis process [51].

2. The Conformational Coupling Hypothesis

This popular hypothesis was first put forward by Paul Boyer in 1964, stating that electron transport generates energized proteins with activated conformations, when interacting with ATP synthase, it will relax back to deactivated states and drive ATP synthesis. This hypothesis eventually evolved into the binding change mechanism in 1970s on the basis of detailed kinetic and binding studies of the reactions catalyzed by F_0F_1 and finally complemented by Boyer with rotational catalysis in 1990s. Two possibilities have been described to support this hypothesis and the classic one suggests that the Ca^{2+} store must be close enough to the plasma membrane to allow a constitutive protein-protein interaction between capacitative Ca^{2+} channels and IP_3 receptors [55]. Boyer's notion is accord with a variety of binding experiments and is essentially consistent with Mitchell's chemiosmotic hypothesis described in the following.

3. The Chemiosmotic Hypothesis

This hypothesis has been accepted as one of the most important contributions to the twentieth century biology proposed by Mitchell in 1961 [56]. This hypothesis states that H^+ is pumped to intermembrane space as electrons go through the electron transfer chain. As a result, an electrical potential $\Delta\psi$ and a proton gradient ΔpH increase across the inner membrane. The membrane potential and proton gradient are the proton-motive force (PMF) which is responsible for driving the ATP synthesis (Figure 5). The following equation obtained from Nernst equation describes calculation of ΔG associated with a proton gradient:

$$\Delta G = RT\ln\frac{[H^+]_{out}}{[H^+]_{in}} + F\Delta\psi \tag{1}$$

where $[H^+]_{in}$ and $[H^+]_{out}$ refer to the proton concentrations inside and outside the cellular membrane respectively; F is Faraday's constant (96,485 J V^{-1} mol^{-1}); R is gas constant (8.31 J mol^{-1} K^{-1}) and $\Delta\psi$ stands for potential difference across the membrane.

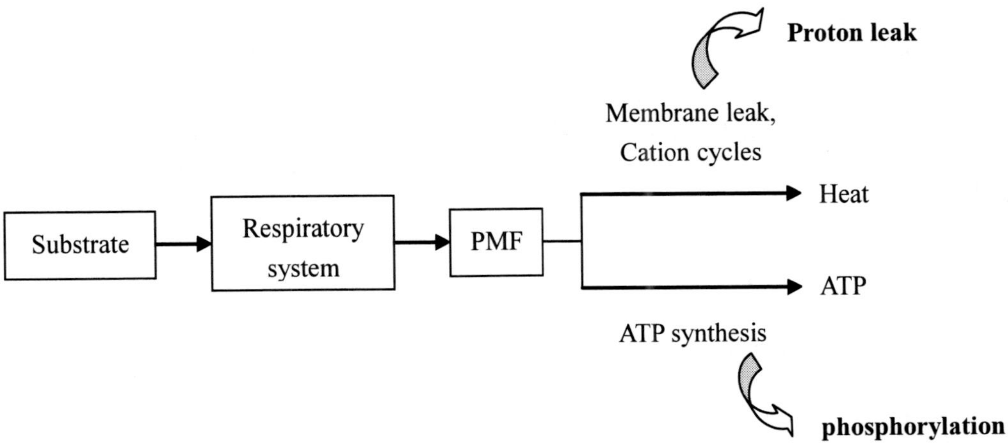

Figure 5. The oxidative phosphorylation system. Substrates are oxidized in respiratory system, which, in turn, produce the proton-motive force (PMF) across the mitochondrial cellular membrane. PMF is subsequently used to drive ATP synthesis (phosphorylation), or membrance leak (proton leak).

3.2. Mechanism of Uncoupling of Oxidative Phosphorylation

Uncoupling of oxidative phosphorylation is an important method of determining the nature of the chemiosmotic hypothesis and provides important information concerning the generation of proton-motive force by the respiratory chain and its utilization for ATP synthesis, as well as the mechanism of uncoupler action [57,58].

In recent years, more and more attentions have been paid on this method in environmental engineering for reduction of excess sludge production in CAS process. Certain conditions and reagents can uncouple oxidation from phosphorylation, such as the presence of excess carbon source and limited nutrients; the presence of anti-metabolites; high temperature; alternative aerobic-anaerobic cycle; etc [59].

1. The Presence of Excess Carbon Source and Limited Nutrients

There is evidence [60] that the observed growth yield decreases significantly with the increase of carbon source. It has been thought that the observed growth yield is a net result of interaction among growth, maintenance metabolism and uncoupling degree between anabolism and catabolism. The observed growth yield decreased with the increase of S_0/X_0 ratio due to energy spilling under substrate sufficient conditions. Bacteria have various metabolic flexibilities with respect not only to available catabolic and anabolic substrates, but also with respect to the continuously changing nutrients [61] which is an indispensable factor for living, such as *Klebsiella pneumonia*, *Escherichia coli*, *Pseudomonas fluorescens*, *Pseudomonas putida*, *Paracoccis denitrificans*, *Bacillus subtilis* and *Bacillus stearothermophilus* [62]. The phenotypic and genotypic responses to low-nutrient growth conditions basically include the regulation of gene expression, variation in the cellular structure and composition, changes in enzyme-specific activity and capacity involving in uptake and/or assimilation of the limiting nutrients, as well as changes in the affinity of these enzymes to different nutrients [61,63]. Under nutrients limited conditions, the uncoupling of oxidative phosphorylation and ATP synthesis occur due to the overproduction of metabolites. One possible mechanism may be the leakage of ions, such as protons or K^+ that in turn weakens the electrical potential across the plasma membrane, resulting in PMF dissipation and finally uncouples oxidative phosphorylation [34]. Such a situation would be encountered during the biological treatment of high strength industrial wastewater in which nutrients are often insufficient.

2. At Unfavorable Temperatures

Senez [49] has reported that there exists a “critical” range of temperature between optimal growth and total inhibition of bacteria. The maximal efficiency of growth may be realized only over a restricted range of temperature, and simultaneous decrease of both the rate and yield of growth would take place outside this range, even with adequate nutrition in the system. Under such conditions, the biosynthesis rate is reduced, while no corresponding fall in the rate of catabolism occurs. Zakharov and Kuzmina [64] found that oxidative phosphorylation to be thermolabile in *Thermus thermophilus* and suggested that elevated temperatures dramatically increased the membrane permeability of the protons. It has also been demonstrated that elevated temperature will cause species shifts toward a thermophilic population while unacclimatized species may achieve reduced biomass production with thermally induced uncoupled oxidative phosphorylation [34].

3. The Presence of Inhibitory Compounds

Many inhibitory compounds of electron transport and oxidative phosphorylation have been employed to demonstrate the pathway of electron flow through the electron transport system, the properties of the PMF, as well as to identify aspects of an enzyme mechanism. So far, the used inhibitory compounds basically can be classified into the following two groups: (i) inhibitors of electron transport at specific sites in the electron transport system, and (ii) chemical uncouplers leading to reduce proton gradient across cellular membrane [65]. Theoretically, inhibitors of electron transport may stop the proton pumping by blocking electron transfer, and uncouplers will minimize PMF through dissipating the proton gradient, leading to stop of ATP synthesis. The specific effects of some inhibitory compounds on electron transport and oxidative phosphorylation are summarized in Table 2.

Uncouplers of oxidative phosphorylation are typically weak acids with substantial lipid solubility and high hydrophobicity to carry proton across the cellular membrane into the matrix (Figure 6) [68,69]. Uncouplers are thought to have the ability to act as lipophilic proton conductors which interact nonspecifically with membranes to dissipate the proton gradient that energizes many membrane bioenergetic functions [57]. These agents cause maximum respiratory rates, electron transport continues, but no ATP synthesis occurs, since the translocated protons do not return to the interior through ATP synthase [65]. As the result, the reduction of ATP production caused by uncoupling oxidative phosphorylation occurs. In such a condition, biomass cells still continue to satisfy their maintenance energy requirement prior to that for anabolism so that the amount of ATP available for biosynthesis would be reduced dramatically which in turn reduce biomass production. Replicating these uncoupling processes in wastewater treatment would therefore be admirable. Oxidative phosphorylation can be effectively uncoupled by the addition of organic protonophores, such as 2,4-dinitrophenol (DNP), para-nitrophenol (PNP), 2,4-dichlorophenol (DCP), 2,4,6-trichlorophenol (TCP), pentachlorophenol (PCP), dicumarol, 3,3',4',5-tetrachlorosalicylanilide (TCS), fluorocarbonyl-cyanide phenylhydrazone (FCCP), etc. The structures of some uncouplers are presented in Figure 7.

3.3. Application of Uncoupling of Oxidative Phosphorylation

Mayhew et al. [7] tested 8 different chemical uncouplers at concentration of 8 μg L^{-1} to 18 mg L^{-1}, and found that erythromycin, 2,4-dNP, rotenone and trypan blue had the highest levels of metabolic uncoupling without the significant effect on microorganisms (Table 3).

The mechanism of the action of 2,4-dinitrophenol (DNP) has been investigated for several decades. It has been shown that this substance acts mainly by inhibiting the uptake of inorganic orthophosphate from the reaction medium, whereas the oxygen uptake is unmodified or increased [70]. Chen et al. [71] studied the response of activated sludge to the presence of DNP in batch cultures, and found that effect of DNP on the substrate removal and growth of activated sludge heavily depended upon its concentration. When DNP concentration was lower than 10 mg L^{-1}, sludge yield and relative specific growth rates (μ/μ_0) were significantly reduced, but its relative specific chemical oxygen demand removal rate (q/q_0) was not substantially affected. DNP addition at 1-20 mg L^{-1} increased the specific oxygen uptake rate of activated sludge, and slightly changed its hydrophobicity.

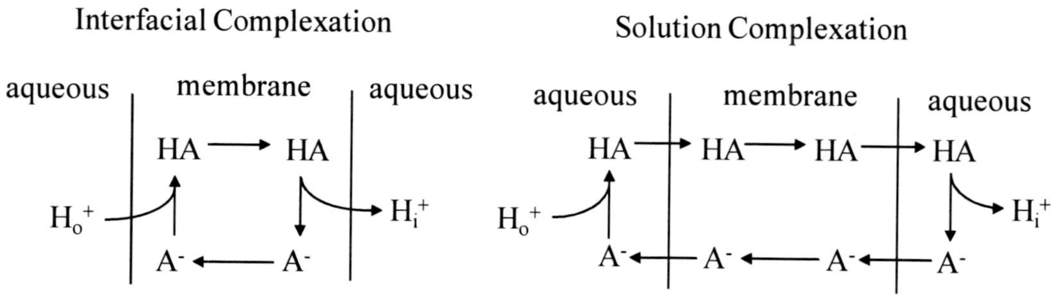

Figure 6. The sketch map of uncoupling process of protonophores. HA: protonated uncoupler; A^-: deprotonated uncoupler; H_o^+, H_i^+: protons in the outer and inner bulk phase, respectively [58].

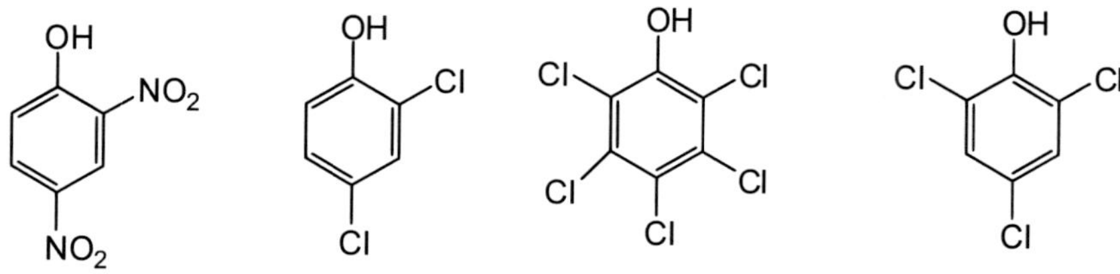

DNP DCP PCP TCP

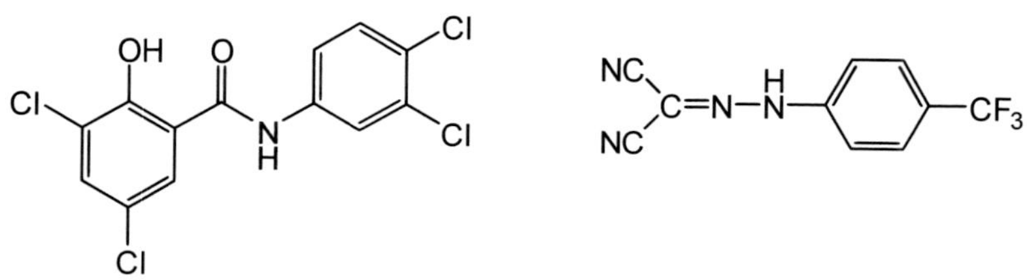

TCS FCCP

Figure 7. Structures of some representative weakly acidic uncouplers.

Table 2. Inhibitory compounds of oxidative phosphorylation [66,67]

Compounds	Effects on oxidative phosphorylation
Rotenone, Amytal	Inhibit electron transfer from Complex I at quinine-binding site
Antimycin A	Inhibit the electron transfer activity of ubiquinol-cytochrome c oxidoreductase
Cyanide, CO, Azide	Inhibit electron transfer to oxygen, preventing the reduction of oxygen
2,4-dinitrophenol, Pentachlorophenol	Disrupt the proton gradient by carrying protons across the membrane
Oligomycin, DCCD	Inhibit ATP synthase by blocking the flow of protons through proton channel F_0

Table 3. The percentage of oxygen utilization rate (OUR) compared to the control [7]

compound	Vanco-mycin	Quina-crine	Congo red	Chlorpro-mazine	Erythro-mycin	2,4 DNP	Rote-none	Trypan blue
Dose (mg L^{-1})	1.0	800	0.008	1000	0.04	4	4	18
Time (h)								
1	54	77	54	100	160	-	-	15
12	100	100	100	100	100	-	-	100
24	65	54	49	42	131	190	131	168
36	78	97	68	53	149	199	216	132
45	101	163	90	69	-	215	311	156
48	85	145	76	-	-	100	144	113

3,3',4',5-tetrachlorosalicylanilide (TCS) which is a component in the formulation of soaps, rinses and shampoos [72] has attracted many researchers to study its effects on the minimization of excess sludge in recent years because of its less toxicity. Chen et al. [73] applied TCS as a metabolic uncoupler to reduce the sludge growth in activated sludge cultures. The results revealed that TCS was an effective chemical agent in limiting the sludge growth and the 0.4 mg L^{-1} concentration was the threshold of inducing a reduction in sludge growth which was associated with the enhancement in microbial activity and the percentage of active cells over the total cell number. It was demonstrated that TCS was able to reduce sludge growth rate by around 40% at a concentration of 0.8 mg L^{-1}. Chen et al. [74] also investigated the effect of TCS on the reduction of sludge production in the activated sludge cultures operated in both batch and continuous modes through energy spilling stimulation, and confirmed that TCS has more significant effect on the reduction of sludge growth, but less influence on COD removal in the continuous culture than in the batch culture. The reduction of the sludge growth yield is correlated with the release of energy, and it was found that addition of TCS did not change the cell storage. Pure culture studies of *E. coli* AB1175 suggested that TCS would not cause the change in cell viability. Meanwhile, Chen et al. [75] in another research group also used TCS as a metabolic uncoupler to investigate its effect on the activated sludge cultures. Batch tests showed that TCS was an effective chemical uncoupler in reducing the sludge yield by approximately 50% when the concentration was greater than 1.0 mg L^{-1}. Moreover, it did not affect substrate removal capability and effluent nitrogen concentration adversely when dosed every other day in a range of 2.0-3.6 mg L^{-1} during the 40-day operation period of the activated sludge batch cultures. However, TCS would cause a shift in the predominant microbial population, e.g. in the reactors with TCS, few filamentous microorganisms were observed, resulting in forming less dense sludge flocs.

Yang et al. [76] tested four metabolic uncouplers (p-chlorophenol, m-chlorophenol, m-nitrophenol and o-nitrophenol) and compared their effectiveness in reducing excess sludge production in the activated sludge process. Different concentrations of individual metabolic uncouplers were added into a series of batch experiments with constant initial biomass and glucose concentrations. Results showed that metabolic uncouplers are very effective in reducing sludge production. Among the four tested metabolic uncouplers, m-chlorophenol was found to be the most effective in reducing sludge production with less effect on the process performance. The acidity constant (pK_a) of a metabolic uncoupler highly influences its effectiveness in sludge reduction, that is to say, stronger lipophilic weak acids may have higher potentials to lower sludge production.

Zheng et al. [4] studied the feasibility of 2,4,6-trichlorophenol (TCP) and malonic acid (MA) as metabolic uncouplers for sludge generation reduction in sequence batch reactors (SBRs) for treating organic wastewater for a long period. Results showed that TCP at the concentrations of 2 mg L^{-1} and 10 mg L^{-1} could reduce sludge production by about 47% and 30%, respectively, while the effluent COD removal efficiency and sludge settleability were not significantly affected. Compared to MA, TCP was found to be a better metabolic uncoupler for sludge reduction for a long period of operation.

Chemical uncoupling may be the simplest method for biomass reduction as it just requires the addition of certain amount of uncouplers and no other plant changes or extra steps are needed. Actually, industrial uncoupling application in two full-scale activated sludge plants in Phoenix Arizona was reported by Okey and Stensel [77]. It appears from the literature that addition of chemical uncouplers to biological wastewater treatment systems can significantly reduce sludge production, and can be regarded as a promising technique for reduction of excess activated sludge generation in CAS process. However, long-term bioacclimation has been observed and can eventually outweigh the advantageous effects of uncoupler addition. Application of chemical uncoupler for sludge reduction may cause reduction of COD removal and even worsen the settleability and dewaterability of activated sludge. It should also be pointed out that most chemical uncouplers are xenobiotics and potentially harmful to human health and the environment, as well as to the microbial cells. Optimization and correct dosing are important to avoid inhibition to microorganisms instead of uncoupling. Thus, thorough consideration and careful plant management should be taken in practice, and further research is needed for the environmental impact of chemical uncouplers in long term [5]. It was also found that addition of metabolic uncouplers would result in higher consumption of dissolved oxygen [78]. In addition, many factors, such as culture medium, pH, temperature, the affinity of microbes to different types of metabolic uncouplers and physiological state and viability of microorganism, may have significant effects on the efficiency of sludge reduction by metabolic uncouplers [1].

Table 4. Eeffects of various uncouplers on sludge reduction and effluent quality

Compound	Dosing & effects	Reference
TCS	0.4 mg L^{-1}: shreshold; 0.8 mg L^{-1}: 40% sludge reduction; 0.5-1.0 mg L^{-1}: no effect on substrate removal efficiency.	[73]
	1.0 mg L^{-1}: 50% sludge reduction; 2.0-3.6 mg L^{-1} dosing every other day for 40 days: no effect on effluent nutrient concentration.	[75]
	40 mg d^{-1}: 30% sludge reduction; 60-day continuous operation: increase effluent nitrogen concentration.	[23]
2,4-DNP	DNP < 10 mg L^{-1}: decrease sludge production significant; DNP□10 mg L^{-1}: inhibit the relative specific COD removal rate.	[71]
TCP	2 mg L^{-1}: 47% sludge reduction, with slight decrease of COD removal efficiency and sludge settleability.	[79]
2,4-DCP	DCP < 5 mg L^{-1}: didn't change the diameter distribution of sludge flocs; 1-20 mg L^{-1}: slightly reduce SOUR and sludge yield.	[80]
PNP	100 mg L^{-1}: 62% sludge reduction with increase in SOUR	[81]

3.4. Modeling of Uncoupled Energy Metabolism

To date, several models have been developed to evaluate the yield of activated sludge under uncoupling conditions. Liu [82] developed a growth yield model to describe the effect of the substrate concentration normalized to the biomass concentration (S_0/X_0) on the observed growth yield in a substrate-sufficient batch culture of activated sludge:

$$\frac{1}{Y_{obs}} = \frac{1}{(Y_{obs})_{\max}} + \frac{1}{(Y_w)_{\min}} \frac{S_0/X_0}{S_0/X_0 + K_{S/X}} \tag{2}$$

Where $(Y_{obs})_{max}$ is the maximum observed growth yield under substrate-limited conditions; $(Y_w)_{min}$ is the minimal energy spilling-related growth yield; S_0 and X_0 are initial substrate concentration and initial biomass concentration, respectively; $K_{S/X}$ is the S_0/X_0 ratio-related saturation constant.

According to the point of view of energy uncoupling, Liu [83] extended the above equation to a general expression :

$$\frac{1}{Y_{obs}} = \frac{1}{(Y_{obs})_{\max}} + \frac{1}{(Y_w)_{\min}} \frac{C_u/X_0}{C_u/X_0 + K_{u/X}} \tag{3}$$

where C_u is initial uncoupler concentration in the culture medium; $K_{u/X}$ is a C_u/X_0-related contant.

Based on growth yield, Liu et al. [84] put forward the concept of energy uncoupling coefficient to describe the observed uncoupling between anabolism and catabolism:

$$E_u = \frac{(Y_{obs})_{\max} - Y_{obs}}{(Y_{obs})_{\max}} \tag{4}$$

where E_u is the energy uncoupling coefficient.

In the study of Chen et al. [85], a model was developed to evaluate the yield of activated sludge in the presence of 2,4-dinitrophenol (DNP):

$$\frac{1}{Y_{obs}} = \frac{1}{Y'_{\max}} + \frac{k_d}{Y'_{\max}} \cdot \theta \tag{5}$$

where Y'_{max} is the maximum observed sludge yield with a consideration of energy requirement maintenance; K_d is specific endogenous respiration rate and θ is SRT.

Chen et al. [80] also analyzed the energy uncoupling and inhibition at the presence of DCP. In regard to energy uncoupling, the difference in the observed growth yield between the DCP-free culture and that dosed with DCP could represent the degree of metabolism uncoupling and be expressed using Equation 4. From the aspect of inhibition, they modified Monod model as the follows:

$$\mu = \mu_{\max} \frac{S}{K_s + S} \frac{K_{i,\mu}}{K_{i,\mu} + I} \tag{6}$$

where μ is specific growth rate; μ_{max} is maximum specific growth rate; S is concentration of the rating-limiting substrate; K_s is concentration giving one-half of the maximum rate; I is inhibitor concentration; $K_{i,\mu}$ is inhibition coefficient of microbial growth.

In the absence of inhibitors (I=0), Equation 6 can be converted as follows:

$$\mu_0 = \mu_{\max} \frac{S}{K_s + S} \tag{7}$$

Where μ_0 is specific growth rate under inhibitor-free condition.

Taking Equation 7 into consideration, then, the specific growth rate described in Equation 6 can be written as:

$$\mu = \mu_0 \frac{K_{i,\mu}}{K_{i,\mu} + I} \tag{8}$$

Equation 8 can be subsequently rearranged as:

$$\frac{\mu_0}{\mu} = 1 + \frac{1}{K_{i,\mu}} I \tag{9}$$

In fact, a larger value of $1/K_{i,\mu}$ would indicate a greater inhibition degree. Similarly, in the presence of inhibitors, the specific COD removal rate can be expressed as:

$$\frac{q_0}{q} = 1 + \frac{1}{K_{i,q}} I \tag{10}$$

where q_0 is specific COD removal rate under inhibitor-free condition. It is clear that the term $1/K_{i,q}$ is a indicator of the inhibition degree on the substrate removal.

Saini et al. [86] used the model organism *Shewanella oneidensis* MR-1 to study the effect of both excess substrate (pyruvate) and chemical uncoupler (TCS) addition on cell growth simultaneously and proposed the following empirical expression:

$$\frac{1}{Y_{obs}} = \frac{1}{(Y_{obs})_{\max}} + \frac{1}{(Y_w)_{\min}} \times \frac{S_0/X_0}{S_0/X_0 + K_{S/X}} + \frac{1}{(Y_{wu})_{\min}} \times \frac{C_u/X}{C_u/X + K_{u/X}} \tag{11}$$

where $(Y_{obs})_{max}$ is observed growth yield under substrate-limited conditions; $(Y_w)_{min}$ is minimal energy spilling related growth yield; $(Y_{wu})_{min}$ is minimal energy spilling related growth yield under uncoupler addition conditions; $K_{s/x}$ is saturation constant related to S_0/X_0; $K_{u/x}$ is saturation constant related to C_u/X; X is biomass concentration at the instant of uncoupler addition; S_0 is initial substrate concentration.

3.5. Molecular Analyses

Microbial populations at the presence of uncouplers can be monitored by several molecular analyses, such as PCR amplification of 16S rRNA genes followed by denaturing gradient gel electrophoresis (DGGE). These two microbiological techniques are always combined together to determine the molecular composition of the microbial community. Besides, Polyacrylamide Gel Electrphoresis (PAGE) is also frequently used to analyze the active biomass with small molecular weight RNA [87].

References

[1] Liu, Y. (2003) Chemically reduced excess sludge production in the activated sludge process. *Chemosphere* **50**(1), 1-7.

[2] Metcalf and Eddy. (1991) *Wastewater engineering, treatment disposal and reuse.* 3rd ed, Mcgraw-Hill, Inc USA.

[3] Andreottola, G. and Foladori, P. (2005) A review and assessment of emerging technologies for the minimization of excess sludge production in wastewater treatment plants. *J Environ Sci Health A Tox Hazard Subst Environ Eng* **41**(9), 1853-1872.

[4] Zheng, G.H., Li, M.N., Wang, L., Chen, Z.Y., Qian, Y.F. and Zhou, Q. (2008) Feasibility of 2,4,6-trichlorophenol and malonic acid as metabolic uncoupler for sludge reduction in the sequence batch reactor for treating organic wastewater. *Appl Biochem Biotechnol* **144**(2), 101-109.

[5] Wei, Y.S., Van Houten, R.T., Borger, A.R., Eikelboom, D.H. and Fan, Y.B. (2003) Minimization of excess sludge production for biological wastewater treatment. *Water Res* **37**(18), 4453-4467.

[6] Liu, Y. and Tay, J.H. (2001) Strategy for minimization of excess sludge production from the activated sludge process. *Biotechnol Adv* **19**(2), 97-107.

[7] Mayhew, M. and Stephenson, T. (1998) Biomass yield reduction: Is biochemical manipulation possible without affecting activated sludge process efficiency?. *Water Sci Technol* **38**(8-9), 137-144.

[8] Saiki, Y., Imabayashi, S., Iwabuchi, C., Kitagawa, Y., Okumura, Y. and Kawamura, H. (1999) Solubilization of excess activated sludge by self-digestion. *Water Res* **33**(8), 1864-1870.

[9] Yasui, H. and Shibata, M. (1994) An innovative approach to reduce excess sludge production in the activated sludge process. *Water Sci Technol* **30**(9), 11-20.

[10] Tsuno, H., Arakawa, K., Kato, Y. and Nagare, H. (2008) Advanced sewage treatment with ozone under excess sludge reduction, disinfection and removal of EDCs. *Ozone-Sci Eng* **30**(3), 238-245.

[11] Saby, S., Djafer, M. and Chen, G.H. (2002) Feasibility of using a chlorination step to reduce excess sludge in activated sludge process. *Water Res* **36**(3), 656-666.

[12] Tokumura, M., Sekine, M., Yoshinari, M., Znad, H.T. and Kawase, Y. (2007) Photo-Fenton process for excess sludge disintegration. *Process Biochem* **42**(4), 627-633.
[13] Guellil, A., Boualam, M., Quiquampoix, H., Ginestet, P., Audic, J.M. and Block, J.C. (2001) Hydrolysis of wastewater colloidal organic matter by extra-cellular enzymes extracted from activated sludge flocs. *Water Sci Technol* **43**(6), 33-40.
[14] Rosenberger, S., Witzig, R., Manz, W., Szewzyk, U. and Kraume, M. (1999) Operation of different membrane bioreactors: experimental results and physiological state of the micro-organisms. *Water Sci Technol* **41**(10-11), 269-277.
[15] Ghyoot, W. and Verstraete, W. (2000) Reduced sludge production in a two-stage membrane-assisted bioreactor. *Water Res* **34**(1), 205-215.
[16] Lee, N.M. and Welander, T. (1996) Use of protozoa and metazoa for decreasing sludge production in aerobic wastewater treatment. *Biotechnol Lett* **18**(4), 429-434.
[17] Lee, N.M. and Welander, T. (1996) Reducing sludge production in aerobic wastewater treatment through manipulation of the ecosystem. *Water Res* **30(**8), 1781-1790.
[18] Ye, F.X. and Li, Y. (2005) Uncoupled metabolism stimulated by chemical uncoupler and oxic-settling-anaerobic combined process to reduce excess sludge production. *Appl Biochem Biotechnol* **127**(3), 187-199.
[19] Chen, G.H., An, K.J., Saby, S., Brois, E. and Djafer, M. (2003) Possible cause of excess sludge reduction in an oxic-settling-anaerobic activated sludge process (OSA process). *Water Res* **37**(16), 3855-3866.
[20] Liu, Y. (2000) Effect of chemical uncoupler on the observed growth yield in batch culture of activated sludge. *Water Res* **34**(7), 2025-2030.
[21] Chen, G.H., Mo, H.K. and Liu, Y. (2002) Utilization of a metabolic uncoupler, 3,3',4',5-tetrachlorosalicylanilide (TCS) to reduce sludge growth in activated sludge culture. *Water Res* **36(**8), 2077-2083.
[22] Mason, C.A., Hamer, G. and Bryers, J.D. (1986) The death and lysis of microorganism in environmental process. *FEMS Microbiol Rev* **39**(4), 373-401.
[23] Ye, F.X. and Li, Y. (2005) Reduction of excess sludge production by 3,3',4',5-tetrachlorosalicylanilide in an activated sludge process. *Appl Microbiol Biotechnol* **67**(2), 267-274.
[24] http://www.molecularstation.com/cell/cell-lysis/
[25] http://www.piercenet.com/Proteomics/browse.cfm?fldID=72F377CD-2581-438C-9B27-5360226EA128
[26] Zhao, Y.X., Yin, J., Yu, H.L., Han, N. and Tian, F.J. (2007) Observations on ozone treatment of excess sludge. *Water Sci Technol* **56**(9), 167-175.
[27] Paul, E. and Debellefontaine, H. (2007) Reduction of excess sludge produced by biological treatment processes: Effect of ozonation on biomass and on sludge. *Ozone Sci Eng* **29**(6), 415-427.
[28] Paul, E., Camacho, P., Sperandio, M. and Ginestet, P. (2006) Technical and economical evaluation of a thermal, and two oxidative techniques for the reduction of excess sludge production. *Trans IChemE, Part B, Process Saf Environ Protect* **84**(B4): 247-252.
[29] Nagare, H., Tsuno, H., Saktaywin, W. and Soyama, T. (2008) Sludge ozonation and its application to a new advanced wastewater treatment process with sludge disintegration. *Ozone-Sci Eng* **30**(2), 136-144.

[30] Tian, S., Lishman, L. and Murphy, K.L. (1994) Investigation Into excess sludge activated sludge accumulation at low temperatures. *Water Res* **28**(3), 501-509.

[31] http://www.learner.org/channel/courses/envsci/unit/text.php?unit=4&secNum=3

[32] Gerlach, S.A. (1978) Food-chain relationships in subtidal silty sand marine sediments and the role of meiofauna in stimulating bacterial productivity. *Oecologia* **33**(1), 55-69.

[33] Ginoris, Y. P., Amaral, A. L., Nicolau, A., Coelho, M.A.Z. and Ferreira E.C. (2007) Recognition of protozoa and metazoa using image analysis tools, discriminant analysis and neural network. *Anal Chim Acta* **595**(1-2), 160-169.

[34] Low, E.W. and Chase, H.A. (1999) Reducing production of excess biomass during wastewater treatment. *Water Res* **33**(5), 1119-1132.

[35] Matz, C., Webb, J.S., Schupp, P.J., Phang, S.Y., Penesyan, A., Egan, S., Steinberg, P. and Kjelleberg, S. (2008) Marine biofilm bacteria evade eukaryotic predation by targeted chemical defense. *PLoS ONE* **3**(7), e2744.

[36] Jürgens, K. and Matz, C. (2002) Predation as a shaping force for the phenotypic and genotypic composition of planktonic bacteria. *Antonie Van Leeuwenhoek* **81**(1-4), 413-434.

[37] Ratsak, C.H. (2000) Effects of Nais elinguis on the performance of an activated sludge plant. *Hydrobiologia* **463**(1-3), 217-222.

[38] Eikelboom, D.H. (2000) Process Control of Activated Sludge Plants by Microscopic Investigation. IWA Publishing, UK, 85-102.

[39] Huang, X., Liang, P. and Qian, Y. (2007) Excess sludge reduction induced by tubifex in a recycled sludge reactor. *J Biotechnol* **127**(3), 443-451.

[40] Wei, Y.S., Wang, Y.W., Guo, X.S. and Liu, J.X. (2009) Sludge reduction potential of the activated sludge process by integrating an oligochaete reactor. *J Hazard Mater* **163**(1), 87-91.

[41] Lapinski, J. and Tunnacliffe, A. (2003) Reduction of suspended biomass in municipal wastewater using bdelloid rotifers. *Water Res* **37**(9), 2027-2034.

[42] Mayhew, M. and Stephenson, T. (1997) Low biomass yield activated sludge: A review. *Environ Technol* **18**(9), 883-892.

[43] Ratsak, C.H. and Verkuijlen, J. (2006) Sludge reduction by predatory activity of aquatic oligochaetes in wastewater treatment plants: Science or fiction? A review. *Hydrobiologia* **564**(1), 197-211.

[44] Low, E.W. and Chase, H.A. (1999) The effect of maintenance energy requirements on biomass production during wastewater treatment. *Water Res* **33**(3), 847-853.

[45] Luong, J.H.T. and Volesky, B. (1983) Heat evolution during the microbial process-Estimation, measurement, and applications. *Adv Biochem Eng Biotechnol* **28**, 1-40.

[46] Dan, N.P. (2001) Biological treatment of high salinity wastewater using yeast and bacterial systems. Ph.D dissertation. Asian Institute of Technology, Bankok, Thailand.

[47] Sun, D.D., Khor, S.L., Hay, C.T. and Leckie, J.O. (2007) Impact of prolonged sludge retention time on the performance of a submerged membrane bioreactor. *Desalination* **208**(1-3), 101-112.

[48] Canales, A., Pareilleux, A., Rols, J.L., Goma, G. and Huyard, A. (1994) Decreased sludge production strategy for domestic wastewater treatment. *Water Sci Technol* **30**(8), 97-106.

[49] Senez, J. C. (1962) Some considerations on the energetic of bacterial growth. *Microbiol Mol Biol Rev* **26**(212), 95-107.

[50] Ye, F.X., Shen, D.S. and Li, Y. (2003) Reduction in excess sludge production by addition of chemical uncouplers in activated sludge batch cultures. *J Appl Microbiol* **95**(4), 781-786.

[51] Russel, J. B. and Cook, G. M. (1995) Energetics of bacterial growth: balance of anabolic and catabolic reactions. *Microbiol Mol Biol Rev* **59**(1), 48-62.

[52] Spycher, S., Escher, B.I. and Gasteiger, J. (2005) A quantitative structure-activity relationship model for the intrinsic activity of uncouplers of oxidative phosphorylation. *Chem Res Toxicol* **18**(12), 1858-1867.

[53] Menendez, R.G. (1996) An electromagnetic coupling hypothesis to explain the proton translocation mechanism in mitochondria, bacteria and chloroplasts. *Med Hypotheses* **47**(3), 179-182.

[54] Hatase, O., Yamamoto, G. and Oda, T. (1969) The competitive effect of adenosine-5'-triphosphate against the stimulating and inhibiting actions of 2, 4-dinitrophenol on the mitochondrial respiration. *Acta Medica Okayama* **23**(3), 227-235.

[55] Rosado, J.A. (2006) Discovering the mechanism of capacitative calcium entry. *Am J Physiol Cell Physiol* **291**(6), 1104-1106.

[56] Mitchell, P. (1961) Coupling of phosphorylation to electron and hydrogen transfer by a chemiosmotic type of mechanism. *Nature* **191**, 144-148.

[57] Decker, S.J. and Lang, D.R. (1978) Membrane bioenergetic parameters in uncoupler-resistant mutants of bacillus megaterium. *J biol chem* **253**(19), 6738-6743.

[58] Brandt, U., Schubert, J., Geck, P. and Vonjagow, G. (1992) Uncoupling activity and physicochemical properties of derivatives of fluazinam. *Biochim Biophys Acta* **1101**(1), 41-47.

[59] Russell, J.B. (2007) The energy spilling reactions of bacteria and other organisms. *J Mol Microbiol Biotechnol* **13**(1-3), 1-11.

[60] Liu, Y. (1996) Bioenergetic interpretation on the S_0/X_0 ratio in substrate-sufficient batch culture. *Water Res* **30**(11), 2766-2770.

[61] de Mattos, M.J. and Neijssel, O.M. (1997) Bioenergetic consequences of microbial adaptation to low-nutrient environments. *J Biotechnol* **59**(1-2), 117-126.

[62] Tempest, D.W. and Niejssel, O.M. (2006) Physiological and energetic aspects of bacterial metabolite overproduction. *FEMS Microbiol Lett* **100**(1-3), 169-176.

[63] Elberse, I.A.M.,Vanhala, T.K., Turin, J.H.B.,Stam, P., van Damme, J.M.M. and van Tienderen, P.H. (2004) Quantitative trait loci affecting growth-related traits in wild barley (Hordeum spontaneum) grown under different levels of nutrient supply. *Heredity* **93**, 22-33.

[64] Zakharov, S.D. and Kuzmina, V.P. (1992) ATP-synthase activity of the thermophilic bacterium Thermus Thermophilus HB-8 membranes. *Biochemistry-Moscow* **57**(4), 365-371.

[65] http://themedicalbiochemistrypage.org/oxidative-phosphorylation.html

[66] http://en.wikipedia.org/wiki/Oxidative_phosphorylation

[67] Kim, H., Esser, L., Hossain, M.B., Xia, D., Yu, C.A., Rizo, J., van der Helm, D. and Deisenhofer, J. (1999) Structure of Antimycin A1, a Specific Electron Transfer Inhibitor of Ubiquinol-Cytochrome c Oxidoreductase. *J Am Chem Soc* **121** (20), 4902-4903.

[68] Lou, P.H., Hansen, B.S., Olsen, P.H., Tullin, S., Murphy, M.P. and Brand, M.D. (2007) Mitochondrial uncouplers with an extraordinary dynamic range. *Biochem J* **407**(Pt 1), 129-140.

[69] Pall, M.L. and Robertson, C.K. (1987) Fatty acid control of cyclic AMP levels in Neurospora crassa. *Curr Microbiol* **15**(1), 25-27.

[70] Dianzani, M.U. and Scuro, S. (1956) The Effects of some inhibitors of oxidative phosphorylation on the morphology and enzymic activities of mitochondria. *Biochem J* **62**(2), 205-215.

[71] Chen, G.W., Yu, H.Q. and Xi, P.G. (2007) Influence of 2,4-dinitrophenol on the characteristics of activated sludge in batch reactors. *Bioresour Technol* **98**(4), 729-733.

[72] Budavari S., O'Neil, M.J., Smith, A. and Heckelman, P.E. (1989) The Merck Index. 11th ed, Rahway, New York.

[73] Chen, G.H., Mo, H.K. and Liu, Y. (2002) Utilization of a metabolic uncoupler, 3,3',4',5-tetrachlorosalicylanilide (TCS) to reduce sludge growth in activated sludge culture. *Water Res* 36(8), 2077-2083.

[74] Chen, G.H., Mo, H.K., Saby, S., Yip, W.K. and Liu, Y. (2000) Minimization of activated sludge production by chemically stimulated energy spilling. *Water Sci Technol* **42**(12), 189-200.

[75] Chen, Y.X., Ye, F.X. and Feng, X.S. (2004) The use of 3,3 ',4 ',5-tetrachlorosalicylanilide as a chemical uncoupler to reduce activated sludge yield. *J Chem Technol Biotechnol* **79**(2), 111-116.

[76] Yang, X.F., Xie, M.L. and Liu, Y. (2003) Metabolic uncouplers reduce excess sludge production in an activated sludge process. *Process Biochem* **38**(9), 1373-1377.

[77] Okey, R.W. and Stensel, H.D. (1993) Uncouplers and activated sludge-the impact on synthesis and respiration. *Toxicol Environ Chem* **40**, 235-254.

[78] Zubay, G.L. (1998) Biochemistry. WCB Publisher, Boston.

[79] Zheng, G.H., Chen, Z.Y., Wang, L., Qian, Y.F. and Zhou, Q. (2007) Treatment of municipal wastewater with the sequence batch reactor under uncoupling metabolic conditions. *J Environ Sci Health A Tox Hazard Subst Environ Eng* **42**(13), 2059-2064.

[80] Chen, G.W., Yu, H.Q., Liu, H.X. and Xu, D.Q. (2006) Response of activated sludge to the presence of 2,4-dichlorophenol in a batch culture system. *Process Biochem* **41**(8), 1758-1763.

[81] Low, E.W. and Chase, H.A. (1997) The use of chemical uncouplers for reducing biomass production during biodegradation. *Water Sci Technol* **37**(4-5), 399-402.

[82] Liu, Y. (1996) Bioenergetic interpretation on the S_0/X_0 ratio in substrate-sufficient batch culture. *Water Res* **30**(11), 2766-2770.

[83] Liu, Y. (2000) Effect of chemical uncoupler on the observed growth yield in batch culture of activated sludge. *Water Res* **34**(7), 2025-2030.

[84] Liu, Y., Chen, G.H. and Paul, E. (1998) Effect of the S_0/X_0 ratio on energy uncoupling in substrate-sufficient batch culture of activated sludge. *Water Res* **32**(10), 2883-2888.

[85] Chen, G.W., Yu, H.Q., Xi, P.G. and Xu, D.Q. (2008) Modeling the yield of activated sludge in the presence of 2,4-dinitrophenol. *Biochem Eng J* **40**(1), 150-156.

[86] Saini, G. and Wood, B.D. (2008) Metabolic uncoupling of *Shewanella oneidensis* MR-1, under the influence of excess substrate and 3,3 ',4 ',5-tetrachlorosalicylanilide (TCS). *Biotechnol Bioeng* **99**(6), 1352-1360.

[87] Low, E.W., Chase, H.A., Milner, M.G. and Curtis, T.P. (2000) Uncoupling of metabolism to reduce biomass production in the activated sludge process. *Water Res* **34**(12), 3204-3212.

In: Biochemical Engineering
Editors: F.E. Dumont and J.A. Sacco, pp. 165-183
ISBN: 978-1-60741-257-1

Chapter 6

MEMBRANE TECHNOLOGY IN THE FISHERY INDUSTRY – A STATE OF THE ART

Wirote Youravong[1,2,*] and Zhen-Yu Li[2]
[1]Department of Food Technology, Faculty of Agro-Industry
[2]Membrane Science and Technology Research Center
Prince of Songkla University, Hat Yai, 90112, Thailand

Abstract

A large amount of raw material is processed with the demand of fishery products. Membrane technology and bioprocessing have increasingly involved in the fishery industry, particularly the fishery waste utilization and the treatment of fishery waste water. A comprehensive review on application of membrane technology in the fishery industry is addressed in this chapter. Fishery wastes in forms of solid and liquid contain high content of organic compounds which may cause the pollution to the environment. Membrane technology can recover valuable compounds from these fishery wastes; therefore, it not only reduces the risk of pollution but also improves economical benefit of the fishery industry. Comparing with other competing methods, membrane technology can serve as a mild temperature, simple and large-scaled method to achieve both high efficiency and maximum preservation of natural properties of recovered compounds. A series of valuable compounds such as protein, enzymes, collagen and marine flavor could be recovered from fishery by-products by membrane technology. Pressure-driven membrane processes which are microfiltration, ultrafiltration, reverse osmosis and nanofiltration have been employed. These membrane processes could work individually or be combined with other biochemical reactions to develop a hybrid multistage membrane process or a membrane reactor for the desirable recovery rate, high purity of recovered compounds and development of new products (e.g. bioactive peptides). In addition to recovery of valuable compounds, membrane process has been also applied for treatment of fish pond and fishing process water to achieve water recycle. In the future, more success of membrane technology in the aspect of the fishery industry is to be expected.

* E-mail address: wirote.y@psu.ac.th. Tel. 66-7428 6321, Fax. 66-7421 2889. Corresponding author's address: Dr. Wirote Youravong. Department of Food Technology, Faculty of Agro-Industry, Prince of Songkla University, Hat Yai 90112, Thailand.

1. Introduction

Fishery has still been a fast-growing industry because of increasing demand of fish and seafood throughout the world. A large amount of raw material is processed for the development of fishery products annually. The main processes presenting in fishery industry are defrosting, cooking, canning and cleaning. For example, the process for canned tuna comprises the following stages: defrosting, cooking, peeling, canning and the finishing operations, which are similar to all lines (sauce or brine filling, sterilization and packing). In the case of mussel, the main steps are washing, trimming, cooking, size classifying, dehydrating and finishing operation. Sardine processing includes scraping, cutting, conveying, canning and finishing operation. The process for other species (octopus, squids, mackerel, etc.) is similar. A more detailed description of all these manufacturing has been previously published (Veiga et al. 1994).

The major part of aquatic animals for fishery processing is meat. Therefore a big part of raw material as well as the effluents from fishery processing plant is wasted. Wastewater from fishery processing contains a high variable content of organic matter of both suspended and dissolved solids, it is thereby characterized as a nontoxic effluent. The main environmental problems of fishery industries are related to the emission of large volume wastewaters. During the processing of fishery products, the amount of raw material which is converted into waste can reach 50 % by weight (Aguilar and Sant Anna 1988). The solid wastes include bone, skin, dark meat, viscera and pieces from flaking, shells, etc. The liquid waste includes the water pumped together with the fish during its unloading, washing water, cooking juice, oil, blood, mucus, etc (Afonso and Borquez, 2002). For example, in the production of surimi, about 80 % of the original raw materials is considered as waste (Dewitt and Morrissey, 2002). The amount of chilled-fresh water used for washing purposes during surimi processing is approximately 27 m^3/ton of surimi product. The organic load of these wastewater is also high. In the case of canned tuna processing, there are 23-25% solid waste (e.g. head, skin, viscera) and nearly 35% liquid waste (e.g. blood, tuna cooking juice, oil) (Prasertsan *et al.*, 1988).

In addition to the process of product development, the increase of the world fishery production can be mainly attributed to aquaculture. Indeed while the total catch from fisheries has been leveling off since the last decade, aquaculture has been constantly growing at a rate of 10% per year due to the fact that the majority of the stocks are being fully exploited (FAO, 2006). Waste discharges from aquaculture facilities can have a pronounced effect on their surroundings (Wu, 1995). Increased sedimentation of organic material below fish cages, nutrient release in receiving waters and the use of antibiotics are a few examples of possible negative impacts of aquaculture on the environment. Unless these waste are properly treated, they can create environmental and ecological hazards associated to their high organic content (Danish Ministry of Fisheries, 1983). The treatment of wastes from fishery industry could be divided into water reuse and by-product recovery. The importance of water reuse has been definitely recognized. By-product recovery is also a cost-effective way to reduce the amount of waste that would, otherwise, be disposed of. There are numerous ways for utilization of fish processing by-product such as production of pet food, fish meal, fertilizers, fish silage, protein hydrolysates, chitin and chitosan, gelatin, collagen, food flavors, bone meal, bait and fish scales. In addition, these by-products can be used as raw materials for producing bioactive compounds such as peptide, collagen hydrolysate, chitooligosaccarides.

Membrane processes have wide industrial applications covering many existing and emerging uses in the chemical, petrochemical environmental, water treatment, pharmaceutical, medical, food, dairy, beverage, paper, textile and electronic industries (Ho and Sirkar, 1992). Membrane filtration process is an approach to separate different materials by semi-permeable membranes. Generally, a membrane process requires two bulk phases to be physically separated by a third phase or a membrane phase. The membrane is an interphase between the two bulk phases. The membrane phase may be any or a combination of the following: nonporous solid, microporous or macroporous solid with a fluid (liquid or gas) in the pores, a liquid phase with or without a second phase, or a gel (Ho and Sirkar, 1992). In all membrane processes, the feed is separated into two fractions, i.e. a permeate which passes through the membrane and a retentate which is rejected by the membrane. Large commercial scaled uses of membrane processes have displaced conventional separation processes and more expectation is supposed in the future. Membrane processes have always been an integral part of biotechnological processes for fermentation, clarification, purification and concentration. Over the last 30 years, the followings are the widely employed membrane processes : reverse osmosis (RO), ultrafiltration (UF), microfiltration (MF), nanofiltration (NF), dialysis (DA), electrodialysis (ED), gas permeation (GP), pervaporation (PV) and liquid membrane. They are used to produce potable water from sea, to clean industrial effluent and recover valuable constituents, to concentrate, purify or fractionate temperature-sensitive solutions (food and drug industry, biotechnology), to remove urea and other toxins from the blood stream in an artificial kidney, to release certain drugs at a predetermined rate (controlled release), and so on (Madaeni, 1999).

In the case of food industry, pressure-driven membrane processes including microfiltration, ultrafiltration, reverse osmosis and nanofiltration have been adopted for both promotion of product and treatment of wastes. Pressure plays a role as driven force.

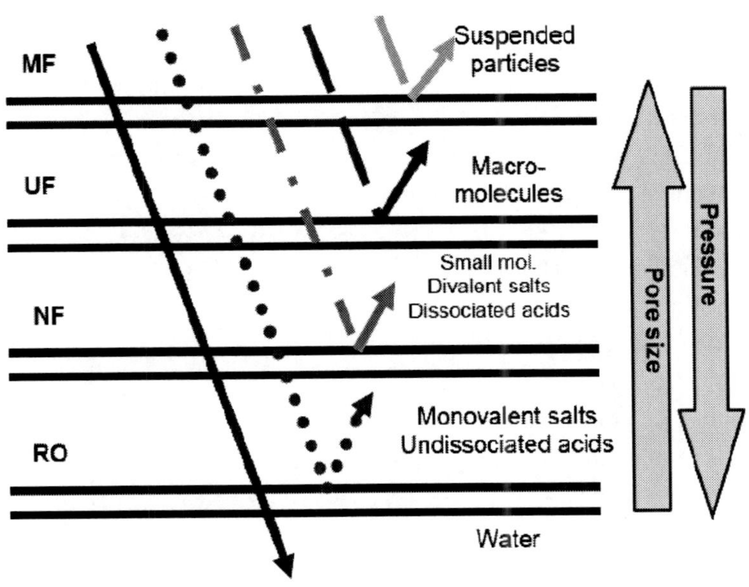

Figure 1. Classification of pressure- driven membrane processes.

Figure 1 and Table 1 summarized the characteristics of pressure-driven membrane processes.

Table 1. Membrane pore size and operational pressure for pressure-driven membrane processes

Process	Pore size (μm)	Applied pressure (bar)
MF	0.05-10	0.1-2
UF	0.05-0.002	1-10
NF	0.002-0.001	5-20
RO	<0.001	10-100

These pressure-driven membrane processes are used in the following exemplificative operations:

Microfiltration:

- Cold sterilization of beverages and pharmaceuticals
- Clarification of fruit juice, wines and beer
- Separation of bacteria from water (biological wastewater treatment)
- Effluent treatment
- Separation of oil/ water emulsions
- Pre-treatment of water for nanofiltration or reverse osmosis

Ultrafiltration:

- Juice clarification
- Enzymatic depectinization
- Fractionation of milk for cheese manufacture
- Fractionation of whey for whey protein concentrates
- Protein recovery in meat, fish and poultry industry
- Blood protein concentration
- Gelatin concentration
- Starch recovery
- High fructose corn syrup from corn starch (enzymatic membrane reactor)
- Solvent recovery and particle removal in oil industry

Nanofiltration:

- Partial demineralization and concentration of whey
- Caustic/acid cleaning recovery solutions
- Brine cleaning in fish industry

Reverse osmosis:

- Fruit juice concentration
- Milk concentration (drying, transport)
- Low-alcohol beer
- Recovery of protiens from soy whey
- Downstream process in corn refining
- Water recycling
- Production of high quality water

Pressure-driven membrane processes are characterized by their low specific energy consumption. Since they constitute a mechanical separation without phase or temperature change, they consume least energy (Mirza, 2008). In addition to energy, these methods have the advantages as follows: (Cheryan and Rajagopalan, 1998):

(1) This technology is more widely applicable across a wide range of industries; moreover, requirement of high temperature can be avoided.
(2) The membrane is a positive barrier to rejected components. Thus, the quality of the treated water (the permeate) is more uniform, regardless of influent variations. These variations may decrease flux, but it does not generally affect the quality of its output.
(3) No extraneous chemicals are needed, making subsequent recovery easier.
(4) Membranes can be used in-process to allow recycling of selected waste streams within a plant.
(5) Membrane equipment has a smaller foot print.
(6) The plant can be highly automated and does not require highly skilled operators.

2. Fundamentals of Pressure-Driven Membrane Process

The pressure-driven membrane process universally relies on the preferential retention and passage of at least one solute and one solvent. Both of these processes often involve selective separation of multiple solutes. The primary objectives of multiple-solute separations are to achieve both a desired yield of the product and an acceptable level of purification, from one or several impurities (Reis and Saksena, 1997). However, one cannot achieve complete retention of retained solute with current membrane technology. There is a compromise between yield and purification. Separating solutes with similar retention properties, such as selective protein filtration, is challenging task (Nakassuka and Michaels, 1992). The function of pressure-driven membrane process is affected by a series of factors. Major factors include pressure, feed velocity, solute concentration and temperature. Interaction between feed solution and membrane, characteristic of solute and cleaning operation should be also considered in many cases.

During pressure-driven membrane process, the flux (J) of permeate is normally described as a function of the driving force and the total resistance in equation (1), called Darcy's law.

$$J = \frac{\Delta P}{\mu R_{tol}} \tag{1}$$

where, ΔP is the differential pressure(or transmembrane pressure); μ is the dynamic viscosity of the permeate; R_{tol} is the total resistance.

In separation of small molecules or salts from water or solvent by reverse osmosis, nanofiltration and ultrafiltration, the effect of osmotic pressure on permeate flux is significant. In order to obtain a flux from bulk side (higher solute concentration) to permeate side (lower solute concentration), the applied pressure have to be greater than the osmotic pressure. Thus an osmotic pressure term should be added to Darcy's law as the following equation:

$$J = \frac{\Delta P - \Delta \pi}{\mu R_{\mathrm{tol}}} \quad (2)$$

where $\Delta\pi$ is osmotic pressure.

According to equation (2), flux is dependent on the driven force, i.e. transmembrane pressure and total resistance. In practice, if feed solution contains particular materials or macromolecules, the permeate will decrease with filtration time to a level which is much lower than initial flux. The decrease of flux is a result of the build-up of a concentration gradient onto membrane surface caused by convective transport of solutes towards membrane (concentration polarization), formation of macromolecule gel layer on membrane surface and other membrane fouling (i.e. narrowing or blocking membrane pore by solutes).

The total resistance increases with filtration time due to the different types of membrane fouling, leading to a decline of flux when a constant pressure is applied. Thus the total resistance is the sum of all sub resistances (equation 3).

$$R_{\mathrm{tol}} = R_{\mathrm{m}} + R_{\mathrm{cp}} + R_{\mathrm{f}} \quad (3)$$

where, R_{m} is the membrane resistance, a membrane constant; R_{cp} is the resistance caused by concentration polarization; R_{f} is the resistance caused by membrane fouling.

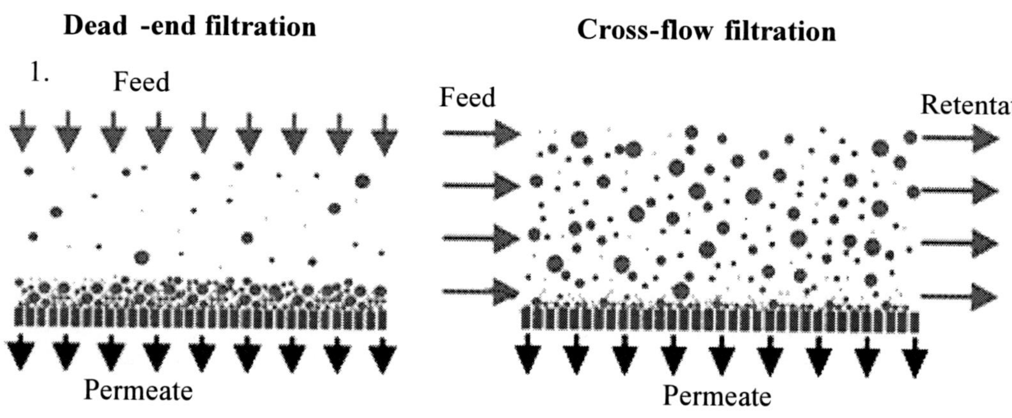

Figure 2. Dead-end and cross-flow filtration..

There are fundamentally two different ways of running pressure-driven membrane process, i.e. dead-end filtration and cross-flow filtration (Figure 2).

In dead-end filtration, feed solution is pressurized through the membrane. The pressure is constant over the whole membrane surface. If the feed solution is contained in a batch cell, the feed volume will decrease with filtration time. In cross-flow filtration, the feed solution is pumped tangentially along the membrane surface. By the pressured applied, the permeate is induced. However, there is a pressure drop along the membrane because of the tangential flow over the membrane. Thus, the pressure is not constant over the whole membrane surface. Cross-flow filtration has several advantages comparing to dead-end filtration. Particularly, the tangential flow in cross-flow filtration can wash through the membrane surface, thus limit the formation of concentration polarization and other membrane fouling. The cross-flow filtration is more widely employed, especially in the industrial applications.

The transmission of a partially rejected solute through a membrane is expressed by the apparent sieving coefficient conveniently (Ghosh and Cui, 2000).

$$S_a = \frac{C_P}{C_b} \tag{4}$$

where C_b is the solute concentration in the bulk feed; C_p is the solute concentration in permeate.

The diamensionless ratio of sieving coefficients has been termed as selectivity (Reis and Saksena, 1997) and it is defined as:

$$\Psi = \frac{S_{a1}}{S_{a2}} \tag{5}$$

where Ψ is the selectivity, and S_{a1} and S_{a2} are the observed sieving coefficients for the lesser- and greater- retained solute, respectively.

From equation (5), a larger value of Ψ is correlated with a better separation between two solutes.

With great potential for concentration, fractionation, and purification of soluble and insoluble materials, pressure-driven membrane processes have been beginning to emerge for applications in fishery industry. As mentioned above, pressure-driven membrane process is based on separation of the substances by passing the liquid under pressure through a membrane. This method provides two essential characteristics: (a) it is a purely physical separation principle; and (b) it is a modular design. Membrane process is therefore an ideal choice for waste treatment in fishery industry, since it allows removing contaminants, recovering dissolved or suspended high value components (protein, pigment and enzyme, etc), and allows water reuse for certain applications (Rautenbach and Linn, 1995).

3. Application of Pressure- Driven Membrane Process in the Fishery Industry

3.1. Microfiltration

Although the exact size range is a matter of debate, microfiltration is generally accepted as a means of filtering colloidal or fine particles having approximate range of 0.02 to 10 μm (Davis, 1992). Microfiltration is not basically different from ultrafiltration, reverse osmosis or nanofiltration, except the size range of retained molecules. Microfiltration is suited to separate larger sizes, such as suspended solids, particulates, and microorganisms. This is accomplished because microfiltration membranes are thought to act like a physical seive. The membranes are highly porous and have discernible pores even the surface skins are asymetric. Therefore, the separation is mainly based on their sizes. Membrane material is usually made up of ceramics, teflon, polypropylene, or other polymers. Applications of microfiltration for water treatment in fishery industry have been reported.

A microfiltration membrane with 0.4 μm nominal pore size has been installed in a membrane bioreactor for treatment of a saline backwash flow from a re-circulating aquaculture system with a total of 35 tonnes of annual rainbow trout (*Oncorhynchus mykiss*) production (Sharrer et al., 2007). The membrane bioreactor performed exceptionally well during this study. The physical exclusion of total suspended solids and bacteria (total heterotrophs and total coliform) from the mixed-liquor suspended solids (MLSS) was nearly completed. Further, the associated biological oxygen demand was almost completely removed. Consistent removal of total nitrogen at all treatment levels was achieved. It provided that sufficient acclimation of each salinity level was also given. A clear water was collected from permeate side. This system can treat approximately 22 m^3/day.

Gemende et al. (2008) tested two microfiltration systems, Zenon ZW500 (PVDF, nominal pore size 0.04 μm, membrane area 10 m^2) and Eido-HF-PP-M6 (PP, nominal pore size 0.1 μm, membrane area 12.75 m^2) in pilot scale (in 6 culture tanks, each with a volume of 1 m^3 and a fish stocking density of about 20 to 60 kg/m^3) at the aquaculture company Fischwirtschaftsbetrieb Andreas von Bresinsky, Oelzschau, Germany. The tests showed the applicability of both microfiltration systems to maintain process stability of the biological water purification and fish breeding. In addition, the biomass was enriched to concentrations that allow a subsequent utilization by fermentation. Furthermore, wastewater and residual loads were reduced remarkably. For long-term operation (more than one year) it could be proved that the integration of microfiltration is not only technically and economically interesting for the special re-circulating aquaculture system, but also results in remarkable reduction in wastewater and residual loads. Though the initial fluxes (about 8 to 15 l/m^2 h) and average fluxes (about 3 to 5 l/m^2 h) in this test were relatively low. This could be attributed to the small transmembrane pressure (less than 0.05 bar) and the high retentate biomass concentrations up to 15 g/l. The flux values for the microfiltration of formulated fish water reported by Viadero et al. (2002), have ranged from 12.5 to 250 l/m^2 h for suspended solid concentration of about 5 to 18 mg/l.

Surimi is a Japanese term of washed and dewatered fish mince widely used as a raw ingredient in manufacturing of artificial crab meats or kamaboko (Lee, 1984). For surimi processors, recovering the suspended solids (myofibrillar proteins) is more commercially interest. Huang and Morrissey (1998) successfully performed a continuous filtration to recover the suspended myofibrillar proteins from minced-fish washing water at refrigerated temperatures (12-15°C) using polysulfone microfiltration membranes with a nominal pore size of 0.2 μm. Lin et al. (1994) used microfiltration for recovering proteins and recycling water from commercial surimi processing wastewater. The solids in the final dewatering process (screw press) were concentrated up to 15.5 folds by a rotary screen (pore size of 100 μm) followed by microfiltration (pore size of 50 and 30 μm, in series). The result demonstrated that the proteins recovered by microfiltration showed highly functional properties in comparison with proteins in regular surimi. Solids recovered from the surimi processing wastewater by microfiltration could be directly added to surimi to increase yield without any effect on its quality.

According to its separation capacity, microfiltration has been more widely applied as a pretreatment method. Afonso and Borquez (2002) applied a microfiltration process (pore size of 5-10 μm) as pretreatment process before ultrafiltration of effluent from fish meal production in a fish meal factory located in Talcahuano, Chile. The result indicated that microfiltration with a pore size of 5-10 μm was an efficient pretreatment process for this effluent. Besides removing a great part of the suspended matter (visual observation), microfiltration drastically reduced the oil and grease content which could be adsorbed onto the ultrafiltration membrane surface, thereby affecting its performance.

During recovery of protease from the spleen of yellowfin tuna which is a solid waste from tuna canning process, microfiltration has been applied as a pretreatment process to remove suspended solids from spleen extract before recovery of protease by ultrafiltration (Li et al., 2008). A hollow fiber membrane with pore size of 0.1 μm was employed in a cross-flow microfiltration system. All visible solids in spleen extract can be removed by microfiltration with given conditions while transmission of protease was higher than 0.90. Microfiltration could remove not only suspended solids but also large molecular weight proteins. As a consequence, protease was also partially purified by microfiltraiton. The clear permeate from microfiltration could be used as a feed for concentration and purification of protease by ultrafiltration. Since suspended solids and a part of proteins were removed by microfiltration, the fouling load in ultrafiltraion process was significantly reduced.

3.2. Ultrafiltration

Ultrafiltration is a selective fractionation process using applied pressures up to 10 bars. It is employed for concentration of suspended solids and solutes having molecular weight greater than 1,000 Da. The permeate contains low-molecular-weight organic solutes and salts. Ultrafiltration has been used for fractionation, concentration and purification of proteins and some bioactive compounds (e.g. bioactive peptides, chitooligosaccharides) produced from fishery processing waste. In comparison with microfiltration, ultrafiltration could perform a higher capacity of separation with a smaller pore size. In comparison with nanofiltration and reverse osmosis, ultrafiltration requires lower applied pressure. Thus, ultrafiltration probably provides the most applications in fishery industry. Within the fish processing industry, ultrafiltration has been mostly applied for brine recovery, wastewater treatments, recovery of marketable compounds from wastes, and applied as a pretreatment process for nanofiltration and reverse osmosis. However, to operate properly, ultrafiltration usually requires the removal of fats, large membrane surface and frequent washing (Marti et al., 1994). Thus there must be a suitable selection of membrane and cleaning systems to prolong its lifetime.

Ultrafiltration treatment has been proposed as an alternative to secondary biological treatment, and has been proved to be effective in the organic matter recovery in fish canning industry (Del-Rosario and Duldulao, 1988). The rejection coefficient for either protein or carbohydrate solute of the tuna broth was approximately 54.5% while solute recovery was about 90%. The ultrafiltered broth had negligible turbidity and odor. It has been also reported that recovery of 90% of proteins from red-meat fish processing wastewater was obtained when ultrafiltration was applied and proteins were concentrated from 0.1-2 % to 0.4-18 % (NovaTec, 1994).

Two ultrafiltration modules equipped with Ceraver and Patterson Candy International (PCI) membranes were tested for recovery and concentration of proteins from the wastewater of a fish plant. Despite of different cut-off values, apparent rejection coefficients were rather comparable (70% and 80% respectively). Optimum economic conditions were established, corresponding to average transmembrane pressures of 2.2 and 3.8 bars and tangential flow velocity of 6.0 and 0.47 m/s for Ceraver and PCI membranes, respectively. The protein concentration in the feed solution was increased from 5 to 35 g/dm^3. The study showed that the method could reduce pollution due to organic matter by decreasing the value of the biological oxygen demand after 5 days (BOD5) by 80% (Mameri et al., 1999).

Collagen and gelatin are currently used in diverse fields including food, cosmetic, and biomedical industries. Fish skin waste could be used as a potential source to isolate collagen and gelatin. Kim and Byun (1994) designed a three-stepped membrane bioreactor for continuous hydrolysis of fish skin gelatin. The system involved three hollow fiber ultrafiltration membranes with molecular weight cut-off of 10 (first step), 5 (second step) and 1 (third step) kDa respectively. The enzymatic hydrolysates of gelatin extracted from fish skin were fractionated and recycled through the membrane bioreactors. Alcalase, pronase E and collagenase were identified as the most suitable enzymes for the hydrolysis of fish skin gelatin. After optimization of operating conditions including temperature, pH, type of enzyme, substrate to enzyme ratio, flow rate and reaction volume, the degree of hydrolysis under the optimum condition in the first, second and third membrane bioreactor were 87%, 77% and 70% respectively. The productivity of hydrolysate in the continuous three-step membrane bioreactor was 430 mg/mg enzyme. These hydrolysates could be not only used in conventional ways as mentioned above, but also applied for production of functional peptides. The fragments arising from the third step were composed of peptides ranging from 0.9 to 1.9 kDa and responsible for an inhibitory activity of angiotensin I converting enzyme. These fragments have great potential application in medical science for regulation of blood pressure (Byun and Kim, 2001).

Alvise et al. (2004) also obtained an antihypertensive peptide from an Alfalfa White protein hydrolysate produced by a continuous enzymatic membrane bioreactor. Insoluble alfalfa protein concentrate (APC) was solubilised by two proteolytic enzymes (pepsin and delvolase) in batch and in continuously flowed membrane bioreactor. The main components of the system included a 8 liter reaction vessel coupled with a membrane module via a recirculation pump. The substrate (APC) was pretreated overnight at pH 2.0 and 40 $^\circ$C and then preset to the required pH (9.5) in the feed vessel and pumped into the reaction vessel at a flow rate equal to the permeate flux (16.6 dm^3/m^2.h). Temperature, pH and reaction volume were continuously regulated. The membrane module was composed of seven tubular ZrO_2 membranes (6 mm inner diameter, 1.2 m length, 0.16 m^2 filtering area) with 10 kDa nominal molecular weight cut-off. The optimum conditions of hydrolysis were found to be 2.4 mg/dm^3 of enzyme, 32 g/dm^3 of APC with a conversion rate of 75.9%. The ultrafiltration was conducted through ZrO_2 membrane of 10 kDa with no significant loss of enzyme. This membrane is approved for application in food industries. Flow rate of 2.22 cm^3/min, reactor volume of 400 cm^3 and resident time of 180 min appeared to be an optimum conditions for ultrafiltration. This process provides a soluble peptidic product with a light brown color. High conversion and complete rejection of the substrate and enzyme are promising. The peptide obtained from this hydrolysis process has been proved to highly inhibit the activity of angiotensin I converting enzyme (Kapel et al., 2006).

NovaTec (1994) provided a comparison between ultrafiltration and other methods for recovery of protein and reduction of BOD from clam processing wastewater. Protein recovery (%) were 60-70, 72, 79, 90, 90, 88 while BOD reduction (%) were 72, 62, 79, 91, 91, 87 for direct acid precipitation, ion-exchange chromatography, single ultrafiltration, ion-exchange chromatography combining with acid precipitation, ultrafiltration combining with acid precipitation and microgas dispersion with flotation respectively. Membrane separation showed its great competitive capacity comparing to other methods.

Except protein separation, enzyme separation is another important and valuable application of ultrafiltration. DeWitt and Morrissey (2002) designed a ultrafiltration pilot plant with membrane molecular weight cut-off 30 and 50 kDa for recovery of catheptic proteases from fish mince washing water. Pretreatment of fish mince washing water by heating at 60 °C, acidification to pH 6, and centrifugation could double ultrafiltration flux and significantly improved protease purity by reducing a majority of the 35–205 kDa proteins. Concentrated crude protease obtained from washing water predominantly contained cathepsin L activity. Enzyme purity was increased about 100-fold, and yield was approximately 80%. Stability (frozen and freeze-dried protease) was maintained for 9 weeks at -80 °C. Freeze-dried preparations were also stable for 9 weeks at 4 and -15 °C. Successful application of pilot plant conditions allows sufficient production of protease for further investigations into their applicability.

Gildberg and Shi (1994) recovered enzyme from fish sauce. A concentrate of tryptic enzymes was obtained by ultrafiltration of fish sauce made from Cod's viscera. The recovery of enzyme activity during ultrafiltration was good (> 80%). It showed that polysulphone membranes can be successfully applied in ultrafiltration even at very high salt concentrations (20-25%). The enzyme concentrate could be stored for 8 months at 3°C or spray dried with only minor activity loss. In addition, nomudfish sauce fermentation was achieved after re-addition of the ultrafiltration permeate to the fish sauce fermentation tank.

Li (2008) applied ultrafiltration to recover protease from yellowfin tuna spleen. After extraction process and pretreatment by microfiltration membrane (Li et al., 2008), ultrafiltration with a hollow fiber membrane (molecular weight cut-off 30 kDa) could separate and purify trypsin and chymotrypsin from this spleen extract. Through a diafiltration mode these protease was purified up to 12-fold. The purified enzyme was a clear solution without odor. The activity of these protease was compared with some commercial enzymes, such as alcalase. The enzymes separated from tuna spleen by ultrafiltion showed competitive cabability of protein hydrolysis. The result showed that a trypsin-like serine protease with low-cost and qualified hydrolysis efficiency could be obtained from tuna canning waste by membrane technology involving microfiltration (pretreatment step) and ultrafiltration (separation and purification steps).

Outer skeleton of crabs, shrimps and lobsters is also a waste from seafood processing. Since it contains abundant chitin, a kind of polysaccharide, this waste has been utilized for production of chitosan which is derived from chitin by deacetylation. Chitosan and its derivatives have shown various functional properties and made them possible to be used in many fields including food, cosmetics, biomedicine, agriculture, environmental protection and wastewater management (Kim and Rajapakse, 2005). Even though chitosan is known to have important functional activities, its poor solubility makes it difficult to use. Recent studies on chitin and chitosan have attracted interest to convert them to chitooligosaccharides, because the chitooligosaccharides are not only water-soluble but also functionally versatile. To increase the solubility of chitosan in an aqueous solution and to facilitate its utilization, the enzymatic production of chitooligosaccharides with a high degree of polymerization (DP) was carried out using an ultrafiltration membrane bioreactor system.

The chitosanase was used for enzymatic hydrolysis of chitosan and able to be repeatedly reused and separated from the final products (Jeon and Kim, 2000a). The system consisted of a supplement tank, a reservoir tank, a water bath for controlling reaction temperature, three peristaltic pumps, a membrane cartridge with molecular weight cut-off 3 kDa and an enzyme reactor vessel. The most important factor in the ultrafiltration bioreactor system was a permeation rate control because the components of the chitooligosaccharides were dependent on the permeation rate. At a permeation rate of 4 ml/min, 80% of produced chitooligosaccharides had 3-6 of DP.

However, ultrafiltration membrane method did not allow continuous production of chitooligosaccharides due to the increased transmembrane pressure during the reaction. This was due to high viscosity of chitosan solution and fouling of membrane by accumulated substrate. Continuous production of chitooligosaccharides was achieved by a combination of a column reactor packed with immobilized enzyme and the ultrafiltration membrane reactor, namely dual membrane reactor system (Jeon and Kim, 2000b). The production of the chitooligosaccharides was performed by two steps. The first step is the preparation of the partially hydrolyzed chitosan from viscous chitosan in the column reactor packed with an enzyme. The second step is the production of the chitooligosaccharides from partially hydrolyzed chitosan in the ultrafiltration membrane bioreactor. Partially hydrolyzed chitosan possessed low viscosity, leading to a decrease in membrane fouling. This dual membrane reactor showed a greater productivity, ability to control molecular weight distribution and more efficient continuous production process than those of conventional methods (Kim and Rajapakse, 2005).

In addition to the treatment of wastes, membrane technology also shows its potential for product promotion in fishery industry. Ultrafiltration was applied for fractionation of Cod frame protein hydrolysate to improve its functional property (Jeon et al., 1999).

Since ultrafiltration membranes are capable of removing suspended solids and colloids, viruses, bacteria, and high-molecular suspended organic material from the water. This technology is very suitable for wastewater treatment or the pretreatment of surface water for reverse osmosis, recycling of backwash from sand filters and recycling of seafood processing and municipal wastewater. (Rosherg, 1997). Ultrafiltratiton can only be used as a pretreatment provided that it is economically feasible.

An innovative membrane-assisted hybrid bioreactor for wastewater treatment in a fish canning factory was developed by Artiga et al. (2005). This system was composed of a hybrid circulating bed reactor and an ultrafiltration membrane module. The aerobic reactor was coupled with an external submerged hollow-fiber ultrafiltration membrane module (average pore size of 0.045 μm, nominal surface area of 0.093 m^2). The operating transmembrane pressure was in the range of 10-50 kPa. The result showed that the combination of a hybrid system with suspended and attached biomass, and an ultrafiltration membrane module may be an alternative for treating industrial wastewaters in compact biological systems with very low footprint requirements. It has been shown that the system can operate simultaneously with high COD and ammonia conversion rates at high nitrogen loading rate and organic loading rate. COD removals around 99% at an organic loading rate of 6.5 kg COD/m^3.d and nitrogen loading rate of 1.8 kg N-NH^+_4/m^3.d were obtained during the treatment of the fish canning wastewater.

Oyanedel et al. (2003) also tested a membrane-assisted hybrid bioreactor with similar membrane to treat a mixture of two streams produced in a fish canning factory. This membrane was a flexible tube resembling a thin macaroni noodle. Filtration was carried out from outside to inside. Air was continuously injected to clean the outer membrane surface. The rising flow of the air-water mixture produced cleaning effects along the membrane surface. The intrinsic characteristics of the system made it feasible to operate the system at high organic loading rate without the problems related with the poor settling properties of the sludge or the drop in nitrogen conversion. There were no solids in the effluent as result of the utilisation of the membrane ultrafiltration module. Ultimately, the membrane-assisted hybrid bioreactor was successfully employed to treat a mixture of two streams generated in a fish canning factory.

3.3. Reverse Osmosis

Reverse osmosis was developed as a water treatment method for more than 40 years ago. The process firstly arose as a technique of desalinating seawater. Once the method's decontaminating capabilities were recognized, reverse osmosis systems began to be commercially produced for home water purification purposes. Such systems were installed in homes as early as the 1970s. Reverse osmosis systems seem to be a viable option to the more costly and energy-wasteful distillation units.

The semi-permeable membrane used in reverse osmosis contains tiny pores through which water can flow. The small pores of this membrane are restrictive to such organic compounds as salt and other natural minerals, which generally have a larger molecular composition than water. These pores are also restrictive to bacteria and disease-causing pathogens. Thus, reverse osmosis is incredibly effective for desalinating water and providing mineral-free as well as pathogen-free water for use. In the areas with no receiving municipally treated water or being at particular risk of waterborne diseases, reverse osmosis is an ideal process of contaminant removal. This process requires a high pressure to be exerted on the high concentration side of the membrane, usually 2-17 bars for fresh and brackish water, and 40-70 bars for seawater, which has around 24 bars of natural osmotic pressure. Nowadays, reverse osmosis has been one of the most advanced techniques for wastewater treatment and reuse. It also widely used to reduce aster salinity for industrial applications (Cao et al., 2007).

Walha et al. (2008) tested a reverse osmosis process for treatment of high salinity drilling water for seafood washing and processing from The Calembo Company (seafood conditioning, Sfax, Tunisia). The company used 50 m^3/d of drilling water to condition cuttlefish before freezing. The water is characterised by its high degree of hardness, high sulphate concentration and salinity (67 g/l). Consequently, generated effluents contain a large amount of salts and a high organic load (COD of 10-20 g/l). Treatment by reverse osmosis at high pressure (70 bars) allows to significantly reduce the salinity of the drilling water (permeate salinity around 2.5 g/l). Mixing reverse osmosis permeate and cuttlefish conditioning effluent together can significantly reduce the salinity of the effluent in the view of biological treatment with a membrane bioreactor (35 g/l or less is required).

Reverse osmosis (TFC membrane, salinity reduction: 96%) has also been employed for recovery of marine flavor from seafood cooking juice (Vandanjon et al., 2002). Cooking juice from buckies, shrimps and tuna have a high level of polluting load (COD of 5-40 g/l) and have to be treated before being rejected in the environment. However, these juices seem to contain interesting flavor compounds. This recovery would allow the industrialists to diminish the wastewater treatment cost and to recover high value molecules. It has been shown that a good retention of marine flavors could be achieved by reverse osmosis membranes. But these membranes present a strong mass transfer limitation during concentration of cooking waters of buckies and shrimps. To maintain acceptable flux, it is necessary to combine a pretreatment by ultrafiltration and followed by reverse osmosis.

However the application of reverse osmosis membrane for aquaculture wastewater treatment has been largely limited. One of the major problems is the energy cost during the reverse osmosis process. In order to reduce the energy cost for reverse osmosis operation, researchers have turned to renewable energy sources for the solutions. A prototype wind-driven reverse osmosis system was developed and constructed in 1997-1998 by University of Hawaii researchers. The system includes a windmill to drive a piston pump for feed supply. The membrane sector involves a pressure tank, a prefilter and reverse osmosis module. A photovoltaic system was installed to utilize solar energy for electric supply (such as PC, electronic sensor). The system can be operated at a moderate wind speed (Liu et al., 2002). Test results of the wind-driven system, which separates the aquaculture wastewater passing through the reverse osmosis membrane into permeate (freshwater) and brine (concentrated wastewater), indicated that freshwater can be processed and recycled.

Liu et al. (2007) applied this wind-driven reverse osmosis system for aquaculture wastewater treatment at the experimental facilities on Coconut Island, Oahu, Hawaii. The wastewater from fish tank is sent to membrane unit by a piston pump driven by windmill. After treatment by reverse osmosis, the permeate, at a quality suitable for fish production is recirculated to the fish tank. The brine is sent back to the storage tank, where it mixes with the wastewater from the fish tank. Occasional discharge of brine is necessary, as the nitrogen concentration in the storage tank builds up over time. The discharged brine is sent to a duckweed-covered reactor for nutrient recovery. By this technology, a fish tank becomes a closed aquaculture production system with zero waste discharge. The system can process aquaculture wastewater at flow rates ranging from 230 to 370 l/h. The ammonia nitrogen concentrations in the permeate remained below 0.02 mg/l, making the permeate suitable for use in fish culture. The permeate (freshwater) from the system can be used as the freshwater supply for fish culture, while the brine (concentrated wastewater) can be further processed into fish feed by a duckweed-covered reactor.

Gang et al. (2005) also proved that the wind-driven reverse osmosis system for nitrogen removal from aquaculture wastewater was technically feasible and environmentally friendly. The most unique and important feature of this system was that it could treat and recycle aquaculture wastewater using renewable energy, making it suitable for use in the remote area where electricity was hard to access. The system had an ability to work at an average wind speed as low as 3.0 m/s. Depending on the wind speed, it can generate and recycle freshwater at a flow rate of 228-366 l/h. The permeate production linearly increased with the wind speed. About 70-84% of aquaculture wastewater can be recycled using this system, which was capable of removing 90-97% of nitrogenous waste presenting in Tilapia culture effluent. The average recovery rate of the membrane used in this system was about 39.2–57.5%.

3.4. Nanofiltration

Nanofiltration is an appreciable special process when reverse osmosis and ultrafiltration are not the ideal choice for separation. Nanofiltration is a relatively recent membrane process frequently used with low total dissolved solids water such as surface water and fresh groundwater, with the purpose of softening (polyvalent cation removal) and removal of disinfection precursors such as natural organic matter and synthetic organic matter. Nanofiltration can perform separation applications that are economically feasible, such as demineralization, color removal, and desalination. Nanofiltration is also becoming more widely used in food processing applications such as dairy for simultaneous concentration and partial (monovalent ion) demineralization (Hillie and Mbhuti, 2007). In concentration of organic solutes, suspended solids, and polyvalent ions, the permeate contains monovalent ions and low-molecular-weight organic solutions, like alcohol. In general, application of nanofiltration could be defined as three areas: (a) removal of monovalent ions from waste water or reaction mixtures; (b) separation ions with different valences; (c) separation of low- and high-molecular weight components.

Ferjani et al. (2005) tested nanofiltration membranes (molecular weight cut-off 550, 342 and 500 Da) for treatment of seafood processing wastewater. The obtained retention rates of COD and protein nitrogen indicated that the nanofiltration process was efficient in terms of retention of organic and biological matters. The retention rates of COD and protein nitrogen increased with pore size diminution. Retention of COD ranged from 87% to 93% while that of protein nitrogen ranged from 52 to 82.5%.

Yeong et al. (2002) tested a nanofiltration membrane (molecular weight cut-off 500 Da) for treatment of fish and surimi wastewater.The permeate flux of nanofiltration remained almost stable for about 270 min. This finding was a very promising, thus polyamide membrane used in this study was suitable for the fish and surimi industries to treat the wastewater. This nanofiltration membrane was capable of reducing COD and total suspended solids up to 93 % and 87 %, respectively. A study on long-term flux decline indicated that polyamide nanofiltration membrane fouled rather slowly.

Blue Whiting fish is mainly used for production of fish meal and fish oil. Recently, it has been used for production of surimi. In order to concentrate the target bioactive peptides and to avoid excessive concentration of salt, Vandanjon et al. (2007) applied a membrane enzymatic bioreactor for concentration and purification of the peptides. The bioreactor involved both ultrafiltration and nanofiltration membranes. The reactional mixture from bioreactor was sent to the ultrafiltration membrane (molecular weight cut-off, 20 kDa). The step was well suited for the separation of peptides and non-hydrolyzed proteins. Enzymes and non-hydrolyzed proteins in the retentate were sent back to the bioreactor. The permeate from first fractionation step was introduced to the ultrafiltration membrane (molecular weight cut-off, 4 kDa). The second step seemed promising for the fractionation of peptides, as their flux and retention were good. At the end, nanofiltration membrane with molecular weight cut-off of 300 Da was applied to concentrate peptides. The fluxes and recovery rate were acceptable and the excessive concentration of salts was avoided.

Nanofiltration membranes have also been evaluated and employed in pre-treatment facilities for both reverse and thermal process (Al-Sofi et al., 1998; Criscuoli and Drioli, 1999). Walha et al. (2008) used a series of membrane including both ultrafiltration and nanofiltration membranes as pretreatment process before reverse osmosis of high salinity drilling water discharged from seafood processing plant. The result showed that the nanofiltration membrane (molecular weight cut-off, 200 Da) was effective for the removal of natural organic matter and reducing the sulphate concentration in drilling water. Approximately 89% of organic materials was rejected with this membrane. Furthermore, as a fact that water composition can vary according to the season and drilling place, the use of nanofiltration as a pre-treatment process also allows to standardize the water quality.

4. Summary

Fishery industry is one of the most important industrial sectors throughout the world. The elimination of waste discharged from various fishery industries leading to environmental pollutions is always unavoidable task. However waste prevention, minimization and valorization and the use of energy-efficient technologies are more and more desirable options in waste management as they are usually rich in organic matters. The controlling of raw material, water and energy consumption, as well as the utilization of by-products and reducing the effluent discharge lead towards exploring new area applications of membrane filtration. In general, membrane filtration is used for concentration, fractionation and purification. In aquaculture production, microfiltration can be employed for recovering suspended particles and removing microorganisms from recycling water while reverse osmosis is used for water desalination. In fishery processing, membrane filtration has been involved in recovering valuable compounds (e.g. proteins, enzymes) and recycling of water. In addition, ultrafiltration coupled with enzymatic reaction, known as membrane bioreactors are successfully employed for converting by-products into high value and functional property products such as chitooligosaccharides, bioactive peptides, etc. Membrane filtration also plays an important role in wastewater treatment for fishery processing plant. In term of membrane configuration or module, either hollow fiber or spiral wound membrane is often preferred depending on the feed characteristics. In the case of strongly fouling or highly viscous feed, tubular or plate and frame systems have to be applied. Generally, specific components in the feed can be related to the fouling severity and mechanism; therefore, the techniques to overcome fouling problem (e.g. by a pre-treatment) should be developed. With the further development of membrane material, module design and great knowledge of membrane fouling, membrane processes are of great potential to achieve desirable requirements of the fishery industry.

References

Afonso, M.D. & Borquez, R. (2002). Review of the treatment of seafood processing wastewater and recovery of proteins by membrane separation process-prospects of the ultrafiltration of wastewaters from the fish meal industry. *Desalination,* **142**, 29-45.

Aguilar A.L.C. & Sant Anna G.L. (1988). Liquid effiuents of the canning industries of Rio de Janeiro state- treatment alternatives. *Environ. Technol Lett,* **9**, 421-428.

Al-Sofi, M., Hassan, A.M., Dalvi, A.G.I., Mustafa, G. & Kither, M.N. (1998). Nanofiltration as a means of achieving higher TBT of >120°C in MSF. *Desalination,* **118**, 123-129.

Alvise, N.P.D., Lesueur-Lambert, C., Fertin-Bazus, A., Fertin, B., Dhulster, P. & Guillochon, D. (2004). Continuous enzymatic solubilization of alfalfa proteins in an ultrafiltration reactor. *Enz. Microbial. Technol*, **34**, 380–91.

Artiga, P., Oyanedel, V., Garrldo, J.M. & Mendez, R. 2005. An innovative biofilm-suspended biomass hybrid membrane bioreactor for wastewater treatment. *Desalination,* **179**, 171-179.

Byun, H. G. & Kim, S. K. (2001). Purification and characterization of angiotensin I converting enzyme (ACE) inhibitory peptides from Allaska Pollack (*Theragra chalcogramma*) skin. *Process Biochem.,* **36**, 1155–1162.

Cao, L., Wang, W.M., Yang, Y., Yang, C.T., Yuan, Z.H., Xiong, S.B. & Diana, J. (2007). Environmental impact of aquaculture and countermeasures to aquaculture pollution in China. *Env. Sci. Pollut. Res.,* **14** (7), 452-462.

Cheryan, M. & Rajapopalan, N. (1998). Membrane processing of oily streams wastewater treatment and waste reduction. *J. Membr. Sci.,* **151**, 13-28.

Criscuoli, A. & Drioli, E. (1999). Energetic and exergetic analysis of an integrated membrane desalination system. *Desalination,* **124**, 243–249.

Danish Ministry of Fisheries. *Technological Laboratory* 1981-1983. Denmark: Danish Ministry of Fisheries; 1983.

Davis, R.H. Microfiltreation. In: Eds. Ho., W.S.W. and Sirkar, K.K. *Membrane Handbook.* New York: Van Nostrand Reinhold; 1992; pp 457-458,

Del-Rosario, E.J. & Duldulao, F.D. (1988). Preparation and characterization of polysulfone ultrafiltration membranes for concentrating waste tuna both. *Desalination*, **70**, 293-298.

Dewitt, C.A.M. & Morrissey, M.T. (2002). Pilot plant recovery of catheptic proteases from surimi wash water. *Bioresour. Technol.*, **82**, 295-301.

FAO. (2006). *The state of the world fisheries and aquaculture.* http://www.fao.org/ docrep/003/ x8002e/x8002e00.htm

Ferjani, E., Ellouze, E., & Amar, R.B. (2005). Treatment of seafood processing wastewaters by ultrafiltration-nanofiltration cellulose acetate membranes. *Desalination,* **177**, 43-49.

Gang, Q., Clark, C.K., Liu, N., Harold, R. & James, E.T. (2005). Aquaculture wastewater treatment and reuse by wind-driven reverse osmosis membrane technology: a pilot study on Coconut Island, *Hawaii. Aqua. Eng.,* 32, 365-378.

Gemende, B., Gerbeth, A., Pausch, N. & Bresinsky, A.V. (2008). Tests for the application of membrane technology in a new method for intensive aquaculture. *Desalination,* **224**, 57–63.

Ghosh, R. & Cui, Z.F. (2000). Simulation study of the fractionation of proteins using ultrafiltration, *J. Membr. Sci.,* **180**, 29-36.

Gildberg, A. & Shi, X.Q. (1994). Recovery of tryptic enzymes from fish sauce. *Process Biochem.,* **29**, 151-155.

Hillie, T., & Mbhuti, H. (2007). Nanotechnology and the challenge of clean water. *Nat. Nanotechnol.,* 2, 663-664.

Ho, W.S.W. & Sirkar, K.K. Chapter 1: overview. In: Eds, Ho. W.S.W. and Sirkar, K.K. *Membrane Handbook.* New York: Van Nostrand Reinold; 1992; pp 3-4.

Huang, L.H. & Morrissey, M.T. (1998). Fouling of membranes during microfiltration of surimi wash water: roles of pore blocking and surface cake formation. *J. Membr. Sci.,* **144**, 113-123.

Jeon, Y.J., Byun, H.G. & Kim, S.K. (1999). Improvement of functional properties of Cod frame protein hydrolysates using ultrafiltration membranes. *Process Biochem.,* **35**, 471–478.

Jeon, Y.J. & Kim, S.K. (2000a). Production of chitooligosaccharides using an ultrafiltration membrane reactor and their antibacterial activity. *Carbohydr. Polym.,* **41**, 133-141.

Jeon, Y.J. & Kim, S.K. (2000b). Continuous production of chitooligosaccharides using a dual reactor system. *Porcess Biochem.,* **35**, 623-632.

Kapel, R., Rahhou, E., Lecouturier, D., Guillochon, D. & Dhulster, P. 2006. Characterization of an antihypertensive peptide from an Alfalfa White protein hydrolysate produced by a continuous enzymatic membrane reactor. *Process Biochem.*, **41**, 1961-1966.

Kim, S.K. & Byun, H.G. (1994). Development of optimum process for continuous hydrolysis of fish skin gelatin using a three-step recycle membrane bioreactor. *J. Korea Ind. Eng. Chem.,* **5**, 681-697.

Kim, S.K. & Rajapakse, N. (2005). Enzymatic production and biological activities of chitosan oligosaccharides (COS): A review. *Carbohydr. Polym.,* 62, 357-368.

Lee, C.M. (1984). Surimi process technology. *Food Tech.,* **38**, 69-80.

Li, Z.Y., H-Kittikun, A. & Youravong, W. (2008). Removal of suspended solids from tuna spleen extract by microfiltration: a batc process design and improvement, *Biochem, Eng. J.,* **38**, 226-233.

Li, Z.Y. 2008. *Development of a multi-stages membrane filtration process for separation and purification of protease from the wastes of fishery industry.* PhD Thesis in Biotechnology. Prince of Songkla University, Thailand.

Lin, T.M., Park, J.W. & Morrissey, M.T. (1994). Recovered protein and reconditioned water from surimi processing waste, *J. Food Sci.,* **60**, 4-9.

Liu, C.C.K., Park, J., Migita, R. & Qin, G. (2002). Experiments of a prototype wind-driven reverse osmosis desalination system with feedback control. *Desalination,* **150**, 277–287.

Liu, C.C.K., Xia, W. & Park, J.W. (2007). A wind-driven reverse osmosis system for aquaculture wastewater reuse and nutrient recovery. *Desalination,* **202**, 24-30.

Madaeni, S.S. (1999). The application of membrane technology for water disinfection. *Waster Res.,* **33**, 301-308.

Mameri , N., Abdessemed, D., Belhocine, D., Lounici, H., Gavach, C., Sandeaux, J. & Sandeaux, R. (1999). Treatment of fishery washing water by ultrafiltration. *J. Chem. Technol. Biotechnol.,* **67**, 169-175.

Marti, C., Roeckel, M., Aspe, E. & Kanda, H. (1994). Recovery of proteins from fish meal factory wastewaters. *Process Biochem.,* **29**, 39-46.

Mirza, S. (2008). Reduction of energy consumption in process plants using nanofiltration and reverse osmosis. *Desalination,* **224**, 132-142.

Nakatsuka, S. & Michaels, A.S. (1992). Transport and separation of proteins by ultrafiltration through sorptive and non-sorptive membranes. *J. Membr. Sci.,* **69**, 189-211.

NovaTec. *Guide for Best Management Practices for Process Water Management at Fish Processing Plants* In British Columbia (Project No: 1043.16). Canada: NovaTec Consultants Inc; 1994.

Oyanedel, V., Garrido, J.M., Lema, J.M. and Mendez, R. (2003). A membrane assisted hybrid bioreactor for the post treatment of an anaerobic effluent from a fish canning factory. *Wat. Sci. Technol.,* **48**, 301-309.

Prasertsan, P., Wuttijumnong, P., Sophanodora, P. & Choorit, W. (1988). Seafood processing industries within Songkhla-Hat Yai region: the survey of basic data emphasis on wastes. Songklanakarin *J. Sci. Technol.,* **10**, 447–451.

Rautenbach, R. & Linn, T. (1995). High-pressure reverse osmosis and nanofiltration, a "zero discharge" process combination for the treatment of waste water with severe fouling/sacling protential. *Desalination,* **105**, 63-70.

Reis, R.V. & Saksena, S. (1997). Optimization diagram for membrane separation. *J. Membr. Sci.,* **129**, 19-29.

Rosherg, R. (1997). Ultrafiltration (new technology), a viable cost-saving pretreatment for reverse osmosis and nanofiltration - a new approach to reduce costs. *Desalination,* **110**, 107-114.

Sharrer, M.J., Tal, Y., Ferrier, D. Hankins, J.A. & Summerfelt, S.T. (2007). Membrane biological reactor treatment of a saline backwash flow from a recirculating aquaculture system. *Aquac. Eng.,* **36**, 159–176.

Vandanjon, L., Johannsson, R., Derouiniot, M., Bourseau, P., & Jaouen, P. (2007). Concentration and purification of Blue Whiting peptide hydrolysates by membrane processes. *J. Food Eng.*, **83**(4), 581-589.

Vandanjon, L., Crosa, S., Jaouen, P., Quemeneur, F. & Bourseau, P. (2002) Recovery by nanofiltration and reverse osmosis of marine flavours from seafood cooking waters. *Desalination* **144**: 379-385.

Viadero, R.C.J and Noblet, J.A. (2002). Membrane filtration for removal of fine solids from aquaculture process water. *Aquac. Eng.,* **26**, 151–169.

Veiga, M.C., Méndez, R. & Lema, J.M. *Wastewater treatment for fisheries operations. In: Ed. Martin,* A. M. Fisheries Processing: Biotechnological Applications. London: Chapman & Hall; 1994, 344-369.

Walha, K., Amar, R.B., Quemeneur, F. & Jaouen, P. (2008). Treatment by nanofiltration and reverse osmosis of high salinity drilling water for seafood washing and processing. *Desalination,* **219**, 231-239.

Wu, R.S.S. (1995). The environmental impact of marine fish culture: towards a sustainable future. *Mar. Pollut. Bull.,* **31**, 159-166.

Yeong, W.T., Mohammad, A.W., Anua, N. & Rahman, R.A. (2002). Potential use of nanofiltration membrane in treatment of wastewater from fish and surimi industries. Songklanakarin *J. Sci. Technol.*, **24**, 977-987.

Reviewed by: Dr. Chakree Thongraung. Department of Food Technology, Faculty of Agro-Industry, Prince of Songkla University, Hat Yai, 90112, Thailand.

In: Biochemical Engineering
Editors: F.E. Dumont and J.A. Sacco, pp. 185-205
ISBN: 978-1-60741-257-1

Chapter 7

AMYLASE PRODUCTION BY *ASPERGILLUS ORYZAE* IN SUBMERGED AND SOLID STATE FERMENTATIONS

Nelson Pérez-Guerra [1,*], ***Lorenzo Pastrana-Castro*** [1,†] ***and Renato Pérez-Rosés*** [2,‡]

[1]Nutrition and Bromatology Group, Department of Analytical and Food Chemistry, Food Science and Technology Faculty, Ourense Campus, University of Vigo, Ourense, Spain

[2]Department of Pharmacy, Natural Science Faculty, University of Oriente, Patricio Lumumba Ave., E-90500, Santiago de Cuba, Cuba

Abstract

Synthesis of amylase by *Aspergillus oryzae* strain FQB-01 was followed in submerged liquid and solid state fermentations. The submerged cultures were carried out in media prepared with brewery (BW) and meat processing (MPW) wastewaters supplemented with different starch concentrations (10, 20 30 and 40 g/L).

Amylase productions (116 and 111 EU/mL) in the BW and MPW media supplemented with 40 g of starch/L of medium were slightly higher than those obtained in the same media supplemented with 30 g of starch/L (113 and 107 EU/mL) after 84 h of fermentation. In addition, the initial chemical oxygen demand in both wastes was reduced by at least 95%. Optimal pH and temperature for amylase activity were estimated at 5.8 and 46.4°C, respectively. In the optimal conditions, the enzyme showed a high stability at 40 or 50°C (pH = 5.8) or at pH values of 5.0 and 6.0 (T = 46.4°C) in the absence of starch.

The optimum conditions for high amylase production (539 EU/g of dry bagasse) under solid stated fermentation were particle size of bagasse in the range of 5-10 mm, incubation temperature of 32.5°C, pH of 5.9, moisture content of bagasse of 75%, starch concentration of 70.5 mg/g of dry bagasse and inoculum size of 1.4×10^7 spores/g of dry bagasse.

Keywords: amylase, brewery wastes, meat processing wastes, solid state fermentation, starch, submerged liquid fermentation, sugar cane bagasse

[*] E-mail address: nelsonpg@uvigo.es. Tel: 34-988-387-062; Fax: 34-988-387-001. Author whom correspondence should be addressed: Nelson Pérez Guerra, Nutrition and Bromatology Group, Department of Analytical and Food Chemistry, Food Science and Technology Faculty, Ourense Campus, University of Vigo, E-32004 Ourense, SPAIN.

[†] E-mail address: pastrana@uvigo.es

[‡] E-mail address: renato@hrpc.uo.edu.cu

Introduction

The meat and brewery industries produce large quantities of wastewaters with high chemical oxygen demand (COD) and biochemical oxygen demand (BOD), thus representing a serious environmental problem in Santiago de Cuba. Therefore, these effluents must be treated by using clean technologies for reducing their initial COD, as reported before for these wastes (Salas et al., 2006) and for other wastes from food industry (Murado et al., 1993; Roukas, 1999; Guerra & Pastrana, 2003).

In this way, different food wastes (*e.g.* mussel processing wastes, brewery wastes, whey) have been used as culture media in submerged cultures to produce different metabolites including amylases (Murado et al., 1993; 1997), pullulan (Roukas, 1999) and bacteriocins (Guerra & Pastrana, 2003; Guerra et al., 2005), among others. However, comparative studies between submerged liquid fermentation (SLF) and solid state fermentation (SSF) claim higher yields and other advantages for products produced by SSF, such as low energy requirements, lower availability of water that reduces the possibilities of contamination by bacteria and yeast, small volumes of polluting effluents and low downstream processing cost (Hesseltine, 1972; Cannel and Moo-Young, 1980; Aidoo et al., 1981; Steinkraus, 1984; Kumar & Lonsane, 1990; Kumaran et al., 1997; Raimbault, 1998; Guerra et al., 2003). Some agro-industrial products (*e.g.* sugar cane bagasse, spent brewing grains, cassava starch, sweet potato residue and peanut meal residue, palm fiber, banana waste) have been used for metabolite production in SSF (Krishna & Chandrasekaran, 1996; Yang and Wang, 1999; Francis et al., 2002; Milagres et al., 2004; John et al., 2006; Spier et al., 2006; Mazutti et al., 2006; 2007; Balkan & Ertan, 2007; Laxmi et al., 2008).

Amylases production by *Aspergillus* species in both submerged and solid state cultivations using different food wastes or agricultural residues has been thoroughly studied (Gigras et al., 2002; Ellaiah et al., 2002; Francis et al., 2002; 2003; Salas et al., 2006). These enzymes have not only been used in fermentation processes, but also in processed food industry and in the textile and paper industries (Pandey et al., 2000; Ellaiah et al., 2002; Gigras et al., 2002). The production of amylases in SLF and SSF is affected by a variety of physicochemical factors, including the composition of the growth medium, incubation temperature, pH, aeration, concentration and type of the carbon, phosphate and nitrogen sources, size and age of the inoculum, particle size and moisture level of the substrate (Gigras et al., 2002; Salas et al., 2006; Balkan & Ertan, 2007). Therefore, after selecting a culture medium for amylase production, it is necessary to determine the optimal culture conditions to improve enzyme production at a low production cost (Salas et al., 2006; Spier et al., 2006; Balkan & Ertan, 2007).

Considering the substantial availability of highly rich wastewaters and sugar cane bagasse at very low prices by local brewery, meat and sugar processing plants of Santiago de Cuba, their use as culture media or support material for different bioproductions could provide profitable substrates for a low cost production. This approach allows the protection of the environment by recycling these residues in production of amylases by *Aspergillus oryzae* strain FQB-01.

In the present study, both brewery and meat processing wastes of Santiago de Cuba were evaluated as culture media for amylase production by *A. oryzae* strain FQB-01 under submerged culture conditions. The reduction in COD in both wastes after biological treatment and the effects of pH and temperature on the stability of amylase produced by *A. oryzae* strain FQB-01were also investigated. Because *A. oryzae* strains are also capable of growing on solid substrates (Murado et al., 1997; Francis et al., 2002), the feasibility of sugar cane bagasse as a support material for the production of amylase by *A. oryzae* FQB-01 was also studied in SSF. To optimize the enzyme production in SSF, the effects of particle size, pH, incubation temperature, moisture content, inoculum size and initial starch levels on the production of amylase were studied.

Materials and Methods

Microorganism

Aspergillus oryzae strain FQB-01, was acquired from the Biotechnology Center of the University of Oriente (Santiago de Cuba, Cuba) and it was maintained on potato dextrose agar slants at 4°C (Ellaiah et al., 2002; Salas et al., 2006).

Inoculum Preparation

Inocula were prepared by transferring 2-mL of 60-h old slant culture in 50 mL of medium (250 mL Erlenmeyer) composed by (g/L): glucose, 20; $(NH_4)_2SO_4$, 6.6; KH_2PO_4, 3.5; $FeSO_4{\cdot}7H_2O$, 0.15; $MgSO_4{\cdot}7H_2O$, 0.10; $MnCl_2{\cdot}2H_2O$, 0.45 and mycological peptone, 3.0; at pH 6.8. The culture was incubated at 30°C/48 h in an orbital shaker at 200 rpm (Salas et al., 2006).

Submerged Fermentations

Brewery (BW) and meat processing (MPW) wastewaters, which were used as a base of the culture media, were obtained from a local brewery and a local meat processing plant. Both wastes were centrifuged at 15000 rpm/15 min to remove the solids in suspension. The supernatant obtained from BW contained (g/L): COD, 3.40; total sugars, 1.98; reducing sugars, 1.46; total nitrogen, 0.095; total phosphorous, 0.034. The composition of the supernatant obtained from MPW (g/L) was: COD, 3.00; total sugars, 1.82; reducing sugars, 0.99; total nitrogen, 0.172; total phosphorous, 0.028 (Salas et al., 2006).

After being used as culture media, both supernatants were supplemented with the following nutrients (g/L): mycological peptone, 3.0; $(NH_4)_2SO_4$, 6.6; $CaCO_3$, 8.0; NaCl, 5; KH_2PO_4, 3.5; $FeSO_4{\cdot}7H_2O$, 0.15; $MgSO_4{\cdot}7H_2O$, 0.10. To study the influence of the initial concentration of starch on amylase production, the media were supplemented with soluble potato starch to obtain initial starch concentrations of 10, 20, 30 and 40 g/L. Media without starch were used as controls. The media were adjusted at pH 6.0, sterilized (121°C/15 min), inoculated with a 2% inoculum level (3×10^7 spores/mL) and incubated at 30°C/96 h. All submerged cultures were carried out in 250-mL Erlenmeyer flasks with 50 mL of production media in an orbital shaker (200 rpm), using eight flasks in each fermentation serie (Salas et al., 2006).

The culture samples, as whole flasks in duplicate, were collected each 12 h. Mycelial mass was harvested by paper filtration using a pre-dried and pre-weighted Whatman filter paper No. 1, washed with distilled water and dried to constant weight at 105°C. The growth of the organism was determined as dry weight. Paper-filtered media were used to perform analytical determinations (pH, enzyme concentration, total sugars and COD).

Method for determination of total sugars was described in a previous work (Guerra & Pastrana, 2003). COD analyses were carried out in both the MPW and BW media before the fermentations and in the paper-filtered media at the end of each fermentation using the closed reflux colorimetric method as described previously (Greenberg et al., 1980). All determinations were carried out in triplicate.

Assay of Amylase Activity

Total amylase activity (*TAA*) in the submerged and solid state cultures was determined as described by Murado et al., (1993). Then, 80 μL of suitably diluted paper-filtered medium (in case of submerged cultures) or crude enzyme extract (in case of solid state cultures) were mixed with 400 μL of 0.15 M citrate-phosphate buffer, pH 5.0 (1 volume) and 4% soluble starch (1.5 volumes) previously maintained at 40°C/15 min. The reaction mixture was incubated at 40°C for 10 min. The reaction was stopped by addition of 480 μL of dinitrosalicylic acid, and the released glucose was determined by 3,5-dinitrosalicylic acid reaction (Bernfeld, 1951). One unit of amylase activity (enzymatic units (EU)/mL) was defined as the amount of enzyme that releases 1 mg/mL of reducing sugars (glucose equivalents) under the assay conditions. In case of SSF in sugar cane bagasse, the amylase activity units were expressed as EU/g of dry support (gds).

Table 1. Experimental domain and codification of the variables used in the factorial design to test the effect of temperature (*T*) and *pH* on amylase activity

Coded values	Natural values	
	***T* (ºC)**	***pH***
- 1.267	26	3.6
-1	30	4.0
0	45	5.5
+ 1	60	7.0
+ 1.267	64	7.4

Effect of Temperature and pH on Amylase Activity

A second-order orthogonal design (Box et al., 1989) based on five levels and two variables was used to study the combined influence of pH and temperature on amylase activity. The design consisted of 13 experiments with four (2^2) factorial points, four axial points to form a central composite design with α = 1.267 and five center points for replication.

Samples of cell-free medium containing the enzymes were buffered at different pHs with the appropriate buffer (potassium hydrogen phthalate-HCl buffer for pH 3.6, 4.0 and 5.5;

sodium phosphate buffer for pH 7.0 and 7.4). The buffering agents were prepared as concentrated stock solutions. Then, appropriate volumes of them were mixed with the samples of cell-free medium to obtain a final buffer concentration of 0.1 M. The samples containing the amylase were incubated for 5 min at the corresponding temperature according to the experimental matrix defined by the design used (Table 1). The enzyme activities obtained were corrected with the corresponding dilution factor.

Results were analyzed by Experimental Design Module of the Statistica software package (Statistica 5.1 for Windows computer program manual; StatSoft Inc. Tulsa, OK, USA). The response surfaces were plotted using the DeltaGraph software, version 4.0 (SPSS, Inc., Chicago, IL, USA).

Solid State Fermentation (SSF)

Sugar cane bagasse was treated with NaOH (0.12 g of NaOH per gram of dry bagasse) and autoclaved at 121°C for 20 min (Gutiérrez-Correa & Tengerdy, 1997) to remove the core and noncore lignin fractions (Doran et al., 1994). Then, the samples of sugar cane bagasse were thoroughly washed with tap water, subsequently with distilled water until neutrality and dried at 80°C (Gutiérrez-Correa & Tengerdy, 1997).

Fermentations were conducted in 250-mL Erlenmeyer flasks containing 5 g of dried bagasse supplemented with the basal salt solution, previously adjusted at the appropriated pH value, and distilled water until reaching the desired moisture level. The salt solution contained (%, g/g of dry support), NH_4NO_3, 1; KH_2PO_4, 1; NaCl, 0.2; $MgSO_47H_2O$, 0.2 (Francis et al., 2003). The contents of the flasks were thoroughly mixed and autoclaved at 121°C for 15 min. After cooling, the flasks were inoculated with 1.25 mL of inoculum containing the appropriate cell suspension. The contents were thoroughly mixed and incubated at the appropriate temperature for 96 h in a chamber with temperature and humidity control. Samples as whole flasks in triplicate were withdrawn after 96 h of fermentation (Francis et al., 2002).

Optimization of Process Parameters in SSF

In a first step, the bagasse was milled to different particle sizes (<1.0, 1.0-2.0, 2.0-5.0, 5.0-10.0 and 10.0-15.0 mm) and used as support materials to determine the appropriate particle size for maximal enzyme production in SSF. The initial moisture content of the growth media was adjusted to 80% (Milagres et al., 2004) with a mixture of 2.5 mL of the basal salt solution (initial pH of 6.0) and 16.25 mL of distilled water previously supplemented with soluble potato starch to obtain the desire level of starch in the solid support (26 mg of TS/g of dry sugar cane bagasse). The media were autoclaved (121°C for 15 min), cooled and inoculated with 1.25 mL of the spores suspension (1×10^7 spores/g of dry sugar cane bagasse) previously prepared and then incubated at 30°C for 96 h.

Secondly, a second-order orthogonal design (Box et al., 1989) was performed to asses the effect of pH of the liquid medium (salt solution plus distilled water) and incubation temperature on amylase production. The design consisted of 13 experiments with four (2^2) factorial points, four axial points to form a central composite design with $\alpha = 1.267$ and five center points for replication. The range and coding values selected for all independent variables in this experimental design are listed in Table 2.

After this previous optimization, a new second-order orthogonal design was used to study the effect of three factors (total sugars concentration, inoculum size and moisture content) on amylase production in the optimal conditions calculated before (particle size, pH and incubation temperature). In this case, the design consisted of 20 experiments with eight (2^3) factorial points, six axial points to form a central composite design with α = 1.525 and six center points for replication.

The medium was supplemented with 2.5 mL of the basal salt solution and an appropriate volume of distilled water containing different amounts of soluble potato starch to give both the different moisture contents and total sugars concentration in the medium (mg of total sugars per g of dry support) defined by the experimental design (Table 3). The media were autoclaved (121°C for 15 min), cooled and inoculated with 1.25 mL of a cell suspension containing the appropriate amount of spores to obtain the inoculum rate (spores/g of dry sugar cane bagasse) defined by the experimental design (Table 3). The media were then incubated at 32.5°C for 96 h.

All the solid state fermentations were done in triplicates and the data presented are mean values of each experiment. Results were analyzed by Experimental Design Module of the Statistica software package (Statistica 5.1 for Windows computer program manual; StatSoft Inc. Tulsa, OK, USA). The response surfaces were plotted using the DeltaGraph software, version 4.0 (SPSS, Inc., Chicago, IL, USA).

Table 2. Experimental domain and codification of the variables used in the factorial design to test the effect of temperature (*T*) and *pH* on amylase production by *A. oryzae* FQB-01 in SSF

Coded values	Natural values	
	***T* (°C)**	***pH***
- 1.267	23.4	3.7
-1	25.0	4.0
0	31.0	5.3
+ 1	37.0	6.5
+ 1.267	38.6	6.8

Table 3. Experimental domain and codification of the variables used in the factorial design to test the effect of total sugars (*TS* in mg/g of dry support (gds)), moisture content (*M* in %) and inoculum size (*IS* in log (number of spores/gds)) on amylase production by A. oryzae FQB-01 in SSF

Coded values	Natural values		
	TS	***M***	***IS***
- 1.525	9.5	58.5	4.2
-1	20.0	63.0	5.0
0	40.0	71.5	6.5
+ 1	65.0	80.0	8.0
+ 1.525	70.5	84.5	8.8

Extraction of the Enzyme

After fermentations the whole sample of each flask was extracted by the addition of distilled water containing 0.1% Tween-80, to a total extract volume of 100 mL. Contents were mixed thoroughly (150 rpm, at room temperature for 1 h) in a rotary shaker and subsequently, the suspension was centrifuged (7000 rpm for 10 min) and the supernatants were used as the crude enzyme extract for further analysis (Francis et al., 2003).

Results and Discussion

Amylases Production in Food Wastes Supplemented with Starch

Amylase production by *Aspergillus oryzae* strain FQB-01 was followed in media prepared with brewery and meat processing wastewaters supplemented with varying concentrations of starch. Figures 1 and 2 show the time course of growth, pH, total sugars (*TS*) and enzyme (*TAA*) production. As it can be observed, the increase in the initial *TS* concentration led to an increase in biomass (as cell dry weight) and amylase synthesis when compared to the control cultures (BW and MPW media without starch). This indicates that the supplements with starch in both media favoured both biomass and enzyme production.

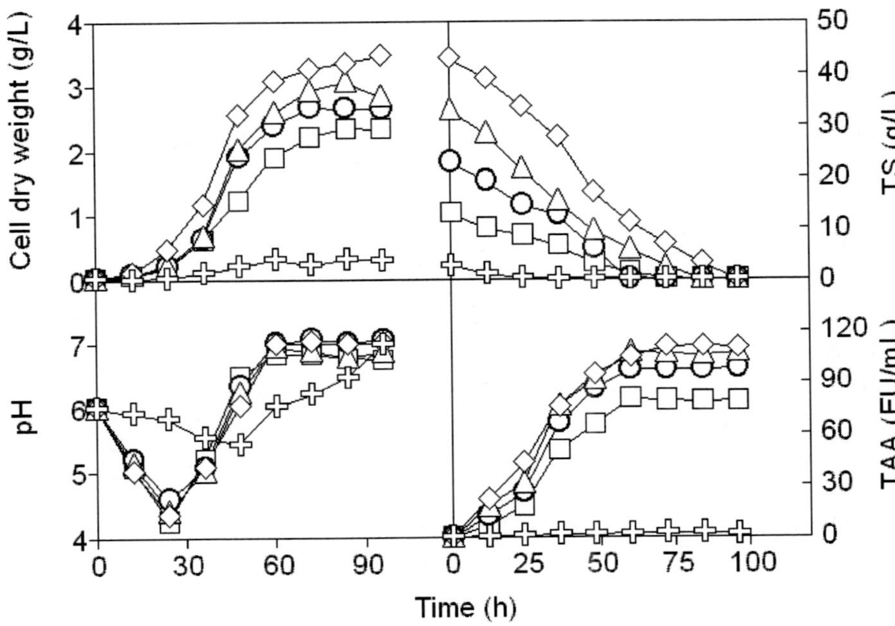

Figure 1. Time course of growth and enzyme synthesis by *A. oryzae* strain FQB-01 grown in media prepared with meat processing wastewaters (MPW) supplemented with different initial concentrations of soluble starch (2: 0, □: 10, O: 20; Δ: 30; ◇: 40 g/L). *TS*: total sugars; *TAA*: total amylolytic activity. The cultures were carried out at 30°C/96 h in an orbital shaker at 200 rpm.

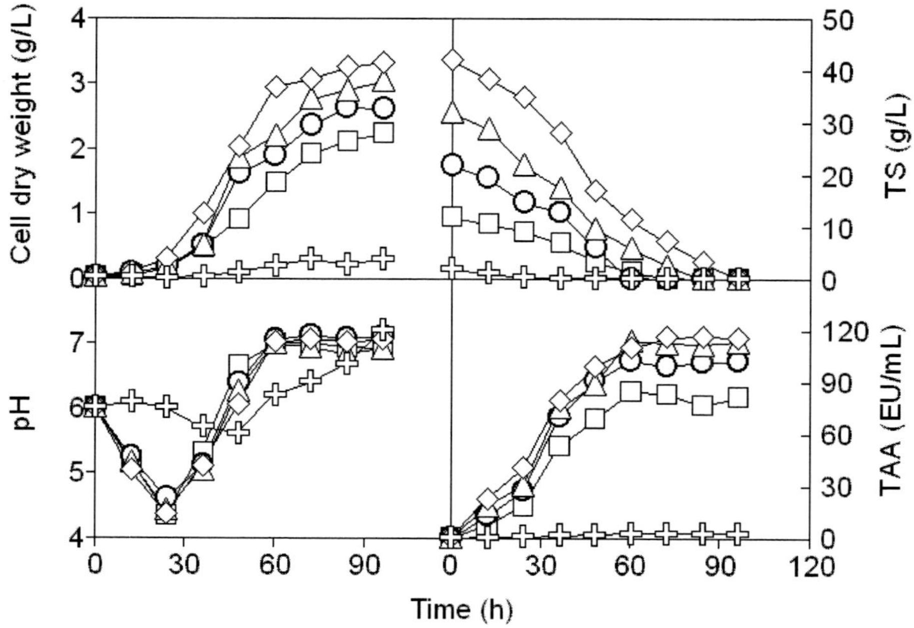

Figure 2. Time course of growth and enzyme synthesis by *A. oryzae* strain FQB-01 grown in media prepared with brewery wastewaters (BW) supplemented with different initial concentrations of soluble starch (2: 0, □: 10, O: 20; Δ: 30; ◇: 40 g/L). *TS*: total sugars; *TAA*: total amylolytic activity. The cultures were carried out at 30°C/96 h in an orbital shaker at 200 rpm.

Thus, the maximum *TAA* levels obtained in the media supplemented with 10, 20, 30 and 40 of g of starch/L of medium were 81, 102, 113 and 116 EU/mL in BW medium and 78, 98, 107 and 111 EU/mL in MPW medium. As it can be noted, increasing the *TS* concentration from 30 to 40 g/L did not produce a significant increase in *TAA* concentrations in both MPW and BW media (Figures 1 and 2).

From the comparison of the results obtained in both media, it can be noted that the maximum *TAA* obtained in BW medium (116 EU/mL) was slightly higher than in MPW (111 EU/mL). In contrast, biomass concentration in MPW medium (3.47 g/L) was slightly higher than that in BW medium (3.32 g/L). Taking into account these results, it can be concluded that the biomass produced in BW was more productive than that in MPW, being the yields of enzyme production (EU on g of cell dry weight) calculated as 34939.76 and 31988.47 EU/g of cell dry weight, respectively.

During fermentation in both the starch-supplemented MPW and BW media, the pH initially dropped from 6.0 to 4.0 after 24 h followed by its increase to 7.0 approximately (60 h), and remained constant thereafter. The *TS* concentration decreased during fermentation in parallel with the increase in biomass and enzyme production. Final COD values of approximately 0.16 g/L or lower were obtained in the media supplemented with 40 g of starch/L of medium and in the media containing lower initial starch concentrations, respectively. This indicates that at least 95% of the initial COD were removed from BW and MPW media.

From a comparative point of view, it can be noted that the *TAA* concentrations produced by *A. oryzae* FQB-01 were considerably higher than those produced by *Aspergillus niger* UO-01 in both BW (70 EU/mL) and MPW (60 EU/mL) media (Salas et al., 2006).

Table 4. Comparison of the observed and the predicted responses (remaining amylase activity) after *pH* and temperature treatment and significance analysis of the model coefficients

Natural values		Remaining TAA (%)	
T	*pH*	Observed response	Predicted value
60	7.0	58.94	*59.04*
60	4.0	39.14	*37.92*
30	7.0	43.16	48.49
30	4.0	28.04	27.37
64	5.5	53.85	55.03
26	5.5	45.06	41.67
45	7.4	75.65	71.66
45	3.6	43.12	44.90
45	5.5	89.63	94.44
45	5.5	92.34	94.44
45	5.5	100.00	94.44
45	5.5	94.00	94.44
45	5.5	95.36	94.44

Significance analysis of coefficients

Factor	*Coefficient*	*t*-value
Constant	**94.4**	**55.34***
T	5.3	3.67*
pH	10.6	7.36*
$T{\cdot}pH$	1.2	0.61
T^2	-28.7	16.92*
pH^2	-22.5	13.28*

* Significant at t ($\alpha < 0.05$; $\upsilon = 4$) > 2.78.

Effect of pH and Temperature on the Stability of Enzymes

Samples of amylase previously produced in BW medium supplemented with 40 g of starch/L, were used to test the effect of temperature and pH on the enzyme activity using a second order orthogonal design. Table 4 summarizes the responses for each individual experiment along with the predicted responses.

The model equation fitted by regression analysis is given by:

$$TAA\,(\%) = 94.4 + 5.3T + 10.6pH - 28.7T^2 - 22.5pH^2 \quad (1)$$

where TAA is the remaining total amylolytic activity and T is the temperature. The model terms T, pH, T^2 and pH^2 were found to be significant according to the Student t-test ($\alpha <$ 0.05), meanwhile the interaction term between T and pH variables ($T{\cdot}pH$) was found to be non significant (Table 4).

The equation (1) was significant in Fisher`s F-tests ($\alpha < 0.05$) applied to both quotients total error/experimental error and lack of fitting/experimental error (Table 5). In addition, the value of the adjusted determination coefficient (adj. r^2) was calculated to be 0.977 (Table 5).

These results indicate that there is a statistically significant relationship between the variables at the 95% confidence level as well as an excellent adequacy of the quadratic models to the experimental data.

The response surface obtained from the empirical model (1) is depicted in Figure 3. Optimal residual amylase activity (95 %) was at pH 5.8 (coded value of 0.23) and 46.4°C (coded value of 0.09), with visible decrease toward low or high values of both variables (Figure 3).

Table 5. Significance analysis of the model (1) obtained for remaining amylase activity after treatment of *pH* and temperature

	SS	*fd*	*QM*
Model	7869.7	4	1967.4
Error	121.8	8	15.2
Exp. error	59.4	4	14.8
Lack of fitting	62.5	4	15.6
Total	7991.5	12	1982.6
QME/QMEe = 1.03		$F^8{}_4(\alpha = 0.05) = 6.04$	
QMLF/QMEe = 1.05		$F^4{}_4(\alpha = 0.05) = 6.39$	
$r^2 = 0.985$		adj $r^2 = 0.977$	

SS: Sum of Squares; df: degrees of freedom; QM: Quadratic Means; E: total error; Ee: Experimental error; LF: Lack of Fitting.

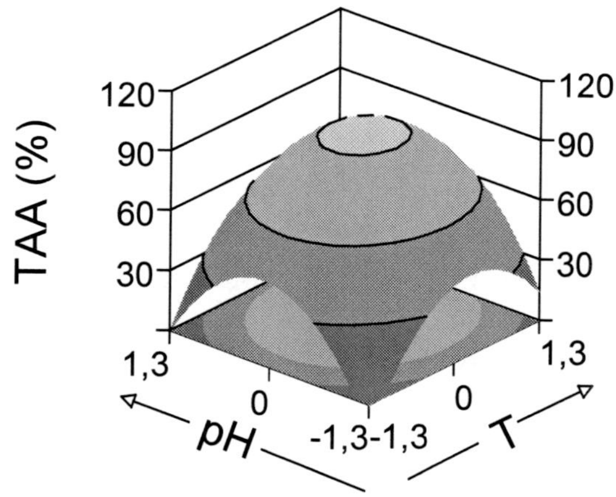

Figure 3. Response surface showing the influence of pH and temperature on the total amylolytic activity (*TAA*) produced by *A. oryzae* FQB-01 in BW medium. *T* and *pH* are in coded values.

On the other hand, when amylase samples were adjusted at optimum pH value of 5.8 and incubated at 40, 50, 60 and 70°C in absence of substrate (starch), the enzyme retained approximately 85% of its initial activity after 1 h of incubation at 40 and 50°C (left part of Figure 4). However, at 60 and 70°C the initial amylase activity was completely lost after 50 and 30 min of incubation, respectively.

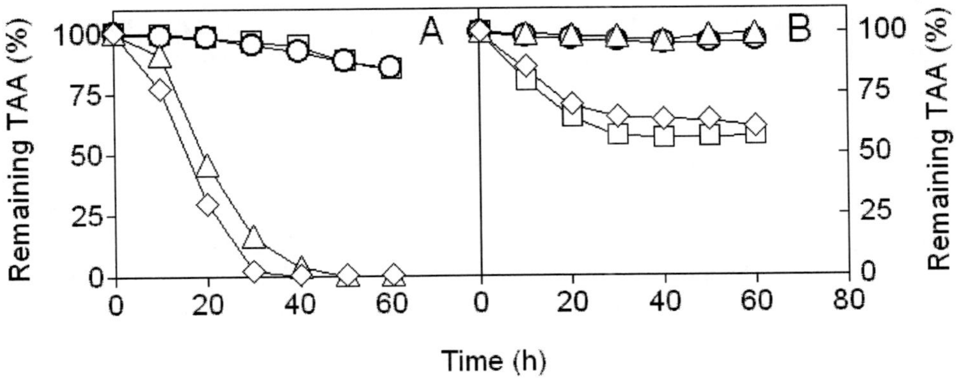

Figure 4. Stability of the total amylolytic activity (*TAA*) produced by *A. oryzae* FQB-01 in BW medium in absence of substrate (starch). A: The amylase samples were buffered at pH 5.8 with 0.1 M potassium hydrogen phthalate-HCl buffer and incubated at 40°C (□), 50°C (O), 60°C (Δ) and 70°C (◇). B: The enzyme preparation was buffered with 0.1 M potassium hydrogen phthalate-HCl buffer (at pH values of 4 (□), 5 (O) and 6 (Δ)) and with 0.1 M sodium phosphate buffer (at pH value of 7 (◇)) and incubated at 46.4°C.

In the same way, when amylase samples were adjusted at pH values of 4.0, 5.0, 6.0 and 7.0 and incubated at 46.4°C for 1h in absence of starch, the highest stabilities were observed at pH values of 5.0 and 6.0 (right part of Figure 4). In these cases, the enzyme retained 96 and 99% of its initial activity, respectively. In contrast, the enzyme lost about 42% and 39% of its initial activity when it was incubated at pH values of 4.0 and 7.0, respectively (right part of Figure 4).

Optimization of *TAA* Production in Solid State Fermentation (SSF)

Particle size, temperature, pH, initial moisture content of support material, inoculum size and starch concentration are important factors that influence amylase production by *A. oryzae* strains in SSF (Gigras et al., 2002; Murado et al., 1997; Francis et al., 2003).

However, the effects of these variables on enzyme production depend on the strain and the solid substrate or the support material used. For example, the use of a large particle size (14 mm) support increased penicillin production by 37% in SSF using bagasse as an inert support (Barrios-González et al., 1993). On the contrary, optimal particle size, moisture level and inoculum concentration for α-amylase production in SSF by a strain of *Penicillium chrysogenum* were >1 mm, 75% and 20% for corncob leaf, >1 mm, 65% and 20% for wheat straw, 1 mm, 65% and 20% for wheat bran and >1 mm, 55% and 30% for rye straw (Balkan & Ertan, 2007). However, an initial moisture content (70%), inoculum rate (1×10^7 spores/gds) and an incubation temperature in the range of 25-30°C were necessary to achieved the maximum α-amylase production by *A. oryzae* NRRL 6270 on spent brewing grains (Francis et al., 2002). The optimal moisture level for glucoamylase production by *Aspergillus* sp. A3 in wheat bran was 80% (Ellaiah et al., 2002), while for α-amylase and glucoamylase production by *A. niger* LPB 28 (Spier et al., 2006) in cassava starch and sugar cane bagasse was 90%.

Taking into account these observations, *TAA* production by *A. oryzae* FQB-01 under SSF using sugar cane bagasse as a support material was optimized by determining the optimal particle size, temperature, pH, inoculum size, initial moisture content and starch concentration in the solid support.

The first optimization experiment was performed to determine the influence of different sugar cane particle sizes (<1.0, 1.0-2.0, 2.0-5.0, 5.0-10.0 and 10.0-15.0 mm) on amylase production by *A. oryzae* FQB-01 in SSF. The results obtained are shown in Figure 5. Maximum enzyme production (191 EU/g of dry support) was observed when the particle size of bagasse was in the range of 5.0-10.0 mm, decreasing for low of high particle sizes. Thus, this particle size probably provides the most effective support material for attachment of the fungal strain (Laxmi et al., 2008), or ***facilitates*** the ***mass transfer*** performance (gas and nutrient diffusion) greatly. This provides better respiration/aeration efficiency (Pandey et al., 1999; Balkan & Ertan, 2007) and an increased availability of nutrients (Mazutti et al., 2007). These facts favour growth and enzyme production (Pandey et al., 1999; Balkan & Ertan, 2007). In subsequent experiments, therefore, a particle size between 5.0-10.0 mm was used for the production of amylase by *A. oryzae* FQB-01.

After optimizing the particle size of bagasse, the following experiment was focused on the determination of the optimal pH and incubation temperature for high amylase production. Table 6 summarizes the response for each individual experiment along with the calculated response and the significance analysis of the model terms. Table 7 shows the significance analysis of the empirical model (2), which gives the level of *TAA* as a function of *pH* and temperature. The value of the determination coefficient (0.940) suggested that the fitted model could explain 94% of the total variation. Thus, it can be concluded that the quadratic model was appropriate to fit satisfactorily the experimental data.

The *TAA* production could be therefore predicted by the model:

$$TAA \text{ (EU/gds)} = 185.5 + 16.8T + 32.9pH - 34.4T^2 - 29.8pH^2 \quad (2)$$

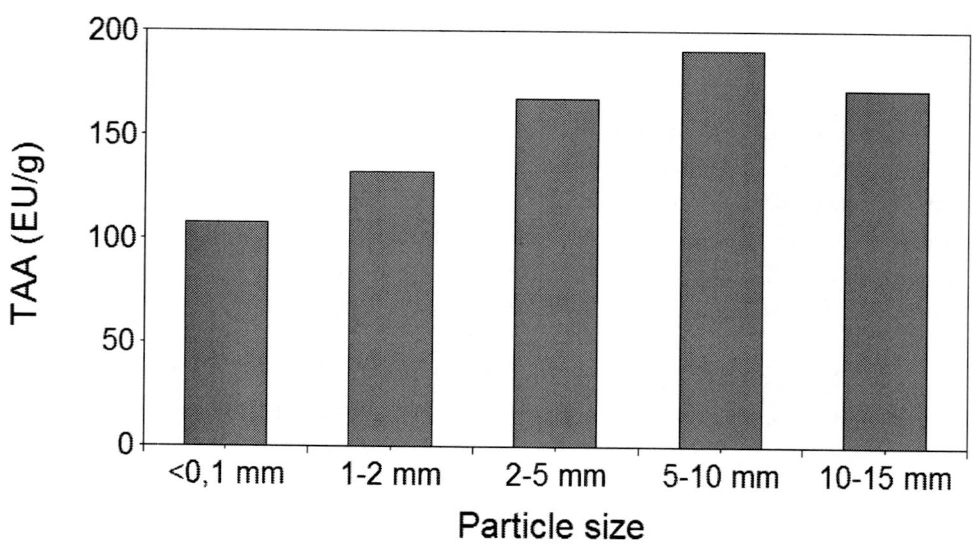

Figure 5. Effect of particle size of the sugar cane bagasse on the production of total amylolytic activity (*TAA*) by *A. oryzae* FQB-01 in solid state fermentation.

Table 6. Results of the experimental design showing both the experimental and the calculated values of *TAA* production levels in response to *pH* and temperature of incubation and significance analysis of the model coefficients

Natural values		*TAA* (EU/gds)	
T	*pH*	Observed response	Predicted value
37.0	6.5	157.90	170.95
37.0	4.0	99.09	105.15
25.0	6.5	129.01	137.41
25.0	4.0	59.64	71.61
38.6	5.3	163.25	151.46
23.4	5.3	121.76	108.96
31.0	6.8	193.02	179.37
31.0	3.7	106.94	96.00
31.0	5.3	176.59	185.47
31.0	5.3	181.94	185.47
31.0	5.3	197.03	185.47
31.0	5.3	183.01	185.47
31.0	5.3	179.09	185.47

Significance analysis of coefficients

Factor	*Coefficient*	*t*-value
Constant	185.5	52.66*
T	16.8	5.66*
pH	32.9	11.11*
T·pH	-2.6	0.66
T^2	-34.4	9.83*
pH^2	-29.8	8.50*

* Significant at t ($\alpha < 0.05$; $\upsilon = 4$) > 2.78.

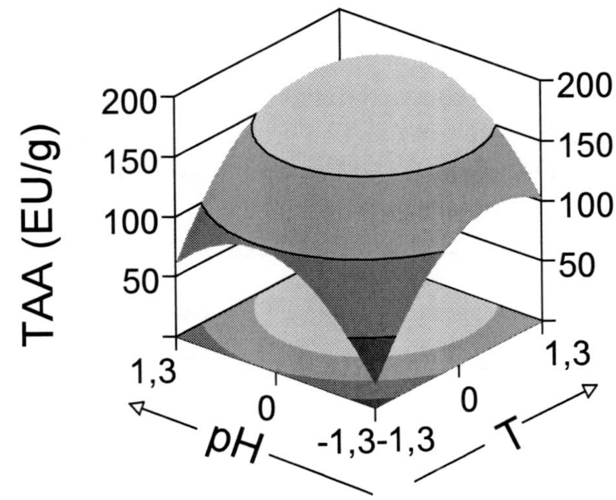

Figure 6. Response surface showing the combined effect of temperature (*T*) and *pH* on the production of total amylolytic activity (*TAA*) by *A. oryzae* FQB-01 in solid state fermentation. *T* and *pH* are in coded values.

The three-dimensional response surface (Figure 6) shows that the maximum *TAA* (197 EU/gds) was obtained at pH 5.9 (coded value of 0.55) and 32.5°C (coded value of 0.24). Thus, *A. oryzae* exhibited its best performance for enzyme production in the mesophilic range (Figure 6), as it was reported before for *A. oryzae* NRRL 6270, which produce the highest α-amylase level in the range from 25 to 30°C (Francis et al., 2003).

High temperature values may lead to enzymatic inactivation (Mazutti et al., 2006) or suppression of cell viability and death (Pandey et al., 2001; Francis et al., 2002), meanwhile low temperature values may reduce the metabolism of the microorganism (Mazutti et al., 2006) and consequently, the enzyme production.

Table 7. Significance analysis of the model (2) obtained for TAA production under different initial pH and temperature conditions

	SS	*fd*	*QM*
Model	20511.9	4	5128.0
Error	1301.2	8	162.6
Exp. error	252.8	4	63.2
Lack of fitting	1048.4	4	262.1
Total	21813.1	12	1817.7
QME/QMEe = 2.57		$F^8{}_4\,(\alpha = 0.05) = 6.04$	
QMLF/QMEe = 4.15		$F^4{}_4(\alpha = 0.05) = 6.39$	
$r^2 = 0.940$		adj $r^2 = 0.911$	

SS: Sum of Squares; df: degrees of freedom; QM: Quadratic Means; E: total error; Ee: Experimental error; LF: Lack of Fitting.

In this study, it was verified that *A. oryzae* FQB-01 has a preference to pH around 6.0 for *TAA* production. However, its production capacity decreased for pH levels higher and lower than the optimum pH probably as a consequence of a reduction in the metabolic activity of the amylase-producing strain at these pH values (Ellaiah et al., 2002).

Table 8. Results of the experimental design showing both the experimental and the calculated values of TAA levels in function of the total sugars (TS in mg/gds), moisture content (M in %) and inoculum size (IS) as log (number of spores/gds) and significance analysis of the model coefficients

Natural values			***TAA* (EU/gds)**	
M	***IS***	***TS***	**Observed response**	**Predicted value**
80	8.0	60.0	470.3	461.0
80	8.0	20.0	347.1	328.1
80	5.0	60.0	400.2	373.1
80	5.0	20.0	218.6	240.2
63	8.0	60.0	342.6	333.2
63	8.0	20.0	228.6	265.7
63	5.0	60.0	243.4	245.3
63	5.0	20.0	183.3	177.8
59	6.5	40.0	189.8	173.0

Table 8. Continued

Natural values			*TAA* (EU/gds)	
M	***IS***	***TS***	**Observed response**	**Predicted value**
84	6.5	40.0	296.9	318.0
72	4.3	40.0	245.0	249.8
72	8.8	40.0	384.4	383.8
72	6.5	9.5	386.1	361.0
72	6.5	70.5	487.8	513.8
72	6.5	40.0	451.1	437.4
72	6.5	40.0	428.7	437.4
72	6.5	40.0	455.6	437.4
72	6.5	40.0	437.6	437.4
72	6.5	40.0	424.2	437.4
72	6.5	40.0	426.8	437.4

Significance analysis of coefficients

Factor	*Coefficient*	*t*-value
Constant	437.0	81.83*
M	47.5	12.71*
IS	43.9	11.74*
TS	50.1	13.39*
$M{\cdot}IS$	6.8	1.44
$M{\cdot}TS$	16.3	3.47*
$IS{\cdot}TS$	-0.6	0.12
M^2	-82.5	20.39*
IS^2	-51.9	12.81*
TS^2	0.7	0.18

* Significant at t ($\alpha < 0.05$; $\upsilon = 5$) > 2.57.

Taking into account the results obtained in this second optimization experiment, a pH value of 5.9 and a temperature of 32.5°C were used for further optimization studies.

The following study was focused on the determination of the effects of three factors (total sugars concentration, inoculum size and moisture content) on amylase production in the optimal conditions calculated before (particle size, pH and temperature) by using a new second-order orthogonal design. Table 8 summarizes the experimental and predicted responses along with significance of the model terms.

With the model coefficients acceptable according to the Student t test ($\alpha < 0.05$), the equation obtained is significant in Fisher´s F tests ($\alpha = 0.05$) applied to the quotients total error/experimental error and lack of fitting/experimental error (Table 9). The value of the coefficient of determination (0.969) ensures that the model fits satisfactorily the data and indicates that 96.9% of the variability in *TAA* production could be explained by the empirical model (3):

$$TAA\ (\text{EU/gds}) = 437.0 + 47.5M + 43.9IS + 50.1TS + 16.3M{\cdot}TS - 82.5M^2 - 51.9IS^2 \quad (3)$$

where *TAA* is the total amylolytic activity, *M* is the moisture content, *IS* is the inoculum size and *TS* is the total sugars concentration. Among the most notable characteristics of the system studied, it should be stated that for enzyme production, all the independent variables (*M*, *IS* and *TS*) and the binary interaction between *M* and *TS* had positive coefficients. In contrast, the coefficients of the quadratic terms M^2 and IS^2 had negative sign and the quadratic term TS^2 as well as the binary interactions *M·IS* and *IS·TS* were found to be non significant (Table 8). The significant coefficients for the quadratic terms (M^2 and IS^2) implied the existence of optimum values of *M* and *IS* inside the experimental domain. Since the quadratic term TS^2 was found to be non-significant and the linear term *T* had a positive coefficient, the optimum value for this variable corresponds to its highest value.

Table 9. Significance analysis of the model (3) obtained for production of TAA in function of the variables M, IS and TS

	SS	*fd*	*QM*
Model	189619.7	6	31603.3
Error	6087.2	13	468.2
Exp. error	885.6	5	177.1
Lack of fitting	5201.6	8	650.2
Total	195707.4	19	32071.5
QME/QMEe = 2.64		$F^{13}{}_5\,(\alpha = 0.05) = 4.66$	
QMLF/QMEe = 3.67		$F^{8}{}_5\,(\alpha = 0.05) = 4.82$	
$r^2 = 0.969$		adj $r^2 = 0.955$	

SS: Sum of Squares; df: degrees of freedom; QM: Quadratic Means; E: total error; Ee: Experimental error; LF: Lack of Fitting.

The response surfaces curves generated according to model (3) show the relative effect of two variables on *TAA* production, when the coded value of the third variable is kept at a constant level of +1 or –1 (Figures 7A, B and C). For example, Figure 7A illustrates the effect of inoculum size and moisture content on *TAA* production at constant starch levels. As it can be observed, the *TAA* concentration increased (for low and high values of *TS*) with the increase in moisture content and inoculum size in the range of –1.267 to 0.44 and –1.267 to 0.42, respectively. Any further increase in the levels of *M* and *IS* led to a sharp decline in *TAA* production. At constant levels of *M* (Figure 7B), a quadratic effect of *IS* on *TAA* production was observed in contrast with the linear effect produced by *TS* on the response. Similarly, Figure 7C shows a quadratic effect of *M* as well as a linear effect of *TS* on *TAA* production for low (–1) and high (+1) values of *IS*.

Thus, a moisture content of 75% (coded value = 0.44), an inoculum size of 1.4×10^7 spores/gds (coded value = 0.42) and starch concentration of 70.5 mg/gds (coded value = 1.525) were the best conditions to produce the high amounts of enzyme (539 EU/gds) with *A. oryzae* FQB-01 on sugar can bagasse.

The effect of initial moisture content on enzyme production in SSF has usually been related to the interference that this variable produces on the physical properties of the solid particles (Ellaiah et al., 2002). Higher moisture content could reduce the porosity of the solid substrate or support material, thus interfering with the oxygen transfer (Feniksova et al., 1960; Lonsane et al., 1985). Lower moisture levels can lead to a reduction in the solubility of

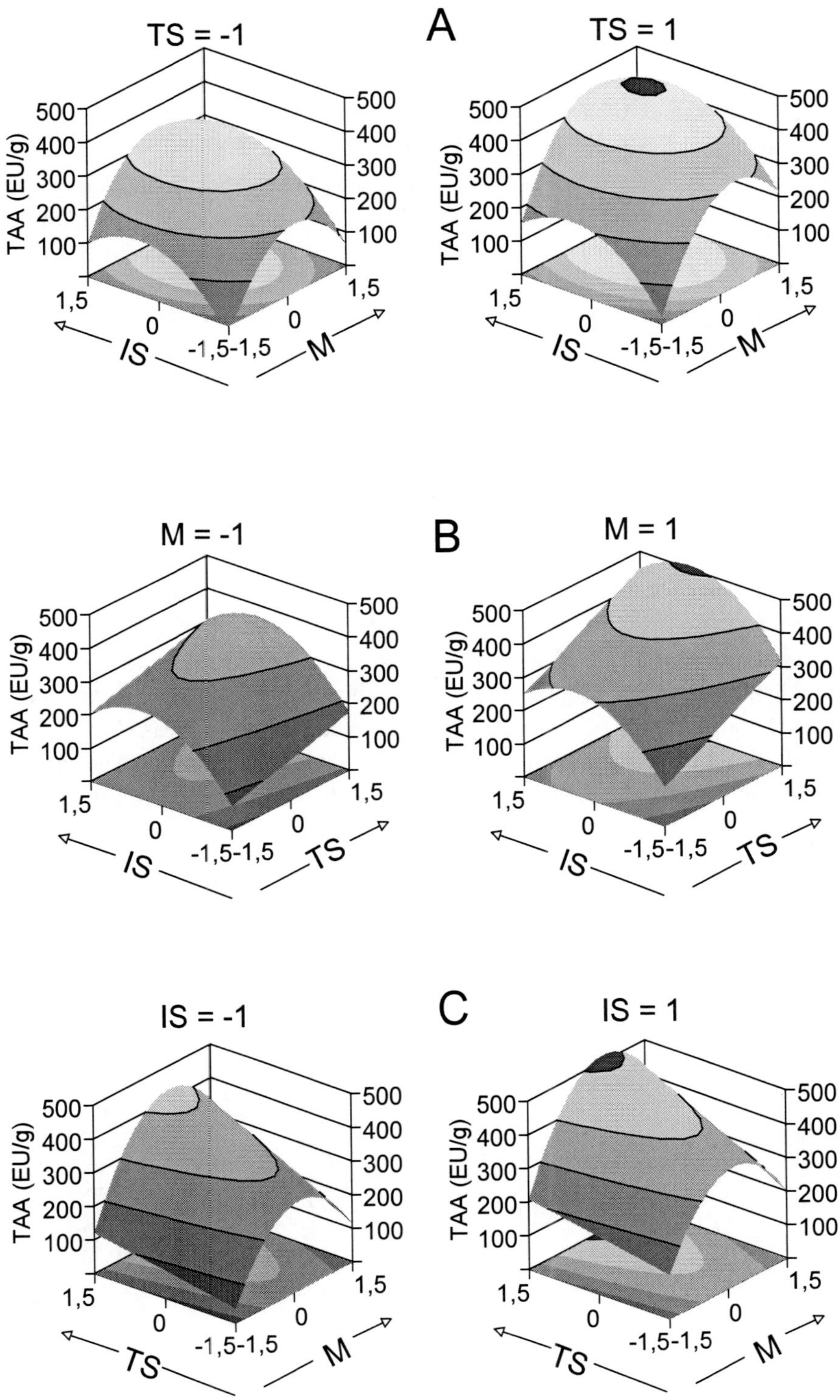

Figure 7. Response surfaces showing the effect of moisture content (*M*), inoculum size (*IS*) and total sugars (*TS*) on the production of total amylolytic activity (*TAA*) by *A. oryzae* FQB-01 in solid state fermentation. Values of *TS*, *M* and *IS* are in coded values.

nutrients of the substrate, low degree of swelling and high water tension (Zadrazil & Brunnert, 1981). These facts explain the presence of the positive interaction between the variables *M* and *TS* in model (3). In fact, the positive effect of variable *M* was more clearly observed for high values of *TS* and similarly, the positive effect of variable *TS* on *TAA* production was higher for high values of *M* (Figure 7A, B and C).

Inoculum size is other important factor that influences the production of metabolites under solid state fermentation (Pandey, 2003; Barrios-González et al., 1993; Mazutti et al., 2007; Balkan & Ertan, 2007). Increase in inoculum size leads to an increase in the moisture content in the solid substrate or support material, which could represent an additional diffusional barrier together with that imposed by the solid particles of the substrate or support material. In addition, a higher inoculum size may produce too much biomass and rapidly deplete the nutrients necessary for growth and product synthesis (Selvakumar & Pandey, 1999). This limits the growth and enzyme production (Baysal et al., 2003). On the other hand, lower inoculum levels may give insufficient biomass and allow the growth of undesirable organisms in the production medium. This increases the necessary time to grow to an optimum number to consume the substrate and synthesise the desired product (Kashyap et al., 2002; Balkan and Ertan, 2007).

From the comparison between the results obtained in submerged (Figures 1 and 2) and solid state fermentation (Figure 7), it can be noted that the levels of *TAA* ceased to increase from a starch concentration of 30 g/L in the submerged cultures (Figure 8A), while in SSF this effect was not observed (Figure 8B). In fact, the empirical model (3) predicted a linear increase in the concentration of *TAA* when the *TS* concentrations increased from 9.5 to 70.5 g *TS*/gds, for different combinations of the variables *IS* and *M* (Figure 8B). The same trend was observed for amylase production by *A. oryzae* CBS 125-59 by SSF on polyurethane foams (Murado et al., 1997).

Thus, the inhibition by substrate in SSF, similar to that described for the production of gibberellic acid by *Gibberella fujikuroi* (Kumar & Lonsane, 1987) was not observed for *TAA* production by *A. oryzae* FQB-01 in SSF on sugar cane bagasse.

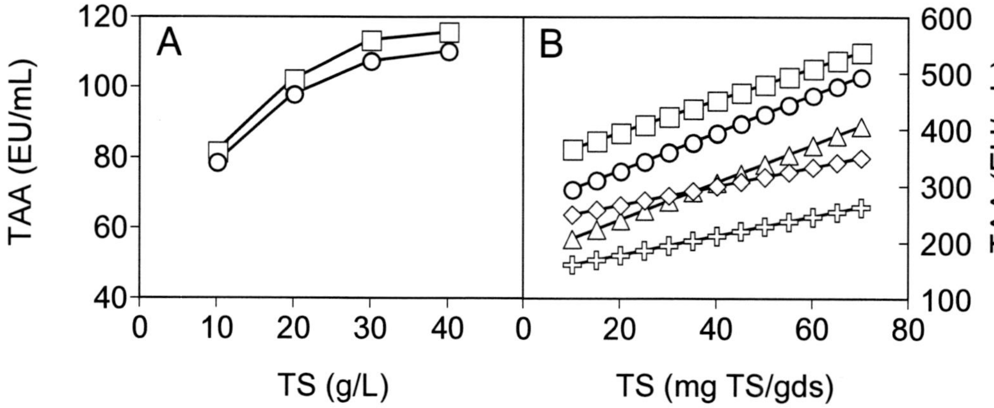

Figure 8. Effect of *TS* concentration on the production of total amylolytic activity (*TAA*) in submerged fermentation (A) on BW (□) and MPW (O) and in solid state fermentation (B) when the levels of the other independent variables are: *M* = 0.44, *IS* = 0.42 (□); *M* = +1.0, *IS* = +1.0 (O); *M* = +1.0, *IS* = -1.0 (◇); *M* = -1.0, *IS* = +1.0 (Δ) and *M* = -1.0, *IS* = -1.0 (2).

Conclusions

Two industrial wastes (brewery and meat processing wastes) and an agro-industrial residue (sugar cane bagasse) showed a great potential as substrates and support material, respectively for *TAA* production by *A. oryzae* FQB-01 at low production costs. In addition, the organic load of the two wastes is substantially reduced at the end of the fermentations.

Increased *TAA* productions (107 and 113 EU/mL) were respectively obtained in the MPW and BW media supplemented with 30 g of starch/L of medium. A further increase in the initial starch concentration in the medium (40 g of starch/L) did not improve significantly *TAA* production in both media (111 and 116 EU/mL, respectively).

The maximum *TAA* concentration (539 EU/gds) in SSF was achieved with optimized process parameters such as particle size of bagasse (in the range of 5-10 mm), fermentation temperature (32.5°C), pH (5.9), moisture content of bagasse (75%), starch concentration (70.5 mg/gds) and inoculum size (1.4×10^7 spores/gds).

However, other studies based on the use of carbon sources from agro-industrial residues (wheat bran, rice bran, rice husk, maize bran, etc) for *TAA* production are necessary to continue optimizing the production of the enzyme in submerged and solid state fermentations.

References

Aidoo, K. E.; Hendry, R. & Wood, B. J. B. (1981). Estimation of fungal growth in a solid state fermentation system *Eur. J. Appl. Microbiol. Biotechnol.*, **12**, 6-9.

Balkan, B. & Ertan, F. (2007). Production of α-amylase from *P. chrysogenum. Food Technol. Biotechnol.*, **45**, 439–442.

Barrios-González, J.; González, H. & Mejía, A. (1993). Effect of size particle, packing density and agitation on penicillin production in solid state fermentation. *Biotechnol. Adv.*, **11**, 539–547.

Baysal, Z.; Uyar, F. & Aytekin, C. (2003). Solid state fermentation for production of α-amylase by a thermotolerant *Bacillus subtilis* from hot-spring water, *Process Biochem.*, **38**, 1665–1668.

Bernfeld, P. (1951). Enzymes of starch degradation and synthesis. *Adv. Enzymol.*, **12**, 397-427.

Box, G. E. P.; Hunter, W. G. & Hunter, J. S. (1989). In Reverte, S.A. (Publ.), *Estadística para investigadores* (pp. 317-361), Barcelona, España.

Cannel, E. & M. Moo-Young. (1980). Solid state fermentation system. *Process Biochem.*, **15**, 2-7.

Doran, J. B.; Aldrich, H. C. & Ingram, L. O. (1994). Saccharification and fermentation of sugarcane bagasse. *Biotechnol. Bioeng.*, **44**, 240–247.

Ellaiah, P.; Adinarayana, K.; Bhavani, Y.; Padmaja, P. & Srinivasulu, B. (2002). Optimization of process parameters for glucoamylase production under solid state fermentation by a new isolated *Aspergillus* species. *Process Biochem.*, **38**, 615-620.

Feniksova, R. V.; Tikhomirova, A. S. & Rakhleeva, E. E. (1960). Conditions for forming amylase and proteinase in surface cultures of *Bacillus subtilis*. *Microbiol.*, **29**, 745–748.

Francis, F.; Sabu, A.; Nampoothiri, K. M.; Ramachandran, S.; Ghosh, S.; Szakacs, G. & Pandey, A. (2003). Use of response surface methodology for optimizing process parameters for the production of α-amylase by *Aspergillus oryzae*. *Biochem. Eng. J.*, **15**, 107-115.

Francis, F.; Sabu, A.; Nampoothiri, K. M.; Szakacs, G. & Pandey, A. (2002). Synthesis of α-amylase by *Aspergillus oryzae* in solid-state fermentation. *J. Basic Microbiol.*, **5**, 320-326.

Gigras, P.; Sahai, V. & Gupta, R. (2002). Statistical media optimization and production of ITS α-amylase from *Aspergillus oryzae* in a bioreactor. *Curr. Microbiol.*, **45**, 203-208.

Greenberg, A. E.; Connors, J. J. & Jenkins, D. (1980). Standard methods for the examination of water and wastewater. 15th ed. American Public Health Association, Washington, D.C.

Guerra, N. P. & Pastrana, L. (2003). Enhancement of nisin production by *Lactococcus lactis* in periodically re-alkalized cultures. *Biotechnol. Appl. Biochem.*, **38**, 157-167.

Guerra, N. P.; Torrado A.; López, C. & Pastrana, L. (2005). Modelling the fed-batch production of pediocin using mussel-processing wastes. *Process Biochem.*, **40**, 1071-1083.

Guerra, N. P.; Torrado-Agrasar, A.; López-Macías, C. & Pastrana, L. (2003). Main characteristics and applications of solid substrate fermentation. *Electron. J. Environ. Agricult. Food Chem.*, **2**, 343-350.

Gutiérrez-Correa, M. & Tengerdy, R. P. (1997). Production of cellulase on sugar cane bagasse by fungal mixed culture solid substrate fermentation. *Biotechnol. Lett.*, **19**, 665–667.

Hesseltine, C. W. (1972). Solid state fermentations. *Biotechnol. Bioeng.*, **14**, 517-532.

Kashyap, P.; Sabu, A.; Pandey, A. & Szakacs, G. (2002). Extra-cellular l-glutaminase production by *Zygosaccharomyces rouxii* under solid-state fermentation. *Process Biochem.*, **38**, 307–312.

Krishna, C. & Chandrasekaran, M. (1996). Banana waste as substrate for α-amylase production by *Bacillus subtilis* (CBTK 106) under solid-state fermentation. *Appl. Microbiol. Biotechnol.*, **46**, 106-111.

Kumar, P. K. R. & Lonsane, B. K. (1987). Potential of fed-batch culture in solid state fermentation for production of gibberellic acid. *Biotechnol. Lett.*, **9**, 179-182.

Kumar, P. R. K. & Lonsane, B. K. (1990). Solid state fermentation: physical and nutritional factors influencing gibberellic acid production *Appl. Microbiol. Biotechnol.*, **34**, 145-148.

Kumaran, S.; Sastuy, C. A. & Vikineswary, S. (1997). Laccase, cellulose, xylanase activities during growth of *Pleurotus sojor-caju* on sago hampas. *World. J. Microb. Biotechnol.*, **13**, 43-49.

Laxmi, G. S.; Sathish, T.; Rao, Ch. S., Brahmaiah, P.; Hymavathi, M. & Prakasham, R. S. (2008). Palm fiber as novel substrate for enhanced xylanase production by isolated *Aspergillus* sp. RSP-6. *Curr.Trends Biotechnol. Pharm.*, **2**, 447-455.

Lonsane, B. K.; Ghildyal, N. P.; Budiatman, S. & Ramakrishna, S. V. (1985). Engineering aspects of solid state fermentation. *Enzyme Microb. Technol.*, **7**, 258–265.

Mazutti, M.; Ceni, G.; Di Luccio, M. & Treichel, H. (2007). Production of inulinase by solid-state fermentation: effect of process parameters on production and preliminary characterization of enzyme preparations. *Bioprocess Biosyst. Eng.*, **30**, 297-304.

Mazutti, M.; Bender, J. P.; Treichel, Helen & Di Luccio, M. (2006). Optimization of inulinase production by solid-state fermentation using sugarcane bagasse as substrate. *Enzyme Microb. Technol.*, **39**, 56-59.

Milagres, A. M. F.; Santos, E.; Piovan, T. & Roberto, I. C. (2004). Production of xylanase by *Thermoascus aurantiacus* from sugar cane bagasse in an aerated growth fermentor. *Process Biochem.*, **39**, 1387-1391.

Murado, M. A.; González, M. P.; Torrado, A. & Pastrana, L. M. (1997). Amylase production by solid state culture of *Aspergillus oryzae* on polyurethane foams. Some mechanistic approaches from an empirical model. *Process Biochem.*, **32**, 35-42.

Murado, M. A.; Siso, M. I. G.; González, M. P.; Montemayor, M. I.; Pastrana, L. & Mirón, J. (1993). Characterization of microbial biomasses and amylolytic preparations obtained from mussel-processing waste treatment. *Biores. Technol.*, **43**, 117-125.

Pandey, A. (2003). Solid state fermentation. *Biochem. Eng. J.*, *13*, 81–84.

Pandey, A.; Nigam, P.; Soccol, C. R.; Soccol, V.T.; Singh, D. & Mohan, R. (2000). Advances in microbial amylases. *Biotechnol. Appl. Biochem.*, **31**, 135-152.

Pandey, A.; Selvakumar, P.; Soccol, R. C. & Nigam, P. (1999). Solid state fermentation for the production of industrial enzyme. *Bioresour. Technol.*, **77**, 149-162.

Pandey, A.; Soccol, C. R.; Rodriguez-Leon, J. A. & Nigam, P. (2001). Solid State Fermentation in Biotechnology: Fundamentals and Applications. Asiatech Publishers, Inc., New Delhi.

Raimbault, M. (1998). General and microbiological aspects of solid substrate fermentation. *Electron. J. Biotechnol.*, *1*, 1-22. Available from http://www.ejb.org/content/vol1/issue3/full/9/index.html.

John, R. P.; Nampoothiri, K. M. & Pandey, A. (2006). Solid-state fermentation for L-lactic acid production from agro wastes using *Lactobacillus delbrueckii. Process Biochem.*, **41**, 759-763.

Roukas, T. (1999). Pullulan production from brewery wastes by *Aureobasidium pullulans. World J. Biotechnol.*, **15**, 447-450.

Salas, M.; Rodríguez, M.; Guerra, N. P. & Pérez R. (2006). Amylase production by *Aspergillus niger* in submerged cultivation on two wastes from food industries. *J. Food Eng.*, **73**, 93-100.

Selvakumar, P. & Pandey, A. (1999). Solid state fermentation for the synthesis of inulinase from *Staphylococcus* sp. and *Kluyveromyces marxianus*. *Process Biochem.*, **34**, 851–858.

Spier, M. R.; Woiciechowski, A. L.; Vandenberghe, L. P. S. & Zoclo, C. R. (2006). Production and characterization of amylases by *Aspergillus niger* under solid state fermentation using agro industrials products. *Int. J. Food Eng.*, **2.** Available at: http://www.bepress.com/ijfe/vol2/iss3/art6.

Steinkraus, K. H. (1984). Solid-state (solid-substrate) food/beverage fermentations involving fungi. *Acta Biotechnol.*, **4,** 83-88.

Yang, S. S. & Wang, J. Y. (1999). Protease and amylase production of *Streptomyces rimosus* in submerged and solid state cultivations. *Bot. Bull. Acad. Sin*, **40**, 259-265.

Zadrazil, F. & Brunnert, H. (1981). Investigation of physical parameters important for the SSF of straw by white rot fungi. *Eur. J. Appl. Microbiol. Biotechnol.*, **11**, 183–188.

In: Biochemical Engineering
Editors: F.E. Dumont and J.A. Sacco, pp. 207-226
ISBN: 978-1-60741-257-1

Chapter 8

MAMMALIAN CELL ENCLOSING CAPSULES AND FIBER PRODUCTION IN A CO-FLOWING AMBIENT LIQUID STREAM

Shinji Sakai*[*] *and Koei Kawakami
Department of Chemical Engineering, Kyushu University, Japan

Abstract

Mammalian cell-enclosing microcapsules have been investigated as devices for bioproduction, cell therapy and stem cell research. Reduction in the diameter of the vehicles is an important issue as it induces beneficial effects such as higher molecular exchangeability between the enclosed cells and the ambient environment, as well as higher mechanical stability and biocompatibility. In this chapter we describe the effectiveness of using a jetting process involving the formation of a stretched thin jet of aqueous polymer solution and its subsequent breakup into droplets in a co-flowing water-immiscible liquid for obtaining droplets of about 100 μm in diameter. The droplet production process and the processes for obtaining gelated microcapsules through a thermal and peroxidase-catalyzed gelation process are also described. In addition, we introduce the production of cell-enclosing hydrogel fibers using the same device developed for the production of cell-enclosing microcapsules.

Introduction

Mammalian cell-enclosing microcapsules have been studied for more than 40 years as research tools, as a device for the production of useful biomolecules such as antibodies and enzymes, and as therapeutic devices for use in cell therapy to treat a variety of diseases such as diabetes, liver failure, hemophilia and cancer [1-3]. More recently, this technology has attracted attention as a feasible tool for investigating a niche of the differentiation pathway of embryonic stem cells between the microenvironment *in vitro* and *in vivo* [4,5]. The injection

[*] E-mail address: sakai@chem-eng.kyushu-u.ac.jp. Corresponding Author. Department of Chemical Engineering, Faculty of Engineering, Kyushu University. 744 Motooka, Nishi-ku, Fukuoka 819–0395, Japan. Tel&Fax: +81-92-802-2768.

of aqueous polymer solution into an ambient co-flowing fluid stream has been used for the preparation of cell-enclosing microcapsules. The oldest but the most widely used ambient co-flowing fluid is air [6-11]. The system using air for producing spherical vehicles for use in biomedical applications has been employed since the 1960s (Figure 1a) [12]. In the first report for the successful transplantation of mammalian cells enclosed in microcapsules for treatment of diseases without dosing immunosuppressive drugs, the system using air as an ambient co-flowing stream was employed for obtaining pancreatic islets-enclosing microcapsules of 300-800 μm in diameter [13]. In the system using air, sodium-alginate is commonly used as a material for surrounding the cells because the aqueous solution of the polymer instantaneously becomes a hydrogel in the presence of multivalent cations, usually calcium, barium or strontium ions [14]. By injecting a cell-suspending sodium-alginate solution into the co-flowing air stream from a needle with a diameter of several hundred micrometers, the breakup into cell-enclosing droplets is caused by the drag force exerted by ambient air flow on the extruded alginate solution. Cell-enclosing calcium-alginate microcapsules can be obtained by collecting the alginate droplets in the aqueous solution containing multivalent cations. The size of the alginate microcapsules strongly depends on the flow rate of the ambient air [15] (the size of the needle from which the cell-suspending aqueous solution is extruded and the flow rate of the aqueous solution are also important factors). The system using air as an ambient co-flowing fluid is effective for obtaining microcapsules of 500-800 μm in diameter with a narrow distribution in size for enclosing pancreatic islets of 50-300 μm in diameter. However, one shortcoming of the system is that it is not possible to obtain droplets which are smaller than the diameter of the needle from which the cell-suspending aqueous polymer solution is extruded [15,16].

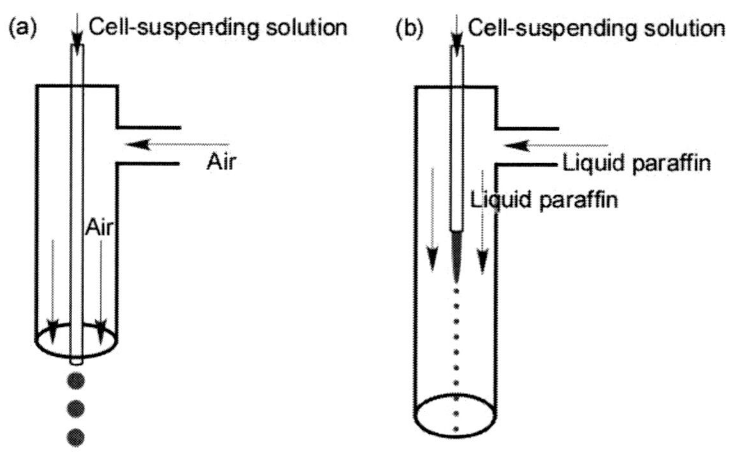

Figure 1. Schematic illustration of droplet production devices using (a) air and (b) liquid paraffin developed as an ambient co-flowing fluid.

Development of smaller microcapsules is a challenging issue in the field of cell encapsulation and is of interest as smaller microcapsules offer many advantages: Exchange of molecules such as oxygen and nutrients is theoretically better in smaller microcapsules [17,18]. Chicheportiche and Reach [17] revealed that the response time of encapsulated pancreatic islets to glucose stimulation decreased with a reduction in microcapsule size. Beneficial effects on activity of enclosed cells were also reported [19]. In addition, reduction in capsule size should allow the use of immune-privileged sites for implantation such as spleen capsules, omental pouch, or liver via the portal vein. Moreover, smaller capsules would reduce the size of the capsule injection device, thereby reducing surgical trauma [20]. It has also been reported that a reduction in microcapsules size suppressed foreign body response to the implanted microcapsules [21].

The reason for the impossibility in the production of droplets smaller than the needle diameter with a narrow distribution in size using air as an ambient fluid is due to the occurrence of fluctuations in the air flow under the flow rate necessary for obtaining such droplets. The fluctuations in air flow result in a heterogeneous drag force which causes the droplets to develop with a heterogeneous distribution in size [22]. For practical applications, the distribution in microcapsule size is an important issue because of the response time toward stimuli inducing secretions of therapeutic products from enclosed cells. Laminar flow with a high flow rate of the ambient co-flowing fluid is essential for the production of smaller droplets than the needle diameter to load a constant force to the cell-suspending aqueous polymer solution for breakup into droplets. From the hydrodynamics point of view, such laminar flow easily results from higher viscous fluid under the same flow rate. It means liquid is more effective for obtaining laminar flow of ambient fluid than gas. In this chapter, cell-enclosing microcapsules of about 100 μm in diameter were prepared using a needle with a diameter of several hundred micrometers with a narrow size distribution by injecting a cell-suspending solution into an ambient co-flowing liquid. The microcapsules produced are too small to enclose pancreatic islets but are large enough to encapsulate single cells, 10-30 μm in diameter. In addition, the novel process of gelation of the droplets in a water-immiscible liquid via an enzymatic reaction, and application of the microcapsule production method for developing cell-enclosing hydrogel fibers are also described based on our recent studies.

Droplets Production via Jetting Process

Using air as an ambient fluid, cell-enclosing microcapsules of about 100 μm in diameter and a narrow distribution in size can be obtained by extruding a cell-suspending solution from a nozzle with a width of several dozen micrometers [23,24]. Such small nozzles are produced by microfabrication techniques. Today, microfabrication techniques are widely known but practical usage of the techniques is less common. In addition, the risk of the nozzle clogging or plugging increases with increasing viscosity of the extruding solution [25] and decreasing nozzle diameter. Without complete dispersion, the suspended cells also become a risk resulting in nozzle clogging and plugging. We attempted to develop the cell-enclosing microcapsules of about 100 μm in diameter using medical syringe needles with a diameter of several hundred micrometers. The small microcapsule preparation technique involves a process of droplet breakup in a co-flowing water-immiscible liquid resulting in an emulsion system. Well-known emulsification techniques include use of a magnetic stirrer and a homogenizer [26,27]. These techniques are effective for obtaining droplets with a diameter of about 100 μm. However, the droplets are heterogeneous in size. In contrast, the droplets produced via the droplet breakup method in co-flowing ambient water-immiscible liquid in a laminar flow fashion show a narrow distribution in size (Figure 2) [28]. The device for the droplet breakup method is simple (Figure 1b) and almost the same structure as that which uses air as an ambient fluid (Figure 1a). The inner needle is positioned upstream in the vicinity of a coaxial outer tubule in which the water-immiscible liquid (liquid paraffin is used in our study) is flowed. Formation of the droplets smaller than the needle diameter is achieved at the tip of a stretched aqueous solution ligament much thinner than the needle diameter, that is, via jetting (Figure 3) [28]. When a disperse phase is injected from a capillary into an immiscible liquid flow at laminar flow, two different droplet breakup modes are observed. Specifically, droplets either form close to the capillary tip, called dripping, or break up at the tip of a stretched liquid jet, called jetting [29]. The formation of a jet thinner than the needle diameter is interpreted as a consequence of being stretched by the drag force exerted by the ambient co-flowing fluid. In general, the behavior of the extruded solution into the co-flowing fluid is affected by a complicated balance of three forces [30-32]: the interfacial tension force,

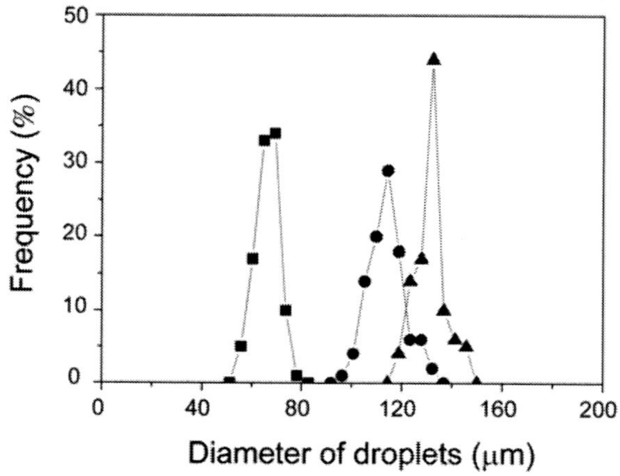

Figure 2. Size distribution of 2.0 wt% sodium-alginate extruded from a needle with 480 μm i.d., 700 μm o.d. at (■) 1.2 cm/sec, (●) 2.6 cm/sec, and (▲) 4.7 cm/sec into a co-flowing liquid paraffin stream with a flow rate of 17.5 cm/sec (Reproduced with permission, from Sakai S et al. Biotechol. Bioeng. [28] @ 2004 Wiley Periodicals, Inc.).

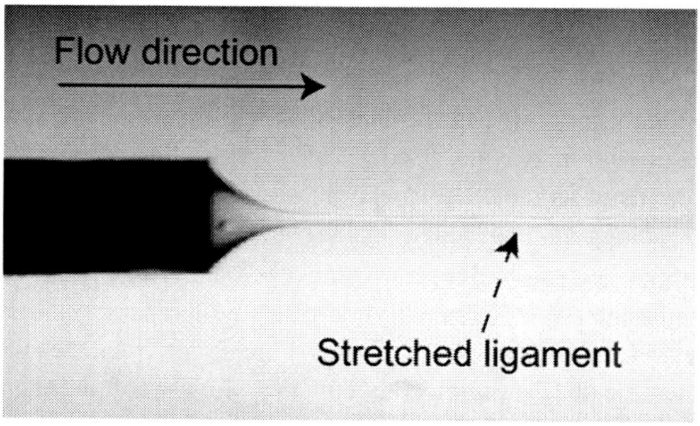

Figure 3. Sodium-alginate aqueous solution (194 mPa·s) extruded from a needle of 300 μm i.d. and 480 μm o.d. into the ambient liquid paraffin stream. The velocities of the sodium-alginate aqueous solution and liquid paraffin were 1.2 cm/sec and 23.5 cm/sec, respectively (Reproduced with permission, from Sakai S et al. Biotechol. Bioeng. [28] @ 2004 Wiley Periodicals, Inc.).

the drag force exerted by the ambient fluid and the viscoelastic force of the aqueous polymer solution. Under the appropriate balance of the forces, a stretched jet of aqueous polymer solution forms in the immiscible liquid. Subsequently, the jet becomes unstable due to the interfacial tension and fragments into individual droplets. The phenomenon of droplet breakup via jetting is explained by Rayleigh-Plateau hydrodynamic instability [33]. As shown in Figures 2 and 4, we can obtain smaller droplets by increasing the velocity of ambient co-flowing liquid, decreasing the velocity of aqueous polymer solution, and decreasing the diameter of needle [28,34]. It is easy to understand the effect of velocities of liquids and diameter of the needle on the diameter of the resultant droplets.

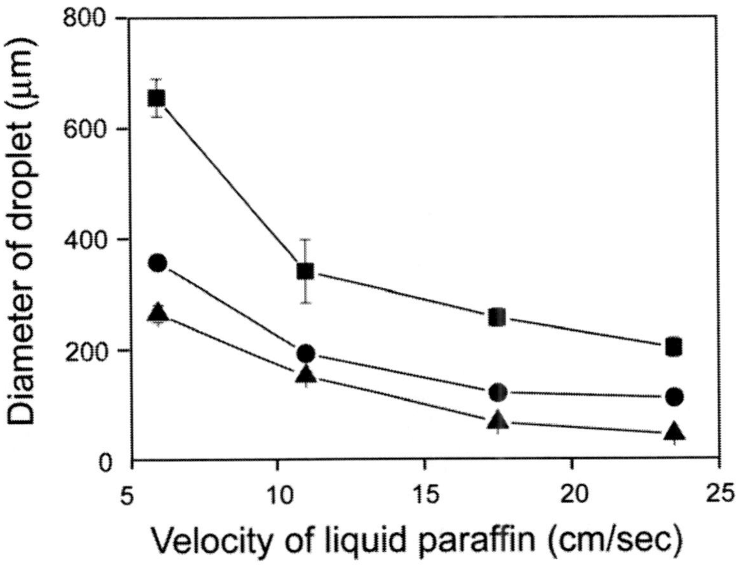

Figure 4. Droplet diameter as a function of liquid paraffin for three aqueous solution differing in viscosity: (■) 1.0, (•) 36, and (▲) 194 mPa·s. Aqueous solutions were extruded at a velocity of 1.2 cm/sec from a needle with a 300 μm i.d. Error bars represent SDs (Reproduced with permission, from Sakai S et al. Biotechnol Prog. [34] @ 2005 American Institute of Chemical Engineers).

In addition to these factors, viscosity of a polymer solution is also a factor influencing the diameter of the resultant droplets [34]. As shown in Figure 4, at the velocities of liquid paraffin and aqueous solution which were 6.0 cm/sec and 1.2 cm/sec, respectively, the droplets obtained from the aqueous solution of 1.0 mPa·s were the largest and their mean diameter was 655 μm. Droplet diameter decreased with the increasing viscosity of the aqueous solution. This tendency was kept even under conditions of a faster flow rate of liquid paraffin. At a liquid paraffin velocity of 23.5 cm/sec, the diameters of droplets were 202, 110, and 44 μm for aqueous solutions of 1.0, 36 and 194 mPa s. The smaller droplets generated from a higher viscous solution results from the formation of a thinner ligament from a higher viscous solution [34]. This means we can obtain the microcapsules with a higher mechanical strength resulting from a synergy of reduction in diameter and a tightening of gel microscopic structure using a smaller quantity of ambient fluid. The droplet breakup method using liquid paraffin as the ambient liquid also is advantageous because it only damages a small percentage of cells. The undamaged percentages of the cells enclosed in the droplets of 110 and 44 μm in diameter [defined as the viability of cells enclosed in the droplet]/[viability of the cells before extruding in co-flowing liquid paraffin × 100] were more than 95% (Figure 5a). In addition, cells retrieved from these droplets showed almost the same growth profiles in cell culture dishes with those seeded using a normal subculture protocol (Figure 5b). It is well known that mammalian cells are easily damaged by the forces exerted by the external environment, such as shear stress. In the droplet breakup process in co-flowing immiscible liquid, drag force is the most influential force to affect the viability of cells suspended in the polymer solution. However, the results for viability and growth profiles of the cells exposed to the droplet breakup process (Figure 5) demonstrate that the drag force necessary for droplet breakup in water-immiscible liquid resulting in droplets of less than 100 μm in diameter from the needle of several hundred micrometers in diameter is insufficient to damage cell viability and growth activity.

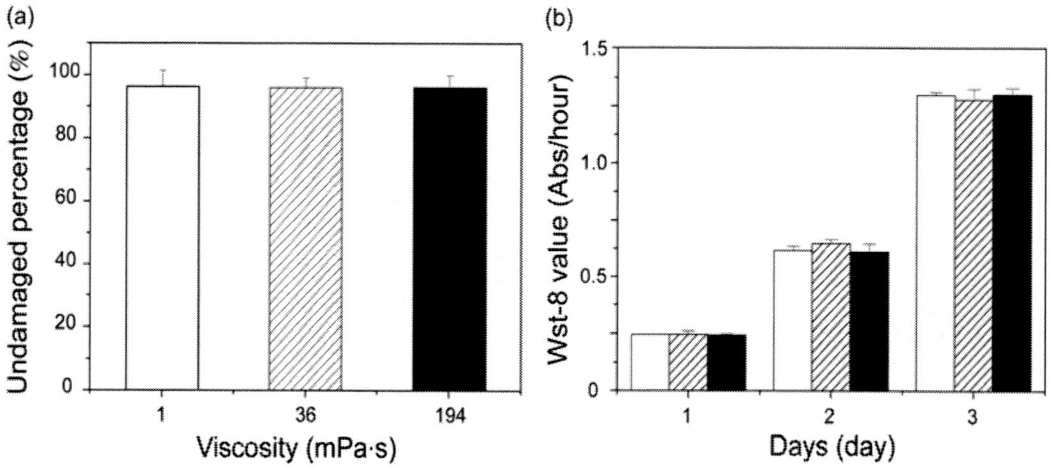

Figure 5. (a) Percentages of undamaged human tongue squamous carcinoma cells enclosed in droplets obtained from an aqueous solution with different viscosities: (white) 1, (diagonal line) 36, and (black) 194 mPa·s. (b) Proliferation profiles of the cells cultured in tissue culture dishes (white) before and after being retrieved from droplets obtained from aqueous solutions of (diagonal line) 36 and (black) 194 mPa·s. Droplets were prepared by extruding cell suspensions from a needle with a 300 μm i.d. at 1.2 cm/sec into an ambient co-flowing liquid paraffin stream of 23.5 cm/sec (Reproduced with permission, from Sakai S et al. Biotechnol Prog. [34] @ 2005 American Institute of Chemical Engineers).

Cell Encapsulation in Microcapsules

As described above, the method using liquid paraffin as an ambient co-flowing fluid is effective for obtaining droplets of polymer solution with a diameter of less than 100 μm while scarcely damaging the cells. The resultant droplets exist in a water-immiscible fluid; thus a smart approach to obtain gelated capsules is to carry out gelation of the droplets. Thermosensitive and photo-curable polymers are good candidates for this process. We prepared capsules based on two types of gelation processes, thermal gelation using unmodified agarose [35,36] and agarose-based materials [37,38], and enzymatic gelation using polymers incorporating phenolic hydroxyl groups [39,40].

Agarose Microcapsules

Agarose is a natural polysaccharide extracted from the cellular walls of agarophyte seaweed and exhibits temperature-sensitive water solubility. Agarose hydrogel has been widely studied for biomedical applications for use in transplantation with cells, such as tissue scaffolds [41-44] and cell encapsulation vehicles [45-47] due to a high biocompatibility. Agarose gel is formed by cooling aqueous agarose solution, which is obtained by heating an aqueous suspension of agarose powder until a clear solution is formed. Using agarose with a low gelling temperature (26-30°C at 1.5%), mammalian cells can be suspended in the solution at 38°C and kept viable. Cell-enclosing agarose microcapsules of about 100 μm in diameter can be obtained by extruding a cell-suspending agarose solution from a needle with an inner diameter of 300 μm into the ambient liquid paraffin stream with laminar flow at 38°C, and subsequently cooling the resultant emulsion system in an ice bath. The gelated microcapsules are collected via centrifugation after adding biological buffer into the emulsion system. The feline kidney cells enclosed in the agarose microcapsules with a diameter of 79 μm had a 89.2% viability [35]. In addition, these cells survived for more than 2 months *in vitro*. These results indicate that use of the thermal gelation process for agarose is effective for obtaining cell-enclosing microcapsules of less than 100 μm in diameter.

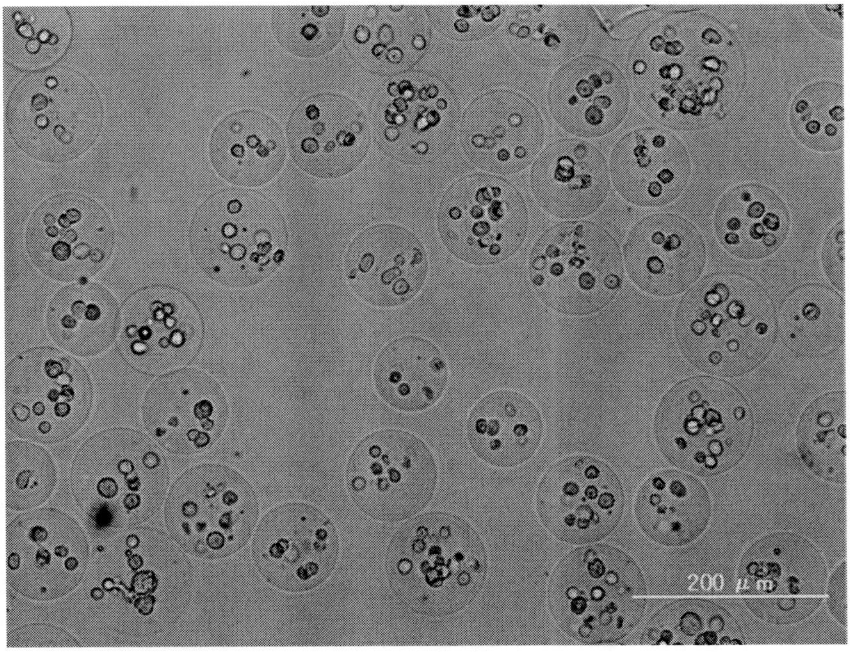

Figure 6. Photograph of CYP2B1-expressing cells-enclosing agarose microcapsules after 1 day of culture *in vitro* (Reproduced with permission, from Sakai S et al. Mol. Cancer Ther. [48] @ 2005 American Association for Cancer Research).

In order to demonstrate a potential application of the agarose microcapsules of less than 100 μm in diameter, we enclosed cells genetically modified to express the cytochrome P450 2B1 enzyme (CYP2B1) in the agarose microcapsules of about 90 μm in diameter (Figure 6).

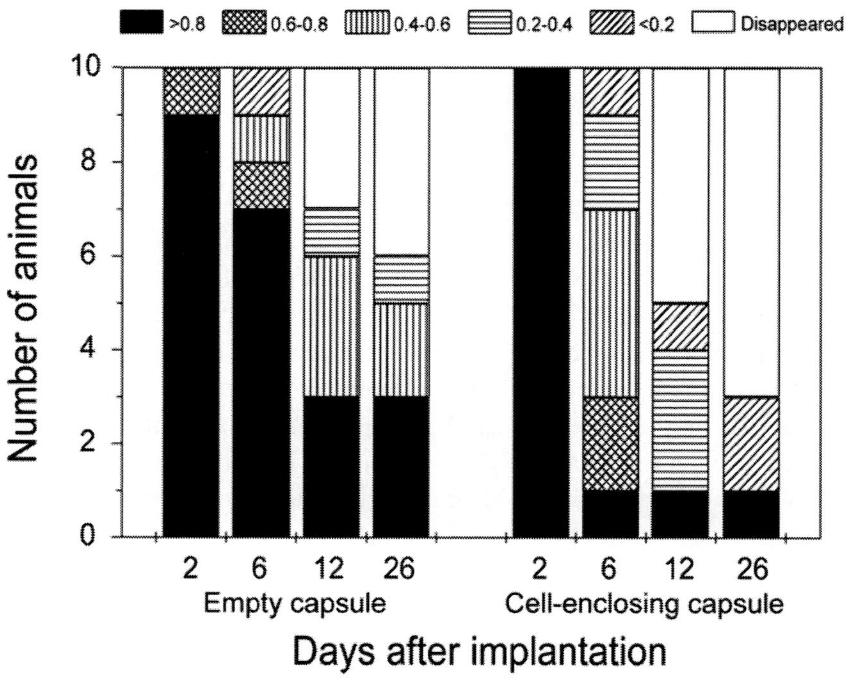

Figure 7. Preformed tumor size changes of the recipients transplanted with empty and CYP2B1 cell-enclosing microcapsules. The nude mice each were used as recipients of empty and cell-enclosing microcapsules. Tumor size just before capsule implantation was set to 1 and each block represents the proportional size of the original tumor (Reproduced with permission, from Sakai S et al. Mol. Cancer Ther. [48] @ 2005 American Association for Cancer Research).

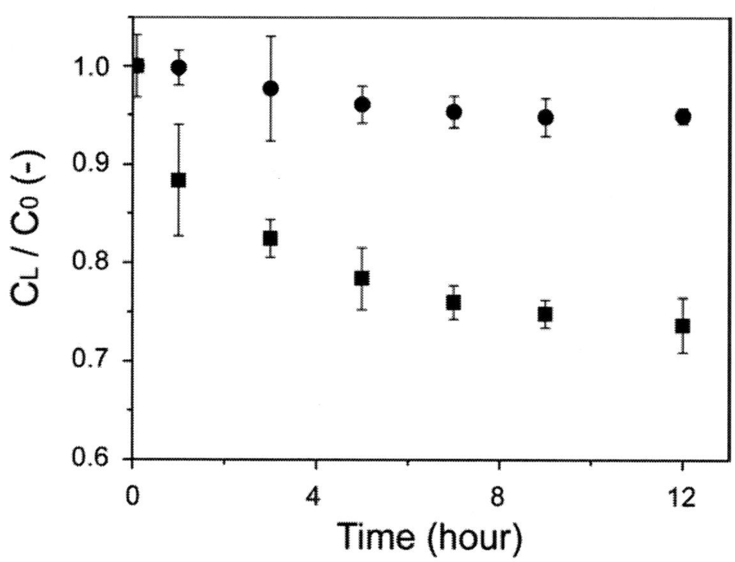

Figure 8. Time-course of diffusion of bovine serum albumin from 10 mL Krebs Ringer Hepes buffer solution (pH 7.4) into 5 mL of microcapsules at 37°C. C_0, C_L: initial concentration and concentration of bovine serum albumin at each time point. (■): Non-treated alginate-agarose microcapsules, (●): Alginate-agarose microcapsules soaked in 0.05 wt% chitosan-acetic acid aqueous solution (pH 6.3) for 10 min (Reproduced with permission, from Sakai S et al. Biochem. Eng. J. [38] @ 2006 Elsevier).

We then implanted them into preformed tumors in nude mice for localized activation of the prodrug ifosfamide in or close to the tumors [48]. Ifosfamide is a prodrug that is metabolized into acrolein by CYP2B1 in the liver [49]. Due to the very short half-life of the activated drug in plasma [50], the prodrug has to be given in a relatively high dose despite severe side effects, such as leucopenia with granulocytopenia. Establishment of a second site of enzyme conversion in or near the tumor using the cells expressing CYP2B1 is recognized as an effective approach to reduce side effects without lowering the response rates to the drug [51-54]. We injected the cell-enclosing microcapsules into tumors using a 26-gauge syringe after suspending them in saline. The effectiveness was demonstrated by the significant regression of tumors in the recipients implanted with the cell-enclosing microcapsules compared with those implanted with the empty agarose microcapsules (Figure 7) [48].

Alginate-agarose Composite and Agarose-Gelatin Conjugate Microcapsules

Two other types of microcapsules were prepared, alginate-agarose composite microcapsules [38] and agarose-gelatin conjugate microcapsules [37], using the thermal gelation process which was previously shown to be harmless towards mammalian cells. The former ones are for subsequent modification with a polyelectrolyte complex membrane. Alginate has been one of the more successful components for bioencapsulation for purposes of cell-encapsulation [13,55-57]. An attractive feature of alginate compared with agarose is that it has anionic charged carboxyl groups in its structure [14]. Thus it is easy to modify the resultant gel using cationic polymers such as poly-L-lysine [13,58-60] and chitosan [7,61] via the formation of electrostatic bonds for controlling biocompatibility and molecular exclusion properties. As described in the Introduction, alginate has a different gelation mechanism from agarose. One well-known method for obtaining alginate microcapsules in a water-immiscible liquid involves the coalescence between droplet(s) of aqueous sodium-alginate solution and droplet(s) of an aqueous divalent cation solution such as calcium ions. Sugiura *et al.* [62] has reported the preparation of cell-enclosing alginate microcapsules of about 150 μm in diameter using the coalescence of droplets prepared with an array of micronozzles of 30 μm × 30 μm square. The prepared droplets were highly spherical before gelation, but the resultant alginate microcapsules were far from a spherical shape. Such an awkward shape results in problems with mechanical stability. In contrast, our method for obtaining the alginate-agarose composite microcapsules involves two steps: First, the droplets are obtained via extrusion of the mixture solution into a co-flowing liquid paraffin flow which is cooled for the gelation of agarose. After formation of spherical microcapsules via gelation of agarose, a 100 mM $CaCl_2$ solution is added to the suspension for the gelation of alginate. Without the gelation of alginate, the alginate molecules are released from the microcapsules into the surrounding aqueous solution. There was no apparent difference between agarose microcapsules and alginate-agarose composite microcapsules. The effectiveness of the incorporation of alginate molecules into microcapsules for the subsequent modification with a polyelectrolyte complex membrane was demonstrated from the obvious suppression of the diffusion of bovine serum albumin into the microcapsules after soaking in aqueous chitosan solution (Figure 8) [38].

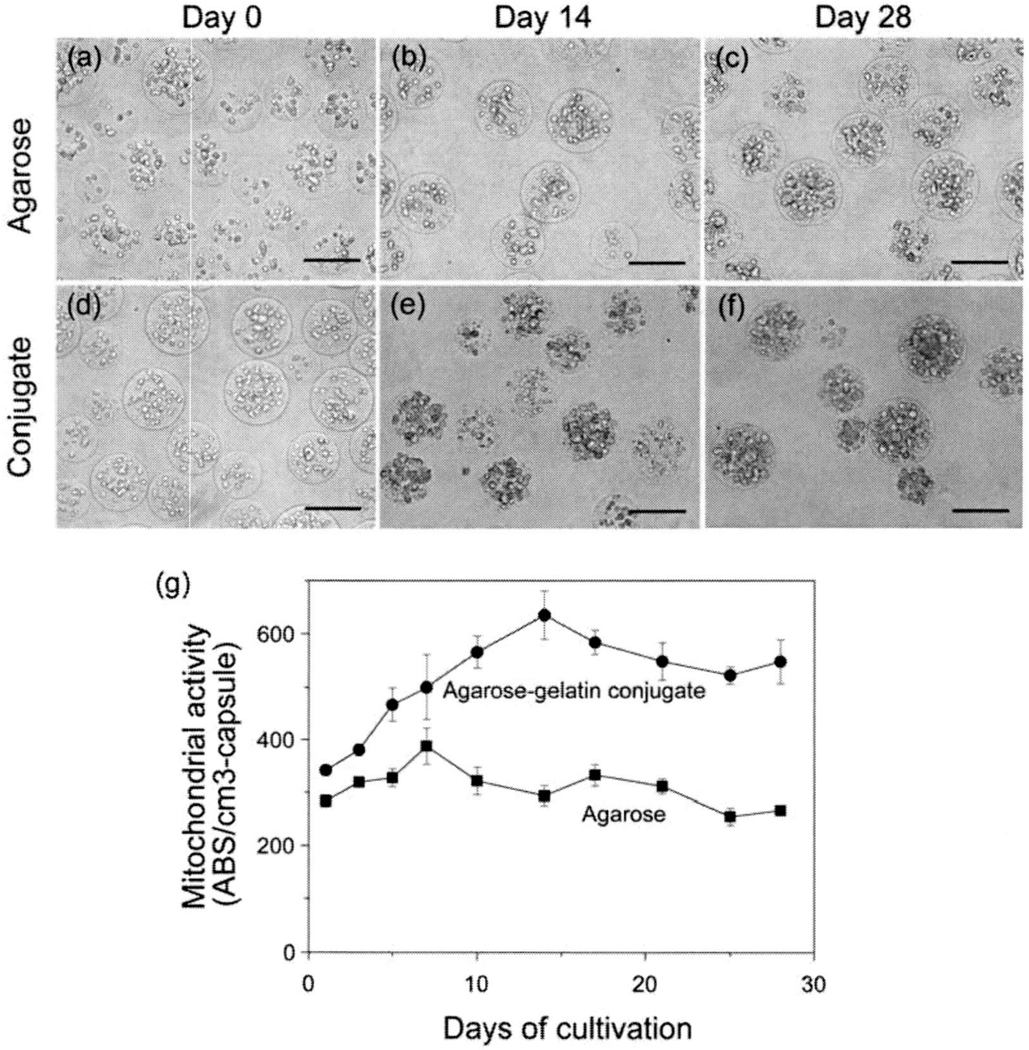

Figure 9. Microphotographs of cell-enclosing (a-c) unmodified agarose and (d-f) agarose-gelatin conjugate microcapsules (a, d) immediately after encapsulation, at (b, e) 14 d, and (c, f) 28 d of cultivation (Bars are 150 μm). Transition of mitochondrial activity of cells enclosed in (■) unmodified agarose microcapsules and (●) agarose-gelatin conjugate microcapsules. (Data are reproduced with permission, from Sakai S et al. Biotechnol. Lett. [37] @ 2007 Springer).

Significant low cell adhesiveness and cell proliferation are two features of unmodified agarose [63]. When transplanting a proliferative cell line *in vivo*, these features are useful for suppressing an excess proliferation of enclosed cells to avoid the risk of tumorigenesis [64]. However, these features are not advantageous for applications such as the bioproduction of therapeutic proteins *in vitro*. For enhancing the growth of enclosed cells, we developed microcapsules from an agarose-gelatin conjugate [37]. The conjugate shows a sol-to-gel transition, but no gel-to-sol transition, around body temperature, similar to native agarose [65]. Gelatin is a natural polymer derived from collagen, and has been used for medical applications such as wound dressings and as scaffolding material for tissue engineering [66-69]. Gelatin hydrogel is obtained by cooling the aqueous gelatin solution obtained by heating a suspension of gelatin in aqueous solution to more than around 40°C. However, this hydrogel can remelt and shows a gel-to-sol transition if subsequently heated at around 40°C [70]. This means the impossibility of using unmodified gelatin as a material for cell-enclosing capsules for transplantation *in vivo* and culture under normal mammalian cell culturing temperatures. Thus we used the gelatin conjugate with agarose for supporting cellular growth in microcapsules. As with the agarose and alginate-gelatin microcapsules, we could prepare cell-enclosing microcapsules via the droplet breakup process using liquid paraffin and the subsequent thermal gelation process [37]. As we expected, the adherent cells enclosed in the agarose-gelatin conjugate microcapsules showed a higher degree of proliferation than those in the unmodified agarose microcapsules (Figure 9).

Enzymatically-crosslinked Alginate Microcapsules

As a novel method for gelation of droplets obtained via the breakup process in ambient co-flowing liquid paraffin, we developed a method using enzymatic polymerization [39]. Enzymatic polymerization *in vitro* has been studied for decades because specific enzyme catalysis provides a novel synthetic route for functional and useful polymeric materials [71,72]. We used peroxidase for obtaining microcapsules. Peroxidases function as oxidoreductases that catalyze the oxidation of donors using H_2O_2 resulting in polyphenols linked at the aromatic ring by a C-C and C-O coupling of phenols. The alginate incorporated hydroxyl phenol moieties into about 3% of the original carboxyl groups (Alg-Ph) and was used as the material of microcapsules. We synthesized this alginate derivative [73] based on the report by Kurisawa *et al.* that a hyaluronic acid conjugate with tyramine forms a hydrogel via a peroxidase-catalyzed oxidative reaction [74]. Because more than 90% of the original carboxyl groups remain, the alginate derivative is gellable via not only the peroxidase catalyzed reaction, but also via the conventional gelation process coming into contact with multivalent cations [73]. One of the problems at the start point of this study was how to supply the H_2O_2 necessary for the enzymatic reaction. One option was to extrude the mixture of Alg-Ph, peroxidase, and H_2O_2 into the ambient co-flowing stream of liquid paraffin. However, this is not practical for continuous production of cell-enclosing microcapsules because of the formation of crosslinks between the Alg-Ph molecules over time in the syringe before being extruded from the needle. Next, we attempted to dissolve H_2O_2 in liquid paraffin. Although liquid paraffin is a 'water-immiscible liquid', it can dissolve a small quantity of water. The liquid paraffin containing H_2O_2 is obtained by vigorously stirring an aqueous H_2O_2 solution with liquid paraffin followed by a centrifugation step to separate the liquid paraffin and non-dissolved aqueous H_2O_2 solution. We added 5 mL of aqueous H_2O_2 solution (31 wt%) into 1000 mL liquid paraffin. Cell-suspending 1.5 w/v% Alg-Ph solution containing 1.6 units/mL of horse radish peroxidase (HRP) was extruded into a co-flowing stream of liquid paraffin containing dissolved H_2O_2 (Figure 10a). The liquid paraffin suspension with partially gelated Alg-Ph microcapsules was collected in a plastic tube. After 10 min of standing to allow for further progress of the enzymatic crosslinking reaction (Figure 10b), Alg-Ph microcapsules were collected via centrifugation. The resultant microcapsules had high sphericity (Figure 11). This high sphericity means that the sufficient degree of enzymatic gelation necessary for fixing the final shape of the gel did not occur instantaneously.

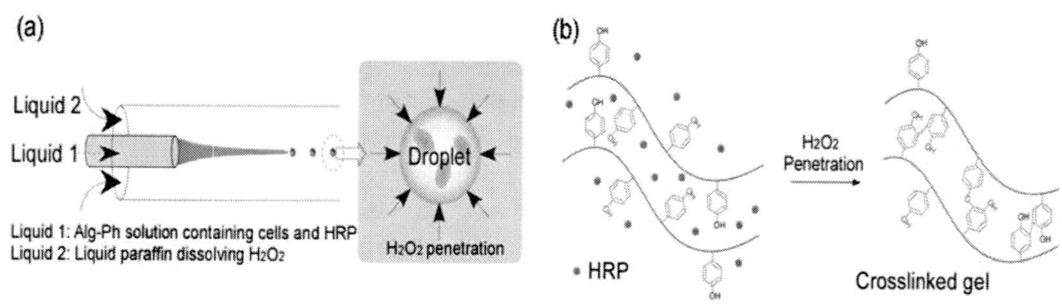

Figure 10. Schematic illustrations of (a) breakup of cell-enclosing droplets in a co-flowing stream of liquid paraffin containing H_2O_2 and HRP-catalyzed gelation and (b) HRP-catalyzed crosslink formation using H_2O_2 penetrating from the ambient liquid paraffin (Reproduced with permission, from Sakai S et al. Biomacromolecules [39] @ 2007 American Chemical Society).

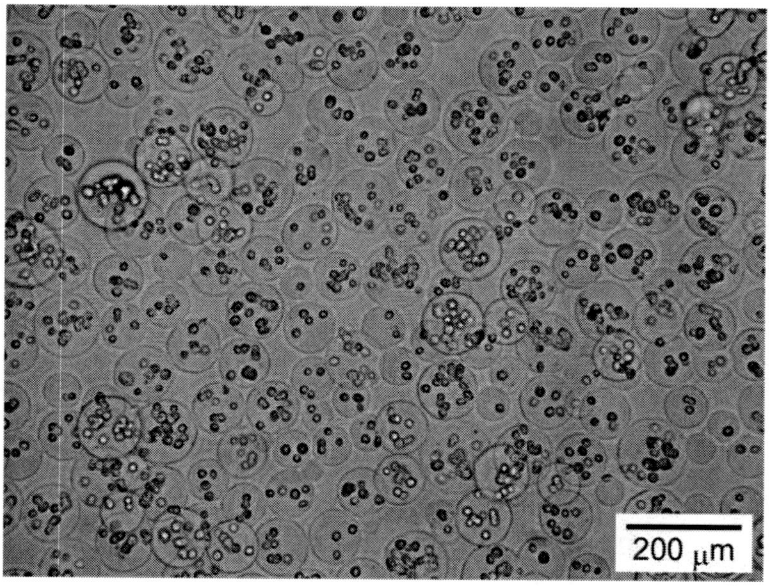

Figure 11. Micrograph of feline kidney cell-enclosing Alg-Ph microcapsules of 84.8 μm in diameter (Reproduced with permission, from Sakai S et al. Biomacromolecules [39] @ 2007 American Chemical Society).

One of the special concerns for the peroxidase-catalyzed encapsulation process would be the potential for a harmful effect of H_2O_2 on mammalian cells, because the cytotoxicity of H_2O_2 is well-known. However, we could not detect any harmful effects of H_2O_2 using feline kidney cells (CRFH cell line) [39]. We investigated the viability and growth profiles of the cells recovered from the Alg-Ph microcapsules by degrading the Alg-Ph gel using alginate lyase. The viability of the cells after 30 min of encapsulation was 91.8%. This value is almost the same as that measured for the cells enclosed in non-gelated droplets of sodium-alginate solution [28]. In addition, the recovered cells adhered, spread and grew on a cell culture dish in the same way as cells seeded using a general subculture protocol (Figure 12). It was unexpected that these promising cytotoxicity results were obtained. A gradual and slight quantity of penetration of H_2O_2 from the ambient liquid paraffin into the droplets with dissolved HRP and the immediate consumption of the H_2O_2 by the HRP-catalyzed oxidative reaction is a possible explanation of these promising results. We also attempted to prepare cell-enclosing microcapsules of more than 400 μm in diameter using the same process. The time necessary for obtaining gelated capsules increased with the increasing diameter of the droplet. This finding shows that diffusion of H_2O_2 from liquid paraffin is limited. Sufficient oxygenation is needed for survival of cells, thus a shorter period of suspension in liquid paraffin is preferable. These findings indicate that the peroxidase-catalyzed cell-encapsulation process using liquid paraffin containing dissolved H_2O_2 is more suitable for the production of the microcapsules of about 100 μm in diameter.

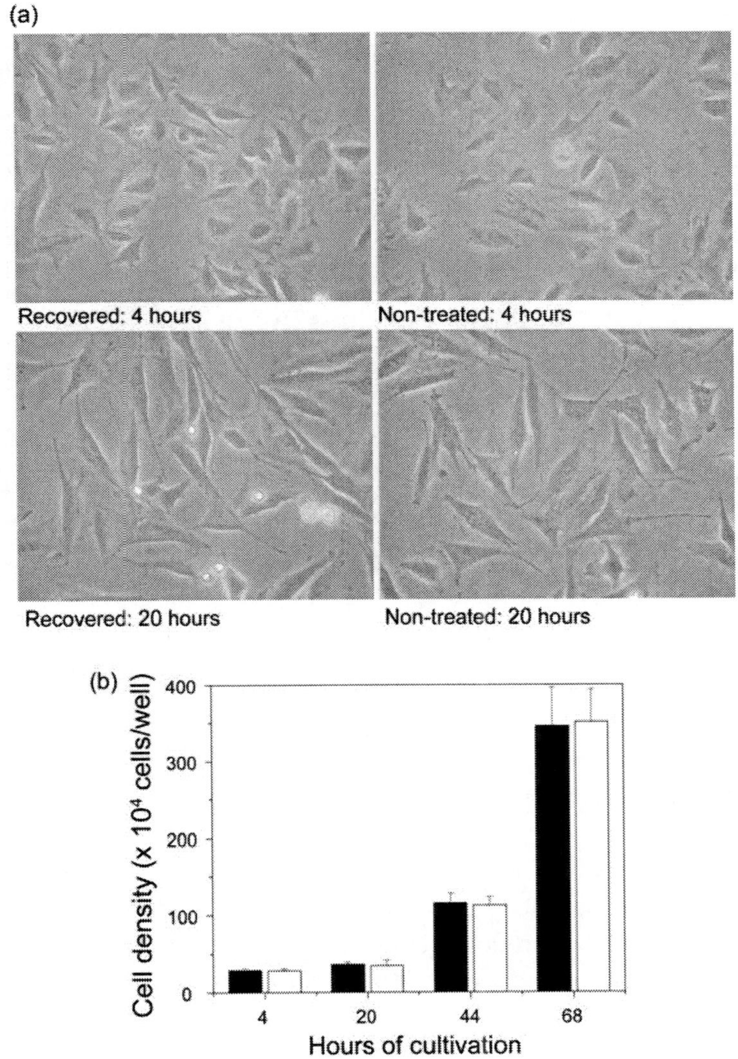

Figure 12. (a) Micrographs of CRFK cells cultured on cell culture dishes of 60 mm in diameter at 4 and 20 hours after seeding. Recovered: cells seeded after being recovered from Alg-Ph capsules by degradation using alginate lyase. Non-treated: cells seeded using a general subculture protocol. Cells were seeded at 3.0×10^5 viable cells/well. (b) Proliferation profiles of CRFK cells on tissue culture dishes. Columns indicate mean densities of (black) recovered cells and (white) non-treated cells. Error bars: SDs (Reproduced with permission, from Sakai S et al. Biomacromolecules [39] @ 2007 American Chemical Society).

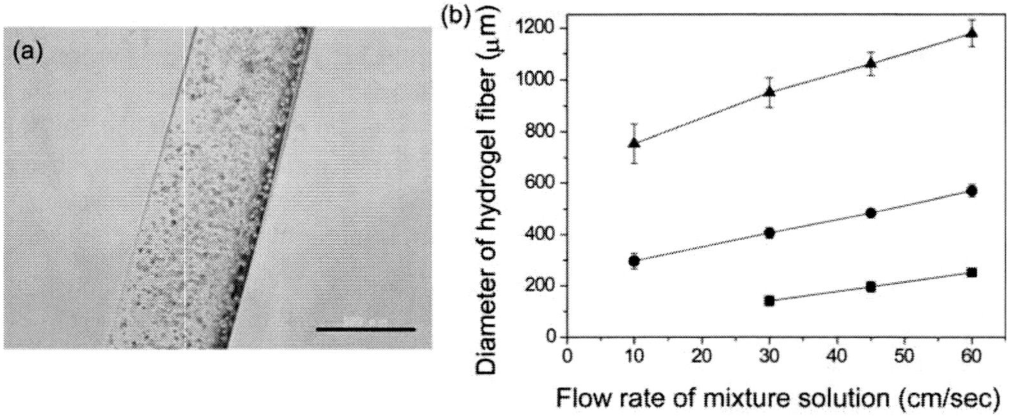

Figure 13. (a) Photograph of a BEC-enclosing Ca-Alg/gelatin fiber. Bar: 500 μm. (b) Diameters of Ca-Alg/gelatin hydrogel fibers as a function of the flow rate of the Na-Alg/gelatin mixture and the diameter of the needles, (■) 270, (●) 480 and (▲) 940 μm in diameter) from which the mixture was extruded. The flow rate of the 100 mM $CaCl_2$ solution was fixed at 33.9 cm/s (Reproduced with permission, from Sakai S et al. Biomacromolecules [75] @ 2008 American Chemical Society).

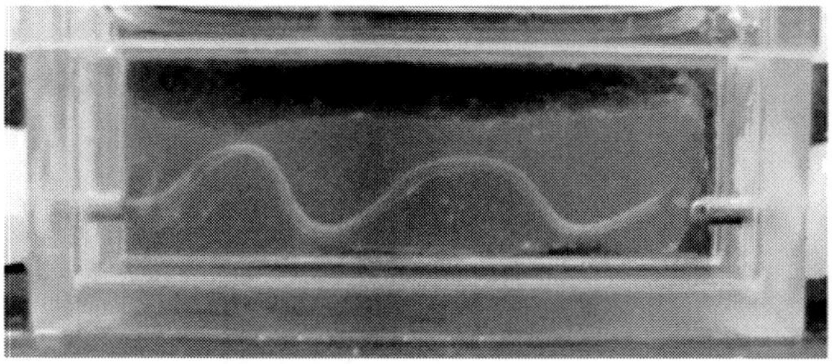

Figure 14. Photograph of the wavy-shaped channel developed in collagen gel by degrading the cell-enclosing alginate fiber (Reproduced with permission, from Sakai S et al. Biomacromolecules [75] @ 2008 American Chemical Society).

Cell Encapsulation in Ca-alginate Gel Fibers

It was also of interest to determine whether it is possible to prepare cell-enclosing microcapsules using the aqueous solution containing dissolved multivalent cations such as calcium ions as an ambient co-flowing liquid. It was not possible because the extruded sodium-alginate solution instantaneously gelates after contacting the multivalent cations in aqueous solution. We obtained cell-enclosing alginate fibers when we injected cell-suspending aqueous alginate solution into a co-flowing 100 mM $CaCl_2$ solution (Figure 13a). Similar to the droplet production in liquid paraffin, the diameter of the fiber is controllable by changing the flow rate of ambient liquid and the diameter of the needle from which the cell-suspending alginate solution is extruded (Figure 13b). One possible application of this cell-enclosing alginate fiber is use it as a template for a vascular-like structure previously developed using biocompatible hydrogels *in vitro* [75]. Due to the flexibility of alginate gel fibers and their ability to degrade using alginate lyase, it is possible to develop a tubular construct with a straight configuration as well as a 3D configuration such as the wavy configuration observed in hydrogels (Figure 14).

Conclusion

In this chapter, we described the methods for the production of cell-enclosing microcapsules of about 100 μm in diameter, much smaller than conventional microcapsules, by extruding a cell-suspending aqueous polymer solution from a needle with a diameter of several hundred micrometers. Droplets of about 100 μm in diameter that subsequently become microcapsules were obtained via a jetting process in a water-immiscible liquid. The gelation of the droplets suspended in a water-immiscible liquid was achieved via a thermal gelation process using agarose and agarose-based materials and an enzyme-catalyzed gelation process. The size of the microcapsules is too small to encapsulate pancreatic islets, but is large enough to encapsulate single cells. Studies using such small microcapsules are still at an early stage in respect of practical applications. However, we believe that these small microcapsules of about 100 μm diameter will become useful devices in the field of cell therapy and bioproduction considering the recent advances made in genetic engineering that enable the use of cells as reactors to produce desired proteins. We also think that microcapsules, smaller than the maximum allowed size for a tightly packed cell mass that does not limit the oxygen supply [76], would be a useful tool for the production of spherical tissues from individual cells.

References

[1] Chang, T.M.S. & Prakash, S. (1998). Therapeutic uses of microencapsulated genetically engineered cells. *Mol. Med. Today*, **4**, 221-227.

[2] Hauser, O. Prieschl Grassauer, E. & Salmons, B. (2004). Encapsulated, genetically modified cells producing in vivo therapeutics. *Curr. Opin. Mol. Ther.*, **6**, 412-420.

[3] Uludag, H. de Vos, P. & Tresco, P.A. (2000). Technology of mammalian cell encapsulation. *Adv. Drug. Deliv. Rev.*, **42**, 29-64.

[4] Maguire, T. Novik, E. Schloss, R. & Yarmush, M. (2006). Alginate-PLL microencapsulation: effect on the differentiation of embryonic stem cells into hepatocytes. *Biotechnol. Bioeng.*, **93**, 581-591.

[5] Wang, X. Wang, W. Ma, J. Guo, X. Yu, X. & Ma, X. (2006). Proliferation and differentiation of mouse embryonic stem cells in APA microcapsule: A model for studying the interaction between stem cells and their niche. *Biotechnol. Prog.*, **22**, 791-800.

[6] Eiselt, P. Yeh, J. Latvala, R.K. Shea, L.D. & Mooney, D.J. (2000). Porous carriers for biomedical applications based on alginate hydrogels. *Biomaterials*, **21**, 1921-1927.

[7] Sakai, S. Ono, T. Ijima, H. & Kawakami, K. (2000). Control of molecular weight cut-off for immunoisolation by multilayering glycol chitosan-alginate polyion complex on alginate-based microcapsule. *J. Microencapsul.*, **17**, 691-699.

[8] Leung, A. Ramaswamy, Y. Munro, P. Lawrie, G. Nielsen, L. & Trau, M. (2005). Emulsion strategies in the microencapsulation of cells: pathways to thin coherent membranes. *Biotechnol. Bioeng.*, **92**, 45-53.

[9] Pajic-Lijakovic, I. Bugarski, D. Plavsic, M. & Bugarski, B. (2007). Influence of microenvironmental conditions on hybridoma cell growth inside the alginate-poly-L-lysine microcapsule. *Process Biochem.*, **42**, 167-174.

[10] Chang, P.L. Hortelano, G. Tse, M. & Awrey, D.E. (1994). Growth of recombinant fibroblasts in alginate microcapsules. *Biotechnol. Bioeng.*, **43**, 925-933.

[11] Zhang, Y. Wang, W. Zhou, J. Yu, W. Zhang, X. Guo, X. & Ma, X. (2007). Tumor anti-angiogenic gene therapy with microencapsulated recombinant CHO cells. *Ann. Biomed. Eng.*, **35**, 605-614.

[12] Sparks, R.E. Salemme, R.M. Meier, P.M. Litt, M.H. & Lindan, O. (1969). Removal of waste metabolites in uremia by microencapsulated reactants. *Trans. Amer. Soc. Artif. Int. Organs*, **15**, 353-359.

[13] Lim, F. & Sun, A.M. (1980). Microencapsulated islets as bioartificial endocrine pancreas. *Science*, **210**, 908-910.

[14] Smidsrod, O. & Skjak-Braek, G. (1990). Alginate as immobilization matrix for cells. *Trends Biotechnol.*, **8**, 71-78.

[15] Wolters, G.H. Fritschy, W.M. Gerrits, D. & van Schilfgaarde, R. (1992). A versatile alginate droplet generator applicable for microencapsulation of pancreatic islets. *J. Appl. Biomater.*, **3**, 281-286.

[16] Ross, C.J. & Chang, P.L. (2002). Development of small alginate microcapsules for recombinant gene product delivery to the rodent brain. *J. Biomater. Sci. Polym. Ed.*, **13**, 953-962.

[17] Chicheportiche, D. & Reach, G. (1988). In vitro kinetics of insulin release by microencapsulated rat islets: effect of the size of the microcapsules. *Diabetologia*, **31**, 54-57.

[18] Schrezenmeir, J. Gero, L. Solhdju, M. Kirchgessner, J. Laue, C. Beyer, J. Stier, H. & Muller Klieser, W. (1994). Relation between secretory function and oxygen supply in isolated islet organs. *Transplant. Proc.*, **26**, 809-813.

[19] Canaple, L. Rehor, A. & Hunkeler, D. (2002). Improving cell encapsulation through size control. *J. Biomater. Sci. Polym. Ed.*, **13**, 783-796.

[20] Leblond, F.A. Simard, G. Henley, N. Rocheleau, B. Huet, P.M. & Halle, J.P. (1999). Studies on smaller (approximately 315 μm) microcapsules: IV. Feasibility and safety of intrahepatic implantations of small alginate poly-L-lysine microcapsules. *Cell Transplant.*, **8**, 327-337.

[21] Robitaille, R. Pariseau, J.-F. Leblong, F.A. Lamoureux, M. Lepage, Y. & Halle, J.-P. (1999). Studies on small (<350μm) alginate-poly-L-lysine microcapsules. III. Biocompatibility of smaller versus standard microcapsules. *J. Biomed. Mater. Res.*, **44**, 116-120.

[22] Ganan-Calvo, A.M. (1998). Generation of steady liquid microthreads and micron-sized monodisperse sprays in gas streams. *Phys. Rev. Lett.*, **80**, 285-288.

[23] Nir, R. Lamed, R. Gueta, L. & Sahar, E. (1990). Single-cell entrapment and microcolony development within uniform microspheres amenable to flow cytometry. *Appl. Environ. Microb.*, **56**, 2870-2875.

[24] Sugiura, S. Oda, T. Aoyagi, Y. Matsuo, R. Enomoto, T. Matsumoto, K. Nakamura, T. Satake, M. Ochiai, A. Ohkohchi, N. & Nakajima, M. (2007). Microfabricated airflow nozzle for microencapsulation of living cells into 150 micrometer microcapsules. *Biomed. Microdevices*, **9**, 91-99.

[25] Zhang, D.F. & Stone, H.A. (1997). Drop formation in viscous flows at a vertical capillary tube. *Phys. Fluids*, **9**, 2234-2242.

[26] Lim, S.T. Martin, G.P. Berry, D.J. & Brown, M.B. (2000). Preparation and evaluation of the in vitro drug release properties and mucoadhesion of novel microspheres of hyaluronic acid and chitosan. *J. Control. Release*, **66**, 281-292.

[27] Ribeiro, A.J. Neufeld, R.J. Arnaud, P. & Chaumeil, J.C. (1999). Microencapsulation of lipophilic drugs in chitosan-coated alginate microspheres. *Int. J. Pharm.*, **187**, 115-123.

[28] Sakai, S. Kawabata, K. Ono, T. Ijima, H. & Kawakami, K. (2004). Preparation of mammalian cell-enclosing subsieve-sized capsules (<100μm) in a coflowing stream. *Biotechnol. Bioeng.*, **86**, 168-173.

[29] Cramer, C. Beruter, B. Fischer, P. & Windhab, E.J. (2002). Liquid jet stability in a laminar flow field. *Chem. Eng. Technol.*, **25**, 499-506.

[30] Zhang, X. & Basaran, O.A. (1995). An experimental study of dynamics of drop formation. *Phys. Fluids*, **7**, 1184-1203.

[31] Cramer, C. Fischer, P. & Windhab, E.J. (2004). Drop formation in a co-flowing ambient fluid. *Chem. Eng. Sci.*, **59**, 3045-3058.

[32] Umbanhowar, P.B. Prasad, V. & Weitz, D.A. (2000). Monodisperse emulsion generation via drop break off in a coflowing stream. *Langmuir*, **16**, 347-351.

[33] Rayleigh, J.W.S. (1878). On the stability of jets. *Proc. Lond. Math. Soc.*, **10**, 4-11.

[34] Sakai, S. Kawabata, K. Ono, T. Ijima, H. & Kawakami, K. (2005). Higher viscous solution induces smaller droplets for cell-enclosing capsules in a co-flowing stream. *Biotechnol. Prog.*, **21**, 994-997.

[35] Sakai, S. Kawabata, K. Ono, T. Ijima, H. & Kawakami, K. (2005). Development of mammalian cell-enclosing subsieve-size agarose capsules (<100μm) for cell therapy. *Biomaterials*, **26**, 4786-4792.

[36] Sakai, S. Hashimoto, I. & Kawakami, K. (2006). Usefulness of flow focusing technology for producing subsieve-size cell enclosing capsules: Application for agarose capsules production. *Biochem. Eng. J.*, **30**, 218-221.

[37] Sakai, S. Hashimoto, I. & Kawakami, K. (2007). Agarose-gelatin conjugate for adherent cell-enclosing capsules. *Biotechnol. Lett.*, **29**, 731-735.

[38] Sakai, S. Hashimoto, I. & Kawakami, K. (2006). Development of alginate-agarose subsieve-size capsules for subsequent modification with a polyelectrolyte complex membrane. *Biochem. Eng. J.*, **30**, 76-81.

[39] Sakai, S. Hashimoto, I. Ogushi, Y. & Kawakami, K. (2007). Peroxidase-catalyzed cell-encapsulation in subsieve-size capsules of alginate with phenol moieties in water-immiscible fluid dissolving H_2O_2. *Biomacromolecules*, **8**, 2622-2626.

[40] Sakai, S. Ito, S. Ogushi, Y. & Kawakami, K. (2008). Feasibility of carboxymethylcellulose with phenol moieties as a material for mammalian cell-enclosing subsieve-size capsules. *Cellulose*, **15**, 723-729.

[41] Rahfoth, B. Weisser, J. Sternkopf, F. Aigner, T. von der Mark, K. & Brauer, R. (1998). Transplantation of allograft chondrocytes embedded in agarose gel into cartilage defects of rabbits. *Osteoarthr. Cartilage*, **6**, 50-65.

[42] Dimicco, M.A. Kisiday, J.D. Gong, H. & Grodzinsky, A.J. (2007). Structure of pericellular matrix around agarose-embedded chondrocytes. *Osteoarthr. Cartilage*, **15**, 1207-1216.

[43] Alaminos, M. Del Carmen Sanchez Quevedo, M. Munoz Avila, J.I. Serrano, D. Medialdea, S. Carreras, I. & Campos, A. (2006). Construction of a complete rabbit cornea substitute using a fibrin-agarose scaffold. *Invest. Ophthalmol. Vis. Sci.*, **47**, 3311-3317.

[44] Rotter, N. Aigner, J. Naumann, A. Hammer, C. & Sittinger, M. (1999). Behavior of tissue-engineered human cartilage after transplantation into nude mice. *J. Mater. Sci. Mater. Med.*, **10**, 689-693.

[45] Tashiro, H. Iwata, H. Warnock, G.L. Ikada, Y. & Tsuji, T. (1998). Viability studies of agarose microencapsulation islets of langerhans from dogs. *Transplant. Proc.*, **30**, 490.

[46] Tun, T. Inoue, K. Hayashi, H. Aung, T. Gu, Y.-J. Doi, R. Kaji, H. Echigo, Y. Wang, W.-J. Setoyama, H. Imamura, M. Maetani, S. Norikawa, N. Iwata, H. & Ikada, Y. (1996). A newly developed three-layer agarose microcapsule for a promising biohybrid artificial pancreas: Rat to mouse xenotransplantation. *Cell Transplant.*, **5**, S59-S63.

[47] Iwata, H. Murakami, Y. & Ikada, Y. (1999). Control of complement activities for immunoisolation. *Ann. N. Y. Acad. Sci.*, **875**, 7-23.

[48] Sakai, S. Kawabata, K. Tanaka, S. Harimoto, N. Hashimoto, I. Mu, C. Salmons, B. Ijima, H. & Kawakami, K. (2005). Subsieve-size agarose capsules enclosing ifosfamide-activating cells: a strategy toward chemotherapeutic targeting to tumors. *Mol. Cancer Ther.*, **4**, 1786-1790.

[49] Dirven, H.A. Megens, L. Oudshoorn, M.J. Dingemanse, M.A. van Ommen, B. & van Bladeren, P.J. (1995). Glutathione conjugation of the cytostatic drug ifosfamide and the role of human glutathione S-transferases. *Chem. Res. Toxicol.*, **8**, 979-986.

[50] Kurowski, V. & Wagner, T. (1993). Comparative pharmacokinetics of ifosfamide, 4-hydroxyifosfamide, chloroacetaldehyde, and 2- and 3-dechloroethylifosfamide in patients on fractionated intravenous ifosfamide therapy. *Cancer Chemother. Pharmacol.*, **33**, 36-42.

[51] Lohr, M. Hoffmeyer, A. Kroger, J.C. Freund, M. Hain, J. Holle, A. Karle, P. Knofel, W.T. Liebe, S. Muller, P. Nizze, H. Renner, M. Saller, R.M. Wagner, T. Hauenstein, K. Gunzburg, W.H. & Salmons, B. (2001). Microencapsulated cell-mediated treatment of inoperable pancreatic carcinoma. *Lancet*, **357**, 1591-1592.

[52] Lohr, M. Muller, P. Karle, P. Stange, J. Mitzner, S. Jesnowski, R. Nizze, H. Nebe, B. Liebe, S. Salmons, B. & Gunzburg, W.H. (1998). Targeted chemotherapy by intratumour injection of encapsulated cells engineered to produce CYP2B1, an ifosfamide activating cytochrome P450. *Gene Ther.*, **5**, 1070-1078.

[53] Lohr, M. Kroger, J.C. Hoffmeyer, A. Freund, M. Hain, J. Holle, A. Knofel, W.T. Liebe, S. Nizze, H. Renner, M. Saller, R. Karle, P. Muller, P. Wagner, T. Hauenstein, K. Salmons, B. & Gunzburg, W.H. (2003). Safety, feasibility and clinical benefit of localized chemotherapy using microencapsulated cells for inoperable pancreatic carcinoma in a phase I/II trial. *Cancer Ther.*, **1**, 121-131.

[54] Kroger, J.C. Benz, S. Hoffmeyer, A. Bago, Z. Bergmeister, H. Gunzburg, W.H. Karle, P. Kloppel, G. Losert, U. Muller, P. Nizze, H. Obermaier, R. Probst, A. Renner, M. Saller, R. Salmons, B. Schwendenwein, I. von Rombs, K. Wiessner, R. Wagner, T. Hauenstein, K. & Lohr, M. (2003). Intra-arterial instillation of microencapsulated, Ifosfamide-activating cells in the pig pancreas for chemotherapeutic targeting. *Pancreatology*, **3**, 55-63.

[55] Martinsen, A. Skjak-Braek, G. & Smidsrod, O. (1989). Alginate as immobilization material: I. Correlation between chemical and physical properties of alginate gel beads. *Biotechnol. Bioeng.*, **33**, 79-89.

[56] Augst, A.D. Kong, H.J. & Mooney, D.J. (2006). Alginate hydrogels as biomaterials. *Macromol. Biosci.*, **6**, 623-633.

[57] Zimmermann, H. Hillgartner, M. Manz, B. Feilen, P. Brunnenmeier, F. Leinfelder, U. Weber, M. Cramer, H. Schneider, S. Hendrich, C. Volke, F. & Zimmermann, U. (2003). Fabrication of homogeneously cross-linked, functional alginate microcapsules validated by NMR-, CLSM- and AFM-imaging. *Biomaterials*, **24**, 2083-2096.

[58] Kung, I.M. Wang, F.F. Chang, Y.C. & Wang, Y.J. (1995). Surface modifications of alginate/poly(L-lysine) microcapsular membranes with poly(ethylene glycol) and poly(vinyl alcohol). *Biomaterials*, **16**, 649-655.

[59] Goosen, M.F.A. O'shea, G.M. Gharapetian, H.M. Chou, S. & Sun, A.M. (1985). Optimization of microencapsulation parameters: Semipermeable microcapsules as a bioartificial pancreas. *Biotechnol. Bioeng.*, **27**, 146-150.

[60] Goosen, M.F.A. King, G.A. McKnight, C.A. & Marcotte, N. (1989). Animal cell culture engineering using alginate polycation microcapsules of controlled membrane molecular weight cut-off. *J. Membr. Sci.*, **41**, 323-343.

[61] Haque, T. Chen, H. Ouyang, W. Martoni, C. Lawuyi, B. Urbanska, A.M. & Prakash, S. (2005). In vitro study of alginate-chitosan microcapsules: an alternative to liver cell transplants for the treatment of liver failure. *Biotechnol. Lett.*, **27**, 317-322.

[62] Sugiura, S. Oda, T. Izumida, Y. Aoyagi, Y. Satake, M. Ochiai, A. Ohkohchi, N. & Nakajima, M. (2005). Size control of calcium alginate beads containing living cells using micro-nozzle array. *Biomaterials*, **26**, 3327-3331.

[63] Gruber, H.E. Hoelscher, G.L. Leslie, K. Ingram, J.A. & Hanley, E.N. Jr (2006). Three-dimensional culture of human disc cells within agarose or a collagen sponge: assessment of proteoglycan production. *Biomaterials*, **27**, 371-376.

[64] Chang, P.L. Van Raamsdonk, J.M. Hortelano, G. Barsoum, S.C. MacDonald, N.C. & Stockley, T.L. (1999). The in vivo delivery of heterologous proteins by microencapsulated recombinant cells. *Trends Biotechnol.*, **17**, 78-83.

[65] Sakai, S. Hashimoto, I. & Kawakami, K. (2007). Synthesis of an agarose-gelatin conjugate for use as a tissue engineering scaffold. *J. Biosci. Bioeng.*, **103**, 22-26.

[66] Ren, L. Tsuru, K. Hayakawa, S. & Osaka, A. (2002). Novel approach to fabricate porous gelatin-siloxane hybrids for bone tissue engineering. *Biomaterials*, **23**, 4765-4773.

[67] Del Guerra, S. Bracci, C. Nilsson, K. Belcourt, A. Kessler, L. Lupi, R. Marselli, L. de Vos, P. & Marchetti, P. (2001). Entrapment of dispersed pancreatic islet cells in CultiSpher-S macroporous gelatin microcarriers: Preparation, in vitro characterization, and microencapsulation. *Biotechnol. Bioeng.*, **75**, 741-744.

[68] Choi, Y.S. Hong, S.R. Lee, Y.M. Song, K.W. Park, M.H. & Nam, Y.S. (1999). Study on gelatin-containing artificial skin: I. Preparation and characteristics of novel gelatin-alginate sponge. *Biomaterials*, **20**, 409-417.

[69] Balakrishnan, B. Mohanty, M. Umashankar, P.R. & Jayakrishnan, A. (2005). Evaluation of an in situ forming hydrogel wound dressing based on oxidized alginate and gelatin. *Biomaterials*, **26**, 6335-6342.

[70] Gillmor, J.R. Connelly, R.W. Colby, R.H. & Tan, J.S. (1999). Effect of sodium poly(styrene sulfonate) on thermoreversible gelation of gelatin. *J. Polym. Sci. B:. Polym. Phys.*, **37**, 2287-2295.

[71] Gross, R.A. Kumar, A. & Kalra, B. (2001). Polymer synthesis by in vitro enzyme catalysis. *Chem. Rev.*, **101**, 2097-2124.

[72] Kobayashi, S. Uyama, H. & Kimura, S. (2001). Enzymatic Polymerization. *Chem. Rev.*, **101**, 3793-3818.

[73] Sakai, S. & Kawakami, K. (2007). Synthesis and characterization of both ionically and enzymatically crosslinkable alginate. *Acta. Biomater.*, **3**, 495-501.

[74] Kurisawa, M. Chung, J.E. Yang, Y.Y. Gao, S.J. & Uyama, H. (2005). Injectable biodegradable hydrogels composed of hyaluronic acid-tyramine conjugates for drug delivery and tissue engineering. *Chem. Comm.*, **34**, 4312-4314.

[75] Sakai, S. Yamaguchi, S. Takei, T. & Kawakami, K. (2008). Oxidized alginate-cross-linked alginate/gelatin hydrogel fibers for fabricating tubular constructs with layered smooth muscle cells and endothelial cells in collagen gels. *Biomacromolecules*, **9**, 2036-2041.

[76] Kim, S.K. Yu, S.H. Son, J.H. Hübner, H. & Buchholz, R. (1998). Calculations on O_2 transfer in capsules with animal cells for the determination of maximum capsule size without O_2 limitation. *Biotechnol. Lett.*, **20**, 549-552.

In: Biochemical Engineering
Editors: F.E. Dumont and J.A. Sacco, pp. 227-244
ISBN: 978-1-60741-257-1

Chapter 9

Effect of Shear Stress on Wastewater Treatment Systems Performance

J.L. Campos[a], B. Arrojo[a], A. Franco[a], M. Belmonte[b], A. Mosquera-Corral[a], E. Roca[a] and R. Méndez[a]

[a]Department of Chemical Engineering. School of Engineering. Rua Lope Gómez de Marzoa s/n. University of Santiago de Compostela, E-15782, Santiago de Compostela, Spain

[b]Environmental Science Center EULA-Chile, University of Concepción, Concepción, Chile

Abstract

Wastewater treatment systems must be operated under hydrodynamic conditions that allow maintaining the biomass in suspension and promoting intimate contact between substrates and biomass. The systems used to maintain the mixture (mechanical stirring, aeration, etc.) exert shear stress on the biomass which can affect its physical properties (density and diameter) and specific activity.

When biomass is subjected to moderate shear stress, stable and dense structures can be formed, improving its retention, and the substrate transfer rates are also favoured. However, high shear stress generally leads to the loss of biomass activity and to the formation of particles with low diameters, which are washed-out of the system. Therefore, it is very important to control shear stress acting on biomass particles in order to optimize the performance of wastewater treatment systems.

The effects of impact stress and hydraulic stress by gas or liquid on the efficiency of different biological systems for carbon and nitrogen removal are discussed in this work.

Keywords: aeration, biological systems, mechanical agitation, pulsed flow, shear stress, wastewater

Introduction

In recent years, the development of different systems for the biological treatment of industrial and urban wastewater have been focused on obtaining biomass with good settling properties (e.g., biofilm and granular sludge). These systems are characterized by their high capacity to retain biomass, which allows increasing the treated volumetric loads and improving the quality of effluents in terms of suspended solids (Beun et al., 2001; Beun et al., 2002; Liu and Tay, 2002).

In these systems shear stress exert important effects on both biomass characteristics and hydrodynamic conditions of the systems and, therefore, on their overall efficiency (Figure 1). The effects on the biomass are related to both distribution of the different microbial populations along the biomass particles and physical properties (size and density) (Liu and Tay, 2002). Shear stress, caused by the operational hydrodynamic conditions, promotes the phenomenon of cellular aggregation, achieving biomass with higher density which favours its retention due to the improvement of its Zone Settling Velocity (ZSV). Moreover, higher biomass concentrations can be accumulated inside the system since its specific volume is lower (low Sludge Volumetric Index, SVI).

Shear stress usually provokes a decrease in particle diameter, which increases the specific surface area of biomass, improving the transfer rate of substrates (Gmachowski, 1995; Liu and Tay, 2002). However, a decrease of diameter has a negative effect on the ZSV and could cause a wash-out of the biomass. Therefore, the system should be operated under those conditions that allow achieving a balance between both transfer rate of substrates and biomass retention.

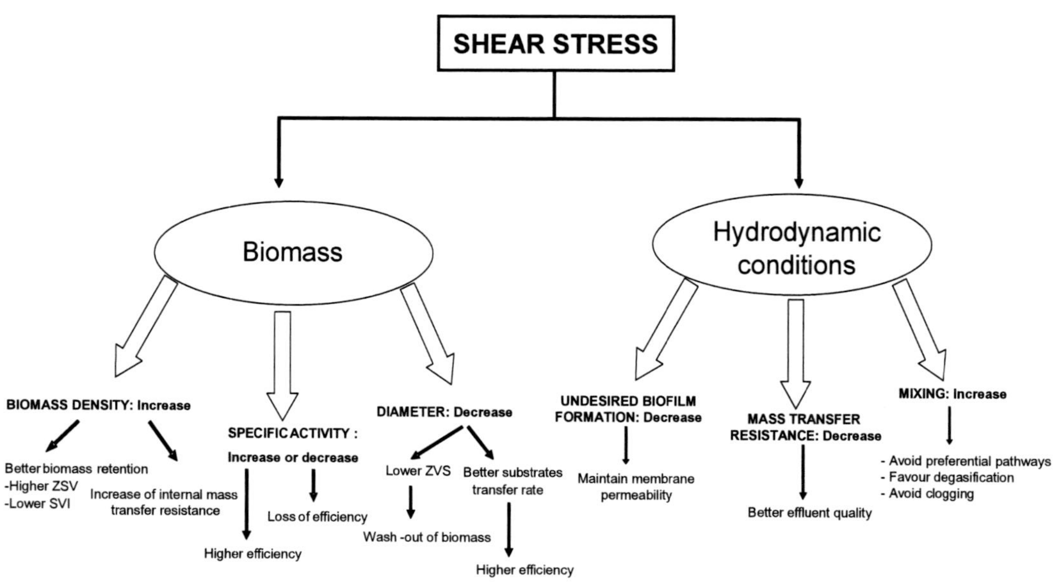

Figure 1. Effects of shear stress on the biomass characteristics and the hydrodynamic conditions and their consequences on the reactor performance.

Shear stress exerted on the biomass also affects its specific activity. Moderate shear stress increases the specific activity due to a decrease of the particle diameter and the generation of turbulent conditions which increase the transfer of substrates, but shear stress higher than a certain value provokes the loss of the activity. The management of shear stress in wastewater treatment systems was also used as a selective pressure to separate the desired biomass. In this way, those biomass particles that are dense can be retained inside the reactor, while the light ones are washed out (Franco et al., 2006).

Reactors used for treating wastewater could be considered heterogeneous catalytic reactors in which mass transfer rate completely or significantly controls the overall rate of the process being promoted in the reactor. Figure 2 shows several physical and biochemical steps involved during the process: oxygen is firstly transferred from the gas phase (in the case of aerobic processes) to the bulk liquid, then through the bulk liquid to the surface of the aggregate (a floc or granule), through the aggregate to the cell surface and, finally, into the cell where biological reaction occurs.

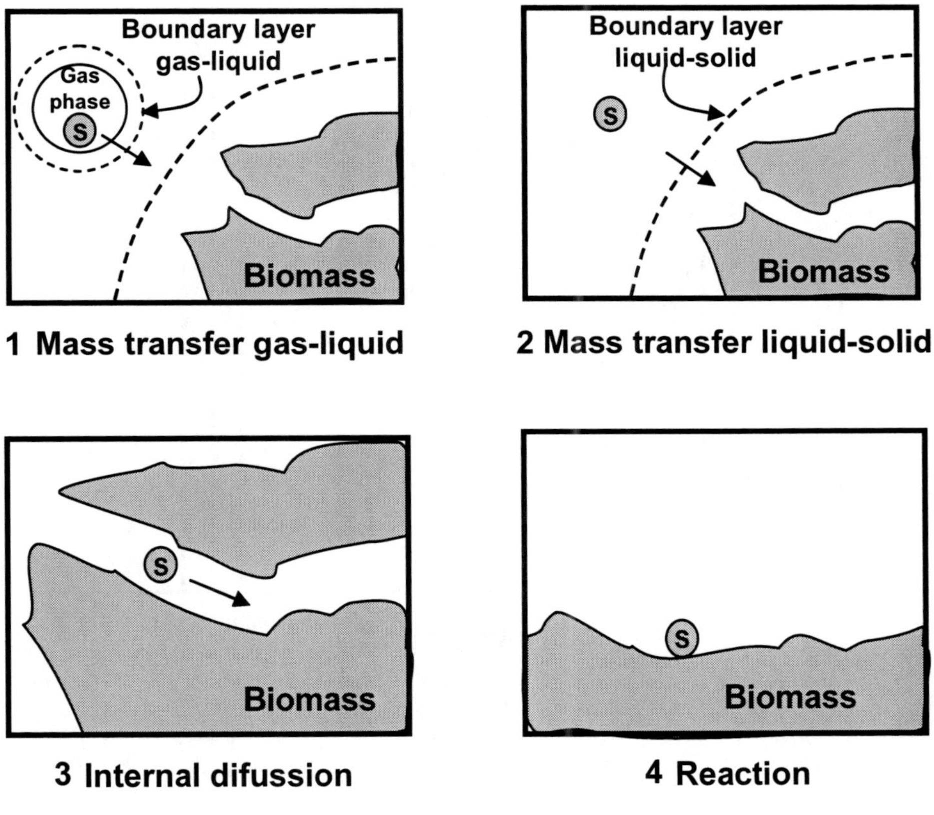

Figure 2. Steps involved during the reaction (S: Substrate).

External mass transfer describes the movement of substrate from the gas phase to the bulk liquid and from the bulk liquid to the aggregate surface. This occurs in the regions where the convective fluid velocity decreases. In this region, often referred to as the external boundary layer, diffusive mass transfer increasingly dominates as the surface is approached, but advection (transport due to fluid motion) can still play a significant role. Internal mass transfer describes the diffusive movement of substrates within the biological aggregate, to the site of the biochemical reaction (at or within the cell).

Resistances associated to either of these transfer steps caused that substrate concentration in the site of biochemical reaction is lower than in the gas phase or in the bulk liquid and, therefore, the overall rate of biological treatment decreased (Figure 3). The relative importance of these resistances in providing a process rate limitation is influenced by numerous factors, including the size and shape of the biomass aggregates, fluid and substrate properties, intrinsic reaction kinetics of the microorganisms and the hydrodynamics of the bulk phase. In this aspect, agitation and aeration are generally used to improve mass transfer rate in reactors due to the hydrodynamic turbulence caused by them. They also can be used to improve the performance of the reactors by enhancing the fluid mixing and controlling the thickness of biofilms (Ebrahimi et al., 2006). However, the energy required to induce and maintain the physical mass transfer pathways in the systems should be also considered.

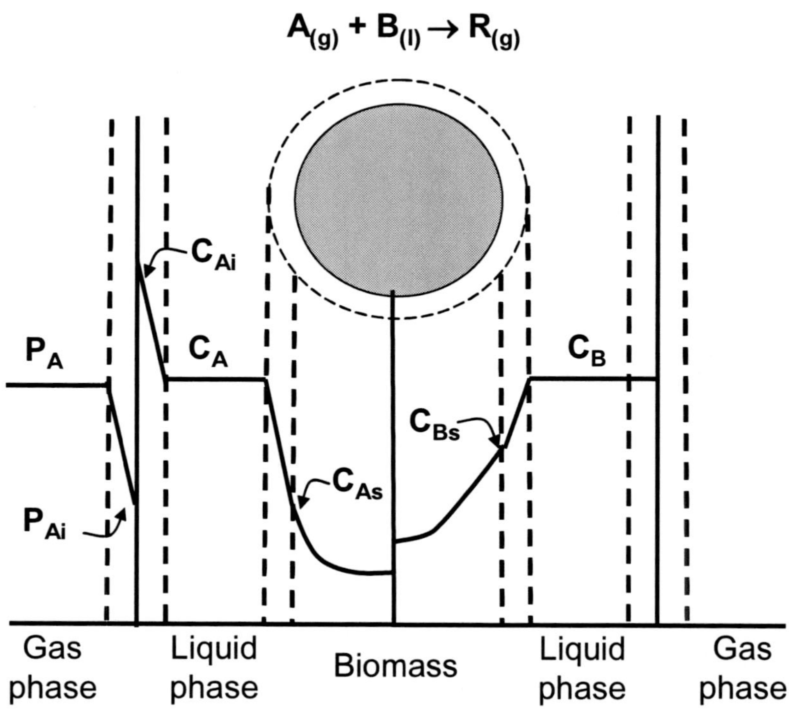

Figure 3. Substrates profiles caused by the mass transfer resistances (P_A: Partial pressure of A in the gas phase; P_{Ai}: Partial pressure of A in the interphase gas-liquid; C_{Ai}: Concentration of A in the interphase gas-liquid (in equilibrium with P_{Ai}); C_A, C_B: Concentration of A and B in the bulk liquid; C_{As}, C_{Bs}: Concentration of A and B inside the biomass aggregate).

Generally, agitation and aeration rates are the most critical parameters used for scale-up biological treatment process and play significant roles in determining the efficiency of the process. For this reason, great efforts were carried out in order to obtain engineering correlations to the scaling-up of biochemical reactors which allows calculating shear stress and mass transfer coefficients as a function of operational parameters such as superficial gas and liquid velocities, stirring velocity or energy input.

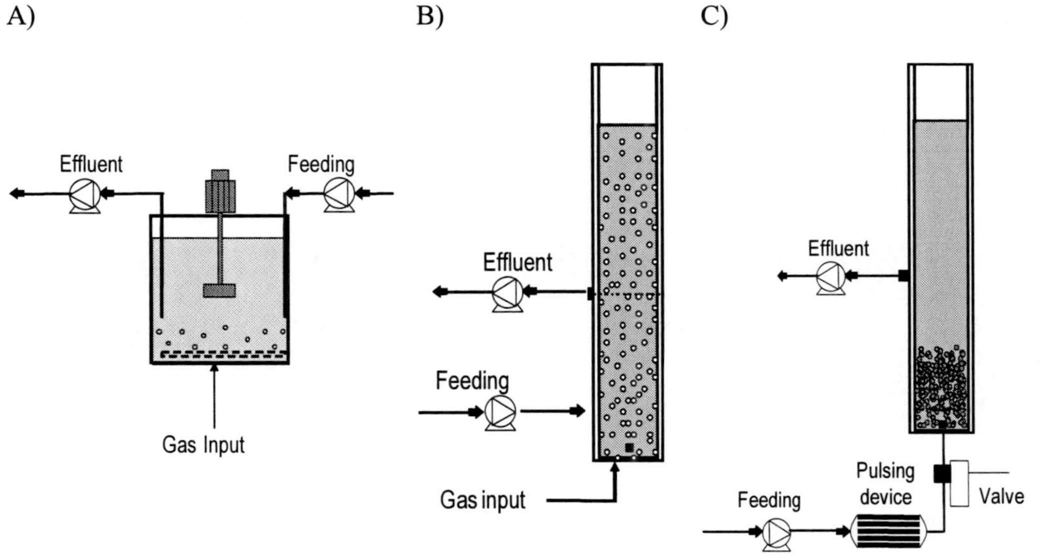

Figure 4. Examples of A) Impact stress: Stirred tank with mechanical stirring; B) Hydraulic stress generated by gas: Fluidized bed; C) Hydraulic stress generated by liquid: Anaerobic reactor with pulsed feeding.

Shear Stress and Mass Transfer Coefficients

In wastewater systems, shear stress can be generated by (Figure 4):

- *Impact stress:* due to collisions between the impeller/baffles/tank shell and the particles or between the particles.
- *Hydraulic stress generated by gas:* due to the aeration (aerobic systems) or to the recycling gas (anaerobic/anoxic systems).
- *Hydraulic stress generated by liquid:* by maintaining a pulsed flow or by a high upflow velocity of influent (by recycling).

The effect of shear stress on the biomass characteristics depends on the type of reactor used (Cao and Alaerts, 1995). Pereboom (1997) found that abrasion rate in stirred tanks was higher than in bubbles columns for similar shear rates. This author observed attrition of granules in bubbles columns while in stirred tanks also granule breakage occurs. Arrojo et al. (2008) also found a higher decrease of biomass specific activity in a recycled gas flow system compared to a mechanical stirred system both operating at similar shear rates values. These authors concluded that the decrease of biomass specific activity could be related not only to the effect of shear stress itself but also to other factors such as the loss of intermediates by gas sparging. The importance of the type of system used to generate shear stress is discussed in the following sections.

Shear stress is an important parameter in bioreactors not only by its effects on biomass characteristics but also by its influence on the average apparent viscosity of non-Newtonian fluids and, therefore, on power absorption, mixing characteristics and mass transfer rate. Shear stress in the systems depends on the shear rate and the specific input power:

$$\tau = \frac{1}{\gamma}\frac{P}{V} \tag{1}$$

where τ is the shear stress (Pa); γ is the shear rate (1/s); P is the power input (W) and V is the volume of fluid (m^3).

For Newtonian fluids, the viscosity is the ratio of shear stress and shear rate, i.e.:

$$\mu = \frac{\tau}{\gamma} \tag{2}$$

being μ is the viscosity (Pa·s).

Therefore, equation (1) can be written as follows:

$$\tau = \left(\mu \frac{P}{V}\right)^{1/2} \tag{3}$$

For non-Newtonian fluids obeying the power law (equation 4):

$$\tau = K \cdot \gamma^n \tag{4}$$

being K the consistency index ($Pa \cdot s^n$) and n the flow index of the fluid. In this case equation (1) can be written as:

$$\tau = \left(K^n \cdot \left(\frac{P}{V}\right)\right)^{n/(n+1)} \tag{5}$$

For agitation under turbulent flow, the power number is almost constant and the specific input power for Newtonian fluids can be expressed as:

$$\left(\frac{P}{V}\right) = \frac{\rho \cdot N^3 \cdot d_i^5}{V} \cdot N_p \tag{6}$$

where ρ is the density of fluid (kg/m^3); N is the agitation rate (s^{-1}); d_i is the diameter of the impeller (m) and N_p is the power number. The power number is defined as a dimensionless number which expresses the ratio between pressure energy and kinetic energy and depends on the various geometrical parameters of the impeller (diameter and height) and reactor (diameter and height), the agitation rate and the fluid properties (density and viscosity). If the stirred tank operates at turbulent conditions, the input power only would depend on the agitation rate.

Substitution of the above in equation (3) leads to the following expression:

$$\tau = \left(\frac{\mu \cdot \rho \cdot N^3 \cdot d_i^5 \cdot N_p}{V} \right)^{1/2} \tag{7}$$

Bubble column and airlift bioreactors are also reactors widely used for the biological treatment of wastewaters because of their high mass transfer rates and simplicity of design. In systems where hydraulic stress is generated by gas, the power input per unit volume of liquid is related with the superficial gas velocity u_g (m/s), as follows (Sánchez Pérez et al., 2006):

$$\left(\frac{P}{V} \right) = g \cdot \rho \cdot u_g \tag{8}$$

Then the combination of equations (5) and (8) gives the shear stress as a function of the superficial gas velocity for no-Newtonian fluids:

$$\tau = \left(K^n \cdot \left(g \cdot \rho \cdot u_g \right) \right)^{n/(n+1)} \tag{9}$$

In packed bed reactors where hydraulic stress is caused by the liquid flow rate, the shear stress can be calculated according to the following equation (Rittmann and McCarty, 2001):

$$\tau = \frac{C_1 \cdot \mu \cdot u_l \cdot (1-\varepsilon)^2}{d_p^2 \cdot \varepsilon^3 \cdot a} \tag{10}$$

where C_1 is an empirical coefficient, u_l is the superficial liquid velocity (m/s), ε is the porosity of the bed, d_p is the equivalent diameter of the carrier (m) and a is the specific surface of the biofilm carrier (1/m).

Iaconi et al. (2005) also developed a simple methodology to calculate hydrodynamic shear forces in a fixed bed system with granular biomass. According to such a procedure, the shear stress was calculated by equation (11):

$$\tau = \frac{C_2 \cdot \mu \cdot u_l \cdot (1-\varepsilon)}{d_p \cdot \varepsilon^2} + \frac{C_3 \cdot \rho \cdot u_l^2}{\varepsilon^2} \tag{11}$$

where C_2 and C_3 are empirical coefficients.

Correlations given in literature for the prediction of mass transfer coefficients are generally expressed as a function of the power input of the impeller and the superficial gas and/or liquid velocities and can be related to shear stress by equations (7), (9), (10) and (11). Several correlations are available for the calculation of the gas-liquid mass transfer coefficients. In mechanically agitated gas-liquid contactors for both Newtonian and non-Newtonian aerated liquids, a typical correlation is expressed in the form (Markopoulos et al., 2007):

$$k_{La} = C_4 \cdot \left(\frac{P}{V} \right)^{\alpha} \cdot u_g^{\beta} \tag{12}$$

where k_{La} is the gas-liquid mass transfer coefficient (s^{-1}), C_4, α and β are empirical coefficients.

For all types of airlift reactors, k_{La} increases with gas velocity (Lazarova et al., 1997; Masoud et al., 2001) and, usually, a power function law of gas velocity is used:

$$k_{La} = C_5 \cdot u_g^{\delta} \tag{13}$$

being C_5 and δ are empirical coefficients.

Nevertheless, in liquid–gas mixing systems when the gas and liquid phases are forced to circulate together, Kouakou et al. (2005) proposed to describe the combined action of air and forced liquid superficial velocities on k_{La} by:

$$k_{La} = (C_6 \cdot u_l + C_7) \cdot u_g^{\gamma} \tag{14}$$

being C_6, C_7 and γ are empirical coefficients.

The liquid-solid mass transfer coefficient is estimated by correlations which give the Sherwood number (Sh), as a function of the Reynolds number (Re) and the Schmidt number (Sc). The equations of Ranz and Marshall (1952) and Brian and Hales (1969) are the most used for the estimation of k_c:

$$Sh = \frac{k_c \cdot d_p}{D} \tag{15}$$

$$\mathrm{Re} = \frac{d_p \cdot u_{s-l} \cdot \rho}{\mu} \tag{16}$$

$$Sc = \frac{\mu}{\rho \cdot D} \tag{17}$$

$$Sh = 2 + 0.6 \cdot \mathrm{Re}^{1/2} \cdot Sc^{1/3} \tag{18}$$

$$Sh = \sqrt{4.0 + 1.21 \cdot (\mathrm{Re} \cdot Sc)^{2/3}} \tag{19}$$

where k_c is the liquid-solid mass transfer coefficient (m/s), u_{s-l} is the terminal settling velocity of the particle (m/s) and D diffusivity of substrate in the liquid (m^2/s).

In the later correlations, it was assumed that the particles move with the rate of free fall. However, when gas is supplied to the system the particles movement is influenced by the turbulent flow in the reactor and other correlations which take into account input energy seem to be more appropriate to calculate k_c (Wijffels et al., 1998):

$$Sh = C_8 + C_9 \cdot Sc^{\eta} \cdot \left(\frac{P}{V} \frac{d_p^4}{\rho \cdot \nu^3} \right)^{\lambda} \tag{20}$$

being C_8, C_9, η and λ are empirical coefficients and ν is the kinematic viscosity (m^2/s).

Impact Stress

Impact stress can be due to the contact between the particles and the tank shell/baffles or the impeller elements, or to the contact between the particles. The later effect is negligible if the density of particles is similar or in the case that the particles concentrations are low. The effect of impact stress due to the impeller elements was widely investigated in fermentation processes (Henzler, 2000; Casas-López et al., 2005). However only few works are reported for wastewater treatment systems (Wilén et al., 2000; Liu et al., 2005; Arrojo et al., 2006). This can be due to the fact that anaerobic systems are generally mixed by recirculating gas or liquid, while air is used in aerobic processes to maintain the mixing inside the system without need of mechanical stirring in both cases.

The effects of agitation rate on the characteristics of flocs of heterotrophic biomass were studied by Liu et al. (2005). These authors found that there was an optimum range of the agitation rate between 400 and 600 rpm which leads to achieve large flocs with a low SVI. They also found a decrease of the sludge yield coefficient with the increase of shear stress. On the contrary, Mikkelsen and Keiding (1999) observed no deflocculation at low agitation rate (100 rpm), whereas an almost linear relation between the input energy and deflocculation was found at high values.

Arrojo et al. (2006) researched, in a stirred tank, the effect of shear stress on the granular Anammox (Anaerobic Ammonia Oxidation) biomass by stepwise increasing the stirring rate. These authors found that an increase of the specific input power from 0.003 to 0.09 kW/m^3, corresponding to an increase of the stirring rate from 60 to 180 rpm, had no effect on the specific biomass activity. However, the Anammox activity decreased a 40% of the original value when the specific power applied to the reactor was 0.23 kW/m^3 (250 rpm). They also observed that the increase of the shear stress also causes a progressive decrease of the average diameter of particles which can be attributed to the breakage of the granules. In this work the loss of activity was not directly correlated to the decrease of the particle diameter.

Impact stress due to collisions between particles is only important when particles with very different densities are present inside the system because the difference of densities causes a difference of their velocities provoking that collisions between them are very frequent. The effect of these collisions has a special importance in biofilm systems during their start-up because in this period significant percentages of bare, partially covered and fully covered particles coexist. In fact several works showed the most important factor which influences the density of the biofilm and its detachment rate, and therefore the biofilm size, is not superficial gas velocity but collisions between bare and covered particles (Gjaltema et al., 1997; Kwok et al., 1998).

Mechanical agitation can be used to improve the mass transfer rate of anaerobic systems (Michelan et al., 2008). These authors determined the apparent kinetic constants of the process and observed that the increase of the rotor speed resulted in an increase in the constant values, by increasing the mass transfer rate, and that axial flow in mechanically stirred reactors was preferable over radial-flow. A similar positive effect of mechanical agitation on the performance of anaerobic systems was found by Pinho et al. (2004) who observed the acceleration of suspended COD degradation with the increase of agitation rate. They attributed these results to the higher velocity of shear of larger particles and major contact between the particulate organic matter and the extra-cellular enzymes. They found the apparent first-order kinetic constant for suspended COD rose approximately 360% when the agitation rate was changed from 500 to 900 rpm, whereas the apparent first-order kinetic constant for soluble COD did not vary significantly. Although the time required to achieve a desirable efficiency can be reduced as the agitation rate is increased, high agitation rates may represent high energy consumption. Therefore, the agitation rate to be applied should be chosen taking into account both economic and environmental aspects (Ratusznei et al., 2001; Rodrigues et al., 2003; Cubas et al., 2004).

Membrane fouling control in membrane bioreactor (MBR) could be achieved by means of mechanical agitation. The major drawback of the membrane bioreactor operation is the decline in membrane filtration performance with time. This decline is due to deposition of particles and soluble materials onto membrane pores, attributed to the interaction of between the biomass components and the membrane. The fouling of membranes is determined, among other factors, by biomass characteristics and the hydrodynamic environment which are closely related to shear stress inside the system. Shear stress can be induced either by cross-flow velocity or by aeration. The cross-flow velocity is key factor in the operation of a side-stream MBR while bubbling has been the strategy of choice for submerged MBRs to induce flow circulation and shear stress on the membrane surface. The actual tendency shows an increase of the submerged membrane systems application in the wastewater treatment field due to their absence of a high-flow recirculation pump and, therefore, their lower power requirements compared to external MBRs (Yang et al., 2006).

The increase in the aeration intensity to avoid membrane fouling can be uneconomical and also affect the biomass growth rate and the microbial community of the sludge (Ji and Zhou, 2006). Therefore, Khan and Visvanathan (2008) proposed the use of mechanical agitation to enhance shear stress in addition to aeration intensity for membrane fouling control in submerged MBRs. These authors observed that the fouling rate in a MBR with mechanical agitation could be up to five times slower than that of a MBR without mechanical agitation in spite of the amount of soluble and colloidal matter in the mixed liquor suspension was higher when mechanical agitation was applied. They attributed this fact to the high stress intensity and to the reduction of the activity of suspended biomass and a subsequent low EPS release of the biofilm formed on the membrane.

Hydraulic Stress Generated by Gas

In aerobic activated sludge systems when the airflow supplied to maintain an adequate dissolved oxygen level is higher than 0.02-0.03 $m^3/(m^3{\cdot}min)$ a complete mixing behaviour can be obtained. In this kind of systems, the loading rate is usually limited by the maximum biomass concentration that can be retained in the reactor. Due to the common settling properties of the sludge, SVI around 50-100 ml/g VSS and ZSV of 1 m/h, the biomass concentration in the reactor is usually between 3 and 6 g VSS/L. Maintenance of higher biomass concentrations in the system requires the increase of the settler volume to separate the biomass from the treated water, involving the need of very large surface areas. In this respect, shear stress due to high airflows could have a positive effect since it causes a decrease of the SVI (Campos et al., 1999) and an increase of the density (Iaconi et al., 2006) which allows the biomass retention to be improved.

An early work of Shin et al. (1992) already showed that a high shear force favours the formation of aerobic granules and granule stability. In fact, Tay et al. (2001) studied the effect of shear stress value in terms of superficial upflow air velocity on the formation of aerobic granules in a bubble column. These authors found that aerobic granules could be formed only above a threshold of 1.2 cm/s. Denser, rounder and compact granules are developed at high hydrodynamic shear force. This fact indicates that the structure of aerobic granules is determined by the hydrodynamic shear force present in the bioreactor. The better stability of the aerobic granules at high shear forces would be related to the higher production of extracellular polysaccharides (EPS) under these conditions which play a crucial role in maintaining the structural integrity in a community of immobilized cells (Sheng et al., 2006). On the contrary, Iaconi et al. (2006) observed the EPS content of granular biomass in a sequencing batch biofilter reactor was not affected by hydrodynamic shear forces. Adav et al. (2007) demonstrated that proteins, rather than polysaccharides, were enriched in the sheared granules, which was consistent with the results of McSwain et al. (2005) and Chen et al. (2007).

Arrojo et al. (2008) performed the Anammox process in a gas lift reactor operated at different upflow velocities (from 3.53 to 9.70 cm/min) in order to expose the system to different shear forces, ranged between 0.0115 and 0.0316 kW/m^3. These authors found the negative effects on the specific activity of biomass started when the system was operated at an upflow velocity higher than 5.29 cm/min, observing a decrease of the activity of 85% at 9.70 cm/min. The average diameter of the granules gradually decreased in 30% during the operational period.

The influence of gas velocity on a methanogenic biofilm was studied by Michaud et al. (2003) in an inverse turbulent bed reactor. They found that the increase of gas velocity caused an increase of the specific activity and the biofilm formed was more dense, stable and resistant to detachment. They also observed higher exocellular polymeric substances production at higher shear stress conditions which could explain the greater cohesion of biological matrix.

Applying high shear force by gas sparging proved also to be effective in improving denitrification rates in a hollow-fiber membrane biofilm reactor by reducing the thickness of the biofilm (Celmer et al., 2008). In submerged MBRs, the membrane fouling can be mitigated with the increase in aeration intensity (Germain et al., 2005). The increase of aeration also causes a decrease of the biomass particle size, favouring cell debris and macromolecules which can have an important contribution to the membrane fouling. However, the net effect observed is the improved membrane filtration with the increased aeration. On the contrary, Ueda et al. (1997) reported that an optimum aeration rate exits beyond which a further increase has no significant effect on membrane fouling suppression.

Hydraulic Stress Generated by Liquid

The upflow velocity of influent could be manipulated to acts as a hydraulic selection pressure on microbial community. A high upflow velocity of influent selects for the growth of fast settling bacteria and the sludge with a poor settleability is washed out. Alphenaar et al. (1993) reported that biomass granulation in an upflow anaerobic sludge blanket (UASB) reactor is favoured by the combination of high-liquid upflow velocity and short hydraulic retention time (HRT). However, for successful start-up and stable operation of UASB reactors, the reactor HRT cannot be reduced below 6 h and upflow velocities must be lower than 1 m/h to avoid both disintegration of granules due to shearing and the wash out of the resulting fragments (Kosaric et al., 1990). Mishima and Nakamura (1991) used the so called Aerobic Upflow Sludge Blanket (AUSB) to obtain granular biomass under aerobic conditions.

In anaerobic immobilized fixed bed reactors shear stress by the liquid flow rate caused the biofilm loss and a relationship between shear stress and both biofilm loss rate and biofilm thickness can be established (Nandy and Kaul, 2002). García-Morales et al. (2003) found that, in spite of the biofilm detachment, the increase of the upflow velocities in anaerobic fluidized bed reactors caused a positive effect on the specific biomass activity due to the reduction of the substrate diffusional limitations to the microorganism growth inside the support pores. Venkata Mohan et al. (2007) also found the recirculation of liquid improved the efficiency of the treatment because it promoted the mass transfer of the substrate from the bulk liquid to the biomass aggregates.

Pulsing flow has been applied to a great number of chemical (Lemay et al., 1975; Lerou et al., 1980; Boelhouwer et al., 2001) and biochemical processes (Serieys et al., 1978; Ghommidh et al., 1982; Murakami et al., 2000) for improving mass transfer, performance and efficiency of the equipment employed (Beeton, 1991; Hewgill et al., 1993; Ni et al., 1995; Dondé et al., 1987). Pulsing flow punctually increases upflow velocity, so shear stress (i.e., selection pressure) may be enhanced under pulsing operation. However, it generates lower mechanical stress than conventional stirring, which prevents disintegration of the granules although it still generates enough to facilitate the formation of aggregates, especially when no carrier is employed. Moreover, the implementation of pulsing systems is easier than other systems and involves a much lower energy cost both at the pilot and industrial scale (Ni et al., 2003).

One type of pulsing bioreactor, the "reciprocating jet bioreactor" was successfully employed for the removal of organic matter and ammonia from wastewater at pilot scale (Brauer and Hennig, 1986) at very short residence times (only 5% of the employed in a conventional system). The turbulence generated by the pulsing movement caused the dispersion of gas bubbles and reduced the size of sludge particles, which implied an increase in the interphase area gas-sludge, favouring mass transfer. Pulsation frequency was in this case directly related to oxygen transfer within biomass particles, and therefore, with its size.

Liquid pulsing flow has been also proposed to improve the performance of anaerobic/anoxic bioreactors. This kind of flow allows removal efficiency and stability to be enhanced in upflow anaerobic filters (UAF) and UASB reactors, as it favours degasification avoiding flotation and wash-out of the biomass, and decreases the formation of preferential pathways (Franco et al., 2002; Franco et al., 2003; Franco et al., 2007). Performance improvement of bioreactors caused by pulsation is due to a combined effect on both biomass characteristics and the hydrodynamic conditions (Franco et al., 2003). From the point of view of hydraulics, pulsation promotes biogas-sludge separation which is positive for controlling biomass wash out at high loading rates, when a quite significant gas flow is produced, and minimises the formation of preferential pathways. Both factors lead to a higher useful volume of the reactor. The high porosity of biomass obtained in pulsed reactors could be one of the causes of the better degasification of the sludge bed due to the enhancement of the internal gas transfer rate (Alphenaar et al., 1992). Improvement in biogas liberation from the inner part of the granule, due to hydraulic stress, may cause a compensation convective flow of liquid within the granule, improving mass transport (van den Heuvel et al., 1997). In addition, a better degasification increases the mixing in anaerobic/anoxic bioreactors (Kato et al., 1994).

Pulsing flow also favours biomass aggregation and granule development in UASB reactors (Franco et al., 2002; Franco et al., 2003). Pulsations promote aggregation of flocculent anaerobic biomass, shortening the granulation process which can last several months in an UASB reactor (Franco et al., 2002). The hydraulic stress or selection pressure generated by pulsation is beneficial to the self-aggregation of particles and also favours the wash out of flocculent particles no susceptible to form aggregates of higher size (Hulshoff Pol, 1989; Alphenaar et al., 1993).

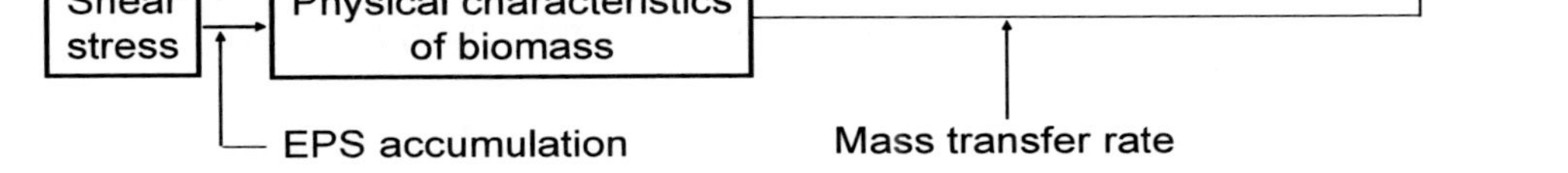

Figure 5. Interactions among wastewater composition, reactor type and shear stress.

The application of pulsing flow to anoxic denitrifying USB reactors results in a more stable and more efficient operation due not only to the granulation of the flocculent sludge but also to the increment in the overall mass transfer rate (Franco et al., 2006). High shear stress produced in pulsed bioreactors generates small-size granules with a high active surface and porosity compared with those formed in non-pulsed reactors. Franco et al. (2008) studied the effect of the pulsing flow on the performance of UAF and UASB reactors. These authors found that in both kinds of systems external mass transfer resistance was decreased by application of the pulsing flow. Nevertheless, Damköhler numbers obtained (maximum kinetic rate to maximum external mass transfer rate ratio) showed that the systems were still limited by the external mass transfer rate due to the formation of granules with a lower size and more active. These authors also observed that internal mass transfer rate was negligible compared to the external one and attributed this fact to the sudden flow regime change during the pulse. In fact, several authors have observed experimental evidence of convective fluxes within biofilms mainly due to pressure gradients (Brito and Melo, 1999; de Ber et al., 1996; van den Heuvel et al., 1997).

Conclusion

Hydrodynamic conditions are proved to highly influence the properties of the obtained biomass and the efficiency of the performed process in biological reactors treating wastewater. Beneficial effects will depend on the values of the input power used up to a limit value. Higher values over this limit are expected to produce a detrimental effect on the biomass characteristics and the efficiency of the process, being possible, in some cases, the complete loss of microbial activity. Shear stress together with the type of reactor and wastewater composition are the main factors that affect to the performance of the bioreactor (Figure 5). It has a significant influence on the structure, mass transfer rate, production of exopolysaccharides, metabolic/genetic behaviour of biofilms, and aerobic and anaerobic granules. From an engineering point of view, hydrodynamic shear stress can be managed, as a control parameter, to enhance the operation and efficiency of wastewater treatment systems.

Acknowledgements

This work was funded by the Spanish Government (Biogramem project CTQ2005-04935/PPQ and NOVEDAR_Consolider project CSD2007-00055).

References

Adav, S.S.; Lee D.J.; Lai, J.Y. (2007). Effects of aeration intensity on formation of phenol-fed aerobic granules and extracellular polymeric substances. *Applied Microbiology and Biotechnology,* **77**, 175–182.

Alphenaar, P.A.; Pérez, M.C.; van Berkel, W.J.H.; Lettinga, G. (1992). Determination of the permeability and porosity of anaerobic sludge granules by size exclusion chromatography. *Applied Microbiology and Biotechnology,* **36**(6), 795-799.

Alphenaar, P.A.; Pérez, M.C.; Lettinga, G. (1993). The influence of substrate transport limitation on porosity and methanogenic activity of anaerobic sludge granules. *Applied Microbiology and Biotechnology,* **39**(2), 276-280.

Arrojo, B.; Mosquera-Corral, A.; Campos, J.L.; Méndez, R. (2006). Effects of mechanical stress on Anammox granules in a sequencing batch reactor (SBR). *Journal of Biotechnology,* **123**, 453-463.

Arrojo, B.; Figueroa, M.; Mosquera-Corral, A.; Campos, J.L.; Méndez R. (2008). Influence of gas flow-induced shear stress on the operation of the Anammox process in a SBR. *Chemosphere,* **72**(11), 1687-1693.

Beeton, S. (1991). Gas transfer characteristics of a novel membrane bioreactor. *Biotechnology and Bioengineering,* **38**(10), 1233-1238.

Beun, J.J.; Heijnen, J.J; van Loosdrecht, M.C.M. (2001). N-Removal in a granular sludge sequencing batch airlift reactor. *Biotechnology and Bioengineering,* **75**, 82-92.

Beun, J.J.; van Loosdrecht, M.C.M.; Heijnen, J.J. (2002). Aerobic granulation in a sequencing batch airlift reactor. *Water Research,* **36**, 702-712.

Boelhouwer, J.G.; Piepers, H.W.; Drinkenburg, A.A.H. (2001). Particle-liquid heat transfer in trickle-bed reactors. *Chemical Engineering Science,* **56**, 1181–1187.

Brauer, H. & Hennig, R. (1986). Aerobic two-stage biological degradation of organic carbon compounds and ammonium. *Bioprocess Engineering,* **1**(1), 13-22.

Brian, P.L.T. & Hales, H.B. (1969). Effects of transpiration and changing diameter on heat and mass transfer to spheres. *AlChE Journal,* **14**, 419-425.

Cao, Y.S. & Alaerts, G.J. (1995). Influence of reactor type and shear stress on aerobic biofilm morphology, population and kinetics. *Water Research,* **29**(1), 107-118.

Campos, J.L.; Garrido-Fernández, J.M.; Méndez, R.; Lema, J.M. (1999). Nitrification at high ammonia loading rates in an activated sludge unit. *Bioresource Technology,* **68**, 141-148.

Casas-López, J.L.; Sánchez-Pérez, J.A.; Fernández-Sevilla, R.; Rodríguez-Porcel, E.M.; Chisti, Y. (2005). Pellet morphology, culture rheology and lovastatin production in cultures of Aspergillus terreus. *Journal of Biotechnology,* **116**, 61-77.

Celmer, D.; Oleszkiewicz, J.A.; Cicek, N. (2008). Impact of shear force on the biofilm structure and performance of a membrane biofilm reactor for tertiary hydrogen-driven denitrification of municipal wastewater. *Water Research,* **42**, 3057–3065.

Chen, M.Y.; Lee, D.J.; Tay, J.H. (2007). Distribution of extracellular polymeric substances in aerobic granules. *Applied Microbiology and Biotechnology,* **73**, 1463–1469.

Cubas, S.A.; Foresti, E.; Rodrigues, J.A.D.; Ratusznei, S.M.; Zaiat, M. (2004). Influence of liquid-phase mass transfer on the performance of a stirred anaerobic sequencing batch reactor containing immobilized biomass. *Biochemical Engineering Journal,* **17**, 99–105.

de Beer, D.; Stoodley, P.; Lewandowski, Z. (1996). Liquid flow and mass transport in heterogeneous biofilms. *Water Research,* **30**, 2761–2765.

Dondé, M.; Goma, G.; Durand, G. (1987). Transfer oxygen potential of an air-pulsed continuous fermentor. *Biotechnology Process,* **4**, 135-141.

Ebrahimi, S.; Kleerebezem, R.; Kreutzer, M.T.; Kapteijn, F.; Moulijn, J.A.; Heijnen, J.J.; van Loosdrecht, M.C.M. (2006). Potential application of monolith packed columns as bioreactors, control of biofilm formation. *Biotechnology and Bioengineering,* **93**(2), 238-245.

Franco, A.; Gresia, G.; Roca, E.; Rozzi, A.; Lema, J.M. (2002). Influence of pulsation on start-up of UASB reactors. *Water Science and Technology,* **45**(10), 163-168.

Franco, A.; Roca, E.; Lema, J.M. (2003). Improvement of the properties of granular sludge in UASB reactors by flow pulsation. *Water Science and Technology,* **48**(6), 51-56.

Franco, A.; Roca, E.; Lema, J.M. (2006). Granulation in high-load denitrifying upflow sludge bed (USB) pulsed reactors. *Water Research,* **40**, 871-880.

Franco, A.; Roca, E.; Lema, J.M. (2007). Enhanced start-up of upflow anaerobic filters by pulsation. *Journal of Environmental Engineering,* **133**, 186–190.

Franco, A.; García, C.; Carrasco, E.F.; Roca, E. (2008). Explaining the enhanced performance of pulsed bioreactors by mechanistic modelling. *AIChE Journal,* **54**(5), 1377-1387.

García-Morales, J.L.; Romero, L.I.; Sales, D. (2003). Influence of operational conditions on biofilm specific activity of an anaerobic fluidized bed reactor. *Water Science and Technology*, **47**(5), 197–200.

Germain, E.; Stephenson, T.; Pearce, P. (2005). Biomass characteristics and membrane aeration: Toward a better understanding of membrane fouling in submerged membrane bioreactors (MBRs). *Biotechnology and Bioengineering,* **90**, 316–322.

Ghommidh, C.; Navarro, J.M.; Durand, G. (1982). A study of acetic acid production by immobilized Acetobacter cells: oxygen transfer. *Biotechnology and Bioengineering,* **24**(3), 605-617.

Gjaltema, A.; van Loosdrecht, M.C.M.; Heijnen, J.J. (1997). Abrasion of suspension biofilm pellets in airlift reactors: effect of particle size. *Biotechnology and Bioengineering,* **55**, 206-215.

Gmachowski, L. (1995). Mechanism of shear aggregation. *Water Research,* **29**, 1815-1820.

Henzler, H.J. (2000). Particle stress in bioreactors. *Advances in Biochemical Engineering and Biotechnoloy,* **67**, 36-82.

Hewgill, M.R.; Mackley, M.R.; Pandit, A.B.; Pannu S.S. (1993). Enhancement of gas-liquid mass transfer using oscillatory flow in a baffled tube. *Chemical Engineering Science,* **48**(4), 799-809.

Hulshoff Pol, L.W. (1989). The phenomenon of granulation of anaerobic sludge. *PhD thesis,* Wageningen Agricultural University, The Netherlands.

Iaconi Di, C.; Ramadori, R.; Lopez, A.; Passino, R. (2005). Hydraulic shear stress calculation in a sequencing batch biofilm reactor with granular biomass. *Environmental Science and Technology,* **39** (3), 889–894.

Iaconi, Di C.; Ramadori, R.; López, A.; Passino, R. (2006). Influence of hydrodynamic shear force on properties of granular biomass in a sequencing batch biofilter reactor. *Biochemical Engineering Journal,* **30**, 152-157.

Ji, L. & Zhou, J. (2006). Influence of aeration on microbial polymers and membrane fouling in submerged membrane bioreactors. *Journal of Membrane Science,* **276**, 168–177.

Khan, S.J. & Visvanathan, C. (2008). Influence of mechanical mixing intensity on a biofilm structure and permeability in a membrane bioreactor. *Desalination,* **231**, 253-267.

Kato, M.T.; Field, J.A.; Versteeg, P.; Lettinga, G. (1994). Feasability of expanded granular sludge bed reactors for the anaerobic treatment of low strength soluble wastewaters. *Biotechnology and Bioengineering,* **44**(4), 469-480.

Kosaric, N.; Blaszczyk, R.; Orphan, L. (1990). Factors influencing formation and maintenance of granules in upflow anaerobic sludge blanket reactors (UASBR). *Water Science and Technology,* **22**, 275–282.

Kouakou, E.; Salmon, T.; Toye, D.; Marchot, P.; Crine, M. (2005). Gas–liquid mass transfer in a circulating jet-loop nitrifying MBR. *Chemical Engineering Science,* **60**, 6346–6353.

Kwok, W.K.; Picioreanu, C.; Ong, S.L.; van Loosdrecht, M.C.M.; Ng, W.J.; Heijnen, J.J. (1998). Influence of biomass production and detachment forces on biofilm structures in a biofilm airlif suspension reactor. *Biotechnology and Bioengineering,* **58**, 400-407.

Lazarova, V. ; Julien, M. ; Laurent, D. ; Jacques, M., (1997). A novel circulating bed reactor: hydrodynamics, mass transfer and nitrification capacity. *Chemical Engineering Science,* **52** (21/22), 3919–3927.

Lemay, Y.; Pineault, G.; Ruether, J.A. (1975). Particle-liquid mass transfer in a three-phase fixed bed reactor with concurrent flow in the pulsing regime. *Industrial and Engineering Chemistry Process Design and Development,* **14**, 280–285.

Lerou, J.J.; Glasser, D.; Luss, D. (1980). Packed bed liquid phase dispersion in pulsed gas-liquid downflow. *Industrial and Engineering Chemistry Fundamentals,* **19**, 66–71.

Liu, Y. & Tay, J.H. (2002). The essential role of hydrodynamic shear force in the formation of biofilm and granular sludge. *Water Research,* **36**, 1653-1665.

Liu, Q.S.; Liu, Y.; Tay, J.H.; Show, K.Y. (2005). Responses of sludge flocs to shear strength. *Process Biochemistry,* **40**, 3213-3217.

Markopoulos, J.; Christofi, C.; Katsinaris, I. (2007). Mass transfer coefficients in mechanically agitated gas-liquid contactors. *Chemical Engineering and Technology,* **30**(7), 829–834.

Masoud, A.J.; Morteza, S.; Farzabeh, V.; Babak, B. (2001). Hydrodynamic and mass transfer characterization of a down flow jet loop bioreactor. *Biochemical Engineering Journal,* **8**, 241–250.

McSwain, B.S.; Irvine, R.L.; Hausner, M.; Wilderer, P.A. (2005). Composition and distribution of extracellular polymeric substances in aerobic flocs and granular sludge. *Applied and Environmental Microbiology,* **71**, 1051–1057.

Michaud, S.; Bernet, N.; Roustan, M.; Delgenès, J.P. (2003). Influence of hydrodynamic conditions on biofilm behavior in a methanogenic inverse turbulent bed reactor. *Biotechnology Progress,* **19**, 858-863.

Michelan, R.; Zimmer, T.R.; Rodrigues, J.A.D.; Ratusznei, S.M.; Moraes, D.; Zaiat, M.; Foresti, E. (2009). Effect of impeller type and mechanical agitation on the mass transfer and power consumption aspects of ASBR operation treating synthetic wastewater. *Journal of Environmental Management,* **90**(3), 1357-1364.

Mikkelsen, L.H.; Keiding, K. (1999). Equilibrium aspects of the effects of shear and solids content on aggregate deflocculating. *Advance in Colloid and Interface Science,* **80**, 151-182.

Mishima, K.; Nakamura, M. (1991). Self-immobilization of aerobic activated sludge-a pilot study of the aerobic upflow sludge blanket process in municipal sewage treatment. *Water Science and Technology,* **23**, 981-990.

Murakami, T.; Usui, J.; Takamura, K.; Yoshikawa, T. (2000). Application of immersed-type membrane separation activated sludge process to municipal wastewater treatment. *Water Science and Technology,* **41**, 295–301.

Nandy, T.; Kaul, S.N. (2002). Biofilm loss in anaerobic immobilized fixed bed reactor system. *Environmental Technology,* **23**(4), 413-419.

Ni, X.; Gao, S.; Cumming, R.H.; Pritchard, D.W. (1995). A comparative study of mass transfer in yeast for a batch pulsed baffled bioreactor and a stirred tank fermenter. *Chemical Engineering Science,* **50**(13), 2127-2136.

Ni, X.; Mackley, M.R.; Harvey, A.P.; Stonestreet, P.; Baird, M.H.I.; Rama Rao, N.V. (2003). Mixing through oscillations and pulsations – a guide to achieving process enhancements in the chemical and process industries. *Transactions of the Institution of Chemical Engineers,* **81**(A3), 373-383.

Pereboom, J.H.F. (1997). Strength characterisation of microbial granules. *Water Science and Technology,* **36**(6-7), 141-148.

Pinho S.C.; Ratusznei S.M.; Rodrigues J.A.D.; Foresti E.; Zaiata M. (2004). Influence of the agitation rate on the treatment of partially soluble wastewater in anaerobic sequencing batch biofilm reactor. *Water Research,* **38**, 4117–4124.

Ranz, W.E.; Marshall W.R. (1952) Evaporation from drops, part II. *Chem. Eng. Progr.,* **48** (1952) 173-80.

Ratusznei, S.M.; Rodrigues, J.A.D.; Camargo, E.F.M.; Zaiat, M.; Borzani, W. (2001). Influence of agitation rate on the performance of a stirred anaerobic sequencing batch reactor containing immobilized biomass. *Water Science and Technology,* **44**(4), 305–412

Rittmann, B.E. & McCarty, P.L. (2001). *Environmental Biotechnology: Principles and Application.* McGraw-Hill, New York.

Rodrigues, J.A.D.; Ratusznei, S.M.; Camargo, E.F.M.; Zaiat, M. (2003). Influence of agitation rate on the performance of an anaerobic sequencing batch reactor containing granulated biomass treating low-strength wastewater. *Advances in Environmental Research,* **7**, 405–410

Sánchez Pérez, J.A.; Rodríguez Porcel, E.M.; Casas López, J.L.; Fernández Sevilla, J.M.; Chisti, Y. (2006). Shear rate in stirred tank and bubble column bioreactors. *Chemical Engineering Journal,* **124**, 1–5.

Serieys, M.; Goma, G.; Durand, G. (1978). Design and oxygen-transfer potential of a pulsed continuous tubular fermentor. *Biotechnology and Bioengineering*, **20**(9), 1393-1406.

Sheng, G.P.; Yu, H.Q.; Li, X.Y. (2006). Stability of Sludge Flocs Under Shear Conditions: Roles of Extracellular Polymeric Substances (EPS). *Biotechnology and Bioengineering,* **93**(6), 1095-1102.

Shin, H.S.; Lim, K.H.; Park, H.S. (1992). Effect of shear stress on granulation in oxygen aerobic upflow sludge reactors. *Water Science and Technology,* **26**, 601–605.

Tay, J.H.; Liu, Q.S.; Liu, Y. (2001). The role of cellular polysaccharides in the formation and stability of aerobic granules. *Letters in Applied Microbiology*, **33**, 222-226.

van den Heuvel, J.C.; Beuling, E.E.; van Dusschoten, D.; Roosenschoon, O.L.; Verschuren, P.G. (1997). Convective flow in methanogenic granules. *Water Science and Technology,* **36**(6-7), 311-316.

Ueda, T.; Hata, K.; Kikuoka, Y.; Seino, O. (1997). Effects of aeration on suction pressure in a submerged membrane bioreactor. *Water Research,* **31**, 489–494.

van den Heuvel, J.C.; Beuling, E.E.; van Dusschoten, D.; Roosenschoon, O.L.; Verschuren, P.G. (1997). Convective flow in methanogenic granules. *Water Science and Technology,* **36**, 311–316.

Venkata Mohan, S.; Lalit Babu, V.; Vijaya Bhaskar Y.; Sarma, P.N. (2007). Influence of recirculation on the performance of anaerobic sequencing batch biofilm reactor (AnSBBR) treating hypersaline composite wastewater. *Bioresource Technology,* **98**(7), 1373-1379.

Wilén, B.M.; Keiding, K.; Nielsen, P.H. (2000). Anaerobic deflocculation and aerobic reflocculation of activated sludge. *Water Research,* **34**(16), 3933-3942.

Wijffels, R.H.; Verheul, M.; Beverloo, W.A.; Tramper, J. (1998). Liquid/solid mass transfer in an air-lift loop reactor with a dispersed solid phase. *Journal of Chemical Technology and Biotechnology,* **71**, 147-154.

Yang, W.; Cicek, N.; Ilg, J. (2006). State-of-the-art of membrane bioreactors: Worldwide research and commercial applications in North America. *Journal of Membrane Science,* **270**, 201–211.

In: Biochemical Engineering
Editors: F.E. Dumont and J.A. Sacco, pp. 245-255
ISBN: 978-1-60741-257-1

Chapter 10

THE ROLE OF BIOFILM AND FLOC STRUCTURE IN BIOLOGICAL WASTEWATER TREATMENT MODELLING

Mario Plattes[*]

Centre de Ressources des Technologies pour l'Environnement, Centre de Recherche Public Henri Tudor, 66, rue de Luxembourg (BP 144), L-4002 Esch-sur–Alzette, Luxembourg

Abstract

Biological wastewater treatment modelling has become and important tool in process engineering. There are state of the art activated sludge models (ASMs) available, which have found wide application in the engineering community. Biofilm models have found less application in engineering practice so far, and a gap has developed between biofilm research and engineering practice in the biofilm modelling community. In this context biofilm and floc structure have played different roles in biological wastewater treatment modelling. Activated sludge models (ASMs) do not explicitly take floc structure into account. In contrast biofilm structure has been strongly emphasized in biofilm models over the past decades. Biofilm models have as a result evolved with increasing complexity from one- to two- to three-dimensional models. One reason for this is that biofilm structure is crucially linked to diffusion by Fick's laws of diffusion, since it is known that diffusion is an important process in biofilm systems. The application of Fick's laws of diffusion has thus been a driving force towards the development of multidimensional biofilm models with increased model complexity, because biofilms have a complex, heterogenous three-dimensional structure. The increasing complexity has not led, however, to increased application of biofilm models in engineering practice, and there is a trend towards simplified (e.g. zero-dimensional) models for this purpose. Further it has been shown that diffusion and structure play an important role in activated sludge systems. The role of activated sludge structure has recently led to the development of multidimensional activated sludge models in activated sludge research, whilst the state of the art ASMs for engineering practice do not take floc structure into account.

[*] Tel. : +352 - 545580 – 612 ; fax. : +352 – 545580 – 601

Introduction

Process modelling has become an important tool in biological wastewater treatment engineering and models are being used for design, control, teaching and research. Biofilm and floc structure have played different roles in biological wastewater treatment modelling. Biofilm structure has been strongly emphasized over the past decades in biofilm research whilst state of the art activated sludge models (ASMs) do not take floc structure into account. It is the aim of this article to analyse and discuss the reasons for the different importance dedicated to biofilm and floc structure in biological wastewater treatment modelling. The role of diffusion as described by Fick's laws of diffusion is discussed and the use and meaning of Monod kinetics in biological wastewater treatment modelling is outlined, since biofilm and floc structure are interrelated to diffusion and Monod kinetics when modelling microbial growth. Both concepts, i.e. Monod kinetics and Fick's laws of diffusion are introduced in the following paragraphs.

Monod Kinetics

Monod kinetics are widely used for modelling microbial growth on a given substrate. The Monod equation (Eq. 1) gives a functional relation between substrate uptake rate (μ), substrate concentration (S), Monod affinity constant or half-saturation coefficient (K_S), and maximum specific uptake rate (μ_{max}) [1]. The functional relation given by the Monod equation is illustrated in figure 1.

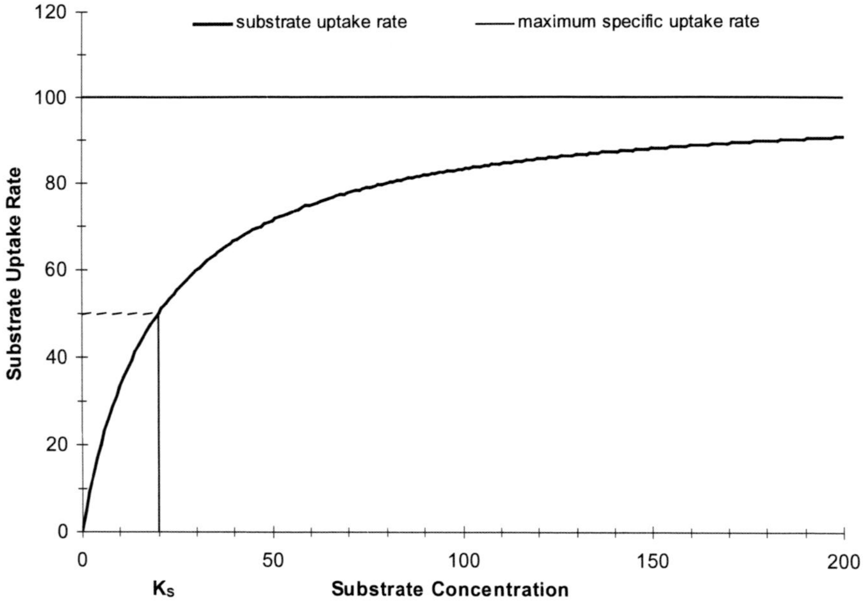

Figure 1. Functional relation between substrate uptake rate and substrate concentration according to the Monod equation ($\mu_{max} = 100$, $K_S = 20$).

$$\mu = \mu_{\max} \cdot \frac{S}{K_S + S} \tag{1}$$

Although there have been attempts to give a mechanistic explanation to the Monod equation [2, 3] and although it has formal similarities to Michaelis-Menten kinetics [4], the Monod equation remains essentially empirical, i.e. it is based on experimental observations and curve fitting of experimental data.

In activated sludge models Monod type kinetics are used for mathematical convenience rather than conformity to any fundamental rate law [5]. This has implications for the interpretation of the kinetic parameters, in particular the Monod affinity constant, since wastewater and activated sludge represents a multisubstrate/multispecies system. The value of Monod's affinity constant in activated sludge systems has been discussed in the literature [6] and some principle implications are given here. In single substrate single species systems the affinity constant takes on values of a few mg/l [7]. In multisubstrate/multispecies systems the value of K_S is typically around 50 mg/l [8]. This difference can be attributed to diffusion within and exterior to flocs [9]. It is thus important to note that diffusion plays an important role in activated sludge processes, and that diffusional mass transport limitations affect the value of Monod's affinity constant when modelling activated sludge systems. Since diffusion is related to floc structure (see 1.2), it becomes apparent that floc structure plays an implicit role when modelling activated sludge systems. The role of diffusion and floc structure in activated sludge modelling needs further discussion (see chapter 2).

In biofilm models Monod kinetics are generally used to model the intrinsic kinetics of biofilm systems, since diffusion is usually explicitly described using Fick's laws of diffusion.

Fick's Laws of Diffusion

Diffusion processes are governed by Fick's laws of diffusion. Fick's first law (Eq. 2) is used in steady-state diffusion, i.e., when the concentration within the diffusion volume does not change with respect to time. Fick's first law states, that the flux (J) of a given substance (S) is proportional to the concentration gradient of the substance along the respective dimension (x), the proportionality factor being the diffusion coefficient (D). Fick's second law (Eq. 3) is used in non-steady or continually changing state diffusion, i.e., when the concentration within the diffusion volume changes with respect to time (t).

$$J = -D\frac{\partial S}{\partial x} \tag{2}$$

$$\frac{\partial S}{\partial t} = D\frac{\partial^2 S}{\partial x^2} \tag{3}$$

In two or more dimensions Fick's second law can be formulated in generalized form using the gradient operator ∇ (Eq. 4).

$$\frac{\partial S}{\partial t} = D\nabla^2 S \qquad Eq.4$$

It is apparent from equation 2 – 4 that biofilm and floc structure must have a crucial importance when applying Fick's laws of diffusion to biological wastewater treatment systems, since the spatial dimensions are part of Fick's laws. It will be shown in the following paragraphs that structure and diffusion play an important role in both biofilm and activated sludge systems. In contrast it will be outlined, how diffusion and biofilm structure have been strongly emphasized in biofilm modelling by using Fick's law of diffusion, whilst structure and diffusion is not explicitly described in state of the art activated sludge models.

The Role of Floc Structure in Activated Sludge Modelling

The composition of activated sludge flocs is complex, but in a simplified way, three types of bacteria can be considered, which determine the structure of an activated sludge floc: floc-forming bacteria, non-floc-forming bacteria and filamentous bacteria. The filamentous bacteria form the backbone of the activated sludge floc whilst the floc forming bacteria function as the glue that holds the floc together. Settleability is an important sludge property since it determines the solid separation in an activated sludge plant and it therefore directly affects effluent quality. The settleability is directly linked to floc structure and the causes of settleability problems can be diagnosed and solved using analysis of floc structure [10, 11]. Large flocs will generally settle faster than small flocs of similar density [12]. Floc strength is thus also important, because activated sludge flocs are exposed to a number of shear stresses that can cause the floc to break apart (see [13] for a review). Weak flocs may break apart resulting in poor settleability and poor dewatering properties [14]. Floc structure is thus an important operational parameter in activated sludge processes, since it determines sludge properties such as flocculation, settling and dewaterability [14].

Diffusion processes and resulting mass transport limitations are linked to floc structure by Fick's laws of diffusion. In activated sludge systems diffusional mass transport limitations are known to occur and influence process quality; they lead to concentration gradients in the sludge floc and play for example an important role in the formation of bulking sludge [15]. The process kinetics in activated sludge systems are also affected by floc structure, diffusion and diffusional mass transport limitations. The nitrification rate for example is strongly dependent up-on the oxygen concentration in the bulk liquid. This is because the nitrifying organisms are agglomerated in large flocs and the dissolved oxygen concentration within the floc may be considerably less than the bulk fluid concentration [16]. The presence of concentration gradients in activated sludge flocs has been confirmed by measurement with microelectrodes [17]. These measurements have revealed the presence of anoxic zones inside sludge flocs under aerobic conditions.

Diffusion and floc structure are hence significant factors in activated sludge systems. However, the state of the art activated sludge models of the ASM model family [5] do not explicitly integrate diffusion of dissolved species into the sludge floc. Instead diffusional mass transport limitations in the sludge floc are modelled using half-saturation coefficients in the Monod expression which are typically one order of magnitude greater than those reported for single suspended cells [18]. Diffusion and floc size play thus an important role when estimating kinetic parameters in activated sludge models, and the relation between floc diffusion and kinetic parameters, in particular half-saturation coefficients, is discussed in the literature [18-20]. Essentially it has been observed and suggested that mass transport limitations are more important in larger flocs than in smaller flocs and that therefore the observed half-saturation coefficients are bigger for larger flocs than for smaller flocs. Kinetic parameters, and in particular half-saturation coefficients, do thus not represent the intrinsic kinetics of activated sludge flocs in activated sludge modelling. Activated sludge models are used to simulate the macro-kinetic behaviour of activated sludge systems, rather than the intrinsic kinetics and micro-environment of the sludge floc. For this reason floc structure is not taken into account in state of the art activated sludge models belonging to the ASM model family.

The Role of Biofilm Structure in Biofilm Modelling

The picture of biofilm structure has changed over the past decades. Initially biofilms were thought to have a homogeneous structure that covers the substratum as a film, i.e. a thin layer (Fig. 2). This picture was supported by low resolution light microscopy and scanning electron microscopy (SEM) [21]. The picture of homogenous biofilm structure has led to the so called continuum approach in biofilm modelling.

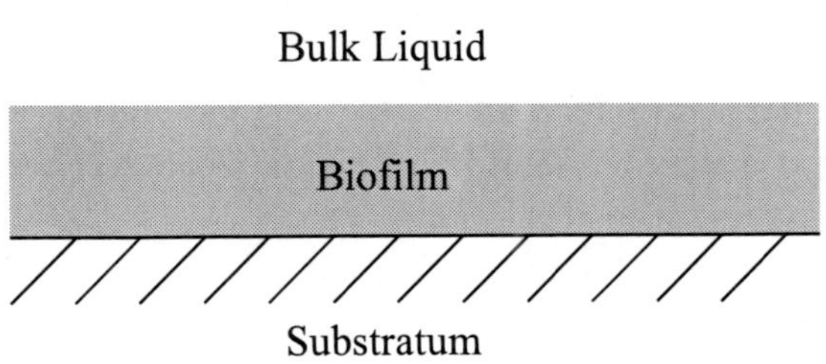

Figure 2. Schematic representation of homogenous biofilm (continuum approach).

In the continuum approach it is assumed that the biomass is evenly distributed in the biofilm; concentration gradients are supposed to occur perpendicular to the substratum only. Biofilm models based on the continuum approach are thus one-dimensional. Until the mid nineties practically all biofilm models were based on the continuum approach [22]. The most prominent example of a one-dimensional biofilm model based on continuum approach is the multispecies biofilm model proposed by Wanner and Gujer [23]. This model was later extended and implemented in the computer program AQUASIM [24].

The picture of a homogenous biofilm structure has changed since the mid-nineties, because advanced microscopic examination tools, namely the confocal scanning laser microscope (CSLM), have revealed that biofilms have a complex, heterogeneous three-dimensional structure with pore channels and cell clusters (see [22] and references there in). The picture of homogenous biofilm structure painted by 1D biofilm models in the continuum approach appears therefore to be inadequate. It also has been shown that mass transport into biofilms is not purely diffusional as assumed in 1D biofilm models; in contrast convective mass transport can also take place through pores and channels of the biofilm [25].

It was thus noted that biofilms also have a complex three-dimensional structure like activated sludge flocs. Biofilms are composed of cell clusters and pore channels and biofilm structure can be affected by filamentous bacteria [26].

This new insights into biofilm structure have led to the development of two- and three dimensional biofilm models [27-34]. Two- and three-dimensional biofilm models are undoubtedly important tools in biofilm research, since they help to understand formation of biofilm structures. However, the picture of biofilm structure painted by 2D and 3D models remains schematic only – the same accounts for the 1D models. On the other hand the development of multi-dimensional biofilm models has not led to more applications of biofilm models in engineering practice.

Problems and Current Trends in Activated Sludge and Biofilm Modelling

Biofilm models have evolved over the past decades with increasing complexity going from one-dimensional to two-dimensional to three-dimensional models. Unlike in activated sludge modelling, structure has been strongly emphasized in the biofilm modelling community. A major reason and driving force for this development was on one hand side the importance of diffusion and the related application of Fick's law, which crucially links diffusion to biofilm structure. On the other hand microscopic advances using the confocal laser scanning microscope (CLSM) have revealed that biofilms have a complex three-dimensional structure. This revelation indicates that one-dimensional (stratified) biofilm models paint an inadequate picture of biofilm structure, which has been an additional driving force towards two- and three-dimensional biofilm models.

However, a gap has developed in the biofilm modelling community between engineering applications and biofilm research, i.e. biofilm models have been primarily used for research whilst they have found little application in engineering practice. The reasons for this have been discussed by *Noguera et al.* [35] and are as follows:

- Biofilm models are perceived as complicated mathematical entities.
- Simplifications and assumptions used in 1D models are often not supported by experimental observations.
- There are many phenomena not considered in the models, such as the fate of particulate substrate, the activity of higher organisms, and the role of exopolymeric substance (EPS) production.
- There is a general lack of trust in the capability of the models to make accurate and reliable predictions.
- The usefulness of biofilm models for the design of full scale systems is not fully appreciated. Many engineers prefer to use simple empirical correlations for design, while models are mostly used as troubleshooting tools when operational problems arise.
- Biofilm models have not been adequately distributed or commercialized.
- Parameters used in biofilm models are sometimes difficult to estimate.

The difficulties encountered when applying biofilm models have been further discussed by *Morgenroth et al.* [36]:

- Biofilm models are too detailed but leave out important factors. Biofilm models have become more and more complex, taking into account an increasing amount of details at the micro-scale. Practitioners are not interested in micro-scale details unless the detailed information becomes directly important for the macro-scale performance of the plant. Thus, for practical applications simpler biofilm models have to be derived from the more complex models. On the other hand, the available models often disregard important processes like attachment and detachment.
- The purpose of mathematical modelling is unclear and many practitioners do not see the need for using mathematical models.
- Too many biofilm models are available. A variety of modelling approaches are available and even biofilm modellers are sometimes in doubt what model to apply for what purpose.
- Model calibration is difficult. A large number of input parameters is required by most models but parameters are often very difficult to determine. Guidance for model calibration is often not provided.

Since biofilm models have found little application in engineering practice so far, certain needs have emerged in the biofilm modelling community [36]:

- Models are needed for predictions of dynamic responses to influent variations.
- Operators of wastewater treatment plants need models for trouble shooting and plant optimization.
- Models are needed that integrate the multiple processes, e.g. particle removal, carbon oxidation, nitrification, denitrification, and biological phosphorous removal, thereby helping to understand the complex interactions between these processes.
- Models are needed for reactor design and testing of reactor configurations. These models could be used to evaluate data from pilot-scale plants and to predict the performance of planned full-scale plants.

There is a trend towards simplified biofilm models with less complexity, which can overcome the difficulties and which can satisfy the needs described above. In the past this trend has pointed from three-dimensional (3D) towards one-dimensional (1D) biofilm models, although it was known, that 1D biofilm models with purely diffusional mass transport may paint an inadequate picture of biofilm structure and performance, since biofilms have a complex 3D structure with pores and channels, which might allow convective mass transport into the biofilm (see chapter 3). In this context the performance of 1D and 3D biofilm models has been evaluated [37]. It was found that a 1D model was suited to approximate average concentration profiles of 3D simulations. This points towards 1D models for engineering applications, because simplified models are needed for this purpose.

In line with the trend towards biofilm models with less complexity a zero-dimensional (0D) biofilm model has recently been developed for dynamic simulation of moving bed bioreactor (MBBR) systems [38-41]. The model was successfully applied to a pilot-scale

MBBR, i.e. the model was able to dynamically predict effluent quality parameters in response to influent variations. The proposed 0D biofilm model is based on the activated sludge model no. 1 (ASM1). The ASM1 processes were extended with attachment of particulates to the biofilm and detachment of biofilm into the bulk liquid. A key feature of the proposed model is that it does not incorporate biofilm structure in any form. The model does not describe a biofilm structure (for example in layers) across which molecular mass transport is explicitly modelled by means of diffusion. Diffusional mass transport limitations are taken into account implicitly by adapted half-saturation coefficients in the Monod terms of the model instead. Kinetic parameters including half-saturation coefficients were obtained from respirometry. In the proposed approach the macro-kinetic behaviour of the biofilm system is modelled rather than the intrinsic kinetics and the micro-environment of the biofilm itself. On a complexity scale the 0D biofilm model situates prior to the biofilm model evolution from 1D to 3D, that is described in the literature [21]. The 0D biofilm model situates in the gap between the bottom edge of the complexity scale and biofilm model evolution from 1D to 3D. The 0D biofilm modelling approach appears to be a complementary alternative to the practicing engineer, not at last because it is in-line with current trends, which point towards biofilm models with less complexity for applications in engineering practice [37, 42] (Fig. 2). Further a fundamental difficulty has been overcome by the 0D biofilm model, i.e. modelling biofilm structure. However, the 0D biofilm model needs validation on real-scale plants, and further testing, including application to other biofilm systems.

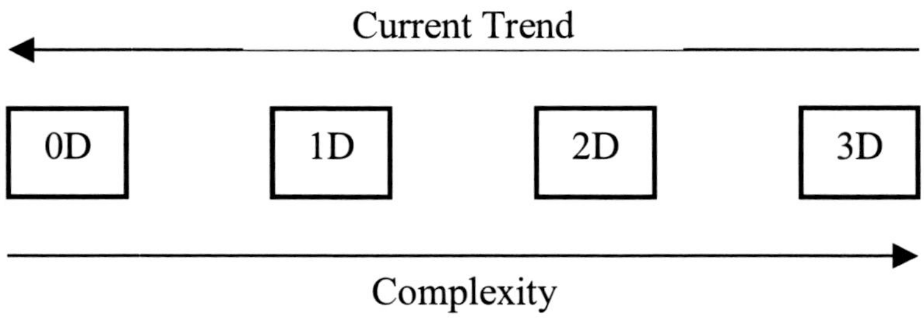

Figure 2. Biofilm model complexity and current trend in the biofilm modelling community

In activated sludge modelling the state of the art models of the ASM model family [5] have proven to be robust tools in engineering practice over the past decades. The ASM models are zero-dimensional since they do not take activated sludge structure into account, although it is known that floc structure and diffusional mass transport limitations play an important role in activated sludge processes (see chapter 2). The ASM models account for the macro-kinetic behaviour of activated sludge systems rather than the micro-environment of activated sludge flocs and activated sludge structure. Only recently a model, originally developed for a biofilm system, was adapted to simulate three-dimensional (3D) formation of activated sludge flocs [43]. This model takes into account floc-forming and filamentous bacteria; diffusion is the only mass transport mechanism. This type of model can be an important tool in activated sludge research, since it helps to understand floc formation. It appears thus, that knowledge from biofilm research, where structure has been strongly emphasized, can be transferred to activated sludge systems. This approach is promising in order to obtain a better fundamental understanding of activated sludge formation and the activated sludge process itself.

Conclusions

The golden rule of modelling as cited by Oskar Wanner [21] says that *models should be as simple as possible and as complex as needed.* This rule holds for both activated sludge and biofilm systems. Activated sludge and biofilm models have evolved with different degrees of complexity over the past decades. This is due to different emphasis on structure, i.e. structure has been strongly emphasized in biofilm modelling whilst state of the art activated sludge models do not take structure into account. Recent developments points towards simplified (zero-dimensional) biofilm models for engineering application whilst three-dimensional activated sludge models appear to be interesting tools in activated sludge research. Most interestingly an exchange of knowledge between activated sludge and biofilm modelling appears to develop: methods from activated sludge modelling (ASM1) have been applied to biofilm systems and methods from biofilm modelling (3D biofilm model) have been applied to activated sludge systems. It is hoped that a further exchange can help both to close the gap that has developed between biofilm research and engineering applications and to obtain a better fundamental understanding of the activated sludge process.

References

[1] J. Monod, La technique de culture continue: Théorie et applications, *Annales de L'Institut Pasteur*, **79** (4) (1950), 390-410.

[2] Y. Liu, Y.-M. Lin and S.-F. Yang, A Thermodynamic Interpretation of the Monod Equation, *Current Microbiology*, **46** (2003), 233-234.

[3] Y. Liu, Overview of some theoretical approaches for derivation of the Monod equation, *Applied Microbiology and Biotechnology*, **73** (2007), 1241-1250.

[4] M. T. Ashby, Appreciating Formal Similarities in the Kinetics of Homogeneous, Heterogeneous, and Enzyme Catalysis, *Journal of Chemical Education*, **84** (9) (2007), 1515-1519.

[5] M. Henze, W. Gujer, T. Mino and M. C. M. Van Loosdrecht, IWA Task Group on Mathematical Modelling for Design and Operation of Biological Wastewater Treatment, Activated Sludge Models ASM1, ASM2, ASM2d and ASM3, Scientific and Technical Report No. 9 (2000).

[6] R. M. Sykes, Value of Monod's Affinity Constant in Activated Sludge, *Journal of Environmental Engineering*, (1999), 780-781.

[7] S. J. Pirt, *Principles of microbe and cell culture*, Wiley, New York, 1975.

[8] C. P. L. Grady, Jr. and H. C. Lim, Biological wastewater treatment*: Theory and applications.,* Marcel Dekker, New York, 1980.

[9] C. R. Baillod and W. C. Boyle, Mass transfer limitations in substrate removal, *Journal of the Sanitary Engineering Devision*, **96** (2) (1970), 525-545.

[10] D. Jenkins, M. Richard and G. Daigger, Manual on the cause and control of activated sludge bulking, foaming and ohter solids separation problems, IWA Publishing, 2003.

[11] M. H. Gerardi, Settleability *Problems and Loss of Solids in the Activated Sludge Process,* Wiley Interscience, Hoboken, New Jersey, 2002.

[12] M. Boller and S. Blaser, Particles under stress, *Water Sci. Technol.,* **37** (10) (1998), 9-29.

[13] P. Jarvis, B. Jefferson, J. Gregory and S. A. Parsons, A review of floc strength and breakage, *Water* Res., **39** (2005), 3121-3137.

[14] P. H. Nielsen, T. R. Thomsen and J. L. Nielsen, Bacterial composition of activated sludge-impotance for floc and sludge properties, *Water Sci. Technol.*, **49** (10) (2004), 51-58.

[15] I. Takács and E. Fleit, Modelling of the Micromorphology of the Activated Sludge Floc: Low DO, Low F/M Bulking, *Water Sci. Technol.*, **31** (2) (1995), 235-243.

[16] M. K. Stenstrom and R. A. Poduska, The Effect of Dissolved Oxygen Concentration on Nitrification, *Water Res.,* **14** (1980), 643-649.

[17] B. Li and P. L. Bishop, Micro-profiles of activated sludge floc determined using microelectrodes, *Water Res.*, **38** (2004), 1248-1258.

[18] J. Pérez, C. Picioreanu and M. C. M. van Loosdrecht, Modeling biofilm and floc diffusion processes based on analytical solutions of reaction-diffusion equations, *Water Res.*, **39** (2005), 1311-1323.

[19] M. Beccari, A. C. Di Pinto, R. Ramadori and M. C. Tomei, Effects of dissolved oxygen and diffusion resistance on nitrification kinetics, *Water Res.*, **26** (8) (1992), 1099-1104.

[20] K. H. Chu, H. van Veldhuizen and M. C. M. van Loosdrecht, Respirometric measurement of kinetic parameters: effect of activated sludge floc size, *Water Sci. Technol.*, **48** (8) (2003), 61-68.

[21] S. Wuertz and C. M. Falkentoft, Modeling and Simulation: Introduction, in: Biofilms in Wastewater Treatment - An Interdisciplinary Approach, S. Wuertz, P. L. Bishop and P. A. Wilderer (ed.), IWA Publishing, London, 2003.

[22] O. Wanner, New Experimental Findings and Biofilm Modelling Concepts, *Water Sci. Technol.*, **32** (8) (1995), 133-140.

[23] O. Wanner and W. Gujer, A Multispecies Biofilm Model, *Biotechnology and Bioengineering*, **28** (1986), 314-328.

[24] O. Wanner and P. Reichert, Mathematical Modeling of Mixed-Culture *Biofilms, Biotechnology and Bioengineering*, **49** (1996), 172-184.

[25] D. de Beer and P. Stoodley, Relation between the structure of an aerobic biofilm and transport phenomena, *Water Sci. Technol.*, **32** (8) (1995), 11-18.

[26] Y. S. Park, J. W. Yun and S. K. Song, Biofilm properties under different substrate loading rates in a rotating biological contactor, *Biotechnology Techniques*, **12** (8) (1998), 587-590.

[27] S. W. Hermanowicz, A model of two-dimensional biofilm morphology, *Water Sci. Technol.*, **37** (4-5) (1998), 219-222.

[28] S. W. Hermanowicz, Two-dimensional simulations of biofilm development: effects of external environmental conditions, *Water Sci. Technol.*, **39** (7) (1999), 107-114.

[29] C. Picioreanu, M. C. M. Van Loosdrecht and J. J. Heijnen, Dicrete-differential modelling of biofilm structure, *Water Sci. Technol.*, **39** (7) (1999), 115-122.

[30] C. Picioreanu, M. C. M. Van Loosdrecht and J. J. Heijnen, Two-Dimensional Model of Biofilm Detachment Caused by Internal Stress from Liquid Flow, *Biotechnology and Bioengineering*, **72** (2) (2001), 205-218.

[31] S. W. Hermanowicz, A simple 2D biofilm model yields a variety of morphological features, *Math. Biosci.*, **169** (2001), 1-14.

[32] D. R. Noguera, G. Pizarro, D. A. Stahl and B. E. Rittmann, Simulation of multispecies biofilm development in three dimensions, *Water Sci. Technol.*, **39** (7) (1999), 123-130.

[33] M. C. M. Van Loosdrecht, J. J. Heijnen, H. Eberl, J. Kreft and C. Picioreanu, Mathematical modelling of biofilm structures, *Antonie van Leeuwenhoek*, **81** (2002), 245-256.

[34] J. B. Xavier, C. Picioreanu and M. C. M. Van Loosdrecht, Assessment of three-dimensional biofilm models through direct comparison with confocal microscopy imaging, *Water Sci. Technol.*, **49** (11) (2004), 177-185.

[35] D. R. Noguera, S. Okabe and C. Picioreanu, Biofilm Modeling: Present Status and Future Directions, *Water Sci. Technol.*, **39** (7) (1999), 273-278.

[36] E. Morgenroth, M. C. M. Van Loosdrecht and O. Wanner, Biofilm models for the practitioner, *Water Sci. Technol.*, **41** (4-5) (2000), 509-512.

[37] E. Morgenroth, H. Eberl and M. C. M. Van Loosdrecht, Evaluating 3-D and 1-D mathematical models for mass transport in heterogeneous biofilms, *Water Sci. Technol.*, **41** (4-5) (2000), 347-356.

[38] M. Plattes, E. Henry, P.-M. Schosseler and A. Weidenhaupt, Modelling and Dynamic Simulation of a Moving Bed Bioreactor for the Treatment of Municipal Wastewater, *Biochem. Eng. J.*, **32** (2006), 61-68.

[39] M. Plattes, D. Fiorelli, S. Gillé, C. Girard, E. Henry, F. Minette, O. O'Nagy and P.-M. Schosseler, Modelling and Dynamic Simulation of a Moving Bed Bioreactor Using Respirometry for the Estimation of Kinetic Parameters, *Biochem. Eng. J.*, **33** (2007), 253-259.

[40] M. Plattes, D. Fiorelli, S. Gillé, C. Girard, E. Henry, F. Minette, O. O'Nagy and P. Schosseler, Modelling and dynamic simulation of a pilot-scale moving bed bioreactor for the treatment of municipal wastewater: model concepts and the use of respirometry for the estimation of kinetic parameters, *Water Sci. Technol.*, **55** (8-9) (2007), 309-316.

[41] M. Plattes, E. Henry and P. M. Schosseler, A zero-dimensional biofilm model for dynamic simulation of moving bed bioreactor systems: Model concepts, Peterson matrix, and application to a pilot-scale plant, *Biochem. Eng. J.*, **40** (2008), 392-398.

[42] S. Wuertz, P. Bishop and P. Wilderer, *Biofilms in Wastewater Treatment.* An interdisciplinary Approach, IWA Publishing, 2003.

[43] A. M. P. Martins, C. Picioreanu, J. J. Heijnen and M. C. M. Van Loosdrecht, Three-Dimensional Dual-Morphotype Species Modeling of Activated Sludge Flocs, *Environmental Science and Technology*, **38** (21) (2004), 5632-5641.

In: Biochemical Engineering
Editors: F.E. Dumont and J.A. Sacco, pp. 257-274
ISBN: 978-1-60741-257-1

Chapter 11

INTERACTION OF CR (VI) WITH GREEN MICROALGAE GROWTH: A COMPARATIVE STUDY

M. Alzira P. Dinis[1,a], Vítor J.P. Vilar[2,b], Álvaro A.M.G. Monteiro[1,2,c], Rui A.R. Boaventura[2,d]

[1]University Fernando Pessoa, Praça 9 de Abril, 349, 4249-004 Porto, Portugal
[2]Laboratory of Separation and Reaction Engineering, Department of Chemical Engineering, Faculty of Engineering, University of Porto, Rua Dr. Roberto Frias, 4200-465 Porto, Portugal

Abstract

Microalgae *Chlorella fusca* ACOI 621, *Chlorella vulgaris* ACOI 879, *Scenedesmus acutus* ACOI 538 and *Scenedesmus obliquus* ACOI 550, all native from Portugal, were characterized in terms of specific growth rate. The effect of pH and the presence of Cr(VI) in concentrations up to 25 mg l^{-1} (50 mg l^{-1} for *Chlorella fusca*) has been evaluated. The logistic equation of population growth $n = 1/\left(\left(1/n_0 - 1/K\right)e^{-\mu t} + 1/K\right)$ adequately describes the cellular growth. Experiments at pH = 6.5 and temperature around 24.5 °C, in the absence of Cr(VI), led to specific growth rates (μ) of 0.0370, 0.0284, 0.0359 and 0.0162 h^{-1} and maximum biomass concentrations (K) of 403.3, 369.2, 542.9 and 604.1 mg l^{-1} for *C. fusca*, *C. vulgaris*, *S. acutus* and *S. obliquus*, respectively. Experiments carried out with the same algae at approximately 21 °C, also in the absence of Cr(VI), gave μ values of 0.0241, 0.0357, 0.0272 and 0.0289 h^{-1} and K values of 292.6, 169.9, 263.1 and 327.8 mg l^{-1} for initial pH = 6.5 and μ values of 0.0115, 0.0177, 0.0137 and 0.0158 h^{-1} and K values of 35.9, 3.0, 32.8 and 54.7 mg l^{-1} for initial pH = 7.9. Higher pH results in a significantly lower growth rate and *C. vulgaris* seems to be the less resistant microalgae to changes in the environmental conditions. Looking simultaneously at μ and K values, the best performance in terms of growth kinetics was obtained for *S. acutus* and *C. fusca*. Growth inhibition is visible for Cr(VI) ≥ 5 mg l^{-1} but concentrations up to 1 mg l^{-1} seem not to seriously affect algal growth, even increasing the *C.*

[a] E-mail address: madinis@ufp.edu.pentru; Tel.: +351-22-5081683; fax: +351-22-5081674 (Corresponding author: M. Alzira P. Dinis).
[b] E-mail address: bventura@fe.up.pt (Rui A.R. Boaventura).
[c] E-mail address: vilar@fe.up.pt (Vítor J.P. Vilar).
[d] E-mail address: aamont@ufp.edu.pt (Álvaro A.M.G. Monteiro).

fusca specific growth rate. For Cr(VI) < 1 mg l^{-1}, μ varies between 0.08 and 0.17 h^{-1}, depending on the algal species. The growth of *C. vulgaris* is severely inhibited by Cr(VI) = 5 mg l^{-1}. The production of metabolites is small compared with biomass production, for all Cr(VI) concentrations. The organic carbon content of algae is about 40%-50% (dry basis), except for *S. obliquus* (around 30%). The biomass of *C. fusca* and *S. acutus* presents the greatest sedimentation rates. The presence of high Cr(VI) concentrations negatively affects the sedimentation.

Keywords: Microalgae; *Chlorella fusca*; *Chlorella vulgaris*; *Scenedesmus acutus*; *Scenedesmus obliquus*; growth kinetics; chromium (VI)

1. Introduction

As population, population density, and development level increase, the issue of wastewater-related nutrient loading has grown as one of the major concerns in environmental management. On the other side, wastewaters resulting from human activities often contain heavy metals [1]. These elements make part of a list of dangerous substances with toxicity characteristics and a high potential of persistence and accumulation.

There has been considerable interest in using aquatic organisms for wastewater treatment. Portugal presents, mainly in Centre and South areas, climatic conditions ecologically favourable to green microalgae development, so the subject has been faced with increasing attention. Because of the fact that, in those areas, the limiting factor to agriculture is water, a judicious management of the hydric resources available is necessary, and so water reuse constitutes an adequate contribution for solving ecological and water shortage problems simultaneously.

Microalgae are often used for wastewater treatment due to their high growth rate and simple growth requirements. Several works refer to nitrogen and phosphorous removal by microalgae [2-6]. Lau et al. [7] show that carrageenan-immobilized *Chlorella vulgaris* cells efficiently removed N and P (95 and 100%, respectively) from wastewater. Meiring et al. [4] reported nitrogen removals of about 85%. Robinson [5] has obtained P removal efficiencies of 80% to 95% using *Chlorella emersonii* in suspension. Continuous-flow culture studies were also carried out using small-scale packed-bed reactors containing *Chlorella* immobilized in alginate beads. Reactors were maintained at 25°C and illumination was provided at 60 μE m^{-2} s^{-1}. Gonzalez et al. [2] run 216 h-experimental cycles of batch cultures of *Chlorella* and *Scenedesmus* and concluded that *Chlorella vulgaris* was less efficient than *Scenedesmus dimorphus* in removing ammonia. Twist et al. [6] used *Scenedesmus subspicatus* cultures incubated at 20°C with an incident photon flux density of 40 μE m^{-2} s^{-1}. The interaction of air flow rate with nutrient supply was tested under controlled temperature (19±1°C) and light (85-95 μE m^{-2} s^{-1}).

The pH of the culture medium determines CO_2 and minerals solubility and influences directly or indirectly the algal metabolism. Becker [8] states that stopping agitation and/or CO_2 supply can cause an increase in pH, leading to the formation of algal flocs, which will settle down. Experiments with *Scenedesmus* sp. to study the effect of pH showed that no flocculation could be obtained for pH values between 5.0 and 7.5, whereas at pH values above 8.5 almost 95% of the algal biomass could be removed.

Light is also of fundamental importance and, in algal cultures, light rapidly becomes limiting due to absorption by the algal biomass. Light intensity also interacts strongly with temperature. Algae growth rate increases with temperature until an optimal value is reached. Further increases in temperature usually lead to a rapid decline in growth rate. At temperatures close to the optimum, algae are also better able to tolerate much higher light intensity before photoinhibition sets in [9].

In a general way, microalgae growth is complex, revealing non linear behaviours in response to the alteration of several environmental parameters such as pH, temperature, light intensity and nutrients, and the interactions between these factors are not well established [8, 10].

Heavy metals removal by the green algae has been widely recognized [11-15]. According to Avery et al. [16], the accumulation of heavy metals comprises a rapid adsorption phase, reversible and metabolism independent and a slower one, metabolism dependent, often irreversible. Batch cultures of *Chlorella fusca* were used for testing the toxicity of Pb(II), Cr(III), Cr(VI) and Cd(II) [17].

Tam et al. [18] investigated the influence of the concentration of *Chlorella vulgaris* dried cells on copper removal from aqueous solution. The evolution of nitrogen and phosphorus uptake in presence and absence of Pb (10^{-6} M), at pH 7 and 4, was studied by Capelo et al. [19] using the microalgae *Selenastrum Capricornutum*. Batch cultures of *Chlorella fusca* were used for testing the toxicity of Cu(II) and Zn(II) [20] and also Pb(II), Cr(III), Cr(VI) and Cd(II) [17]. Lam et al. [13] evaluated the effect of Cd and Cu on *Chlorella vulgaris* growth. The direct toxic effect due to copper addition and the indirect impact arising from a decrease in pH following an increase in the metal ions concentration were investigated. Light intensity was in the range 85-90 μE s^{-1} m^{-2}. Lupi et al. [21] studied the inhibition of *Chlorella vulgaris* by Cu, at 27 °C and 100 or 150 W m^{-2} light intensity. The effect of the contact time on the binding capacity of five toxic metals (Cd, Cu, Ni, Pb and Zn) in solution by *Chlorella vulgaris* at pH 8.5 and constant temperature was also evaluated [14].

For an efficient biomass separation, good flocculation must occur [4, 22]. Algae autoflocculation may result from a pH rise (>8), promoting an initial nucleation of calcium phosphate crystals. In the presence of Ca^{2+}, the calcium phosphate, positively charged, precipitates, adsorbing the negatively charged cells, agglomerating them and promoting their flocculation [8]. Under these conditions, polyvalent cations of Cr, Cu, Fe, Mn, Pb, Sr and Zn, together with Ca and Mg, tend to become incorporated into the flocs, which settle down and promote water clarification [23].

This work aims to describe the growth kinetics of pure cultures of four microalgae, *Chlorella fusca* ACOI 621, *Chlorella vulgaris* ACOI 879, *Scenedesmus acutus* ACOI 538 and *Scenedesmus obliquus* ACOI 550. Results are compared with those obtained in the presence of Cr(VI). The effect of temperature and initial pH upon algal growth was also investigated.

2. Modelling

Growth of unicellular organisms like microalgae is often considered as a first-order autocatalytic reaction [24-26]. So, the increase in biomass concentration, n (mg l^{-1}), as a function of time, t (h), can be expressed as:

$$\frac{dn}{dt} = \mu \times n \tag{1}$$

where μ is the specific growth rate of biomass (h^{-1}), characteristic of each organism and culture medium, and is primarily governed by the growth capacity of the organism and the environmental conditions [8, 25]. This model assumes unlimited exponential growth. However, in batch systems every population is limited either by lack of nutrients or accumulation of toxic metabolites and then some allowance must be made to restrict the growth [25]. Verhulst derived the logistic equation of population growth, sometimes called Verhulst model, in 1845 [24], which is frequently presented as the initial form of a Riccati equation:

$$\frac{dn}{dt} = \mu \times n \times \left(1 - \frac{1}{K} \times n\right) \tag{2}$$

This equation can be easily integrated to obtain the logistic curve:

$$n = \frac{1}{\left(\frac{1}{n_0} - \frac{1}{K}\right) e^{-\mu t} + \frac{1}{K}} \tag{3}$$

where n_0 is the initial biomass concentration (mg l^{-1}) and K the maximum biomass concentration (mg l^{-1}). The logistic curve is sigmoidal and leads to a stationary population of size K, i.e., for $t = \infty$, $n = K$.

The logistic equation perfectly describes the exponential and stationary growth phases, but it does not include the initial lag phase and the decline phase. According to Krebs [27], there are two ways of viewing the logistic curve. One is to view it as an empirical description of how populations tend to grow in number when conditions are initially favourable. This is the more general, more flexible viewpoint. The other way is to view the logistic model as an implicit strict theory of population growth, as a "law" of population growth. This last one is, perhaps, the one that interests us most.

3. Materials and Methods

3.1. Microalgae

Samples of *Chlorella fusca* ACOI 621, *Chlorella vulgaris* ACOI 879, *Scenedesmus acutus* ACOI 538 and *Scededesmus obliquus* ACOI 550 strains, all chlorophyceae, were obtained from stock cultures maintained by the Algal Culture Collection ACOI, Department of Botany, University of Coimbra, Portugal. The genera *Chlorella* and *Scenedesmus* belong to the order *Chlorellales* [22]. All species are unicellular green algae, without motion capacity. Species from the *Chlorella* gender are spherical or ellipsoidal, exhibiting a simple life cycle and also simple nutritional needs. They are small cells, 1 to 5 μm in diameter, which reproduce asexually, each mature cell producing 6, 8, or, more rarely, 16 autospores. Species from the *Scenedesmus* gender are largely distributed in water and soil. Cells are cylindrical with rounded or sharp extremities. They can be found laterally united in 4, 8 or 16 – more rarely – cells. They reproduce by autocolonies formation [22].

Algal cultures were stored at 4°C, in sterilized test tubes, and renewed with algae proceeding from successive experiments performed in sterile conditions.

3.2. Growth Medium

Table 1 presents the composition of the culture medium. As autoclaving leads to the precipitation of certain mineral salts [8], four concentrated solutions (Table 2) were prepared separately and aseptically filtered through 0.45 μm Gellman Sciences membranes. All solutions were prepared with dechlorinated tap water, which was previously sterilized by filtration. The only source of specific inorganic carbon was $NaHCO_3$, besides the CO_2 present in the sterile air bubbled into the medium.

Table 1. Composition of the synthetic medium

Constituent	Concentration (mg l^{-1})
$NaNO_3$	525
$NaHCO_3$	50
KH_2PO_4	626
K_2HPO_4	266
$MgSO_4 \cdot 7H_2O$	500
$CaCl_2 \cdot 2H_2O$	55.87
Na_2EDTA	50
NaOH	22.1
$FeSO_4 \cdot 7H_2O$	4.98

Table 2. Composition of the concentrated inorganic solutions

Constituent	Concentration (g l^{-1})	Solution
$NaNO_3$	5.25	1
$NaHCO_3$	0.5	1
KH_2PO_4	6.26	1
K_2HPO_4	2.66	1
$MgSO_4 \cdot 7H_2O$	5.0	2
$CaCl_2 \cdot 2H_2O$	0.559	2
Na_2EDTA	50	3
NaOH	22.1	3
$FeSO_4 \cdot 7H_2O$	4.98	4

One of the factors affecting microorganism growth is pH. Experiments aiming at determining kinetic parameters must be performed at constant or approximately constant pH. According to Mitchell and Slaughter [25], control of pH is also very important since in many cases it changes during the development of a microbial population. Phosphate buffers are especially useful to control pH, since phosphorus is required as nutrient and they buffer the solution at around neutral pH, where microorganisms like microalgae grow optimally. In this work, the culture medium was buffered with hydrogen / di-hydrogen phosphates.

3.3. Cr(VI) Solutions

Hexavalent chromium solution was prepared by dissolving $K_2Cr_2O_7$ in distilled and demineralised water. Test cultures were spiked with pre-defined volumes of this solution in order to get different Cr(VI) concentrations.

3.4. Analytical Techniques

Light intensity was measured using a luxmeter (Luxmeter HD 8366 Delta OHM).

The cell concentration was determined by optical density measurements at 680 nm, as used by Schelenz [28]. Optical densities were measured with a UV-VIS Jenway 6405 spectrophotometer, using the sterile medium culture as blank. In the range of optical densities measured, cell concentration is proportional to optical density. A calibration curve was prepared by plotting dry biomass weight against optical density. In order to eliminate the interference of cell sedimentation in optical density values, samples were previously stirred in a Vortex Mixtub, Raypa. The pH was daily measured with a 540 GLP/ WTW pH meter. Cells were counted in an improved Neubauer hemacyclometer.

Total organic carbon (TOC) and dissolved organic carbon (DOC) concentrations were measured using a Rosemount Analytical Dohrman DC-190 analyser. Prior to DOC measurements, samples were centrifuged at 5470 rpm, in an Alresa Mod Digicen centrifuge, for 15 minutes and then filtered through 0.45 μm porosity membranes (Gellman Sciences).

3.5. Experimental Procedures

Four 250-ml Erlenmeyer flasks containing 250 ml of medium were inoculated with the above referred microalgae species to obtain an initial cell concentration around 100 000 cells ml^{-1}, which has given the best results in preliminary experiments. A flask containing only culture medium was also used as blank for optical density measurements, sterilization control and pH variation. Experiments were carried out at temperature approximately constant. Mixing was accomplished and some CO_2 was provided by bubbling sterile air into the culture medium. As source light, special Osram L36W/72 Biolux lamps, giving a spectrum similar to sunlight, were used. Experiments were performed under a 24 h light photoperiod. All glass material was sterilized at 180°C for 1 h 30 min.

In the first set of experiments, room temperature was controlled at 24.6 ± 1.0°C (mean ± standard deviation). Daily pH measurements were carried out but no pH corrections were performed.

Mixing was maintained approximately uniform in all flasks by adjusting the diffused air flowrate. Besides, flasks were also manually stirred several times a day, to minimize the formation of a biofilm onto the flask wall.

Light intensity values ranged from 2000 to 3000 lux (32-48 μE m^{-2} s^{-1}). Twist et al. [6, 29], Minowa and Sawayama [30] used similar values, 40, 30 and 40-60 μE m^{-2} s^{-1}, respectively.

A second set of experiments was carried out to evaluate the effect of pH on microalgae growth rate. Experiments performed with buffered culture medium at pH ≈ 8 were compared with experiments where initial pH = 6.5. All the operating conditions were as before, except temperature, which was around 20.8 ± 1.8°C.

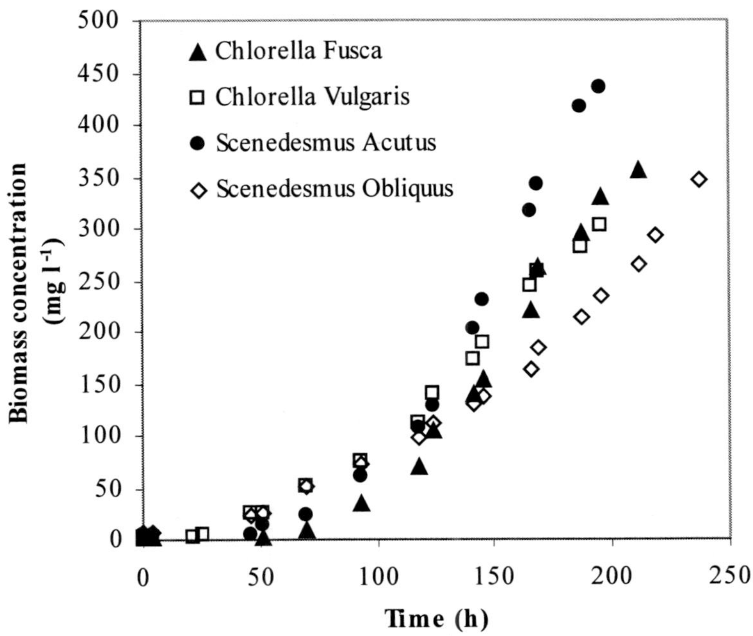

Figure 1. Microalgal growth as a function of time. (T = 24.6 °C; pH = 6.5).

To study the effect of the addition of Cr(VI), pre-defined volumes of metal stock solution were added simultaneously to 2000 ml-glass flasks containing 1500 ml of each algal culture once a density of 100 000 cells ml^{-1} was reached. Experiments with Cr(VI) concentrations of 0, 1, 5 and 25 mg l^{-1} (50 mg l^{-1} for *Chlorella fusca* culture) were carried out at room temperature approximately constant: 21.6±1.1, 22.4±0.6, 21.0±0.71 and 21.0±0.3 °C (mean ± standard deviation) for *Chlorella fusca*, *Chlorella vulgaris*, *Scenedesmus acutus* and *Scenedesmus obliquus* cultures, respectively. No pH corrections were performed. An Erlenmeyer flask containing only sterile culture medium was submitted to the same operating conditions, for each set of experiments, and used as blank to correct optical density measurements over time. Daily samples were collected both in test and control flasks for pH, biomass (optical density) and total, dissolved and suspended organic carbon.

Flocculation properties of algal cultures were evaluated through settling tests. Initial optical densities were compared with those of the supernatants after a 6 h-settling period.

4. Results and Discussion

4.1. Experiments under the Same Initial pH Value (T = 24.6 °C, pH = 6.5)

For each algae, the parameters μ, K and the initial biomass concentration, n_0, were determined by fitting the logistic equation to the experimental results, using the non-linear least squares technique. Adjusting n_0 allows overcoming high experimental errors associated with the measurement of low biomass concentrations. The growth curve obtained for all algae species involved in this study are shown in Figure 1. All curves present a similar growth pattern, although *Scenedesmus acutus* growth seems to increase after 150 h operation and *Chlorella fusca* shows a slower growth rate in the beginning.

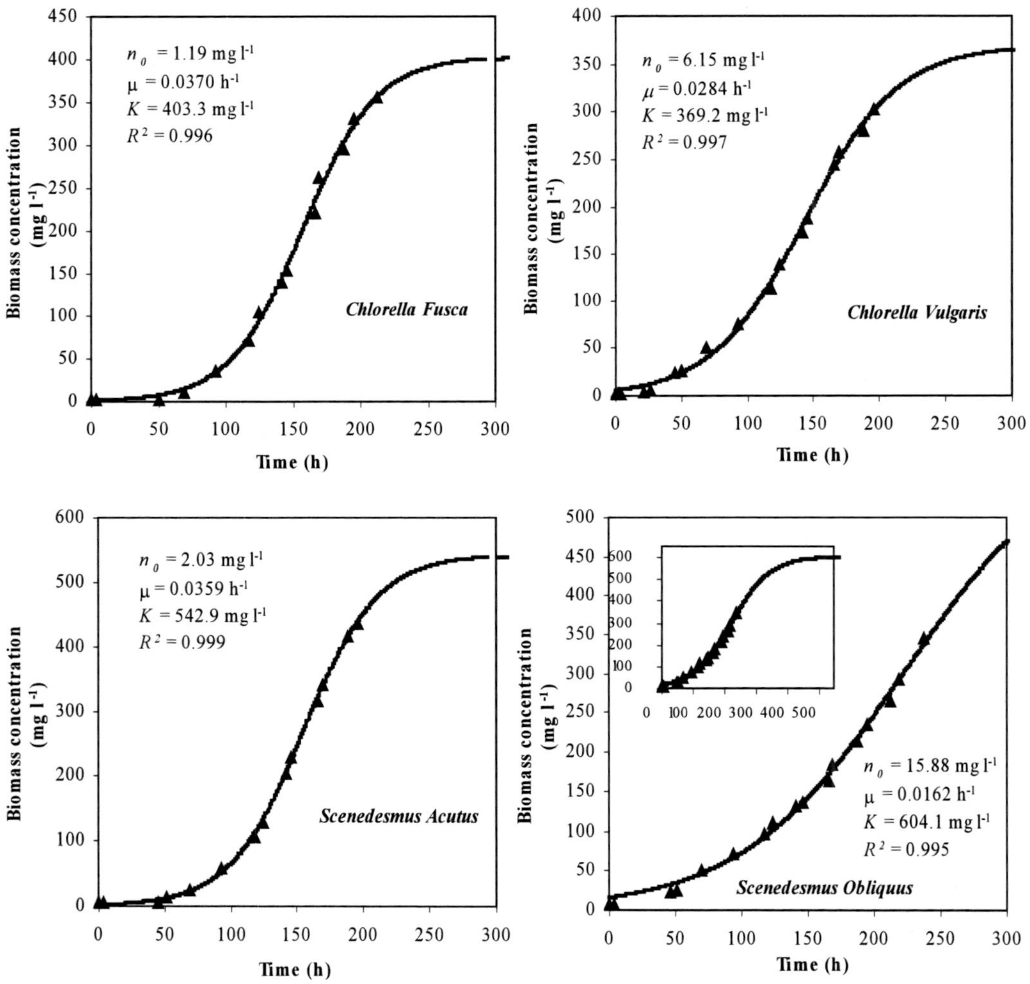

Figure 2. Microalgal growth (▲) as a function of time. (—) Logistic model fittings (T = 24.6 °C; pH = 6.5).

Growth curves obtained for each algae by fitting the logistic equation to the experimental data are also presented in Figure 2. *Scenedesmus acutus* and *Chlorella fusca* show higher lag phases. Maximum biomass concentration, K, ranges from 369 mg l^{-1} for *Chlorella vulgaris* to 604 mg l^{-1} for *Scenedesmus obliquus*. The pH increased from 6.4-6.5 to 7.2-7.7, after about 190 h operation, as observed in Figure 3. As it could be expected, algae species growing faster are those presenting the highest pH variation, which is clearly visible for *Chlorella fusca*, by comparing Figure 3 with Figure 2.

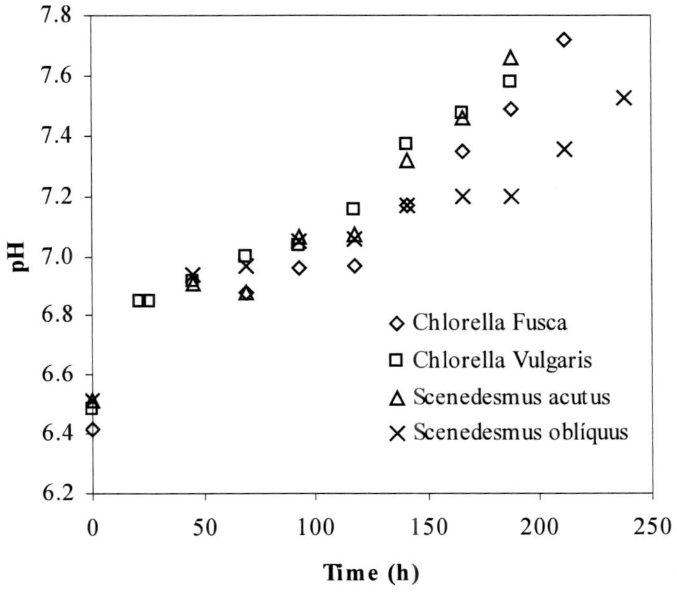

Figure 3. pH evolution over time for each alga (T = 24.6 °C; pH = 6.5).

4.2. Experiments with Different Initial pH Values (T = 20.8 °C, pH = 6.5 and pH = 7.9)

4.2.1. Effect of pH

The influence of the initial pH on algal growth was investigated, although maintaining the temperature at a lower value than in the previous experiments. The initial pH was kept around 6.5 in some experiments, whereas in others it was increased up to 7.9. Figure 4 shows the experimental and predicted growth curves obtained and the parameters μ, K and n_0, determined by fitting the logistic equation to the experimental results.

As indicated by K values, there is no significant growth of biomass when the initial pH was adjusted to around 7.9: the highest K value is ≈ 54 mg l^{-1} for *Scenedesmus obliquus*, whereas for *Chlorella vulgaris* is only ≈ 3 mg l^{-1}. The maximum biomass concentration follows the same order as for pH = 6.5: *Scenedesmus obliquus* > *Chlorella fusca* > *Scenedesmus acutus* > *Chlorella vulgaris*. The specific growth rate of biomass μ is also reduced by about 50% for all species, when compared to the values obtained at initial pH = 6.5.

In the experiments performed at a temperature of 20.8 °C and initial pH = 6.5, it was obtained an increase in pH of around 0.5 after about 230 h incubation. Starting with a pH = 7.9, no visible alterations were visible during the same operating time. At higher pH values, the amount of CO_2 available is smaller, thus affecting algal growth. Results confirm that a high initial pH acts as an inhibitory factor preventing algae from developing normally [8].

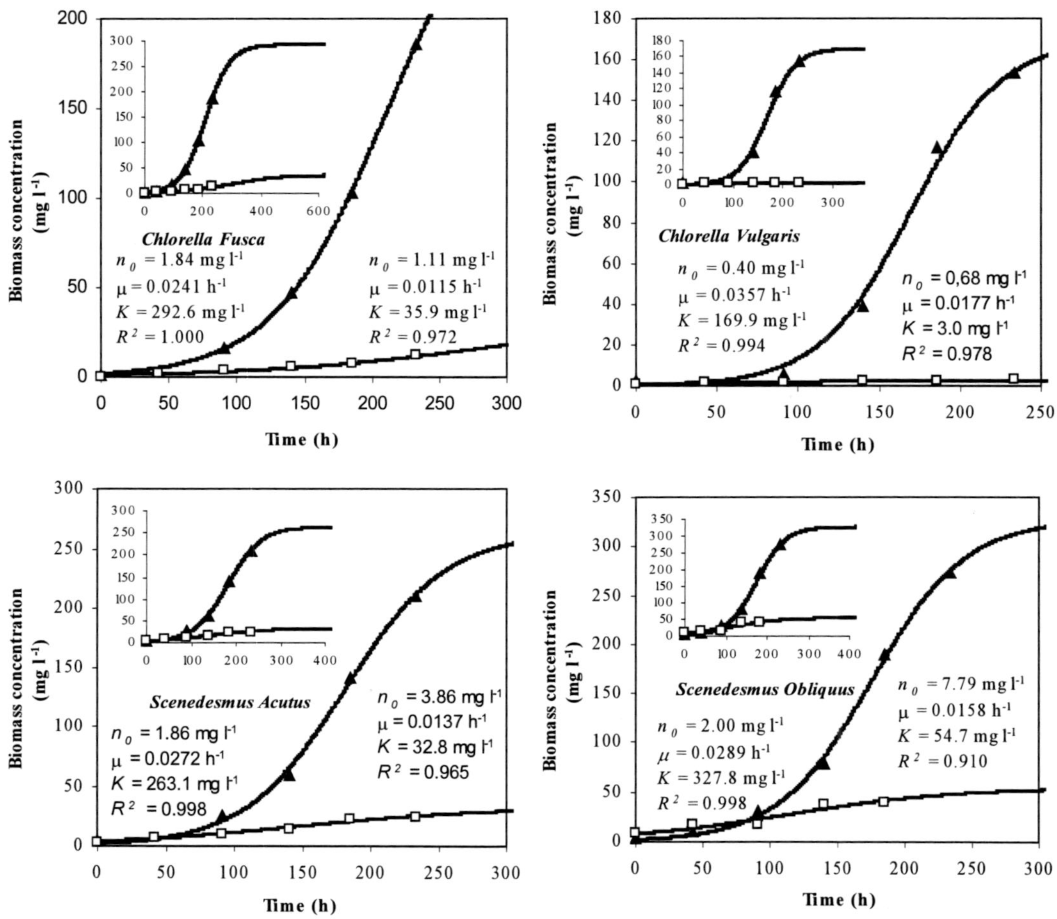

Figure 4. Microalgal growth as a function of time for different initial pH values. (▲) pH = 6.5; (□) pH = 7.9. (—) Logistic model fittings (T = 20.8 °C).

4.2.2. Temperature Influence

Comparing the values of the model parameters for the experiments performed at pH = 6.5 and two different temperatures (24.6 and 20.8 °C), it can be concluded (Figure 2 and Figure 4) that although the variation of the specific growth rate of biomass, μ, is casual, the maximum biomass concentration, *K*, increases with temperature.

At 20.8 °C, it was possible to observe that the pH increased from 6.5-6.6 to 7.0-7.1, after about 190 h operation, not reaching values so high as at 24.6 °C (Figure 3). The lower production of biomass at 20.8 °C originates a lower impact on the initial pH.

The variation of kinetic parameters due to temperature variation can be also associated to environmental stresses, as, according to Becker [8], living organisms like algae are extremely susceptible to changes occurring in the environment.

4.3. Effect of Cr(VI) on Growth Kinetics

Figure 5 shows the experimental results of biomass growth for each algal species. Growth in the presence of different Cr(VI) concentrations is compared with that obtained using Cr(VI)-free culture medium. Cr(VI) concentrations of 1 mg l^{-1} do not affect or slightly improve algal growth. However, higher concentrations cause an inhibitory influence on growth kinetics. The more marked effect is visible in *C. vulgaris* culture, since growth is almost interrupted for Cr(VI) concentration as low as 5 mg l^{-1}. In the absence of Cr(VI) or for low Cr(VI) concentrations (1 mg l^{-1}), pH increases over time from the initial value around 6.5 to about 7.4-7.8 (Figure 6). As it could be expected, pH rises more slowly or even remains constant when inhibition occurs.

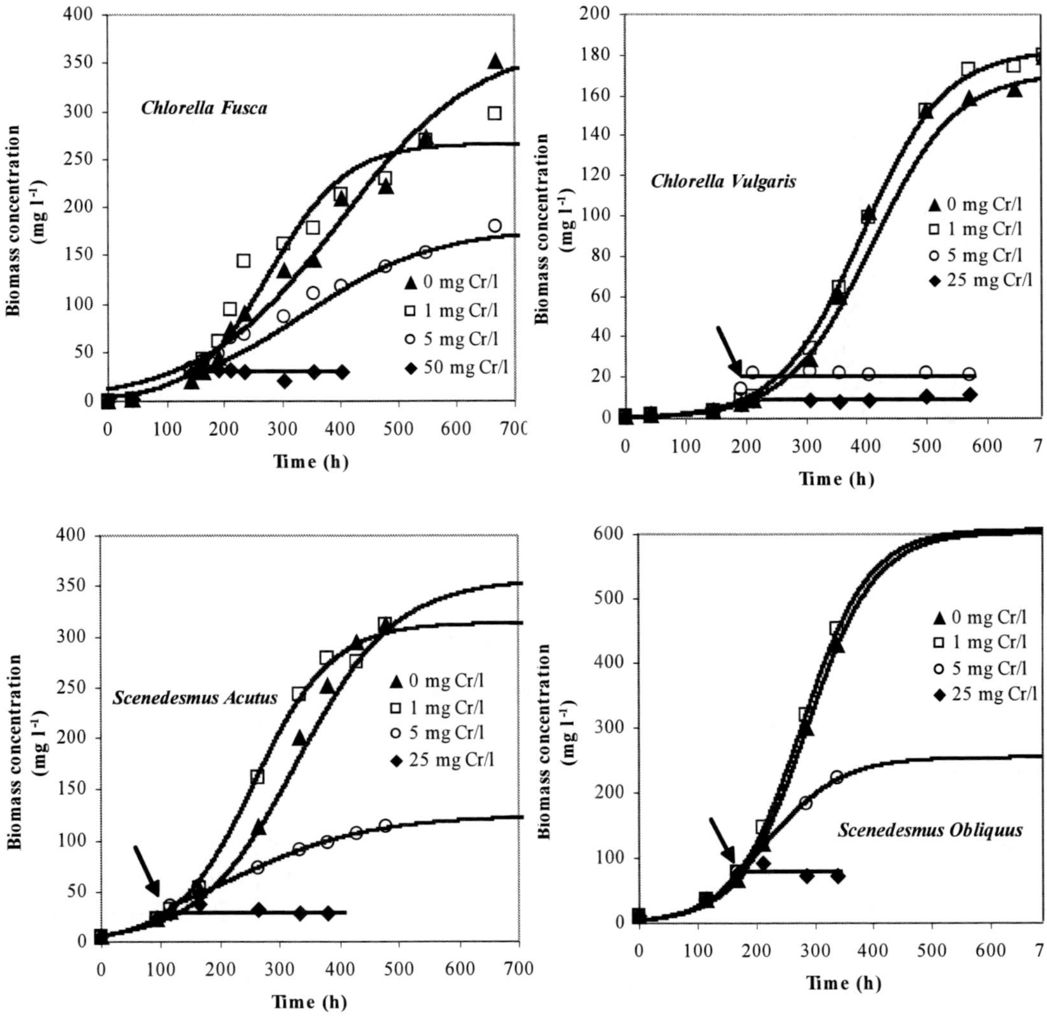

Figure 5. Effect of Cr(VI) on algal growth as a function of time (T = 21-22 °C; initial pH = 6.6-6.7; arrow indicates the moment of Cr addition). (—) Logistic model fittings.

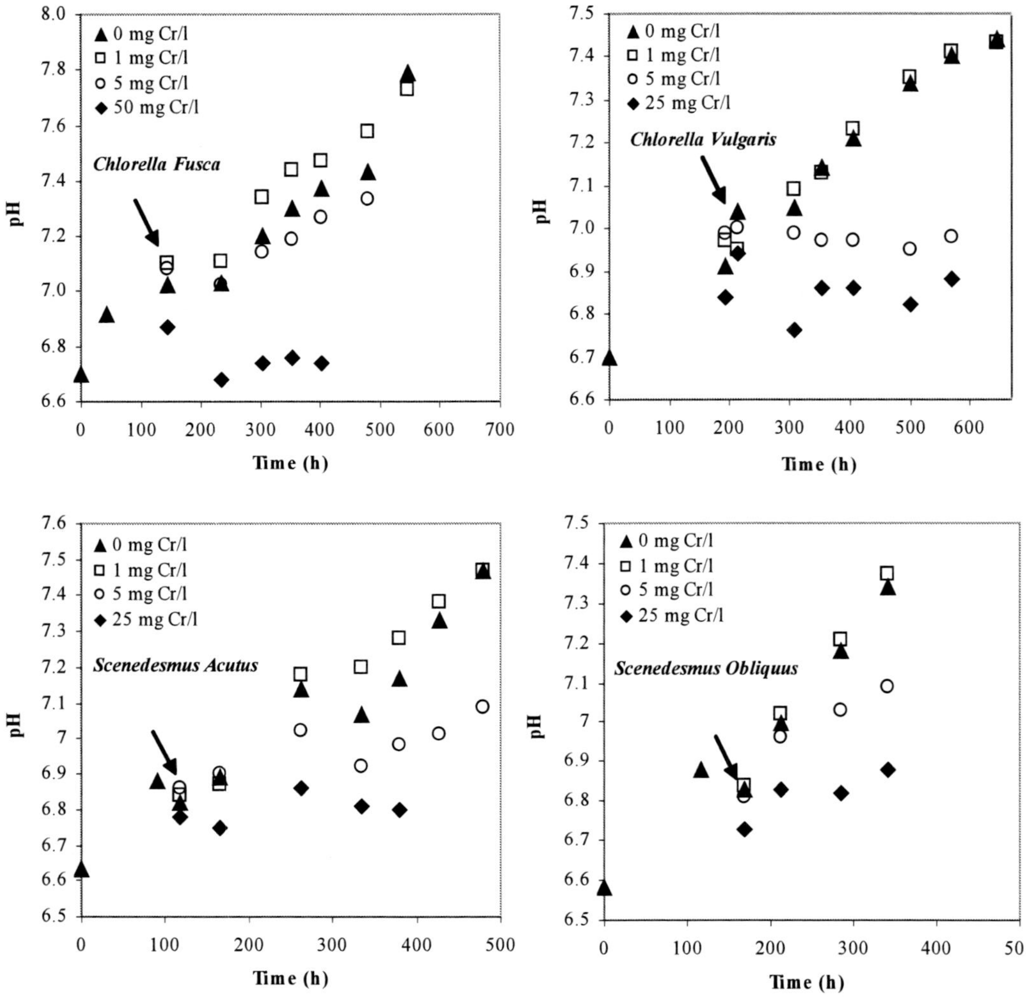

Figure 6. Evolution of pH over time (T = 21-22 °C; initial pH = 6.6-6.7; arrow indicates the moment of Cr addition).

The logistic fittings to the experimental results are also displayed in Figure 5 and the corresponding parameter values are presented in Table 3. Hexavalent chromium concentrations up to 5 mg l^{-1} seem not to negatively affect the initial specific growth rate of biomass (μ) of *C. fusca* and the presence of 1 mg l^{-1} even increases μ. However, the maximum biomass concentration (K) continuously decreases as Cr(VI) concentration increases. According to Wong et al. [20], *C. fusca* is known for its tolerance to heavy metals, which explains this behaviour. On the other hand, *C. vulgaris* (Figure 5) is less tolerant to Cr(VI) than *C. fusca*. Although the addition of 1 mg l^{-1} does not affect the normal development of the population, growth is interrupted within a few hours for Cr(VI) = 5 mg l^{-1}. The effect of the presence of Cr(VI) on growth kinetics of *S. acutus* is similar to that observed for *C. fusca*, even though a lower maximum biomass concentration (K) has been reached for Cr(VI) = 5 mg l^{-1}. The presence of Cr(VI) up to 5 mg l^{-1} practically does not affect the biomass specific growth rate of *S. obliquus*, but the maximum biomass concentration (K)

markedly decreases. However, *K* values are significantly higher than those obtained for the other species. For Cr(VI) = 25 mg l^{-1} the algal growth is completely inhibited.

Comparing the results obtained without Cr(VI) addition (Figure 4) with those presented in Figure 5, one can conclude that growth rates are lower in the first experiments. The difference may be due to the higher medium volume used in the second study (1500 ml instead of 250 ml). As light intensity over the flasks was the same, the available light for algal growth was lower. As light is one of the most important environmental factors influencing the growth and development of photosynthetic organisms like chlorophyceae [21], the maximum concentration of growing algae in a continuously mixed culture is determined by the degree of penetration of light into the culture.

4.4. Effect of Cr(VI) on Metabolites Production

Total, dissolved and suspended organic carbon was measured over time in each algal culture. Table 4 presents the mean production of metabolites (in terms of dissolved organic carbon) for different Cr(VI) concentrations. Dissolved organic carbon concentration is small when compared with total organic carbon concentration. There are no indications that cellular growth by non photosynthetic routes may have occurred. Results also indicate that an outstanding dissolved organic carbon production due to the metal presence seems not to occur. It is also possible to observe that *Chlorella fusca* yields the lowest dissolved organic carbon concentration.

Table 3. Specific growth rates of biomass (μ) and maximum biomass concentrations (K) obtained by fitting the experimental data to the logistic equation

Concentration of Cr(VI) (mg l^{-1})	Microalgae	μ (h^{-1})	K (mg l^{-1})	R^2
0	***Chlorella fusca***	**0.008**	**372.4**	**0.97756**
	Chlorella vulgaris	0.014	171.4	0.98586
	Scenedesmus acutus	0.013	356.1	0.99839
	Scenedesmus obliquus	0.017	603.0	0.99914
1	***Chlorella fusca***	**0.015**	**267.9**	**0.93015**
	Chlorella vulgaris	0.015	182.7	0.99860
	Scenedesmus acutus	0.015	313.6	0.99351
	Scenedesmus obliquus	0.017	607.7	0.99973
5	***Chlorella fusca***	**0.009**	**178.1**	**0.97919**
	Chlorella vulgaris	0	-	-
	Scenedesmus acutus	0.009	124.0	0.98879
	Scenedesmus obliquus	0.016	254.6	0.99525
25 (50 for *CF*)	***Chlorella fusca (CF)***	**0**	-	-
	Chlorella vulgaris	0	-	-
	Scenedesmus acutus	0	-	-
	Scenedesmus obliquus	0	-	-

Table 4. Effect of Cr(VI) addition on the production of metabolites (expressed as dissolved organic carbon, DOC) by the algal cells

Microalgae	Concentration of Cr(VI) (mg l^{-1})			
	0	1	5	25 (50 for *CF*)
	Metabolites (DOC average ± standard deviation) (mg l^{-1})			
Chlorella fusca (CP)	11.0±4.9	8.0±4.1	8.9±3.9	16.2±6.7
Chlorella vulgaris	29.1±9.7	26.8±5.3	24.6±1.0	24.2±2.4
Scenedesmus acutus	29.2±4.3	40.9±29.6	38.4±34.6	28.9±5.2
Scenedesmus obliquus	26.2±3.7	28.3±8.6	35.3±21.8	26.2±2.6

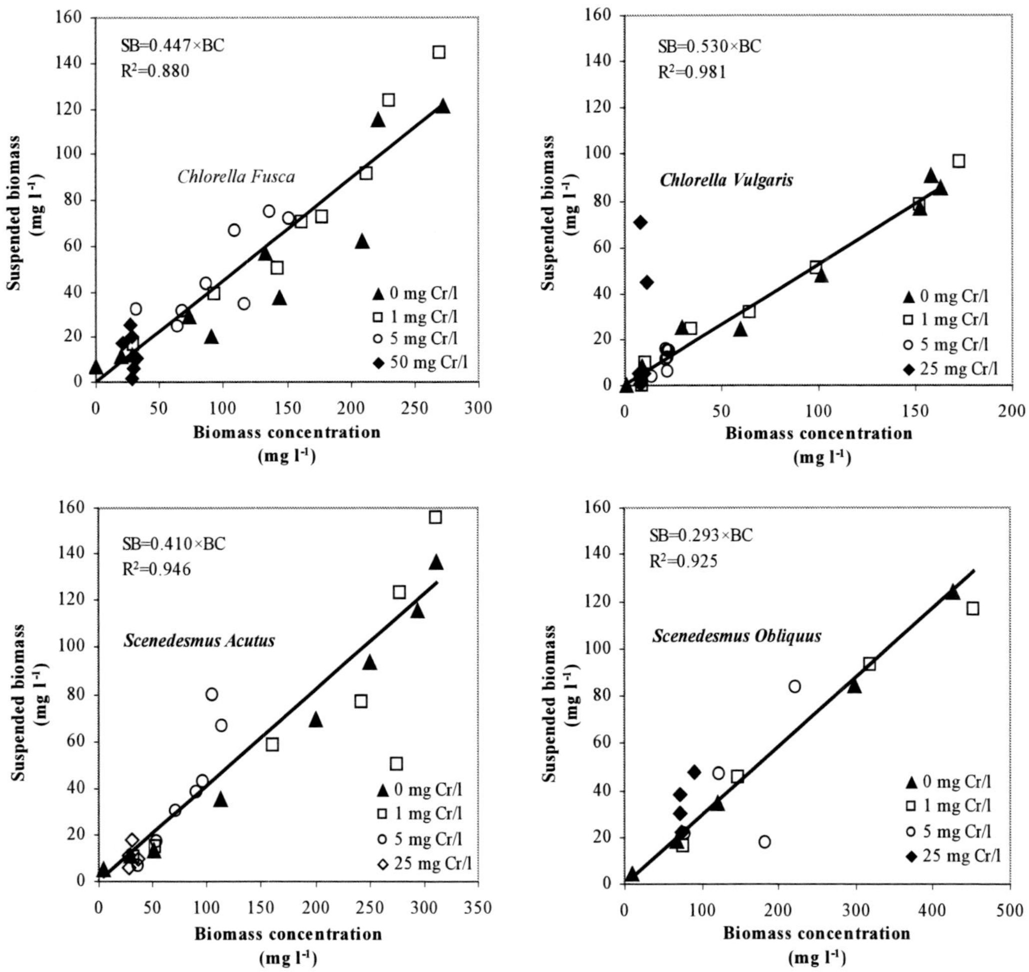

Figure 7. Suspended biomass as a function of dry biomass concentration. (—) Linear regression. SB = Suspended Biomass; BC = Biomass Concentration.

Suspended organic carbon was plotted against biomass for each Cr(VI) concentration and no significant dependence on Cr(VI) was observed. A global linear correlation between biomass concentration and suspended organic carbon for each algal culture was found, as shown in Figure 7. The percentage of organic carbon in the algal biomass varies from 29% (*S. obliquus*) to 53% (*C. vulgaris*). These results are in agreement with those reported by several authors that present values of the same order of magnitude by using empirical formulas [23, 30-33].

4.5. Effect of Cr(VI) on Biomass Settling

The ability of algal biomass to settle down has been expressed in terms of the decrease of the suspended biomass concentration after a 6 h-settling period. Results of the biomass-settling rate for each algae species and different Cr(VI) concentrations are presented in Table 5. In the absence of Cr(VI) and for Cr(VI) = 1 mg l^{-1}, the settling rate is similar for *C. fusca*, *S. acutus* and *S. obliquus* but much lower for *C. vulgaris*. As Cr(VI) concentration increases, the settling rate decreases for all algal cultures but the effect is particularly evident for *C. vulgaris* and *S. acutus*. Results suggest that Cr(VI) ≥ 5 mg l^{-1} impairs algae autoflocculation.

Table 5. Biomass settling rate (mg l^{-1} h^{-1})

Concentration of Cr(VI) (mg l^{-1})	*Chlorella fusca (CF)*	*Chlorella vulgaris*	*Scenedesmus acutus*	*Scenedesmus obliquus*
0	55.1	13.4	41.1	42.7
1	45.8	12.3	46.4	54.6
5	26.5	1.4	16.0	33.2
25 (50 for *CF*)	nd	0.52	nd	6.5

nd – not determined.

5. Conclusions

Microalgae growth kinetics is adequately described by the logistic equation. Experiments carried out at 24.6 °C and initial pH = 6.5 led to values of the specific growth rate of biomass, μ, of 0.0162, 0.0284, 0.0359 and 0.0370 h^{-1} for *Scenedesmus obliquus*, *Chlorella vulgaris*, *Scenedesmus acutus* and *Chlorella fusca*, respectively. The corresponding maximum biomass concentration, *K*, was 604.1, 369.2, 542.9 and 403.3 mg l^{-1} for the same algal species. In spite of the low growth rate, *Scenedesmus obliquus* culture reaches the highest biomass concentration. A decrease in temperature operation to around 21 °C originated lower μ (between 0.0241 and 0.0357 h^{-1}) and *K* (between 169.9 and 327.8 mg l^{-1}) values. An increase in pH to about 7.9 led to drastic decreases in the specific growth rate of biomass (below 0.0177 h^{-1}) and the maximum biomass concentration (below 54.7 mg l^{-1}). Results show that pH plays a central role in microalgae growth. *Chlorella vulgaris* seems to be the less resistant microalgae to changes in pH and temperature. Looking simultaneously at μ and *K* values, the best performance was obtained with *Scenedesmus acutus* and *Chlorella fusca.*

Growth inhibition of *Chlorella fusca*, *Chlorella vulgaris*, *Scenedesmus acutus* and *Scenedesmus obliquus* by Cr(VI) occurs for concentrations ≥ 5 mg l^{-1}, but concentrations up to 1 mg l^{-1} do not seem to seriously affect growth or even may increase it slightly. The more marked effect of Cr(VI) on algal growth is observed in *C. vulgaris* culture. Metabolites production has proved to be in small quantity either in presence or absence of chromium, thus indicating that there was no relevant growth based on a non photosynthetic process. Microalgal biomass produced in all experiments is composed of approximately 40 - 50% organic carbon. *C. fusca* and *S. obliquus* are the algal species that present a higher degree of autofloculation, then being more easily separated from the liquid phase. However, Cr(VI) ≥ 5 mg l^{-1} impairs algae autoflocculation particularly of *C. vulgaris* and *S. acutus* cultures. For Cr(VI) ≤ 1 mg l^{-1} only *C. vulgaris* flocs present a low settling rate.

Acknowledgements

Financial and logistic support for this work was in part provided by Fernando Pessoa Foundation Teaching and Culture and by LSRE financing by FEDER/POCI/2010, for which the authors are thankful.

References

[1] Patterson JW. *Industrial Wastewater Treatment Technology.* Boston, Butterworth, 1985.

[2] González LE, Cañizares RO, Baena S. Efficiency of ammonia and phosphorus removal from a colombian agroindustrial wastewater by the microalgae Chlorella vulgaris and Scenedesmus dimorphus. *Bioresource Technology* **60**(3): 259-262, 1997.

[3] Lau PS, Tam NFY, Wong YS. Operational optimization of batchwise nutrient removal from wastewater by carrageenan immobilized Chlorella vulgaris. *Water Science and Technology* **38**(1): 185-192, 1998.

[4] Meiring PGJ, Rose PD, Shipin OV. Algal aid puts a sparkle on effluent. *WQI* **2**(30-32, 1994.

[5] Robinson PK. Immobilized algal technology for wastewater treatment purposes. In: Wong Y-S, Tam NFY, eds. *Wastewater Treatment with Algae,*vol USA, Springer-Verlag, 1-16; 1998.

[6] Twist H, Edwards AC, Codd GA. A novel in-situ biomonitor using alginate immobilised algae (Scenedesmus subspicatus) for the assessment of eutrophication in flowing surface waters. *Water Research* **31**(8): 2066-2072, 1997.

[7] Lau PS, Tam NFY, Wong Y-S. Carrageean as a matrix for immobilizing microalgal cells for wastewater nutrients removal. In: Wong Y-S, Tam NFY, eds. *Wastewater Treatment with Algae,*vol USA, Springer-Verlag, 1998.

[8] Becker EW. *Biotechnology and Microbiology.* Great Britain, Cambridge University Press, 1994.

[9] Borowitzka M. Limits to growth. In: Wong Y-S, Tam NFY, eds. *Wastewater Treatment with Algae,*vol USA, Springer-Verlag, 203-226; 1998.

[10] Sterner RW, Grover JP. Algal growth in warm temperate reservoirs: kinetic examination of nitrogen, temperature, light, and other nutrients. *Water Research* 32(12): 3539-3548, 1998.

[11] Cervantes C, Campos-Garcia J, Devars S, Gutierrez-Corona F, Loza-Tavera H, Torres-Guzman JC, Moreno-Sanchez R. Interactions of chromium with microorganisms and plants. *FEMS Microbiology Reviews* **25**(3): 335-347, 2001.

[12] Corradi MG, Gorbi G, Morsi A-E-MH, Torelli A, Bassi M. Exudates from The Wild Type and a Cr Tolerant Strain of Scenedesmus acutus Influence Differently Cr(VI) Toxicity to Algae. *Chemosphere* **37**(14-15): 3019-3025, 1998.

[13] Lam PKS, Wut PF, Chan ACW, Wu RSS. Individual and combined effects of cadmium and copper on the growth response of Chlorella vulgaris. *Environmental Toxicology* 14(347-353, 1999.

[14] Soto EG, Rodríguez MVG, Suárez EL, Romero JMC, Iglesias JP, Solís JMF. Binding of metals in solution by Chlorella vulgaris. *Annali di Chimica* **89**(323-326): 1999.

[15] Travieso L, Pellón A, Benitéz F, Sánchez E, Borja R, O´Farril N, Weiland P. BIOALGA reactor: preliminary studies for heavy metals removal. *Biochemical Engineering Journal* **12**(87.91, 2002.

[16] Avery SV, Codd GA, Gadd GM. Microalgal removal of organic and inorganic metal species from aqueous solution. In: Wong Y-S, Tam NFY, eds. *Wastewater Treatment with Algae,*vol USA, Springer-Verlag, 55-72; 1988.

[17] Bundy K, Mowat F. Speciation studies and toxicity assessment of complex heavy metal mixtures. In *HSRC/WERC Joint Conference on the Environment,*vol 1996.

[18] Tam NFY, Wong YS, Simpson CG. Removal of copper and immobilized microalgae, Chlorella vulgaris. In: Wong Y-S, Tam NFY, eds. *Wastewater Treatment with Algae,*vol Springer-Verlag, 17-36; 1998.

[19] Capelo S, Vilhena MF, Gonçalves MLSS, Sampayo MA. Effect of lead on the uptake of nutrients by unicellular algae. *Water Research* **27**(1563-1568, 1993.

[20] Wong SL, Wainwright JF, Pimenta J. Quantification of total and metal toxicity in wastewater using algal biomass. *Aquatic Toxicology* **31**(57-75): 1995.

[21] Lupi FM, Fernandes HML, Sá-Correia I. Increase of copper toxicity to growth of Chlorella vulgaris with increase of light intensity. *Microbial Ecology* **35**(193-198, 1998.

[22] Richmond A. Large scale microalgal culture and applications. In: Chapman DG, Round FE, eds. *Progress in Phycological Research,*vol 7. USA, Biopress Ltd., 1-62; 1990.

[23] Oswald WJ. Micro-algae and wastewater treatment. In: Borowitzka MA, Borowitzka LJ, eds. *MicroAlgal Biotechnology,*vol Great Britain: Cambridge, University Press, 305-328; 1988.

[24] Bailey JE, Ollis DF. *Biochemical Engineering Fundamentals.* USA, McGraw-Hill Book Company, 1977.

[25] Mitchell MJ, Slaughter JC. *Biology and Biochemistry for Chemists and Chemical Engineers.* USA, Ellis Horwood Limited, 1989.

[26] Wilson EO, Bossert WH. *A Primer of Population Biology.* USA, Sinauer Associates Inc, 1971.

[27] Krebs CJ. *Ecology: The Experimental Analysis of Distribution and Abundance.* Menlo Park, California, Benjamin Cummings, 2001.

[28] Schelenz T. Relationships between algistatic activity and physicochemical properties of 5-amino.1-aryl-1H-tetrazoles. . *Journal fur Praktishe Chemie Chemiker Zeitung* **342**(2): 205-210, 2000.

[29] Twist H, Edwards AC, Codd GA. Algal growth responses to waters of contrasting tributaries of the River Dee, North-East Scotland. *Water Research* **32**(8): 2471-2479, 1998.

[30] Minowa T, Sawayama S. A novel microalgal system for energy production with nitrogen cycling. *Fuel and Energy Abstracts* **78**(1213-1215, 1999.

[31] Fogg GE. *The Metabolism of Algae.* USA, John Wiley & Sons Inc., 1953.
[32] Gloyna EF. *Bassins de Stabilization des Eaux Usées.* OMS 1972.
[33] Stumm W, Morgan JJ. *Aquatic Chemistry.* New York, John Wiley & Sons, 1996.

In: Biochemical Engineering
Editors: F.E. Dumont and J.A. Sacco, pp. 275-291
ISBN: 978-1-60741-257-1

Chapter 12

SHORT-TERM EFFECTS OF GLUCOSE ADDITION ON NITRIFICATION AND ACTIVATED SLUDGE SETTLEMENT IN SEQUENCING BATCH REACTORS

Guangxue Wu[1] and Yuntao Guan[2]
[1]Department of Civil Engineering, National University of Ireland, Galway, Ireland
[2]Research Center of Environmental Engineering and Management, Graduate School of Shenzhen, Tsinghua University, Shenzhen 518055, China

Abstract

The short-term effects of glucose addition on nitrification and activated sludge settlement were investigated in two laboratory-scale sequencing batch reactors (SBRs): one with the addition of glucose (G-Reactor) and the other without the addition of glucose (N-Reactor). The characteristics of nitrification activity, nitrite accumulation, and activated sludge settlement were examined. A high specific nitrification rate was obtained in the N-Reactor, while a high volumetric nitrification rate was obtained in the G-Reactor. Nitrite accumulation occurred in both reactors, and the nitrite/total oxidized nitrogen ratio in both reactors was over 67%. Nitrite accumulation in both reactors was due to low pH caused by the processing of nitrification. In the G-Reactor, the biomass concentration did not change much; in the N-Reactor, the biomass concentration decreased with time. The reason for decreasing biomass concentration in the N-Reactor was as follows: (1) high extracellular polymeric substances (EPS) produced in the N-Reactor due to shortage of organic carbon substrate, resulting in poor settlement of activated sludge flocs; (2) poor settlement of activated sludge flocs causing activated sludge wash out of the system, and, consequently, a low sludge retention time occurred; and, finally, (3) the low sludge retention time further encouraged the poor settlement of activated sludge flocs.

1. Introduction

Nitrification plays an important role in the nitrogen cycle. Nitrification includes two steps: (i) ammonia (NH_4-N) is oxidized to nitrite (NO_2-N) in the nitritation step, and then (ii) NO_2-N is oxidized to nitrate (NO_3-N) in the nitratation step. These two steps are carried out sequentially by ammonia oxidizing bacteria (AOB) and nitrite oxidizing bacteria (NOB). Both AOB and NOB are named as "nitrifiers". Nitrifiers are slow-growing bacteria and have yield rates of 0.21 g biomass/g N (Rittmann and McCarty, 2001). In addition, nitrifiers are sensitive to shock-loading and toxins (Wagner et al., 1995; Okabe et al., 2004). Therefore, nitrification is usually the limiting process in wastewater treatment. As a result, more attention is being paid to the improvement of the efficiency and stability of nitrification. Biofilm systems or bio-augmentation processes, wherein the sludge retention time (SRT) is separated from the hydraulic retention time (HRT) to maintain low-growth nitrifiers, are commonly used to improve nitrification (Park et al., 2002; Head and Oleszkiewicz, 2004).

Nitrification can be hindered by heterotrophic activities, because heterotrophs out-compete nitrifiers for space and nutrients (Hanaki et al., 1990; Nogueira et al., 2002). The carbon/nitrogen (C/N) ratio is one of the factors that determine the heterotroph/autotroph population ratio in wastewater treatment systems. In general, a high carbon substrate concentration can cause the inhibition of nitrification in activated sludge systems and biofilm systems (Hanaki et al., 1990; Nogueira et al., 2002). Lindemann and Wiesmann (2000) found that aerobic nitrifiers were out-competed by heterotrophs at a C/N ratio of 1.8. Komorowska-Kaufman et al. (2006) observed that a chemical oxygen demand/nitrogen (COD/N) ratio below 4 encouraged nitrification and a value above 4 caused instability of nitrification, especially at low temperatures. Nogueira et al. (2002) found that, at long HRTs, nitrification efficiency decreased drastically when acetate was added. However, some studies have shown that nitrification may be improved with the existence of a heterotrophic protective layer (Germain et al., 2007). Compared with a biological oxygen demand (BOD) loading rate of 410–450 mg/l·d, at the BOD loading rate of 290 mg/l·d, nitrification activity was reduced, which was caused by the diminishment of the protective heterotrophic layer for nitrifiers at low temperatures ranging from 9 to 14°C (Germain et al., 2007). Oyanedel et al. (2005) observed that, for the suspension biomass, the nitrification capacity was improved with COD addition compared with no COD addition.

The settlement characteristics of activated sludge flocs are very important for the design and operation of wastewater treatment processes. Several studies have examined the settlement characteristics of activated sludge flocs in nitrifying activated sludge systems (Yoshinori et al., 1999; Martinez et al., 2004; Texier and Gemez, 2004). In these studies, the sludge volume index (SVI) values were below 22 ml/g, indicating good settleability of nitrifying activated sludge. In addition, the accelerated nitrification activity could improve the activated sludge settleability. However, few studies have investigated the effect of carbon substrate on the settleability of nitrifying activated sludge flocs. In the study of Wu et al. (2008a), the biomass concentration decreased drastically in a nitrifying reactor without the addition of organic carbon substrate and there was little increase in the biomass concentration even without desludging; however, the settlement characteristics of activated sludge were not investigated and no explanation was given for the decrease in the biomass concentration.

Therefore, in this study, the effects of glucose addition on nitrification and activated sludge settlement were investigated in a short-term experiment. The activity of nitrification and the characteristics of activated sludge settlement were examined under conditions with and without glucose addition. The underlying mechanisms were analyzed.

2. Materials and Methods

2.1. Nitrification Enrichment

Two identical laboratory-scale sequencing batch reactors (SBRs), each with an effective volume of 2 litres, were used to investigate the short-term effects of glucose addition on nitrification and settlement of activated sludge flocs. One reactor (the G-Reactor) was operated with the addition of glucose, and the other reactor (the N-Reactor) without the addition of glucose. The SBRs were operated as once-a-day feeding and drawing-off reactors at temperatures ranging from 18 to 20°C. Each day, 1 litre of treated wastewater in both reactors were exchanged, resulting in a HRT of 2 days. Pulse feeding and drawing-off were applied, which occurred within approximately 2-minute durations. The period for aeration was 23 hours and the settlement period was approximately 58 minutes. Air was supplied by air pumps during the aerobic reaction phase and the concentration of dissolved oxygen (DO) in the bulk water was maintained above 2 mg/l (usually around 8 mg/l). Mixed liquor (100 ml) was withdrawn from both reactors once a day just before the end of the aerobic react phase. Both SBRs were seeded with activated sludge taken from Tuam Municipal Wastewater Treatment Plant, Galway, Ireland.

The components of the synthetic wastewater were: 420 mg/l of NH_4Cl, 200 mg/l of $Na_2HPO_4 \cdot 12H_2O$, 12 mg/l of yeast extract, 500 mg/l of $NaHCO_3$, 200 mg/l of $KHCO_3$, 200 mg/l of $MgSO_4 \cdot 7H_2O$, 8 mg/l of $FeCl_3 \cdot 6H_2O$, 12 mg/l of $CaCl_2 \cdot 6H_2O$ and 8 mg/l of $MnSO_4 \cdot H_2O$. In addition, glucose was added to the G-Reactor at a concentration of 200 mg/l.

2.2. Analytical Methods

Suspended solids (SS), volatile suspended solids (VSS) and SVI were determined according to standard methods (APHA, 1995). SVI was measured using a 100 ml graduated cylinder. NH_4-N, NO_2-N and NO_3-N were tested using a Konelab 20 analyzer (Thermo Clinical Labsystems, Vantaa, Finland). DO and pH were measured using ***WTW*** portable ***DO*** and pH probes.

The free ammonia (NH_3) and free nitrous acid (HNO_2) concentrations were calculated using (Anthonisen et al., 1976; Ford, 1980):

$$NH_3 = \frac{NH_4^+ 10^{pH}}{10^{pH} + e^{6344/(273+t)}} \quad (1)$$

$$HNO_2 = \frac{NO_2^-}{10^{pH} e^{-2300/(273+t)}} \quad (2)$$

To examine extracellular polymeric substances (EPS), the activated sludge mixed liquor, taken from the two reactors, was centrifuged at 3900 rpm for 10 minutes and then the biomass was re-suspended in a 0.85% NaCl solution. EPS was extracted from the re-suspended activated sludge mixed liquor using the formaldehyde–NaOH method (Liu and Fang, 2002). After extraction, the mixed liquor was centrifuged at 4°C at 17,000 g for 20 minutes. The carbohydrate and protein concentrations in the supernatant were measured by the sulfuric-phenol method (Dubois et al., 1956) and the Lowry method (Lowry et al., 1951), respectively.

The dynamics of NH_4-N, NO_2-N and NO_3-N concentrations were measured in typical cycles in the two reactors. The data from the periods when the parameters had linear relationships with time were used to calculate the nitritation and nitratation rates based on the reactor volume (the volumetric rate) or the unit of biomass concentration (the specific rate), respectively.

The morphology of activated sludge was observed with a phase contract light microscopy (OPTIKA Microscopes, Italy) and photos were taken with a digital camera (INOVO USB2.0-CAM) connected to the microscopy.

3. Results

3.1. Short-term Dynamics of Various Nitrogen Forms in the Two SBRs

The dynamics of various nitrogen forms in the N-Reactor and the G-Reactor during the 32-day short-term study period are shown in Figures 1 and 2.

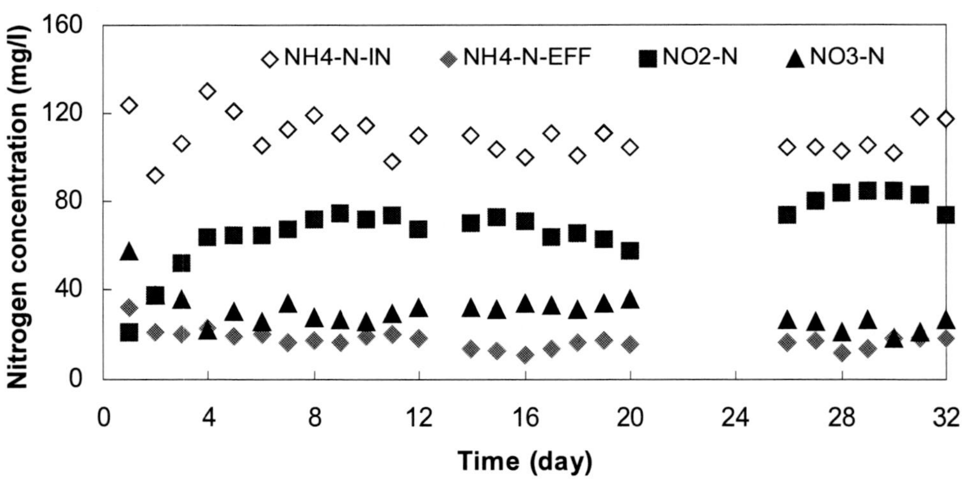

Figure 1. Nitrogen dynamics in the N-Reactor during the short-term study period. NH4-N-IN: NH_4-N in the influent; NH4-N-EFF: NH_4-N in the effluent; NO2-N: NO_2-N in the effluent; and NO3-N: NO_3-N in the effluent.

At the influent NH_4-N concentrations of 109.3 ± 8.7 mg/l in the N-Reactor and 109.1 ± 9.1 mg/l in the G-Reactor, the respective NH_4-N removal efficiencies were 84% in the N-Reactor and 93% in the G-Reactor. The total oxidized nitrogen (NOx-N: NO_2-N + NO_3-N) concentrations were 97.2 ± 8.4 mg/l in the N-Reactor and 86.1 ± 7.5 mg/l in the G-Reactor. The high NH_4-N removal efficiency and the lower effluent NOx-N concentration in the G-Reactor showed that a high amount of NH_4-N was used for biomass synthesis by heterotrophs due to the addition of glucose. In the N-Reactor, the influent total nitrogen (influent NH_4-N) was 109 mg/l and the effluent total nitrogen (effluent NH_4-N + NOx-N) was 115 mg/l. The increased total nitrogen concentration in the N-Reactor could be attributed to the lysis of bacteria in the reactor due to shortage of carbon substrates. NO_2-N was the main component of the total NOx-N in the effluent in both reactors, with ratios (NO_2-N/NOx-N) of 0.69 in the N-Reactor and 0.67 in the G-Reactor.

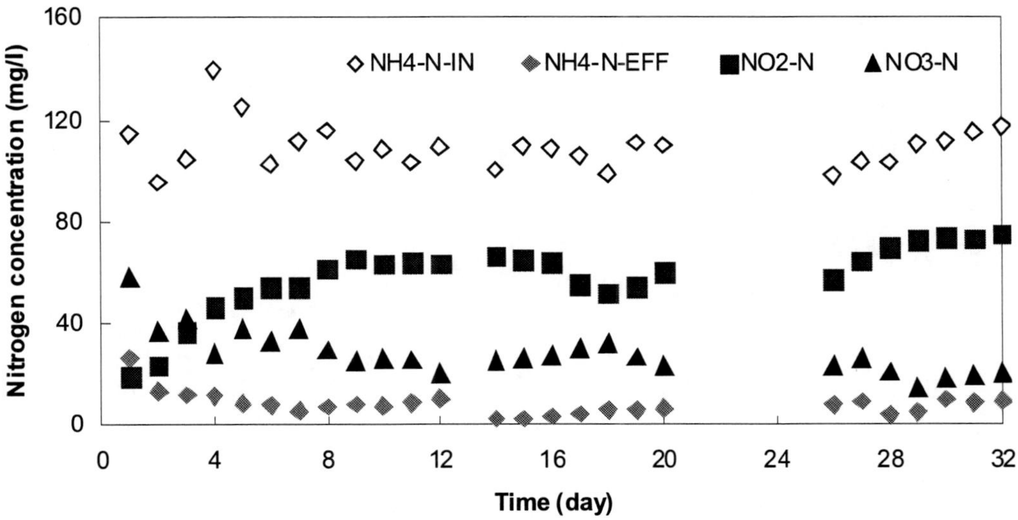

Figure 2. Nitrogen dynamics in the G-Reactor during the short-term study period. NH4-N-IN: NH_4-N in the influent; NH4-N-EFF: NH_4-N in the effluent; NO2-N: NO_2-N in the effluent; and NO3-N: NO_3-N in the effluent.

3.2. Typical Cycle Study in the Two SBRs

In order to examine the nitrification activity and the nitrite accumulation in detail, on Day 19, the dynamics of various nitrogen forms and pH in a react cycle in both reactors were examined and the results are shown in Figures 3 and 4.

Nitrification mainly occurred within initial 14 hours and there was little change thereafter. Therefore, only parameters obtained within the initial 14 hours are presented. Using linear regression, the volumetric nitritation and nitratation rates were 5.4 mg/l·h and 1.6 mg/l·h in the N-Reactor, and 6.4 mg/l·h and 2.6 mg/l·h in the G-Reactor. However, the specific nitritation and nitratation rates based on unit biomass concentrations were 9.4 mg/g VSS·h and 3.9 mg/g VSS·h in the N-Reactor, and 4.2 mg/g VSS·h and 1.7 mg/g VSS·h in the G-Reactor.

The NO_2-N/NOx-N ratios, obtained at Hour 14 in both reactors, were 0.69. Based on Equation 1, the NH_3 concentration ranged from 0 to 4.2 mg/l in the N-Reactor and from 0 to 1.5 mg/l in the G-Reactor. According to Equation 2, the HNO_2 concentration ranged from 0 to 0.25 mg/l in the N-Reactor and from 0 to 0.24 mg/l in the G-Reactor. With the processing of nitrification, pH decreased with time in both reactors. pH varied from 8.4 to 5.9 in the N-Reactor and from 8.0 to 5.9 in the G-Reactor. When the pH was above 6.0, the nitrification rate was constant, and this reflected from the good linear regression of the produced NO_2-N and NO_3-N. When pH decreased to below 6.0, there was no further nitritation or nitratation processed.

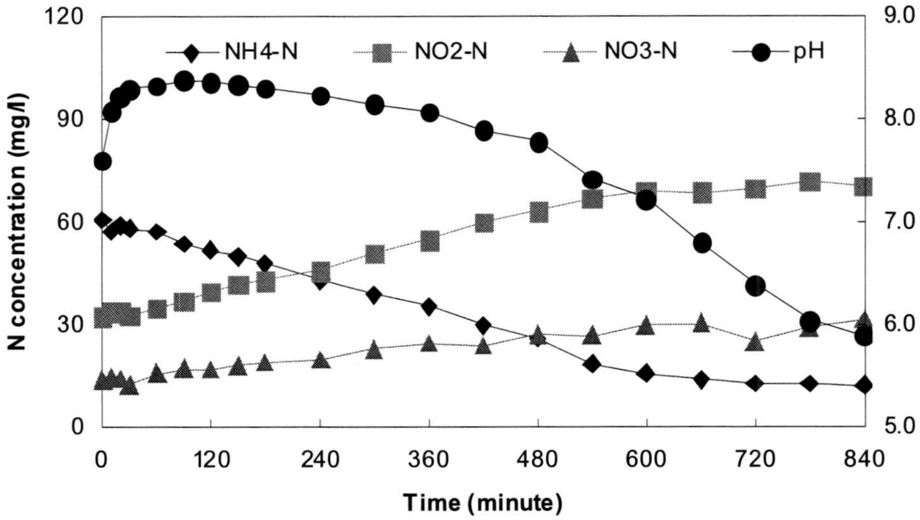

Figure 3. Nitrogen and pH dynamics in a typical cycle in the N-Reactor.

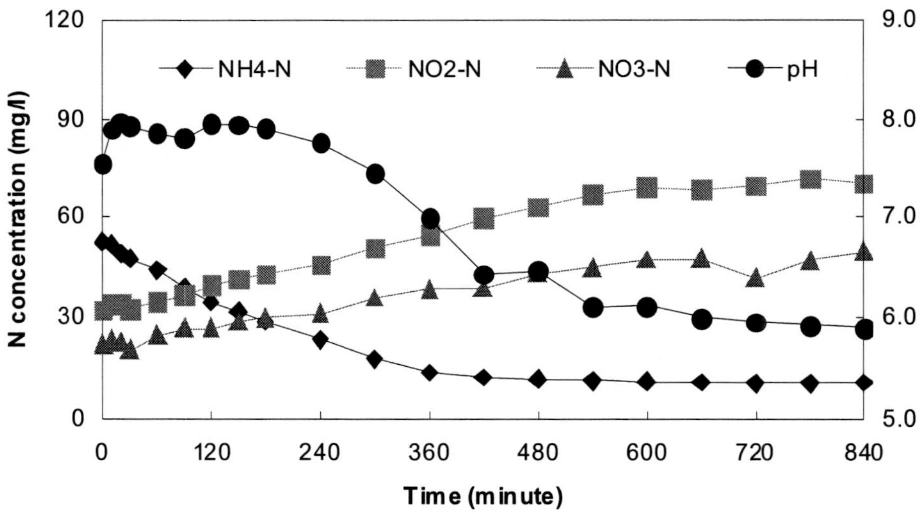

Figure 4. Nitrogen and pH dynamics in a typical cycle in the G-Reactor.

3.3. Short-term Dynamics of Biomass Concentration in the Two SBRs

The dynamics of the biomass concentration in both reactors and in the effluent exiting from both reactors are given in Figure 5.

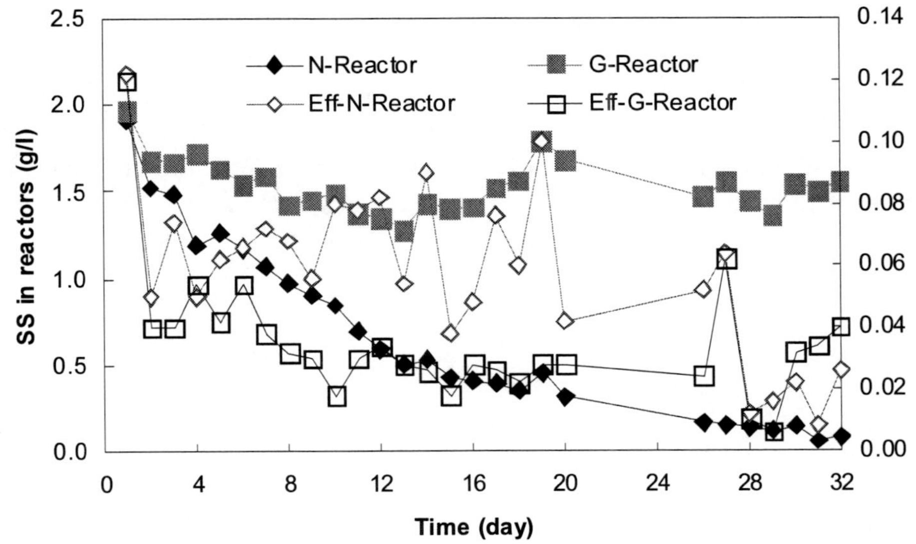

Figure 5. Biomass (SS) concentrations in the reactor (N-Reactor and G-Reactor) and in the effluent (Eff-N-Reactor and Eff-G-Reactor) during the study period.

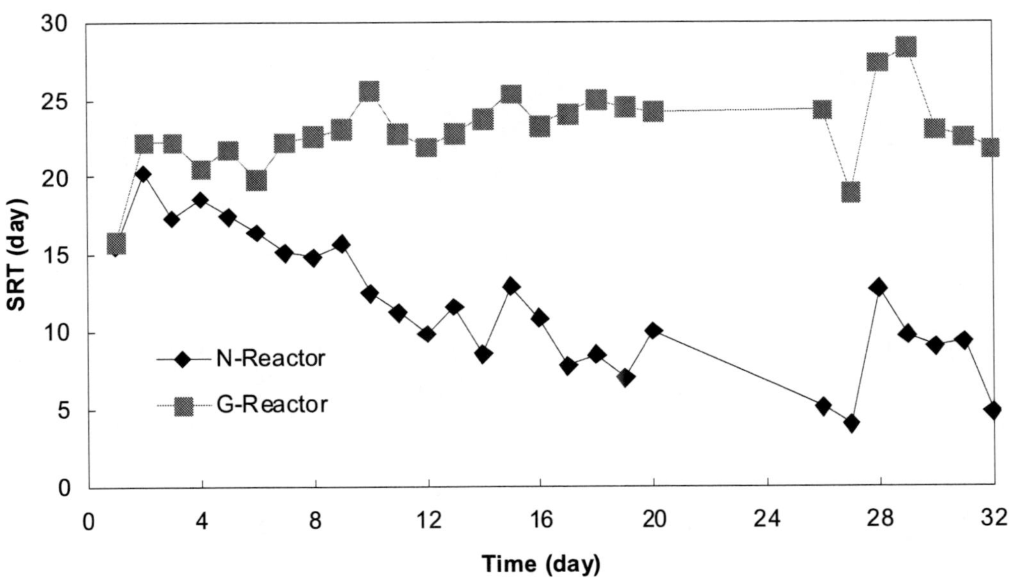

Figure 6. Dynamics of sludge retention time (SRT) in the N-Reactor and the G-Reactor.

The biomass concentration (*SS*) in the N-Reactor decreased drastically from 1.80 g/l to less than 0.10 g/l within 32 days. This is described by:

$$SS = -0.59Ln(t) + 2.11 \ (R^2 = 0.97) \tag{3}$$

where t is time in days. The biomass in the G-Reactor decreased from 1.87 to 1.55 g/l within approximately 6 days and remained constant thereafter. A slightly higher effluent SS of 58 ± 27 mg/l existed in the N-Reactor than that of 35 ± 21 mg/l in the G-Reactor.

The higher effluent biomass concentrations and lower biomass concentrations inside the reactor caused the lower SRT in the N-Reactor compared to that in the G-Reactor (Figure 6). The SRT in the N-Reactor ranged from 4 to 20 days, while that in the G-Reactor ranged from 19 to 25 days. The relationships between SS and SRT in the N-Reactor are presented in Figure 7, which shows that the lower biomass concentration in the N-Reactor was mainly caused by the low SRT.

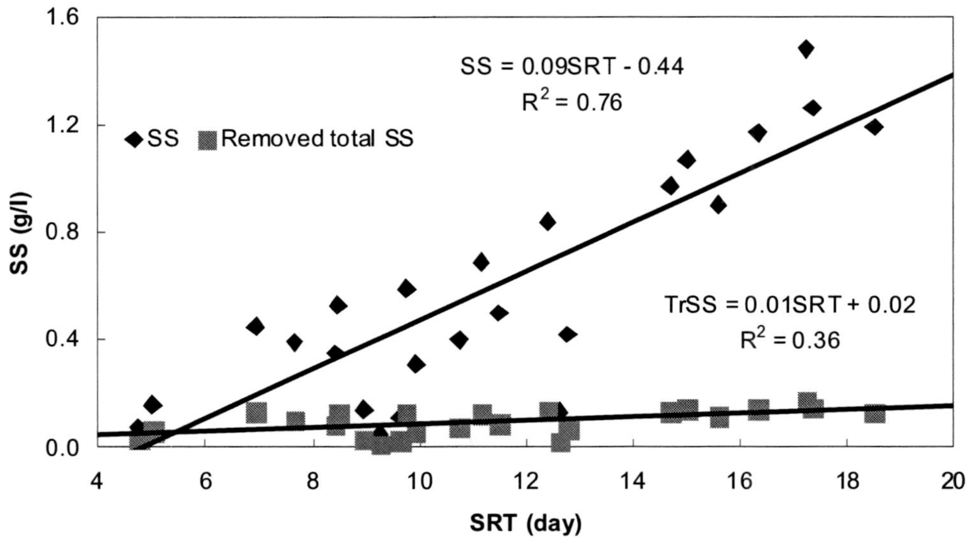

Figure 7. Relationships between the sludge retention time (SRT) and the biomass in reactor (SS) and total removed biomass per day (TrSS) in the N-Reactor.

The high effluent SS from the N-Reactor could be due to the poor settlement of activated sludge flocs in the N-Reactor compared to the G-Reactor, which was reflected by the slightly higher SVI of 218 ± 24 ml/g in the N-Reactor than that of 171 ± 52 ml/g in the G-Reactor.

3.4. Morphology and Extracellular polymeric Substances of Activated Sludge Flocs

The morphology of activated sludge flocs taken on Day 32 is shown in Figure 8. There were two types of floc forms presenting in the N-Reactor: one type was granular and the other type comprised small-dispersed particles. In the G-Reactor, most flocs were bridged together and formed larger particles.

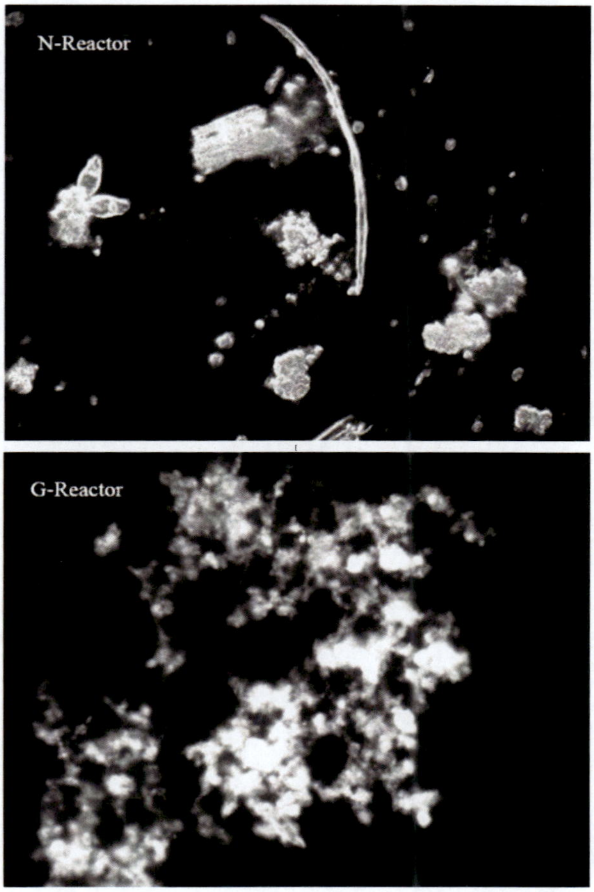

Figure 8. Activated sludge floc structure in the N-Reactor and G-Reactor taken by the phase contrast microscopy at 100× on Day 32. The scale of the picture is 1024 × 768 μm.

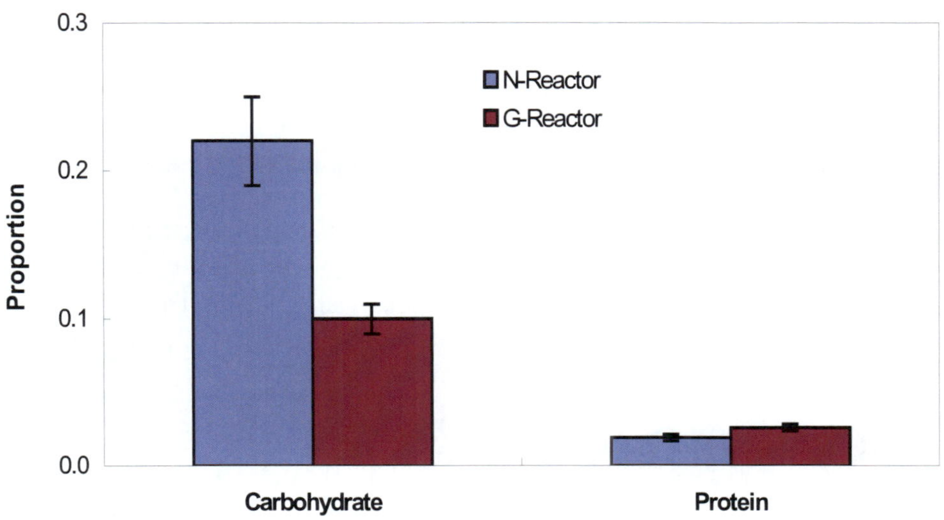

Figure 9. Proportions of extracellular polymeric substances (carbohydrate and protein) in the biomass in the N-Reactor and the G-Reactor.

The EPS content in the biomass taken from both reactors is shown in Figure 9. The total EPS (carbohydrate + protein) proportion was 24.0% in the N-Reactor and 12.6% in the G-Reactor. There was little difference in the protein proportion in both reactors: 1.9% in the N-Reactor and 2.6% in the G-Reactor. However, the carbohydrate proportion was significantly different in both reactors: 22% in the N-Reactor and 10% in the G-Reactor.

The average carbohydrate/protein ratios in the EPS were 12.0 ± 1.1 in the N-Reactor and 3.8 ± 0.4 in the G-Reactor. The carbohydrate/protein ratio could be correlated to SVI in both reactors (Figure 10).

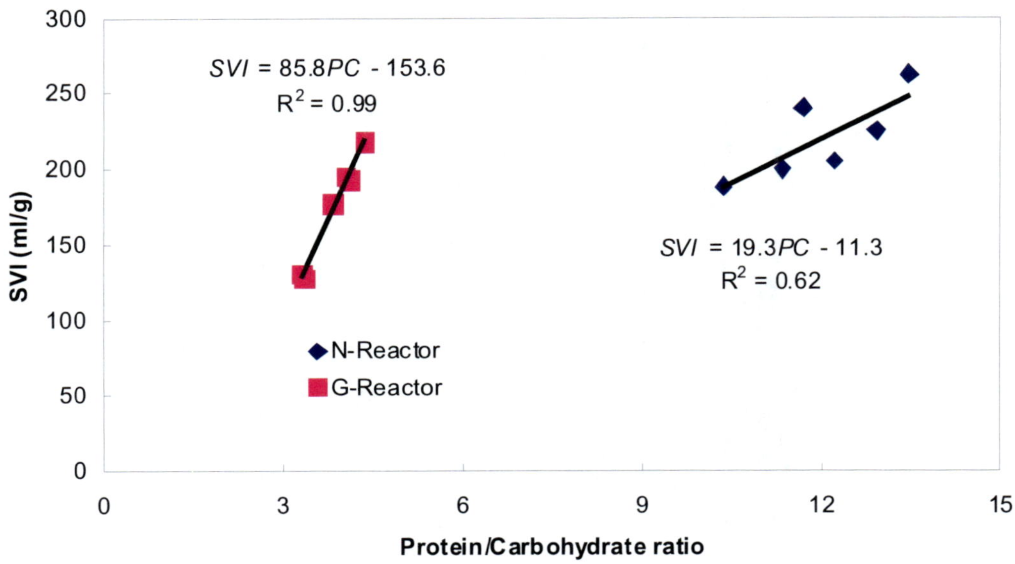

Figure 10. Relationships between the SVI and the protein/carbohydrate ratio (PC).

4. Discussion

4.1. Effect of the Glucose Addition on the Activity of Nitrification

Compared with the addition of glucose in the G-Reactor, a high specific nitrification rate and a low volumetric nitrification rate were obtained in the N-Reactor without the addition of glucose. This is similar to the result obtained by Wu et al. (2008a). The possible reasons causing this phenomenon are: (1) the effect of carbon substrates; (2) the effect of the distribution of microbial communities; and (3) the effect of SRT.

4.1.1. Effect of Carbon Substrates

The effect of carbon substrate on nitrification has been intensively investigated (Hanaki et al., 1990; Nogueira et al., 2002). However, different results have been obtained from different studies.

The ratio of COD/N and the applied loading rate should be both considered when discussing the effect of carbon substrate on nitrification. For example, even at high COD concentrations, if there still has adequate time for nitrification to process after carbon oxidation, nitrification will not be much affected. Therefore, in order to achieve stable nitrification, an adequate nitrification duration should be maintained and also an enough loading rate should be adopted to maintain the growth of nitrifiers (Rothman, 1998; Altinbas, 2001; Gieseke et al., 2002).

With glucose addition, a higher biomass concentration (including heterotrophs and protozoa concentrations) was obtained compared with the condition without the addition of glucose. Nitrification capacity was found to increase with increasing protozoa concentrations, which could be due to the following protozoa activities: (1) supplying nutrients through excretion; and (2) predating heterotrophs (Petropoulos and Gilbride, 2005). Therefore, by ensuring a high biomass concentration and an adequate nitrification duration, nitrification could be improved with the addition of organic carbon substrates.

4.1.2. Effect of Microbial Distribution

The distributions (suspended or attached) of heterotrophs and autotrophs can be affected by the C/N ratio. At conditions without organic carbon addition, heterotrophs and nitrifiers co-existed in the outermost biofilm layers and heterotrophs dominated in the innermost biofilm layers; while, at the C/N ratio of 1.5, heterotrophs dominated in the outermost biofilm layers and nitrifiers were only present in the innermost biofilm layers (Okabe et al., 1996). In addition, other environmental conditions like HRT can also affect the distribution of microbial communities. When HRT is about an order of magnitude larger than the reciprocal of the net maximum growth rate, the majority of the active biomass will grow suspended in the mixed liquid rather than in the biofilm or granules (Van Benthum et al., 1997; Villaverde et al., 2000). Nitrifiers existing at the top layer of biofilm or flocs can increase the nutrient supply and enhance the nitrification activity.

The distribution of nitrifiers (within the same flocs or biofilms) can contribute to the different nitrification rates at conditions with and without the addition of glucose. AOB were mainly in a form of clusters in the activated sludge flocs or biofilm systems without the addition of carbon substrate, but were well dispersed throughout the flocs or biofilms with the addition of carbon substrate (Aoi et al., 2000; Wu et al., 2008a). The dispersed distribution of AOB can improve the nutrient availability for these microbial communities within activated sludge flocs, where nutrient diffusion may be limited. As a result, nitrification activity can be improved by dispersed distribution of AOB with the addition of organic carbon substrate.

4.1.3. Effect of SRT

Several other conditions that contribute to the different nitrification activities may be due to the different SRTs caused by different environmental conditions, such as temperature and settlement characteristics of activated sludge flocs.

Temperature affects the microbial activity, doubling with increasing temperatures in 10°C increments. Temperatures ranging from 14–27°C had no significant influence on the nitrification activity (Zhu and Chen, 2002; Germain et al., 2007). With a temperature above 15°C, nitrification performed efficiently when the SRT was above 10 days (Komorowska-

Kaufman et al., 2006). The specific nitrification activity decreased with increasing SRT (Germain et al., 2007). Oyanedel et al. (2005) found that nitrification activity kept constant when protein was used as the biomass unit, whereas it changed with the SS or VSS as the biomass unit. This should be considered in future studies.

In the present study, the settlement of activated sludge was found to affect the microbial activity. With a good settlement in the G-Reactor, a relatively high SRT was maintained and a stable nitrification was achieved. However, in the N-Reactor, a relatively high effluent SS (as a consequence of poor settlement of activated sludge flocs) caused a low SRT in this reactor, finally a very low biomass was maintained. Therefore, the settlement characteristics of activated sludge flocs can affect the SRT and further affect the nitrification activity at conditions with or without the addition of glucose.

4.2. Nitrite Accumulation Analysis

Nitrite accumulation in the nitrification process can reduce 25% of the operational cost by lowering aeration; in addition, with a reduction of 40% in organic carbon demand of denitrifiers for denitrifying nitrite rather than nitrate in the following denitrification process (Hellinga et al., 1998; Blackburne et al., 2008; Dytczak et al., 2008). In our study, nitrite accumulation was observed in both reactors with and without the addition of glucose. The possible reasons for this are discussed as follows.

NOB is more sensitive to NH_3 than AOB (Anthonisen et al., 1976). The NH_3 inhibition concentration for AOB ranges from 7–10 mg/l, whereas it is 0.1–5 mg/l for NOB (Anthonisen et al., 1976; Abeling and Seyfried, 1992; Turk and Mavinic, 1989; Villaverde et al., 1997). The inhibition of NH_3 could be also described based on the specific biomass concentration - mg NH_3/g biomass. Villaverde et al. (1997) and Fdz-Polanco et al. (1994) found that the threshold concentration was 0.5 mg NH_3/g biomass due to the nutrient limitation, while a concentration above 2.5 mg NH_3/g biomass caused inhibition of AOB. The NH_3 concentration above 1.5 mg/g VSS was shown to inhibit the activity of NOB, causing the accumulation of nitrite (Villaverde et al., 1997). In the present study, the NH_3 concentration ranged from 0 to 4.2 mg/l in the N-Reactor, and from 0 to 1.5 mg/l in the G-Reactor. The biomass based concentration was 0 to 10.2 mg/g VSS in the N-Reactor and 0 to 1 mg/g VSS in the G-Reactor. This showed that a higher NH_3 concentration occurred in the N-Reactor than in the G-Reactor. However, the higher NH_3 concentration occurred during the early aerobic stage, where a good linear regression of nitrification was obtained. Therefore, the NH_3 was not the main reason responsible for nitrite accumulation.

pH has a great influence on nitrification by: (i) activation-deactivation of enzymes in nitrifiers; (ii) affecting alkalinity through changing the forms of inorganic carbon; and (iii) inhibition by substrate of NH_3 or HNO_2 (Villaverde et al., 1997). Villaverde et al. (1997) showed that nitrification improved 13% with a pH increase from pH 5.0 to 8.5 in a submerged biofilm system. According to the discussion by Villaverde et al. (1997), in our study, the possible reason for nitrite accumulation could be due to the affected enzyme systems in nitrifiers and/or alkalinity with the pH decreased from above 8 to around 6.0.

4.3. Effect of Glucose Addition on the Settlement of Activated Sludge Flocs

A higher amount of EPS, specifically carbohydrate, was produced by biomass in the N-Reactor than in the G-Reactor, which could be due to shortage of carbon substrates in the N-Reactor, and the excretion of more carbohydrate could be a survival strategy.

The higher total EPS can cause a high amount of vicinal water, which may further cause less efficiency in flocculation and a high SVI (Liao et al., 2000; Liss et al., 1996). Protein content contributes to the hydrophobic characteristics of activated sludge flocs and causes good settlement, while carbohydrate contributes to the neutral or hydrophilic characteristics of activated sludge flocs and leads to poor settlement (Jorand et al., 1998; Liao et al., 2001). Due to the similar protein proportion in both reactors, a high carbohydrate proportion in the EPS in the N-Reactor caused poor settlement of activated sludge, resulting in washing out more sludge from this reactor; this led to drastic decreases in both the biomass concentration and the SRT.

Martinez et al. (2004) found that the settlement of the activated sludge was correlated to the protein concentration in the EPS and the type of protein varied with time. However, in their study, EDTA was used to extract EPS and the total EPS proportion in the biomass was only around 1.2%. In the present study, the ratio of carbohydrate/protein was better correlated to the SVI, which is similar to other studies where the ratio rather than the individual component contributed to the surface characteristics of flocs (Morgan et al., 1990; Liao et al., 2001).

SRT was shown to affect the sludge floc structure with more dense and physically stable flocs at SRTs of 16 to 20 days than at SRTs of 4 to 9 days (Liss et al., 2002). In the N-Reactor, due to low SRTs, loose flocs with less settlement ability existed in this reactor and more biomass was washed out.

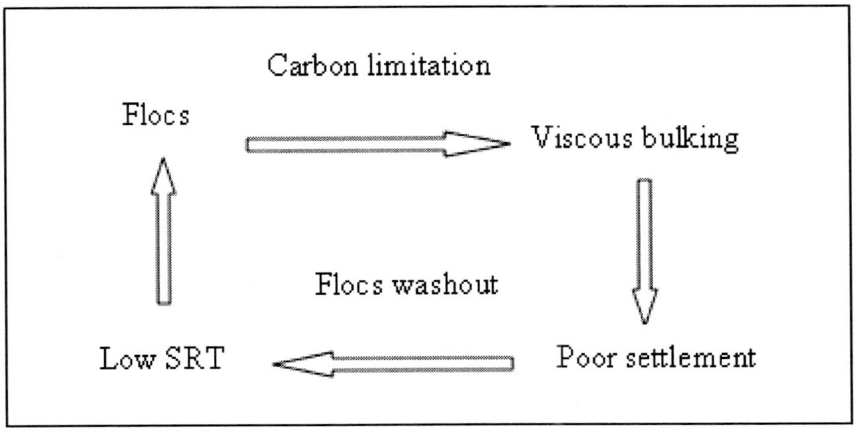

Figure 11. Mechanisms causing the viscous bulking of activated sludge flocs in the N-Reactor.

In the N-Reactor, the possible mechanism for sludge settlement is illustrated in Figure 11. Due to shortage of carbon substrates, microorganisms excrete a high amount of carbohydrate, which can act as both a carbon substrate reserve and an energy source. The high carbohydrate content causes poor settlement of activated sludge flocs (viscous bulking; Jenkins, 1992) and more biomass is washed out from the system. Consequently, a low SRT exists and further causes the poor settlement of activated sludge flocs (Liss et al., 2002). This mechanism can also be used to explain the results obtained from other studies in which heterotrophs dominated, or saline wastewater was treated and a low SRT occurred (Wu et al., 2008b; Liss et al., 2002).

Based on this study, it is recommended that biofilm systems or membrane systems can be used to nitrify wastewaters with low carbon substrate concentrations. By this means, the effect of poor settleability of activated sludge flocs can be avoided.

5. Conclusion

Short-term effects of glucose addition on nitrification and sludge settlement were investigated. High specific nitrification rates were observed in the N-Reactor without the addition of glucose, while high volumetric nitrification rates occurred in the G-Reactor with the addition of glucose. Nitrite accumulation was due to low pH caused by the processing of nitrification in both reactors. In the G-Reactor the biomass concentration remained relatively stable, while in the N-Reactor the biomass concentration decreased with time. The reason for the decrease in the biomass concentration in the N-Reactor was due to the high amount of EPS produced, which caused the poor settlement of activated sludge flocs; consequently, more activated sludge was washed out from the N-Reactor and a very low SRT occurred; the low SRT further encouraged poor settlement of activated sludge flocs.

References

Abeling, U; Seyfried, CF. Anaerobic-aerobic treatment of high-strength ammonium wastewater – nitrogen removal via nitrite. *Water Science and Technology,* 1992, 26, 1007-1015.

Altinbas, U. Nutrient removal from low strength domestic wastewater in sequencing batch biofilm reactor. *Water Science and Technology,* 2001, 44, 181-186.

Anthonisen, AC; Loehr, RC; Prakasam, TBS; Srinath, EG. Inhibition of nitrification by ammonia and nitrous acid. *Journal of the Water Pollution Control Federation,* 1976, 48, 835-852.

Aoi, Y; Miyoshi, T; Okamoto, T; Tsuneda, S; Hirata, A; Kitayama, A; Nagamune T. Microbial ecology of nitrifying bacteria in wastewater treatment process examined by fluorescence *in situ* hybridization. *Journal of Bioscience and Bioengineering,* 2000, 90, 234–240

APHA; AWWA; WPCF. Standard methods for the examination of water and wastewater. 19th Edition. Washington DC, USA: American Public Health Association. 1995.

Blackburne, R; Yuan, Z; Keller, J. Demonstration of nitrogen removal via nitrite in a sequencing batch reactor treating domestic wastewater. *Water Research,* 2008, 42, 2166-2176.

Dubois, M; Gilles, KA; Hamilton, JK; Robers, PA; Smith, F. Colorimetric method for determination of sugars and related substances. *Analytical Chemistry,* 1956, 28, 350-356.

Dytczak, MA; Londry, KL; Oleszkiewicz, JA. Activated sludge operational regime has significant impact on the type of nitrifying community and its nitrification rates. *Water Research,* 2008, 42, 2320-2328.

Fdz-Polanco, F; Garcia, P; Villaverda, S. Temperature effect on nitrifying bacteria activity in biofilters: activation and free ammonia inhibition. *Water Science and Technology,* 1994, 30, 121-130.

Ford, DL. Comprehensive analysis of nitrification of chemical processing wastewaters. *Journal of the Water Pollution Control Federation,* 1980, 52, 2726-2746.

Germain, E; Bancroft, L; Dawson, A; Hinrichs, C; Fricker, L; Pearce, P. Evaluation of hybrid processes for nitrification by comparing MBBR/AS and IFAS configurations. *Water Science and Technology,* 2007, 55, 43-49.

Gieseke, A; Arnz, P; Amann, R; Schramm, A. Simultaneous P and N removal in a sequencing batch biofilm reactor: insights from reactor- and microscale investigations. *Water Research,* 2002, 36, 501-509.

Hanaki, K; Wanlermraj, C; Ohgaki, S. Effects of the activity of heterotrophs on nitrification in a suspended-growth reactor. *Water Research,* 1990, 24, 289-296.

Head, MA; Oleszkiewicz, JA. Bioaugmentation for nitrification at cold temperatures. *Water Research,* 2004, 38, 523-530.

Hellinga, C; Schellen, AAJC; Mulder, JW; Van Loosdrecht, MCM; Heijnen, JJ. The SHARON process: an innovative method for nitrogen removal from ammonium-rich waste water. *Water Science and Technology,* 1998, 37, 135–142.

Jenkins, D. "Towards a comprehensive model of activated sludge bulking and foaming." *Water Science and Technology,* 1992, 25(6), 215-230.

Jorand, F; Boue-Bigne, F; Block, JC; Urbain, V. Hydrophobic/hydrophilic properties of activated sludge exopolymeric substances. *Water Science and Technology,* 1998, 37, 307-315.

Komorowska-Kaufman, M; Majcherek, H; Klaczynski, E. Factors affecting the biological nitrogen removal from wastewater. *Process Biochemistry,* 2006, 41, 1015-1021.

Liao, BQ; Allen, DG; Droppo, IG; Leppard, GG; Liss, SN. Bound water content of activated sludge and its relationship to solids retention time, floc structure, and surface properties. *Water Environmental Research,* 2000, 72, 722-730.

Liao, BQ; Allen, DG; Droppo, IG; Leppard, GG; Liss, SN. Surface properties of sludge and their role in bioflocculation and settleability. *Water Research,* 2001, 35, 339-350.

Lindemann, J; Wiesmann, U. Single-disc investigations on nitrogen removal of higher loads in sequencing batch and continuously operated RDR system. *Water Science and Technology,* 2000, 41, 77-84.

Liss, SN; Liao, BQ; Droppo, IG; Allen, DG; Leppard, GG. Effect of solids retention time on floc structure. *Water Science and Technology,* 2002, 46, 431-438.

Liss, SN; Droppo, IG; Flannigan, DT; Leppard, GG. Floc architechture in wastewater and natural riverine systems. *Environmental Science and Technology,* 1996, 30, 680-686.

Liu, H; Fang, HHP. Extraction of extracellular polymeric substances (EPS) of sludges. *Journal of Biotechnology,* 2002, 95, 249-256.

Lowry, OH; Rosebrough, NJ; Farr, AL; Randall, RJ. Protein measurement with the folin phenol reagent. *Journal of Biological Chemistry,* 1951, 193, 265-275.

Martinez, FO; Lema, J; Mendez, R; Cuervo-Lopez, F; Gomez, J. Role of exopolymeric protein on the settleability of nitrifying sludges. *Bioresource Technology,* 2004, 94, 43-48.

Morgan, JW; Forster, CF; Evison, L. A comparative study of the nature of biopolymers extracted from anaerobic and activated sludge. *Water Research,* 1990, 24, 743-750.

Nogueira, R; Melo, LF; Purkhold, U; Wuertz, S; Wagner, M. Nitrifying and heterotrophic population dynamics in biofilm reactors: effects of hydraulic retention time and the presence of organic carbon. *Water Research,* 2002, 36, 469-481.

Okabe, S; Hiratia, K; Ozawa, Y; Watanabe, Y. Spatial-microbial distribution of nitrifiers and heterotrophs in mixed-population biofilms. *Biotechnology and Bioengineering,* 1996, 50, 24-35.

Okabe, S; Kindaichi, T; Ito, T; Satoh, H. Analysis of size distribution and areal cell density of ammonia-oxidizing bacterial microcolonies in relation to substrate microprofiles in biofilms. *Biotechnology and Bioengineering,* 2004, 85, 86-95.

Oyanedel, V; Campos, JL; Garrido, JM; Lazarova, V; Mendez, R. Development of a membrane-assisted hybrid bioreactor for ammonia and COD removal in wastewaters. *Journal of Chemical Technology and Biotechnology,* 2005, 80, 206-215.

Park, SJ; Lee, HS; Yoon, TI. The evaluation of enhanced nitrification by immobilized biofilm on a clinoptilolite carrier. *Bioresource Technology,* 2002, 82, 183-189.

Petropoulos, P; Gilbride, KA. Nitrification in activated sludge batch reactors is linked to protozoan grazing of the bacterial population. *Canadian Journal of Microbiology,* 2005, 51, 791–799.

Rittmann, BE; McCarty, PL. *Environmental biotechnology: principles and applications.* New York: McGraw-Hill Book Co. 2001.

Rothman, M. Operation with biological nutrient removal with stable nitrification and control of filamentous growth. *Water Science and Technology,* 1998, 37, 549-554.

Texier, AC; Gomez, J. Settleability and kinetics of a nitrifying sludge in a sequencing batch reactor. *Canadian Journal of Microbiology,* 2004, 50, 943-949.

Turk, O; Mavinic, DS. Benefits of using selective inhibition to remove nitrogen from highly nitrogenous wastes. *Environmental Technology Letters,* 1987, 8, 419-426.

Van Benthum, WAJ; Van Loosdrecht, MCM; Heijnen, JJ. Control of heterotrophic layer formation on nitrifying biofilm in a biofilm airlift suspension reactor. *Biotechnology and Bioengineering,* 1997, 53, 397–405.

Villaverde, S; Garcia-Encina, PA; Fdz-Polanco, F. Influence of pH over nitrifying biofilm activity in submerged biofilters. *Water Research,* 1997, 31, 1180-1186.

Villaverde, S; Fdz-Polanco, F; Lacalle, ML; Garcia, PA. Influence of the suspended and attached biomass on the nitrification in a two submerged biofilters in series system. *Water Science and Technology*, 2000, 41, 169-176.

Wagner, M; Rath, G; Amann, R; Koops, HP; Schleifer, KH. In situ identification of ammonia-oxidizing bacteria. *Systematic and Applied Microbiology,* 1995, 18, 251-264.

Wu, G; Rodgers, M; Zhan, X. Nitrification in sequencing batch reactors with and without glucose addition at 11 °C. *Biochemical Engineering Journal,* 2008a, 40, 373-378.

Wu, G; Guan, Y; Zhan, X. Effect of salinity on the activity, settling and microbial community of activated sludge in sequencing batch reactors treating synthetic saline wastewater. *Water Science and Technology,* 2008b, 58, 351-358.

Yoshinori, S; Minoru, N; Hiroshi, Y; Shoji, W. An improvement of the settleability of the activated sludge by the acceleration of the nitrification with the anaerobic-oxidic (AO) process. *Journal of Japan Sewage Works Association,* 1999, 36, 147-157.

Zhu, S; Chen, S. The impact of temperature on nitrification rate of fixed film biofilters. *Aquacultural Engineering,* 2002, 26, 221-237.

In: Biochemical Engineering
Editors: F.E. Dumont and J.A. Sacco, pp. 293-306
ISBN: 978-1-60741-257-1

Chapter 13

ACACIA CAVEN (MOL.) MOLINA POLLEN PROTEASES. APPLICATION TO THE PEPTIDE SYNTHESIS AND TO LAUNDRY DETERGENTS

***Cristina Barcia*[1], *Evelina Quiroga*[1,3,a], *Carlos Ardanaz*[2], *Gustavo Quiroga*[1] *and Sonia Barberis*[1,3,b]**

[1]Laboratorio de Bromatología

[2]Área de Química Orgánica. Facultad de Química, Bioquímica y Farmacia. Universidad Nacional de San Luis. Chacabuco y Pedernera (5700) San Luis, Argentina

[3]Instituto de Física Aplicada (INFAP). Universidad Nacional de San Luis, Argentina

Abstract

It is known that the proteases have applications in several industrial processes such us leather processing, laundry detergents, producing of protein hydrolysates and food processing, as well as in the peptide synthesis in non conventional media. The application of proteases as catalyst of short oligopeptides in aqueous-organic media, have received a great deal attention as a viable alternative to chemical approach because of their remarkable characteristics. On the other hand, alkaline proteases have also been used to improve the cleaning efficiency of detergents. Detergent enzymes account for about 30% of the total worldwide enzyme production and represent one of the largest and most successful applications of modern industrial biotechnology. The aim of this work was to study the performance of proteolytic enzymes of *Acacia caven* (Mol.) Molina pollen for its potential application as an additive in various laundry detergents formulations and as catalyst of the peptide synthesis in aqueous-organic media. Pollen grains (35 mg/ml) were suspended in 0.1M Tris-HCl buffer pH 7.4 and slowly shaken for 2 h at 25° C. Then, the slurry was centrifugated for 30 min at 8000 rpm and the supernatant (crude enzyme extract, CE) was tested in protein content (Bradford's method) and proteolytic activity (using BAPNA and Z-Ala-pNO as substrates). A partial characterization of *Acacia caven* CE was carried out: enzyme extract displayed maximum proteolytic activity at pH 8 and 35-40° C; it showed remarkable thermal stability after 1.5 h at 25-40° C but it decreased as long as temperature increased to 60° C. On the other hand, the enzyme extract was incubated with different surfactants and commercial laundry detergents at

[a] E-mail address: equiroga@unsl.edu.ar. (Corresponding author: Evelina Quiroga)
[b] E-mail address: sbarberi@unsl.edu.ar (Corresponding author: Sonia Barberis).

25-60° C during 30 min and 1h; and it showed high stability and compatibility with them. The peptide synthesis catalyzed by *Acacia caven* CE was carried out in a mixture of 0.1M Tris-HCl buffer pH 8.5 and ethyl acetate (50:50 ratio) at 37° C using 2-mercaptoethanol as activator and TEA as neutralizing agent of the amino component (Phe-OMe.HCl). Carboxylic components were selected in base of the highest preference of CE. The identification of synthesized peptide products was carried out by HPLC-MS. According to the obtained results, this work contributes with a new variety of phytoprotease useful as catalyst of the peptide synthesis and as additive of laundry detergents.

Introduction

Enzymes are used in a wide range of industrial processes and in consumer products. The technical industries segment comprises the detergent, starch, textile, fuel alcohol, leather, and pulp and paper industries [1].

Proteases are one of the three largest groups of industrial enzymes and represent about 60% of the total worldwide sales of enzymes [2]. Several of these are used in a wide range of industrial processes and in consumer products such as detergents, brewing, leather, dairy and food-processing industries [3, 4]. Proteases are also useful as components in biopharmaceutical products such as contact-lens enzyme cleaners and enzymatic debriders [5].

Alkaline proteases also have a long history in the food and detergent industries. Their application in the leather industry for dehairing and bating of hides to substitute the traditional methods involving sodium sulfide (toxic chemical) is a relatively new development and has conferred added biotechnological importance [6].

Enzymes have also been used to improve the cleaning efficiency of detergents. Detergent enzymes account for about 30% of the total worldwide enzyme production and represent one of the largest and most successful applications of modern industrial biotechnology [7]. In detergent formulations, the proteases act on protein-based stains (blood, grass and food stains, for example); the amylases assist in the removal of starch-based stains from many types of food products, and the lipases facilitates the removal of fat and oil-based stains from greasy food and human sebum [8]. Enzymes function optimally in detergents at temperatures of 20-60° C and within a pH range of 7.5 – 10.5. The performance of enzymes in detergents depends on a number of factors, including the detergent composition, type of stains to be removed, wash temperature, washing procedure and wash-water quality.

Detergents available on the international market, such as Dynamo®, Era Plus® and Tide® (Procter and Gamble), contain proteolytic enzymes, the majority of which are produced by members of the genus Bacillus. Subtilisins have been the enzymes of choice in detergent formulations (U.S. Patent N° 1240058, 374971, 370482 and 4266031, and UK Patent N° 13155937), despite their low stability in the presence of commercial detergents and their short half-time on the shelf [9, 10].

For an enzyme to be used as a detergent additive it should be stable and active in the presence of typical detergent ingredients, such as surfactants, builders, bleaching agents, bleach activators, fillers, fabric softeners and various other formulation aids. The food process industry has come to rely on improved detergent efficiency in their cleaning programs [11].

On the other hand, the application of proteases to the production of short oligopeptides in aqueous-organic media, have received a great deal attention as a viable alternative to chemical approach and for their remarkable characteristics. Biocatalysis in non-conventional (non-aqueous) media has expanded the spectrum of application of proteases to those reactions that cannot proceed effectively in aqueous environments, this is, the synthesis of peptide bonds instead of their hydrolysis [12-15].

Enzymatically synthesized small peptides (usually di or tripetides) are being used successfully for human and animal nutrition and also as pharmaceuticals and agrochemicals. Some relevant examples are the synthesis of the leading non-caloric sweetener aspartame, the lysine sweet peptide, kyotorphin, angiotensin, enkephalin and dynorphin [16-22], and some nutritional dipeptides and tripeptides [23-25]. Some small peptides have been synthesized at commercial scale in continuously operated enzyme reactors [26, 27]. Kimura et al. [23] have proposed various kinds of reactors to synthesize peptides from essential aminoacids using papain, α-chymotripsin, and thermolysin.

The aim of this work was to study the performance of proteolytic enzymes of *Acacia caven* (Mol.) Molina pollen for its potential application as an additive in various laundry detergents formulations and as catalyzer of the peptide synthesis in aqueous-organic media.

Materials and Methods

Plant Material

Pollen grains of flowers of *Acacia caven* (Mol.) Molina *Fabaceae* (argentine folk names: "espinillo") was obtained from San Luis range, Argentina. *Acacia caven* plant was classified at Universidad Nacional de San Luis and registered as Voucher N° 462, III, 1998. The pollen sample was maintained safe from the light, dried to room temperature and stored in dessicator.

Elution of the Enzyme from the Pollen Wall

Pollen grains (35 mg/ml) were suspended in 0.1M Tris-HCl buffer pH 7.4 and slowly shaken for 2 h at 25° C. Then, the slurry was centrifugated for 30 min at 8000 rpm. The supernatant (crude enzyme extract, CE) was tested in proteolytic activity and protein content by Bradford's method [28].

Chemical Reagents

Synthetic substrates L-phenylalanine methyl ester (Phe-OMe.HCl), N-□-benzyloxycarbonyl-L-amino acids (Z-Ala, Z-Gln and Z-Phe) and N-□-benzyloxycarbonyl-L-Alanine p-nitrophenyl ester (Z-Ala-pNO) were supplied by Bachem (California, USA). The synthetic substrate N-□-benzoyl-DL-Arginine 4-nitroanilide (BAPNA) and HPLC-grade organic solvents were purchased from Sigma (St. Louis, USA).

Proteolytic Activity Assays

The measurement of the proteolytic activity was carried out using BAPNA as substrate (5 mg/ml). After 5 min of incubation at 37° C, the absorbance of the p-nitroaniline released was measured spectrophotometrically at 410 nm. Enzymatic units (IU) were obtained by performing a standard curve of p-nitroaniline [29].

Effect of ph and Temperature on Proteolytic Activity

The dependence of proteolytic activity of *Acacia caven* CE on pH and temperature was evaluated using (5 mg/ml) BAPNA as substrate pH was varied from 6.0 to 11.0 at 25° C, and temperature was modified from 25 to 60° C at pH 8, using 0.1M Tris-HCl buffer.

Thermostability of Acacia Caven Crude Extract

Thermal stability of *Acacia caven* CE in 0.1M buffer Tris-HCl pH 8 was evaluated as residual proteolytic activity (%), using BAPNA (5 mg/ml) as substrate, after incubation between 0 and 90 min at 25, 40, 50 and 60° C. Enzyme activity prior to incubation was taken as 100 % at each assayed temperature.

Effect of Surfactants on the Proteolytic Activity of Acacia Caven Crude Extract

The effect of some non-ionic surfactants (Triton X100, Tween® 20 and Tween® 80) and ionic surfactants: -cationic (Phospholipid® EFA) and -anionic (Arlatone® MAP 230) on the proteolytic activity of *Acacia caven* CE, was assessed at different concentrations (0.1, 0.4 and 1% (v/v)). The crude extract was incubated in the surfactant solutions at 25, 40 and 60° C, for 30 min and 1h. Then, the enzyme residual activity was determined using (5 mg/ml) Z-Ala-pNO as substrate and the absorbance of the released p-nitrophenol was measured spectrophotometrically at 405 nm [30].

Effect of Commercial Detergent on the Proteolityc Activity of Acacia Caven Crude Extract

Compatibility of *Acacia caven* CE with different commercial laundry detergents (Woolite®, Ariel Hydro Gel Max and Ace (Procter & Gamble), Skip Intelligent and Ala Matic (Unilever) and Enzimax (TVB)) was assessed. The detergents were diluted in distilled water (7 mg/ml) to simulate washing conditions [31]. The CE was incubated in the detergent solutions for 30 min and 1h at 37° C. Then, the residual proteolytic activity was measured using Z-Ala-pNO (5 mg/ml) as substrate under similar conditions that the above described. Control samples without any detergent and incubated under similar conditions were taken as 100 %.

Enzymatic Synthesis Reaction Conditions

The enzymatic synthesis was carried out in a mixture of Tris-HCl buffer (0.1M, pH 8.5) and ethyl acetate in 50:50 ratios. The condensation reaction was initiated by mixing the aqueous phase containing the enzyme (0.11 IU/ml), 4mM Phe.OMe, 20mM 2-mercaptoethanol as activator and 4mM TEA (since the amino component is commercially available as hydrochloride derivative), with the organic phase containing 4mM carboxylic component (Z-Ala, Z-Phe or Z-Gln). The reaction was conducted at 37° C in stopper flask under magnetic stirring at 160 rpm during 24 h at pH 8.5. At time intervals, aliquots were taken and analyzed by HPLC-MS. Simultaneously, blanks with identical composition but without the enzyme and with only the enzyme in the organic system, were carried out.

Analytical Control of Enzymatic Synthesis

RP-HPLC

The enzymatic synthesis was followed by RP-HPLC. Analyses were performed on a Gilson System (Model 712) equipped with a C-18 Luna (5 □), 250mm x 4.60mm column (Varian). For all the experiences, the volume of injection was of 20 μL, the speed of flow was 0.8 ml/min and the eluted material was monitored spectrophotometrically at 254 nm and 25° C. The mobile phase consisted of 50% acetonitrile and pH was adjusted to 3 by the aggregate of 0.1% trifluoroacetic acid (TFA). The quantification of products and substrates was carried out using the corresponding patterns.

HPLC-MS

Reaction products were identified by HPLC-MS. Analyses were performed on VG-Quattro (Micromass Instruments S.A.), with a C-18 Nucleosil (120-5) (5 □m), 250mm × 40mm column (Scharlan). The technique of electrospray was used with positive ion reading (100–1000 uma). Nebulizer gas: N_2 (flow: 10 l/h). Drying gas: N_2 (flow: 450 l/h). Temperature of the source: 80° C. Voltage of the capillary: 3.5 kV. Voltage focus: 55V. Flow: 15 (ml/min). Elution: solvent A, H_2O containing 1% formic acid and solvent B, CH_3CN; A:B = 50:50. Volume of injection: 10-20 □l.

Results and Discussion

In the present study, the performance of a proteolytic crude extract (CE) of *Acacia caven* (Mol.) Molina was investigated as both, -additive of laundry detergents formulations and -catalyst for the peptide synthesis in aqueous-organic media.

Initially, a partial characterization of *Acacia caven* CE was carried out and it showed maxima proteolytic activities (3 IU/mg of protein) at pH 8 and 40° C (Figure 1 and 2a).

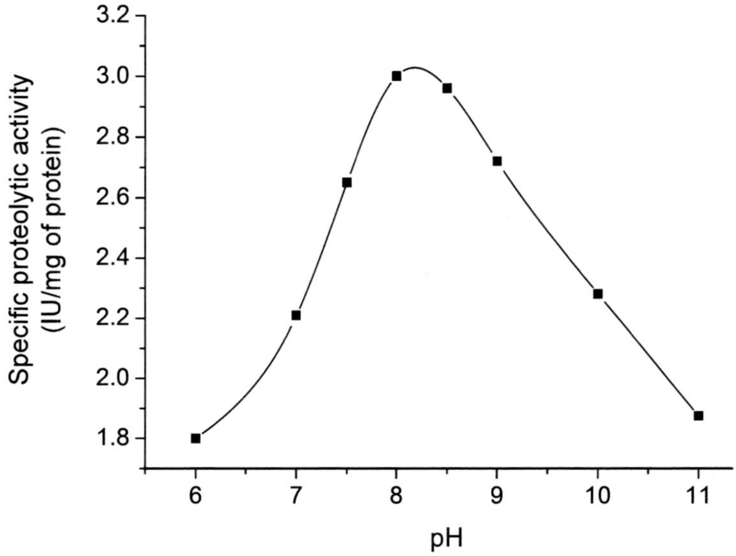

Figure 1. Effect of pH on the proteolytic activity of *Acacia caven* CE, after 5 min at 37° C.

According to Figure 2a, temperature profile of *Acacia caven* CE showed a sharp peak of maximal proteolytic activity at 35-40° C but high activities were obtained between 25 and 50° C. Besides, the *Acacia caven* CE displayed a remarkable thermal stability after 1.5 h at 25-40° C retaining 100 % of its initial activity, but it decreased as long as temperature increased to 60° C (Figure 2b). Although the enzyme was very stable below 40° C after 1.5 h, only 25 % of the initial proteolytic activity was retained at 60° C under similar incubation conditions. The high thermostability of *Acacia caven* proteases below 40° C is desirable for laundry purposes and from the ecological and economical point of view, mainly, because saving of energy.

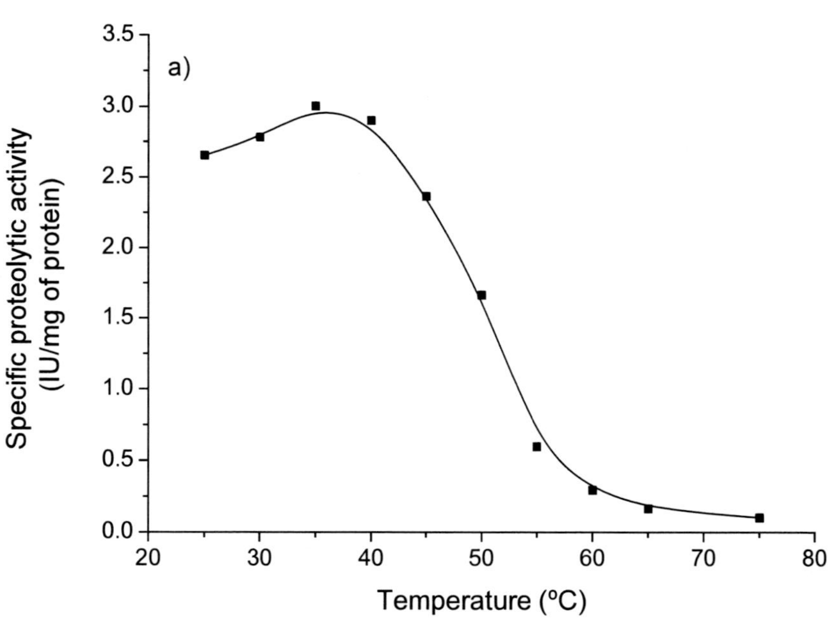

Figure 2. Continued on next page.

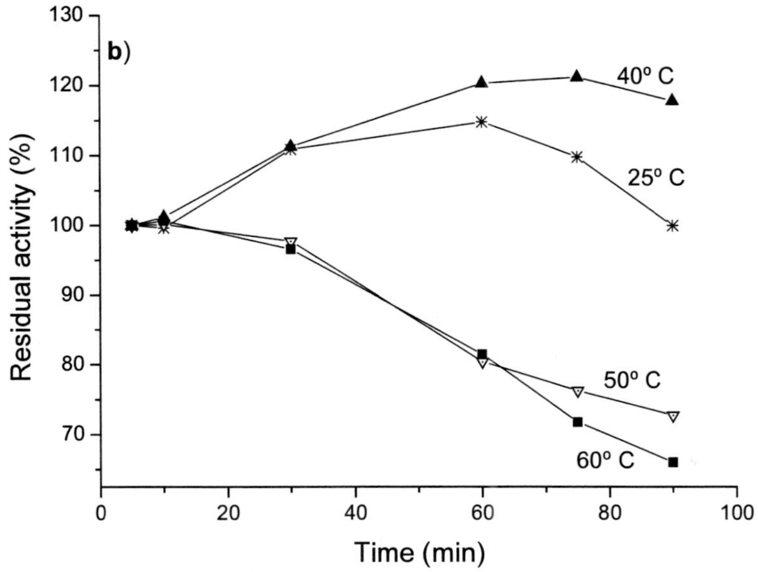

Figure 2. a) Effect of the temperature on the proteolytic activity of *Acacia caven* CE in 0.1M Tris-HCl buffer pH 8, after 5 min of incubation. b) Thermostability of *Acacia caven* CE in 0.1 M Tris-HCl buffer pH 8. BAPNA was used as substrate.

On the other hand, *Acacia caven* CE demonstrated to be active over a broad pH range and more than 40 % of its maximum activity was exhibited between pH 7.5 and 10. One of the most important parameters for selecting proteases for both laundry detergents and catalyst of peptide synthesis is the optimum pH. The desired pH of a detergent solution in which proteases work should be approximately the same as the optimum pH for the enzyme [32]. Since the pH of laundry detergents is commonly alkaline protease and other enzymes currently used in detergent compositions should be alkaline in nature with a high optimum pH [33]. Besides, peptide synthesis reactions should be carried out at high pH in order to maximize the concentration of the unionized amine form of the nucleophile. These properties were displayed by *Acacia caven* proteases.

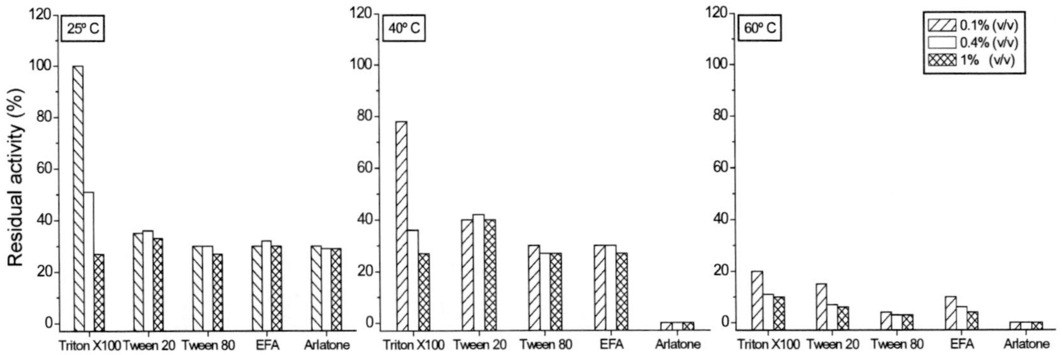

Figure 3. Effect of different concentration of some ionic and non-ionic surfactants (0.1, 0.4 and 1% (v/v)) on the proteolytic activity of *Acacia caven* CE, after 1 h at 25, 40 and 60° C.

The proteolytic activity of *Acacia caven* CE was notably enhanced by the addition of 12mM cysteine, but it was completely inactivated by incubation with 0.1 and 10mM of E64, $HgCl_2$ and 0.1mM of PMSF after 30 min. This behavior is characteristic of cystein proteases, although serine proteases can not be discarded before further purification steps were carried out. Preferences of the *Acacia caven* CE on the p-nitrophenyl esters of different amino acids were also tested. *Acacia caven* CE showed high activity for Phe, Ala, Gln and Pro derivatives (unpublished).

In addition of high optimum pH and thermostability, many parameters are involved in selecting a protease for detergents, such as compatibility with detergent components, e.g. surfactants, perfumes and bleaches [34, 35].

Acacia caven proteases (CE) were assayed in the presence of non-ionic and ionic surfactant at different concentrations and temperatures as were previously described in Material and methods. Proteolytic activities were all retained after being incubated 30 min under above conditions. Nevertheless, proteolytic activities were modified when the time of incubation increased at 60 min (Figure 3). After 1 h at 60° C, the residual activity of *Acacia caven* CE showed a sharp decreasing with every studied surfactant because of the thermal inactivation of the proteases. After 1 h of incubation below 40° C, the residual activities of *Acacia caven* CE ranged between 27 and 100 % and the effect of the surfactant concentration was negligible with the exception of Triton X100. This surfactant was the only one capable of reducing the proteolytic activity when the temperature and the concentration increased.

In spite of the above results, *Acacia caven* CE was able to retain more than 55 % of initial activity when it was incubated with six commercial detergents, after 1 h at 37° C (Figure 4). The observed decreasing on CE activity was attributed to other detergent components than the detergent own proteases because of these were previously inactivated freezing (24 h) and overheating (95° C, 1h).

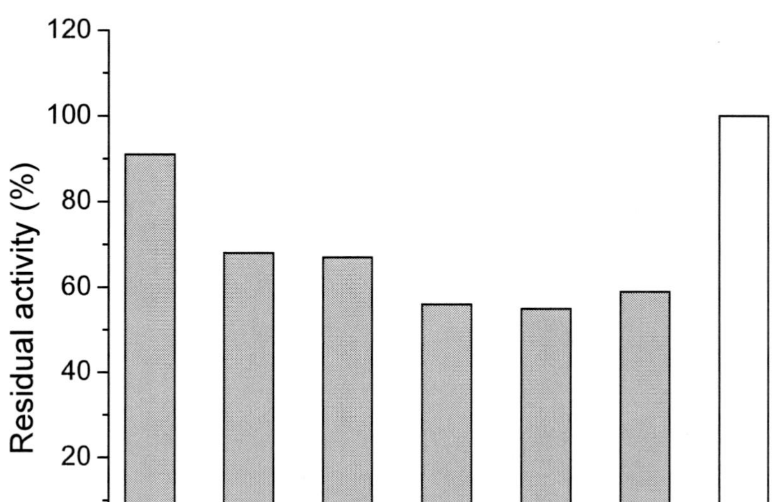

Figure 4. Effect of different commercial detergent on the proteolityc activity of *Acacia caven* CE, after 1h of incubation at 37° C. (5 mg/ml) Z-Ala p-nitrofenil ester was used as substrate. (1. Woolite, 2. Ariel, 3. Ace, 4. Skip, 5. Ala Matic, 6. Enzimax, 7. None)

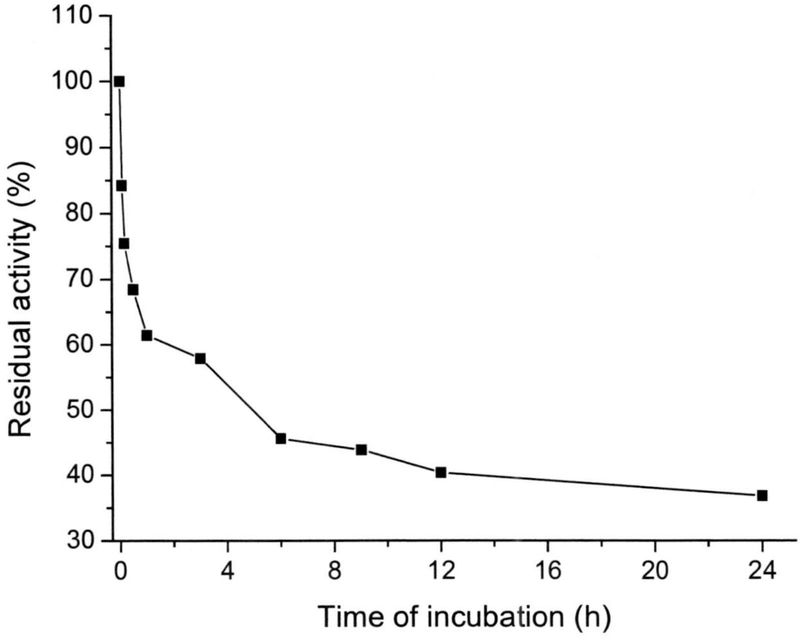

Figure 5. Effect of incubation time in 50 % (v/v) ethyl acetate on the residual activity of de *Acacia caven* CE, at 37° C.

In order to maximize the concentration of the unionized amine form of the phenyl nucleophile, the peptide synthesis was performed at pH 8.5. In this respect, pH values above 9 might have been optimum. However, such conditions reduced the overall activity of the enzyme (Figure 1). The chosen temperature of reaction (37° C) was based on the optimum temperature of *Acacia caven* CE and the remarkable thermostability of the enzyme extract at that temperature (Figure 2).

The peptide synthesis catalyzed by *Acacia caven* CE was carried out in the biphasic medium of 0.1M Tris-HCl buffer pH 8.5 and ethyl acetate (50:50 ratio). In these conditions, a clear decreasing of 60 % on the activity enzyme was observed after 24h at 37° C and 160 rpm (Figure 5). Nevertheless, enzyme inactivation by the organic solvent did not influence on the synthesis yield and high dipeptide yields were obtained before 1 h.

N-protected (N-benzyloxycarbonyl-L-amino acids (Phe, Ala, Gln)) acyl donors were selected in base of the highest preferences showed by *Acacia caven* CE by those amino acids derivatives.

Under such reaction conditions, *Acacia caven* CE was able to catalyze the synthesis of different bioactive peptides: Z-Gln-Phe.OMe, Z-Phe-Phe-OMe (substrate for cathepsin A [36] and Z-Ala-Phe.OMe (bitter dipeptide precursor with commercial interest for the food industry [37]).

When Z-Ala and Phe.OMe were used as carboxylic and nucleophilic components in the assayed biphasic medium, *Acacia caven* CE catalyzed the synthesis of the bitter dipeptide precursor Z-Ala-Phe.OMe, and a maximum product conversion (Xp) of 50% was obtained after 10 min of reaction. Xp parameter represents the amount of product obtained as a function of the limitant reactive and it was defined as the relation between the dipeptide concentration (mM) and the initial concentration (mM) of the limitant reactive (Phe.OMe).

Figure 6 shows the time-course of the synthesis of Z-Ala-Phe.OMe in 0.1M Tris-HCl buffer pH 8.5 and ethyl acetate (50:50 ratio). As can be observed, the dipeptide concentration decreased because of unwished by-products (Z-Ala.OMe and Z-Ala-Phe) were formed. It is apparent that the o-methyl derivated used as nucleophile was hydrolyzed by the esterolitic ativity of *Acacia caven* CE and then, both by-products were achieved. Further studies of media engineering should be carried out in order to increase aminolysis/hydrolysis ratio and other nucleophiles should also be assayed.

When Z-Gln (or Z-Phe) and Phe.OMe were used as carboxylic and nucleophilic components, respectively, HPLC-mass spectrometry analysis of the samples indicated the formation of dipeptide Z-Gln-Phe.OMe (or Z-Phe-Phe.OMe) after 30 min (or 1h) of reaction (Figure 7).

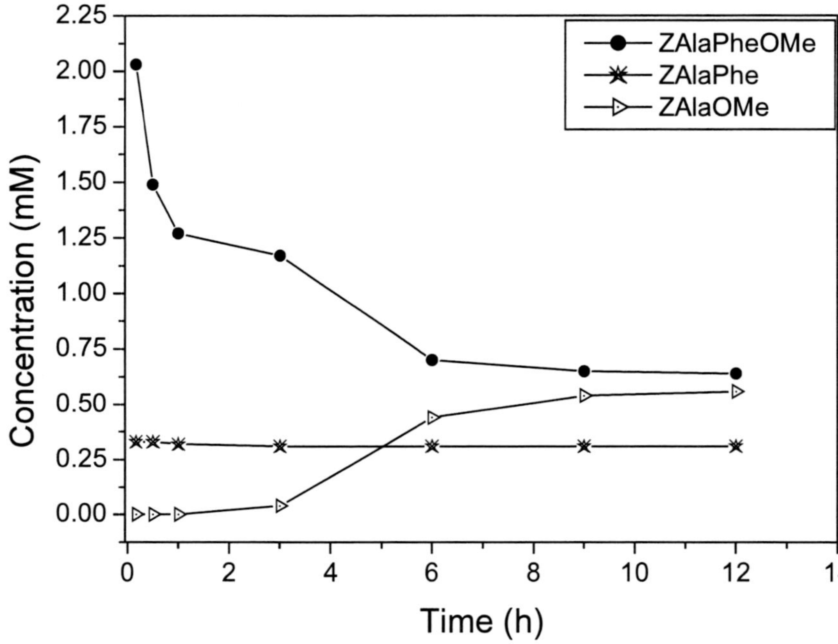

Figure 6. Time-course production of dipeptide Z-Ala-Phe.OMe and by-products (Z-Ala.OMe and Z-Ala-Phe) in 50 % (v/v) ethyl acetate, using (4mM) Z-Ala as carboxylic component and (4mM) Phe.OMe as amino nucleophilic component.

As a general conclusion, this work provides a new variety of phytoprotease as catalyst of the peptide synthesis in aqueous-organic media. The successful synthesis of some dipeptides has been achieved with *Acacia caven* CE although synthesis reactions need further optimization. At present, the effect of several factors (temperature, pH, molar ratio of enzyme/substrate, aqueous-organic media and external diffusional restrictions) on the yield of dipeptides is being evaluated. It is interesting to emphasize that the proteolytic extract of *Acacia caven* was unable to form oligopeptides derivatives; and it showed both, a high selectivity to form short peptide and a good specificity to react only with some amino acids. Consequently, this enzyme constitutes a promissory catalyst for the synthesis of small peptides.

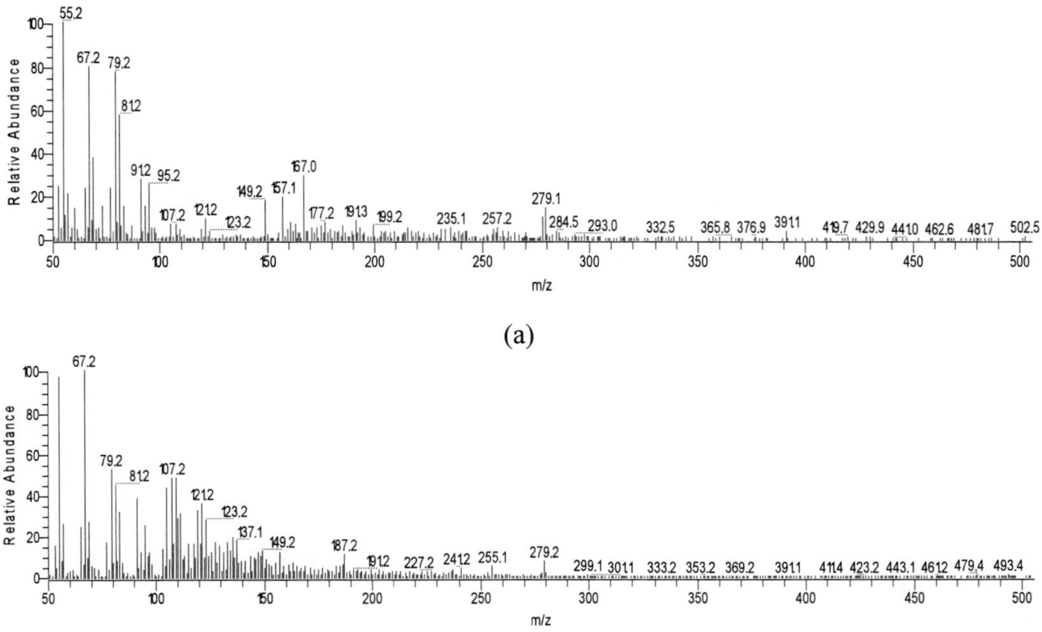

(a)

(b)

Figure 7. Mass spectrum of material eluted of RP-HPLC corresponded to the enzymatic synthesis catalyzed by *Acacia caven* CE in a 50 % (v/v) ethyl acetate, using: a) (4mM) Z-Gln and (4mM) Phe.OMe, after 30 min of reaction (Z-Gln-Phe.OMe ESMS [*M*+ 1] 441.0; Z-Gln ESMS [*M*+ 1] 279.1; Phe.OMe ESMS [*M*+ 1] 177.2); b) (4mM) Z-Phe and (4mM) Phe.OMe, after 1h of reaction (Z-Phe-Phe.OMe ESMS [*M*+ 1] 461.2; Z-Phe ESMS [*M*+ 1] 299.1; Phe.OMe ESMS [*M*+ 1] 177.2).

Conclusion

Enormous efforts over the past two decades to find detergent proteases and to improve known ones have not succeeded in bringing forth a single new enzyme as versatile and universally applicable as the high-alkaline subtilisin. At this moment, the investigation is focused in finding new suitable enzymes to be used in specific environments such as low-temperature automatic washing detergents, essentially due to the necessity to save energy and to adapt to new fragile materials in clothing. Proteolytic extract of *Acacia* caven pollen fulfilled all the requirements for its application in those detergent formulations: high stability in a broad range of temperature (25-50° C), high activity in alkaline pH (7.5-10) and very good compatibility with the commercial detergent additives.

By last, but not least important, this work reports a new plant protease with potential application to the peptide synthesis and the laundry detergent industry, from a renewable natural resource (an autochthonous plant) of the Latin American region. In this sense, we believe that this research has novelty and in addition, the crude extract obtaining is very simple and cheap, and it avoids the use of fermentation and genetic engineering.

Acknowledgements

This work was supported by grants from ANPCyT and Universidad Nacional de San Luis, Argentina. E. Quiroga and S. Barberis are career members of CONICET.

References

[1] Nogueira E. (2004) Study of the Novel Protease secreted by Teredinobacter turnirae - Diffusion of the Protease in Immobilized Cells, *Application of the Protease as a Detergent Additive and Proteases Market Study.* PhD Thesis. Baltimore, University of Maryland Baltimore County, USA.

[2] Sandhya, C., Nampoothiri, M.K. & Pandey, A. (2005) Microbial Proteases. In: Barredo, J.L. (Ed.), *Microbial Enzyme and Biotransformations,* Humana Press Inc. Vol 10 (pp. 165-179) Totowa, Nj.

[3] Layman, P.L. (1986) *Industrial Enzymes:* Battling to Remain Specialties. *Chem. Eng. News.* **64**, 11-14.

[4] Kalisz, H.M. (1988) Microbial Proteinases. In: Fietcher A. (Ed.), *Advances in Biochemical engineering/biotechnology* (enzymes studies), Springer-Verlag. Vol 36 (1-65) Berlin, Germany.

[5] Anwar, A. & M. Saleemuddin. (2000) Alkaline Protease from Spilosoma oblique: Potential Applications in Bio-formulations. *Biotechnol. Appl. Biochem.* **31**, 85-89.

[6] Kamini, N.R., Hemachander, C., Mala, J.G.S. & Puvanakrishnan, R. (1999) *Microbial Enzyme Technology as an Alternative to Conventional Chemicals in Leather Industry.*

[7] http://www.specialtyenzymes.com/pubs.shtml

[8] Ferreira, G. (2001) Characterization of Proteolytic Activity Secreted by Different Morphologies of Teredinobacter turnirae. *PhD Thesis.* Department of Chemical and Biochemical Engineering, University of Maryland Baltimore County.

[9] Samal, B.B., Karan, B. & Stabinsky, Y. (1989) Stability of Two Novel Proteinases in Commercial Laundry Detergents. *Biotechnol. Bioeng.* **35**, 650-652.

[10] Aunstrup, K. (1979) Production, isolation, and economics of extracellular enzymes. *Appl. Biochem. Bioeng.* **2**, 27-69.

[11] Inhs, D. A., W. Schmidt & F. R. Richter. 1999. Enzyme Product, *U.S. Patent.* 5 861 366.

[12] Barberis, S.E. & Illanes, A. (1994) Catálisis enzimática en fase orgánica. In: O.E.A. (Ed.), Enzyme Biotechnology, Biology Serie, *Monography* No. **35** (pp. 225-254). Washington, DC.

[13] Barberis, S., Quiroga, E., Arribére, M.C. & Priolo, N. (2002) Peptide synthesis in aqueous–organic biphasic systems catalyzed by a protease isolated from Morrenia brachystephana (Asclepiadaceae). *J. Mol. Catal. B: Enzym.* **17**, 39-47.

[14] Morcelle, S.R., Barberis, S., Priolo, N., Caffini, N.O. & Clapés, P. (2006) Comparative behaviour of proteinases from the latex of Carica papaya and Funastrum clausum as catalysts for the synthesis of Z-Ala-Phe-OMe, *J. Mol. Catal. B: Enzym.* **41**, 117-124.

[15] Quiroga, E., Priolo, N., Obregón, D., Marchese, J. & Barberis, S. (2008) Peptide synthesis in aqueous–organic media catalyzed by proteases from latex of *Araujia hortorum* (*Asclepiadaceae*) fruits. *Biochem.l Eng. J.* **39**, 115-120.

[16] Kullman, W. (1982) Protease-catalyzed peptide bond formation: application to synthesis of the COOH-terminal octapeptide fragment of cholecystokinin. *Proceedings of the National Academy of Sciences of the United States of America.* **79**, 2840-2844.

[17] Kullman, W. (1979) Enzymatic synthesis of Leu- and Metenkephalin. *Biochem. Biophys. Research Comm.* **91**, 693-698.

[18] Takai, H., Sakato, K., Nakamizo, N. & Isowa, Y. (1981) Enzymatic synthesis of caerulein peptide. In: Oyama, K. & Niimura, S. (Eds.) *Peptide Chemistry* (pp. 213-214). Protein Research Foundation, Osaka.

[19] Oyama, K., Irino, S. & Hagi, N. (1987) Production of aspartame by immobilized thermoase. *Methods in Enzymology.* **136**, 503-516.

[20] Aso, K. (1989) Enzymatic approach to the synthesis of a lysinecontaining sweet peptide, N-acetyl-L-phenylalanyl-Llysine. *Agric. Biol. Chem.* **53**, no. 3, p. 729-733.

[21] Clapés, P., Mata-Alvarez, J., Valencia, G., Reig, F., Torres, J.L. & García-Antón, J.-M. (1989) Continuous enzymatic synthesis of Z-kyotorphin amide in an enzyme-immobilized fixed-bed reactor. *J Chem. Technol. Biotech.* **45**, 191-202.

[22] Nakanishi, K., Takeuchi, A. & Matsuno, R. (1990) Long-term continuous synthesis of aspartame precursor in a column reactor with an immobilized thermolysin. *App. Microbiol. Biot.* **32**, 633-636.

[23] Kimura, Y., Muraya, K., Araki, Y., Matsuoka, H., Nakanishi, K. & Matsuno, R. (1990) Synthesis of peptides consisting of essential amino acids by a reactor system using three proteinases and an organic solvent. *Agric. Biol. Chem.* **54**, 3331-3333.

[24] Kimura, Y., Nakanishi, K. & Matsuno, R. (1990) Enzymatic synthesis of the precursor of Leu-enkephalin in water-immiscible organic solvent systems. *Enzyme Microb. Technol.* 12, 272-280.

[25] Monter, B., Herzog, B., Stehle, P. & Fürst, P. (1991) Kinetically controlled synthesis of dipeptides using ficin as biocatalyst. *Biotech. Appl. Biochem.* **14**, 183-191.

[26] Herrmann, G., Schwarz, A., Wandrey, C., Kula, M.R., Knaup, G., Drauz, K.H. & Berndt, H. (1991) Scale up of enzymatic peptide synthesis in an enzyme membrane reactor. *Biotech. Appl. Biochem.* **13**, 346-353.

[27] Serralheiro, M.L.M., Prazeres, D.M.F. & Cabral, J.M.S. (1994) Dipeptide synthesis and separation in a reversed micellar membrane reactor. *Enzyme Microb. Technol.* **16**, no. 12, p. 064-1073.

[28] Bradford, M.M. (1976) A rapid and sensitive method for the quantitation of microgram quantities of protein utilizing the principle of protein-dye binding. *Anal. Biochem.* **72**, 248-254.

[29] Mole, J.E. & Horton, R. (1973) Kinetic of Papain-Catalyzed Hydrolysis of α-N-Benzoyl-L-arginine-p-nitroanilide. *Biochem.* **12**, 816-822.

[30] Priolo, N., Morcelle del Valle, S., Arribére, M., López, M. & Caffini, N. (2000) Isolation and characterization of cysteine protease from the latex of *Araujia hortorum* fruits. *J. Protein Chem.* **19**, 39-49.

[31] Phadatare, S.U., Srinivasan, M.C. & Deshpande, V.V. (1993) High Activity Alkaline Protease from *Cornidiobolus coronatus* (NCL 86.8.20): Enzyme Production and Compatibility with commercial Detergents. *Enzyme Microb. Technol.* **15**, 72-76.

[32] Gupta R., Beg Q.K. & Lorenz P. (2002) Bacterial alkaline proteases: molecular approaches and industrial application. *Appl Microbiol Biotechnol.* **59**,15-32.

[33] Banerjee U.C., Sani R.K., Azmi W. & Sani R. (1999) Thermostable alkaline protease from Bacillus brevis and its characterization as a laundry detergent additive. *Process Biochem.* **35**, 213-9.

[34] Gupta R., Gupta K., Saxena R.K. & Khan S. (1999) *Bleach-stable alkaline protease from Bacillus sp. Biotechnol. Lett.* **21**, 135-8.

[35] Kumar C.G. & Takagi H. (1999) Microbial alkaline proteases from a bioindustrial viewpoint. *Biotechnol. Adv.* **17**, 561-94.

[36] McDonald, J.K. & Barrett, A.J. (1986) *Mammalian Proteases: a Glossary and Bibliography.* Vol. 2: Exopeptidases, Academic Press, London.

[37] Trusek-Holownia, A. Synthesis of ZAlaPheOMe, the precursor of bitter dipeptide in the two-phase ethyl acetate-water system catalysed by thermolysin, *J. Biotechnol.* 2003, 1021, 53-163.

In: Biochemical Engineering
Editors: F.E. Dumont and J.A. Sacco, pp. 307-319

ISBN: 978-1-60741-257-1

Chapter 14

DEACTIVATION AND REJUVENATION OF PHOSPHORUS ACCUMULATING ORGANISMS IN THE PARALLEL AN/AO PROCESS

***Hong-bo Liu*[1,2,*] *and Si-qing Xia*[2]**

[1]Rhine-Main Water Research Institute, Justus-von-Liebig-Str.10, 64584 Biebesheim am Rhein, Germany

[2]State Key Laboratory of Pollution Control and Resource Reuse, Tongji University, 200092 Shanghai, P. R. China

Abstract

The parallel AN/AO process, first proposed by the authors to efficiently use denitrifying phosphorus removing bacteria, was briefly introduced in the paper. Deactivation of phosphorus-accumulating organisms (PAO) occurred in the process when its SRT (Sludge Retention Time) and HRT (Hydraulic Retention Time) were too long, i.e. SRT and HRT were 30d and 18h respectively. PAO deactivation was observed also in three anaerobic-anoxic SBR reactors fed with different COD/NO_3^--N synthetic wastewater using seed sludge from the parallel AN/AO process. Possible factors that could cause PAO deactivation such as pH, temperature, internal/external return ratio, SRT, HRT, DOC (Dissolved Organic Carbon) at the beginning of anoxic stages and NO_3^--N concentration at the beginning of anaerobic stages were studied. Results showed that SRT and HRT were main factors accounting for PAO deactivation occurrence in the parallel AN/AO process while DOC concentration at the beginning of anoxic stages and NO_3^--N concentration at the beginning of anaerobic stages were main factors influencing PAO activity in the anaerobic-anoxic SBR reactors. PAO rejuvenation occurred in both configurations shortly after main influencing factors were reset to right values: PAO was rejuvenated by adjusting SRT and HRT to 15d and 9h respectively for the parallel AN/AO process; by controlling DOC at the beginning of anoxic stages and NO_3^--N concentration at the beginning of anaerobic stages lower than 3 mgL-1 and 2.3 mgL-1 respectively could rejuvenate PAO in anaerobic-anoxic SBR reactors.

Keywords: Enhanced Biological Phosphorus Removal (EBPR); Parallel AN/AO process; Phosphorus-accumulating organisms (PAO); PAO inactivation; PAO rejuvenation

* E-mail address: lhb_7906@163.com Tel: +49(0)6925490-8015 Fax: +49(0)6925490-8009. (Corresponding author)

1. Introduction

EBPR (Enhanced Biological Phosphorus Removal) technology is promising for phosphorus removal since it is more environment friendly and more cost-effective than chemical phosphorus removal processes (Robert et al., 2003). The reliability of EBPR processes, however, is notorious and prevents wide spreading of the technology (Csiti, 1991; Robert et al., 2003). Failure or deteriorated performance of EBPR processes are inherently related to PAO deactivation, which are triggered by improper application of easily adjustable operational parameters such as sludge retention time (SRT) (Akar et al., 2006), hydraulic retention time (HRT) (Csiti, 1991) , internal/external return ratio (Ochmen et al., 2005; Beatons et al., 1999; Csiti 1991), and costly adjustable parameters such as temperature (T) (Beatons et al, 1999), pH (Carsson et al., 1997; Kuba et al., 1997; Ochmen et al., 2005; Roske and Schonborn, 1994; Smolders et al., 1994;) , and wastewater characteristics (Cech and Hartman, 1993; Meinhold et al., 1998; Ahn et al., 2002; Maurer et al., 1997) . Numerous studies have been done regarding to each single factor listed above, but mainly with synthetic wastewater in laboratory scales (Csiti. 1991; Meinhold et al., 2005; Ochmen et al., 1998; Smolders et al.,1994).

The parallel AN/AO process was first proposed by the authors to efficiently use denitrifying phosphorus removing bacteria (DPB) (Xia and Liu, 2006), a kind of facultative bacteria performing phosphorus and nitrogen removal simultaneously (Meinhold et al., 1999; Kerrn et al., 1994; Kuba 1997; Kuba 1996; Chuang et al., 1998;). The process performed well with average COD, TN, TP removal efficiency 74%, 65%, and 84% respectively before occurrence of PAO deactivation. However, in some operations of the parallel AN/AO process, PAO deactivation occurred and decreased phosphorus removal efficiency of the process to 55%; PAO deactivation occurred also in some operations of three anaerobic-anoxic SBR reactors using seed sludge from the parallel AN/AO process.

Aim of this study was to investigate factors that trigger PAO deactivation in both configurations, determine interaction effects of different factors and find solutions to rejuvenate PAO from deactivated EBPR systems.

2 Materials and Methods

2.1. The Parallel AN/AO Process

Schematic diagram of the parallel AN/AO process was demonstrated in figure 1 together with mass balance of main pollutants for a reoported operation (Xia and Liu, 2006).. Main body of the process consists of an AN unit and a parallel AO unit. The AO unit consists of an anaerobic compartment and two aerobic compartments (abrreviated as Ana1, Aer1 and Aer2 respectively in the schematic diagram) and the AN unit consists of an anaerobic compartment and two anoxic compartments (abrreviated as Ana2, Ano1 and Ano2 respectively in the schematic diagram). The 6 compartments are uniform in size and each has a 20L working volume. Oxygen was provided by air diffusors in Aer1 and Aer2; nitrate needed in anoxic compartments was provided by internal return stream from Aer2 to Ano1. Mixers were installed in Ana1 , Ana2 , Ano1, and Ano2 to stir and suspend biomass. Extenal sludge return ratios were identical in both AO and AN units, where mixed liquid solids (MLSS) in the AN unit can flow into the AO unit through holes (20mm in diameter) in shared walls betwwen Aer1 and Ano1 as well as betwwen Aer2 and Ano2. The working volume of the final settler was 50L. Detailed description of the proces and operations can reffer to Xia and Liu 2006.

2.2. Anaerobic-anoxic SBR Reactors

Three anaerobic-anoxic SBR reactors used in this study were identical in dimensions with effective volume of 1.5L and feeding/discharging ratio of 80%. For each SBR reactor, 400mL sludge was seeded from wasted sludge of the parallel AN/AO process and cultured in the following 6h-period sequencing operation until a quasi-steady effluent was achieved: feeding of synthetic wastewater (15min)----anaerobic stage (2h)----sedimentation (30min)----decanting (15min)----feeding of nitrate solution (15min)----anoxic stages (2h)----sedimentation (30min)----effluent (15min). Flow rate of feeding and decanting (effluent) were accurately controlled by peristaltic pumps (Lange, China); magnetic stirrers (Meiyinpu, China) were used to suspend biomass in anaerobic and anoxic stages. Nitrate concentration at the beginning of anaerobic stages and DOC concentration at the beginning of anoxic stages were controlled by washing MLSS of the corresponding above mentioned stages before starting of their next sequential stages.

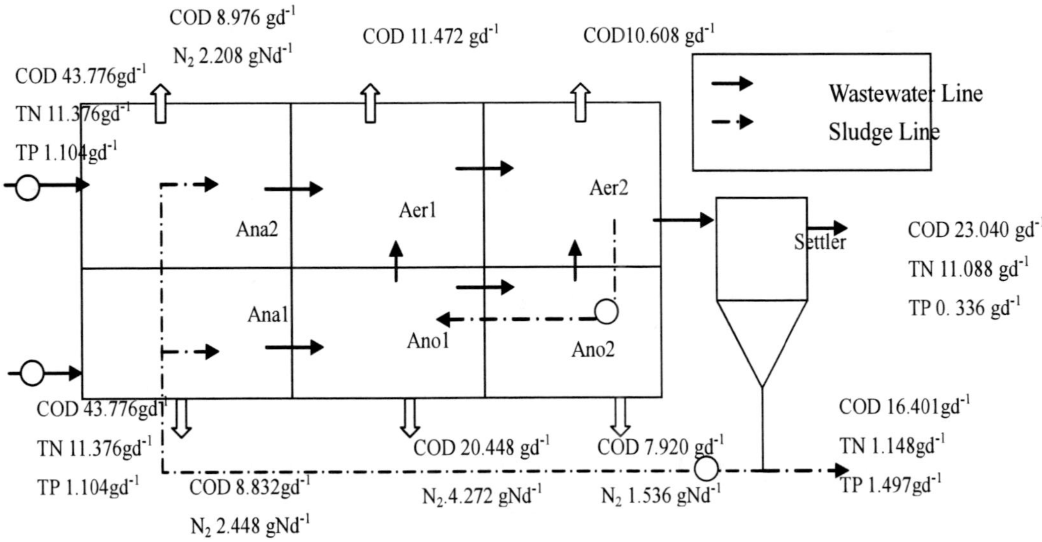

Figure 1. Schematic diagram of the parallel AN/AO process.

2.3. Wastewater and Analysis

Domestic wastewater from Tongji Newly Village Pump Station (Shanghai China) was used in experimental operations of the parallel AN/AO process; characteristics of the wastewater were listed in table 1.

Table 1. Characters of the domestic wastewater

Item	Concentration (mgL^{-1})
COD	160-430[a]
TN	40-60
TP	5-10
NH_4^+-N	30-50
NO_3^--N	<1
NO_2^--N	<1

[a] COD concentration ranged from 160-430 mgL^{-1} with 80% possibility lower than 200 mgL^{-1}

Synthetic wastewater was used in experimental operations of anaerobic-anoxic SBR reactors. Characteristics of the wastewater were listed in table 2-4, where different concentrations of NO_3^--N were used to control influent COD/ NO_3^--N as 3, 6, and 12 respectively.

pH was measured with glass electrode (Leici, Shanghai China); Temperature (T) was measured with mercury temperature meter (Ruiming, China); COD, TN, TP, NH_4^+-N, NO_3^--N, and NO_2^--N were measured with standard methods (APHA, 1998).

Table 2. Main components of the synthetic wastewater

Components	Concentration (mgL^{-1})	Chemicals
COD	200	NaAc
NH_3-N	10	NH_4Cl
TP	6	KH_2PO_4 and K_2HPO_4 (1:1 in mass)
Trace solution	1mLL^{-1}	Refer to table 4

Table 3. Main components of the nitrate solution

Components	Concentration (mgL^{-1})	Chemicals
NO_3^--N	17/34/68	KNO_3
TP	6	KH_2PO_4 and K_2HPO_4 (1:1 in mass)
Trace solution	1 mLL^{-1}	Refer to table 4

Table 4. Main components of the trace solution

Components	Concentration (mgL^{-1})
$FeCl_3$	0.900
H_3BO_3	0.150
$CaCl_2·6H_2O$	0.150
$CuSO_4·5H_2O$	0.030
KI	0.180
$MnCl_2·4H_2O$	0.060
Na2Mo$SO_4·2H_2O$	0.060
$ZnSO_4·7H_2O$	0.120

3. Results and Discussions

3.1. PAO Deactivation

PAO deactivation was observed in both field based parallel AN/AO reactors and laboratory based anaerobic-anoxic SBR reactors. Activity of PAO was evaluated in terms of phosphorus removal efficiency in the processes and/or reactors.

3.1.1. PAO Deactivation in Parallel AN/AO Reactors

PAO deactivation was observed when either SRT or HRT was too long (30d and 18h respectively). Performances of the parallel AN/AO process under different SRTs and HRTs were illustrated in figure 2 and figure 3.

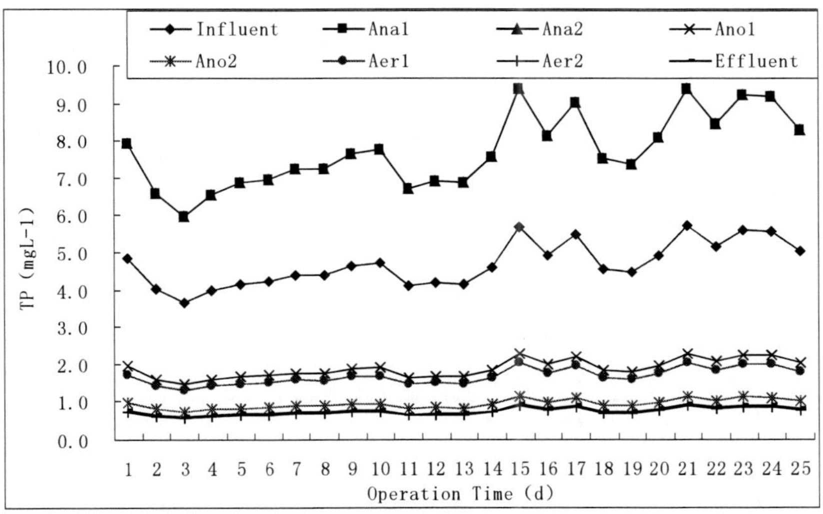

(a) SRT 10d

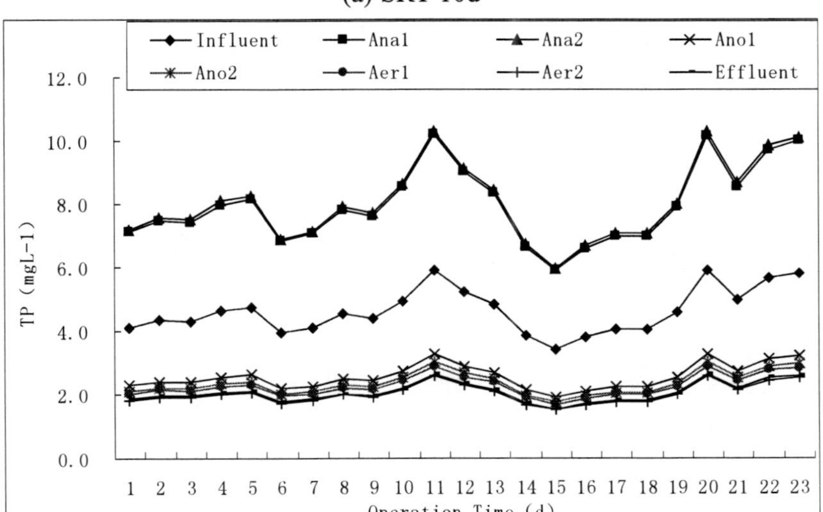

(b) SRT 30d

Figure 2. Continued on next page.

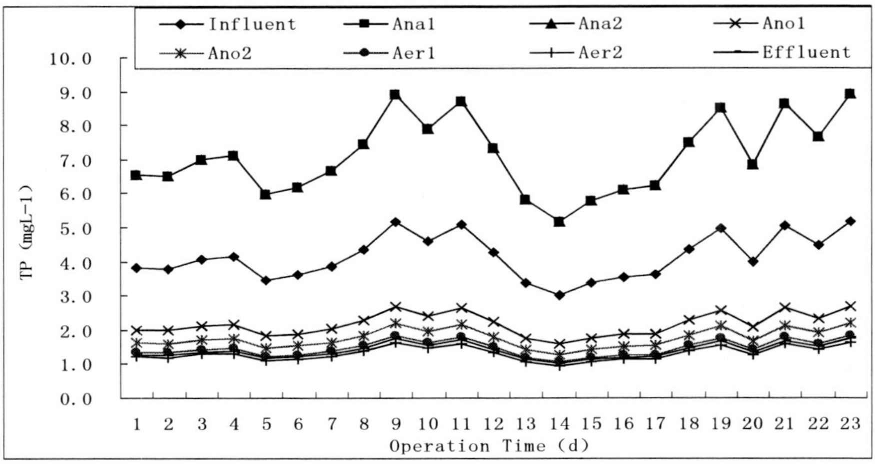

(c) SRT 15d

Figure 2. TP profiles under different SRT operations; HRT=6h.

The process performed normally when SRT was 10d, during which operation time average COD, TN, TP removal efficiency were 64%, 65% and 84% respectively and average effluent TP concentration was 0.8 mgL^{-1} (figure 2(a)). PAO deactivation was observed when SRT was 30d, where TP removal efficiency was down to 55% and average effluent TP concentration was increased to 2.1 mgL^{-1} (figure 2(b)), which was much higher than discharging limit (1.0 mgL^{-1}) according to *Discharging Standards for Chinese Urban WWTPs* (GB18918-2002) first grade (B). Falling temperature accounted for part of PAO deactivation

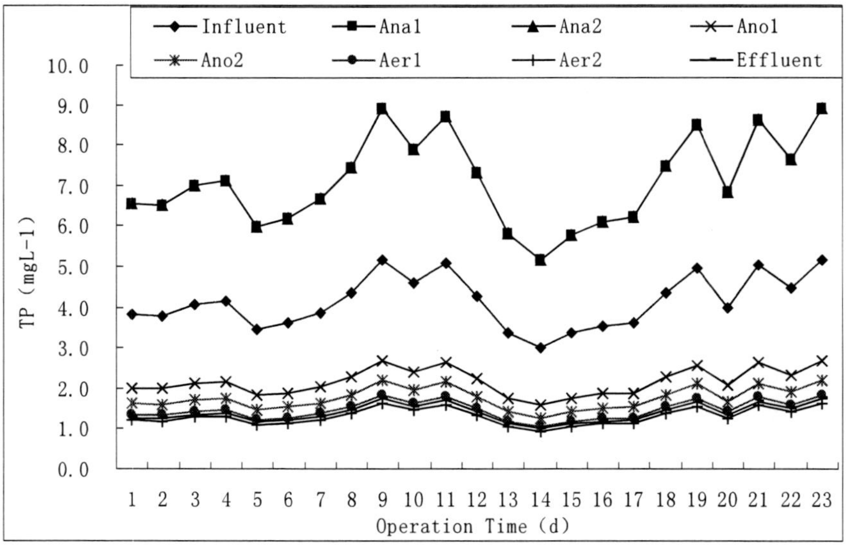

(a) HRT 6h

Figure 3. Continued on next page.

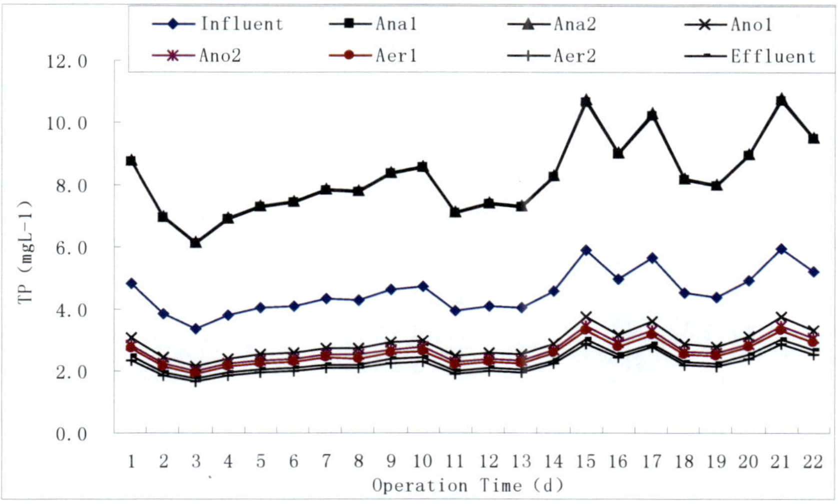

(b) HRT 18h

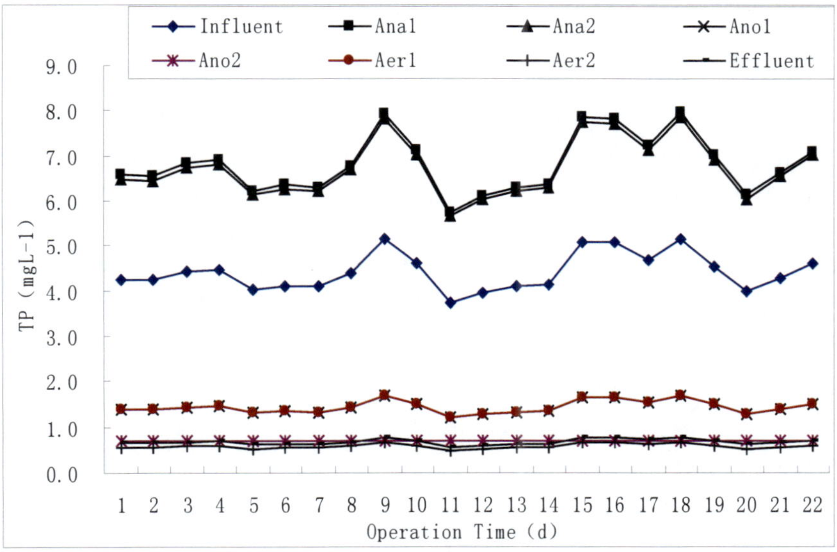

(c) HRT 9h

Figure 3. TP profiles under different HRT operations; SRT=15d.

since operation temperature during SRT=30d was lower than that of during SRT=10d (15°C vs. 18°C). Considering average operation temperature was only 13℃ when SRT=15d but TP removal efficiency was up to 67% and average effluent TP concentration was decreased to 1.4 mgL^{-1} (figure 2(c)) , SRT was the main factor that cause PAO deactivation in figure 2(b).

PAO deactivation was observed when HRT=18h with effluent TP 2.3mgL-1 and TP removal efficiency 49% (figure 3). Since atmosphere temperature was even higher when HRT=18h than that when HRT=6h in this group of operations (15.6℃ vs. 13℃), HRT was the main factor that causes PAO deactivation in figure 3(b). Anaerobic phosphorus release was less than 4mgL-1 in both figure 2 and figure 3 because of limited COD content in the influent.

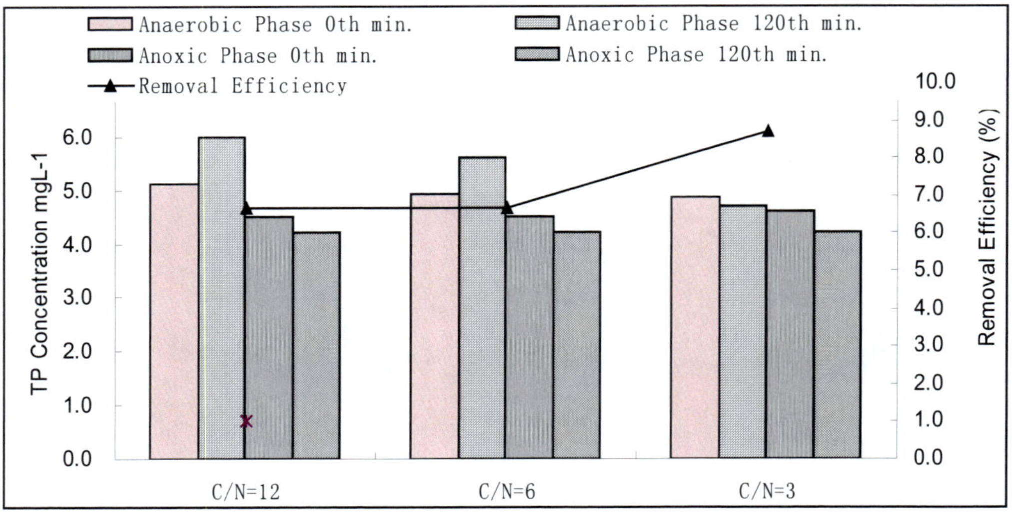

Figure 4. TP removal in anaerobic-anoxic SBR reactors during PAO deactivation.

3.1.2. PAO Deactivation in SBR Reactors

PAO deactivation was observed in three anaerobic-anoxic SBR reactors feeding with different C/N ratio synthetic wastewater and seeded with waste sludge from the parallel AN/AO process; results were shown in figure 4. No significant phosphorus release occurred in reactors 1 and 2 (0.9mgL-1 and 0.7mgL-1 respectively; C/N=12 and 6 respectively); negative phosphorus release was even observed in reactor 3 (C/N=3). For all the reactors, negative phosphorus uptake occurred, which indicated complete activity loss of PAO in the reactors.

Measured DOC at the beginning of anoxic stages was 15.9-24.1 mgL-1 and measured NO_3^--N concentration at the beginning of anaerobic stages was 2.8-7.2 mgL^{-1}.

3.2. PAO Rejuvenation

3.2.1. PAO Rejuvenation in Parallel AN/AO Reactors

PAO rejuvenation occurred when SRT was changed from 30d to 15d (figure 2(c)) and when HRT was changed from 18h to 9h (figure 3(c)). The fast rejuvenation process indicated that SRT and HRT were main operational factors that influence phosphorus removal capacity of the process.

3.2.2. PAO Rejuvenation in SBR Reactors

PAO rejuvenation occurred in anaerobic-anoxic SBR reactors by reducing DOC concentration at the beginning of anoxic stages (anoxic phase 0th min.) and NO_3^--N concentration at the beginning of anaerobic stages (anaerobic phase 0th min.) to as low as 3.0 mgL-1 and 2.3 mgL-1 respectively, which was done by washing the remaining MLSS with distilled water at the end of anoxic stages and anaerobic stages respectively. Results were shown in figure 5.

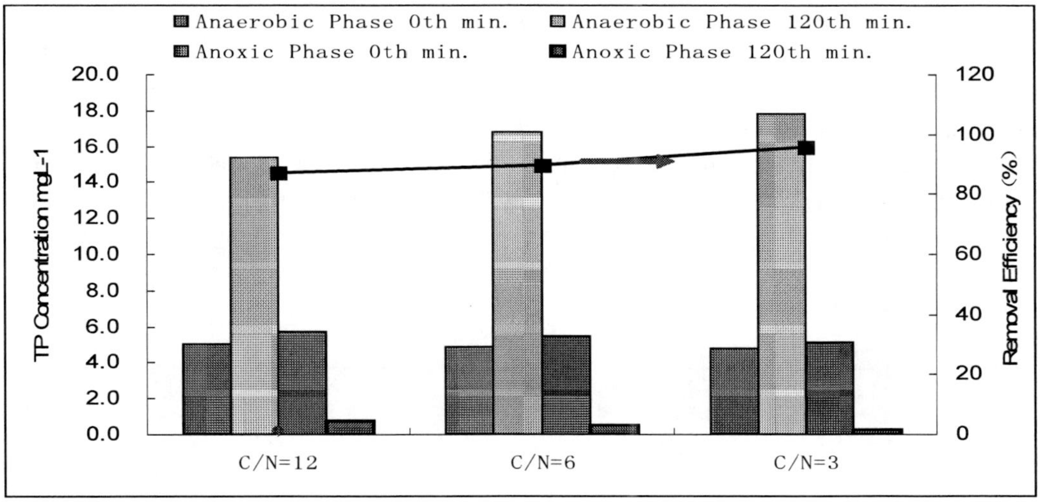

Figure 5. TP removals of anaerobic-anoxic SBR reactors after PAO rejuvenation.

Obvious phosphorus release was observed in all the three SBR reactors; and phosphorus removal efficiencies of the SBR reactors were more than 85%. The lower DOC concentration at the beginning of anoxic stages and NO_3^--N concentration at the beginning of anaerobic stages, the better phosphorus removal efficiency the process achieved (figure 6).

3.3. Factors Influencing PAO Activity

Influent wastewater characters, SRT, HRT, internal/external return ratios, temperature, pH are the main factors that influence PAO activity in any EBPR process (Csiti 1991; Hu et al., 2003).

Temperature accounted for part of PAO deactivation in the parallel AN/AO process since it happened during winter time; but as stated above it was not the main factor and its effects are very complicated since lower temperature can also help PAO out competing non-PAO in the process (Cech and Hartman, 1993; Ahn et al., 2002; Raymond et al., 2003). Considering operation temperature was even lower in the PAO rejuvenation run than that in the PAO deactivation run, temperature was not the main factor that deactivated PAO in the parallel AN/AO process. No temperature effect should be considered in the anaerobic-anoxic SBR reactors as they were operated in temperature controlled room.

Influent pH for the parallel AN/AO process was 6.5-8.5; Influent pH of synthetic wastewater for the anaerobic-anoxic reactors was around 7.2. When phosphorus release and uptake functioned normally, pH in anaerobic tanks and/or anaerobic stages were higher than those in anoxic/aerobic tanks and/or anoxic stages (data not shown); less pH difference between anaerobic stages and anoxic/aerobic stages was observed when PAO deactivation accrued. pH control can help improve phosphorus removal efficiency, which was not studied in this study but was done in other researches (Carsson et al., 1997; Kuba et al., 1997; Ochmen et al., 2005).

It was found in anaerobic-anoxic and anaerobic-aerobic SBR reactors that ready biodegradable organic at the beginning of anoxic stages and NO_3^--N at the beginning of anaerobic stages were main factors that influence PAO activity; DOC at the beginning of anoxic stages and NO_3^--N at the beginning of anaerobic stages as low as 3 mgL-1 and 2.3

mgL-1 could avoid PAO from deactivation in this study. Considering the fact that there was almost no NO_3^--N detected in anaerobic reactors of the parallel AN/AO process when PAO deactivation occurred (data not shown), and NO_3^--N at the beginning of anaerobic stage in one of the PAO deactivated anaerobic-anoxic SBR run was very close to 2.3 mgL-1 (2.8 mgL-1, figure 4), ready biodegradable organic at the beginning of anoxic stages was probably more essential to activity of PAO exposing to cyclic anaerobic-anoxic circumstances. In this study threshold DOC content in the anoxic stage was found to be 3.0 mgL-1; varying DOC from 3.0 mgL-1 to 0.3 mgL-1 improved effluent phosphorus concentration only from 0.7 mgL-1 to 0.2 mgL-1.

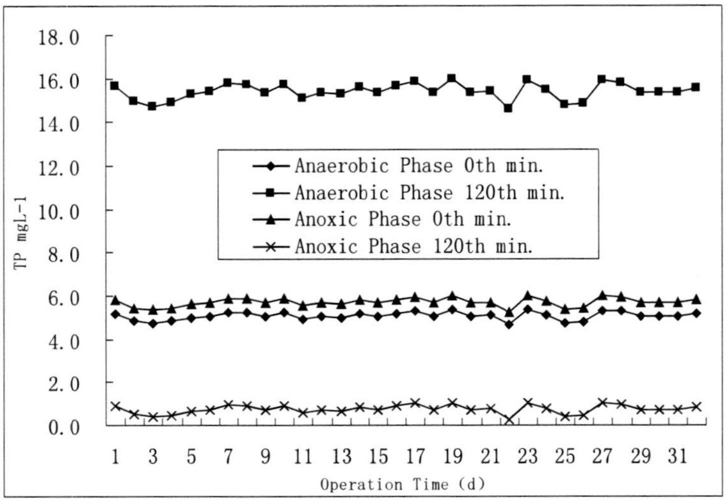

(a) DOC=3.0mgL-1

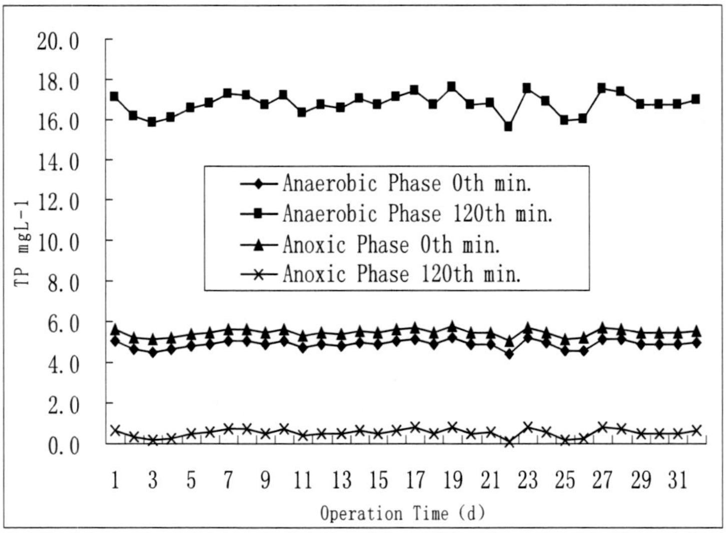

(b) DOC=0.9mgL-1

Figure 6. Continued on next page.

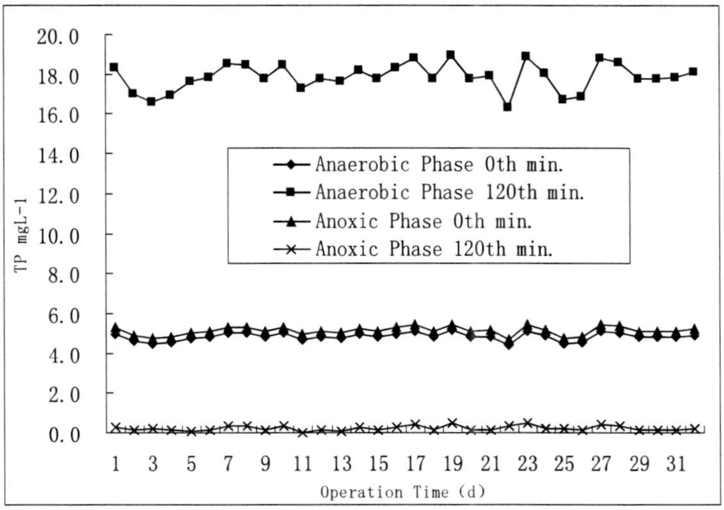

(c) DOC=0.3mgL-1

Figure 6. TP profiles under different DOC concentrations at the beginning of anaerobic stages.

Since seed sludge of anaerobic-anoxic SBR reactors were taken from the parallel AN/AO process, which was also operated partly in anaerobic-anoxic mode, they should share some common characters. Operation results showed however, that PAO deactivation was caused by different factors in two configurations, which indicated the complicated nature of EBPR systems. NO_3^--N concentration at the beginning of anaerobic stages in the anaerobic-anoxic configuration can be linked to external sludge return ratio in the parallel AN/AO process. During operation of the parallel AN/AO process, PAO deactivation was not observed when external sludge return ratio was 50% (mass content of phosphorus in biomass was 4% , at which time NO_3^--N in the anaerobic reactor was 0.1 mgL-1), but occurred when external sludge return ratio was 100% (mass content of phosphorus in biomass was 2.6%, at which time NO_3^--N in the anaerobic reactor was 0.4 mgL-1). Ready biodegradable organic contents at the beginning of anoxic stages in the anaerobic-anoxic SBR configuration is related to internal MLSS return ratio, influent wastewater characters, and performance of the anaerobic compartments; the latter two are very difficult to be evaluated quantitatively. Internal MLSS return ratio was controlled as 200% in this study. No PAO deactivation was observed by varying internal MLSS return ratio from 100% to 300% in the study.

4. Conclusions

PAO deactivation was observed in both the parallel AN/AO process and the anaerobic-anoxic SBR reactors. Excessive HRT and SRT were main factors that caused PAO deactivation in the parallel AN/AO process; ready biodegradable organic content at the beginning of anoxic stages and NO_3^--N concentration at the beginning of anaerobic stages were main factors that influenced PAO activity in anaerobic-anoxic SBR reactors. Ready biodegradable organic content was probably more essential to PAO activity in anaerobic-anoxic SBR reactors. Other factors such as pH, temperature, internal MLSS return ratio and external sludge return ratio also influenced PAO activity in the parallel AN/AO process and were linked to factors that influenced PAO activity in anaerobic-anoxic SBR reactors.

SRT as long as 30d and HRT as long as 18h triggered PAO deactivation in the parallel AN/AO process, which could be rejuvenated by adjusting SRT and HRT to 15d and 9h respectively. By controlling DOC concentration at the beginning of anoxic stages and NO_3^--N concentration at the beginning of anaerobic stages lower than 3 mgL-1 and 2.3 mgL-1 respectively could rejuvenate deactivated PAO in the anaerobic-anoxic SBR reactors.

Acknowledgements

This study was financially co-supported by New Century Educational Talent Program (NCET-05-0387, Commission of China Education Ministry) and High School Fund for Doctors(No.20050247016).

References

A., Ochmen, M.T., Vives, H.B., Lu, Z.G., Yuan, J., Keller , 2005. The effect of pH on the competition between polyphosphateaccumulating organisms and glycogen-accumulating organisms. *Water Res.* **39**, 3727-3737.

American Public Health Association (APHA), American Water Works Association (AWWA), Water Environment Federation (WEFW) , 1998. *Standard Methods for the Examination of Water and Wastewater,* 20th edition. Washington, DC.

A., Akar, E.U., Akkaya, S. Koray, Yesiladali, G.C., E.U., Cokgor, C.T., Derin, Orhon, Z., P., Cakar , 2006. Accumulation of polyhydroxyalkanoates by microlunatus phosphovorus under various growth conditions. *J. Ind. Microbial Biotechnol.* **33**, 215-220.

Beatons, D., Vanrolleghem, P.A., van Loosdrecht, M.C.M., Hosten, L.H., 1999. *Temperature effects in bio-P removal. Wat. Sci. Tech.* **39** (1), 215-225.

Carsson, H., Aspegren, H., Lee, N., Hilmer, A., 1997. Calcium phosphorus precipitation in biological phosphorus removal systems. *Water Res.* **31**(5), 1047-1055.

Cech, J.S., Hartman, P., 1993. Competition between polyphosphate and polysaccharide accumulating bacteria in enhanced biological phosphorus removal systems. *Water Res.* **27**, 1219-1225.

Csiti, A., 1991. Factors affecting the biological phosphorus removal efficiency. *Periodica Polytechnica Ser. Civil Eng.* **35**(1-2), 97-106.

J., Meinhold, Carlos, D.M. Filipe, G., T., Daigger, 1999. Characterization of the denitrifying fraction of phosphate accumulating organisms in biological phosphate removal. *Wat. Sci. Tech.* **39**(1), 31-42.

J., Meinhold, H., Pedersen, E., Arnold, 1998. Effect of continuous addition of an organic substrate to the anoxic phase on biological phosphorus removal. *Wat. Sci. Tech.* **38** (1), 97-105.

J., Ahn, Tomotaka D., Satoshi T., 2002. Transformation of phosphorus and relevant intracellular compounds by a phosphorus-accumulating enrichment culture in the presence of both the electron acceptor and electron donor. *Biotechnology and Bioengineering.* **79** (1), 84-92.

Kerrn, J.P., Henze, M., Strube, R., 1994. Biological phosphorus release and uptake under alternating anaerobic and anoxic conditions in a fixed-film reactor. *Water Res.* **28** (5), 1253-1255.

Raymond, J., M.C.M., van Loosdrecht, Z.G., Yuan, J., Keller, 2003. Metabolic model for glycogen-accumulating organisms in anaerobic/aerobic activated sludge systems. *Biotechnology and Bioengineering.* **81**(1), 92-106.

Maurer, M., Gujer, W., Hany, R., Bachmann, S., 1997. Intracellular carbon flow in phosphorus-accumulating organisms from activated sludge systems. *Water Res.* **31**, 907-917.

Robert, J.S., T., Mino, M., Onuki, 2003. The microbiology of biology phosphorus removal in activated sludge systems. *FEMS Microbiology Reviews.* **27**, 99-127.

Roske, L.N., Schonborn, C., 1994. Interactions between chemical and advanced biological phosphorus elimination. *Water Res.*, **28**, 1103-1109.

Smolders, G., J.F., Meij, Loosdrecht, M.C.M., Heijnen, J.J. 1994. Model of the anaerobic metabolism of the biological phosphorus removal process: stoichiometry and pH influence. *Biotech. and Bioengin.* **44**, 837-848.

S.H., Chuang, C.F., Ouyang, H.C., Yuang, S.J., You, 1998. Evaluation of phosphorus removal in anaerobic-anoxic-aerobic system-via polyhydroxyalkonoates measurements. *Wat. Sci. Tech.,* **38** (1), 107-114.

T., Kuba, M.C.M., van Loosdrecht, J.J., Heijnen, 1997. Biological dephosphatation by activated sludge under denitrifying conditions: pH influence and occurrence of denitrifying dephosphatation in a full scale waste water treatment plant. *Wat Sci. Tech.* **36** (12), 75-82.

T., Kuba, M.C.M., van Loosdrecht, J.J., Heijnen, 1996. Phosphorus and nitrogen removal with minimal COD requirement by integration of nitrification in a two sludge system. *Water Res.* **30** (7),1702-1710.

Xia, S.Q., Liu, H.B., 2006. Operation of three parallel AN/AO processes to enrich denitrifying phosphorus removing bacteria. *J. Environ. Sci.* (China). **18**(3): 433-438.

Z.R., Hu, M.C., Wentzel, G.A., Ekama, 2003. Modeling biological nutrient removal activated sludge systems-a review. *Water Res.* **37**, 3430-3444.

In: Biochemical Engineering
Editors: F.E. Dumont and J.A. Sacco, pp. 321-339
ISBN 978-1-60741-257-1

Chapter 15

ALBUMIN-BOUND TOXIN REMOVAL IN LIVER SUPPORT DEVICES: CASE STUDY OF BILIRUBIN ADSORPTION AND DIALYSIS

M. Cristina Annesini, Vincenzo Piemonte and Luca Turchetti
Department of Chemical Engineering Materials & Environment
University of Rome
"La Sapienza" via Eudossiana 18, 00184 Roma (Italy)

Abstract

Dialysis and adsorption units are commonly used in liver support devices for the removal of albumin-bound toxins such as bilirubin. In this chapter, an engineering approach to the analysis of a liver support device implementing these units is presented. Starting from the physico-chemical description of the basic phenomena involved in the detoxification process, a mathematical model of a recirculating albumin dialysis liver support device was built and used to calculate bilirubin clearances obtained by the device with different operating conditions.

The results highlight the possible existence of an optimum dialysate albumin concentration; furthermore, the overall bilirubin clearances obtained in the simulations did not exceed 4% of the blood flow-rate fed to the device, this poor performance being limited by the slow bilirubin mass transfer across the membrane. The information presented in this chapter can be helpful for the optimization of existing liver support devices and for the design of new ones; nevertheless, for a complete assessment of the device performance, a similar analysis should be extended to the clearance of other toxins and some of the model parameters should be also checked against clinical data.

1. Introduction

The liver is a vital organ that performs a wide range of life-important functions, among which is blood detoxification. Therefore, liver failure is associated with high plasmatic levels of different types of toxins, some of which are involved in the aethiology of secondary life-threatening multi-organ pathologies. For this reason, if not correctly and timely managed, liver failure can lead to a sudden worsening of the clinical state of the patient, and easily result in a fatal outcome.

At present, the clinical approach to acute and acute-on-chronic liver failure is mainly based on blood detoxification as a temporary solution aimed at keeping patients alive in wait of a recovery of liver functionality or an organ transplantation. Blood detoxification is performed by means of extracorporeal liver support devices (LSDs), whose primary aim is to remove selectively the noxious substances, while avoiding the depletion of biologically valuable macromolecules, like proteins, hormones and growth factors.

This task is successfully performed in the treatment of renal failure with hemodialysis, for the removal of small, water-soluble toxins such as urea and creatinine. However, toxins not cleared by the failing liver include also molecules that are tightly bound to plasma proteins such as albumin, and more complex processes are required for blood detoxification in this case.

The first detoxification process specifically designed for the removal of albumin-bound toxins is Albumin Dialysis (AD, Stange et al. 1993). Basically, this process consists in blood dialysis across a special albumin impregnated membrane against a concentrated albumin solution (albumin dialysate). The particular structure of the membrane and the presence of a binder in the dialysate allow for albumin-bound toxin transfer across the membrane, while the cut-off of the membrane is chosen so as to avoid transfer of albumin and higher molecular weight substances.

AD can be either applied without recirculation of the albumin dialysate, as in Single Pass Albumin Dialysis (SPAD, Sauer et al., 2004), or with regeneration and recirculation of the dialysate, as in the MARS (Molecular Adsorbents Recirculating System, Gambro, Lund, Sweden). This latter device, that is one of the best-known LSDs used in clinical practice, implements AD with on-line regeneration of the albumin dialysate by conventional dialysis and adsorption on activated carbon and anionic resin (see Fig. 1.).

Another detoxification process designed for albumin-bound toxin removal is Fractionated Plasma Separation and Adsorption (FPSA, Falkenhagen et al., 1999). The clinically used LSD that implements this process is the Prometheus (Rifai et al., 2005, Evenopoel et al., 2005), which is schematically represented in Fig. 1.. In this device, plasma is filtered from blood along with albumin and smaller molecules, circulated in closed loop through one anionic resin and one non-ionic resin column in series, and then sent back into the patient blood circuit.

Other processes have been proposed (for a review see Stegmayr et al. 2005 and Rozga et al. 2006), anyway, under a general point of view, it can be affirmed that all of them are based on some combination of dialysis and adsorption unit operations.

Analysis and comparison of the performance of the different LSDs is currently based mainly on clinical criteria (Mitzner et al. 2006 and Evenopoel et al. 2006). It is clear that in-vivo clinical tests of LSDs are essential and can certainly not be substituted with in-vitro experiments and theoretical considerations only; nevertheless, the patient-device system is extremely complex and it is hard to gain insight into the detoxification process by the mere analysis of clinical data. As a consequence, at present it is difficult to define rationally the advantages and drawbacks of different device configurations.

In this chapter, an engineering approach is applied to the analysis of LSDs. Starting from a description of the physico-chemical phenomena that characterize each unit operation of the detoxification process, mathematical models of dialysis and adsorption units are developed and combined into a LSD model. The analysis refers to a MARS-type LSD and is

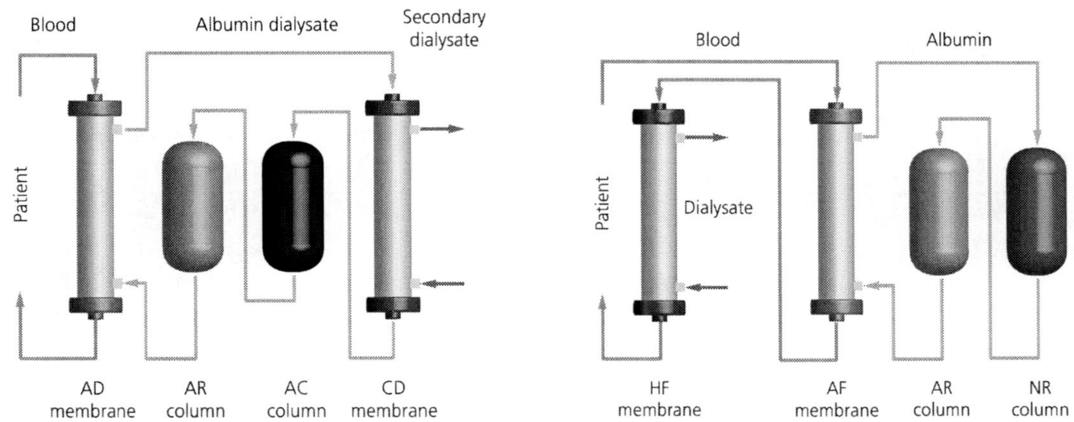

Figure 1. Liver support devices. On the left: MARS (AD: albumin dialyzer; CD: conventional dialyzer; AR: anionic resin column, AC: activated carbon column); a similar scheme, excluding dialysate regeneration and recirculation, applies to SPAD. On the right: Prometheus (NR: polymeric non-ionic resin column; AF: plasma fractionator, HF: high-flux dialyzer); adsorption is performed directly after plasma filtration on a polysulphone membrane.

focused on removal of bilirubin, that is a standard marker of the clinical state of liver-failure patients and can be considered as representative of a wide class of strongly albumin-bound toxins. Nevertheless, the results and the considerations reported in this chapter can be extended to other strongly albumin-bound toxins and other albumin dialysis LSDs.

2. Thermodynamic Considerations

2.1. Bilirubin-Albumin Binding

One of the key role of serum albumin is binding and transport in blood of a variety of endogenous and exogenous molecules that include, among others, fatty acids, bile acids, heterocyclic compounds as bilirubin, aromatic aminoacids as tryptophan and other indols.

As for bilirubin, it is generally agreed that the albumin molecule has one high affinity binding site (Sudlow I) and one (or more) low affinity site(s) for this metabolite: neglecting, for the sake of simplicity, the formation of complexes with more than two bilirubin molecules, the following binding equilibria can be considered in a liquid phase containing albumin and bilirubin

$$\mathrm{A+B} \rightleftharpoons \mathrm{AB} \tag{1}$$

$$\mathrm{AB+B} \rightleftharpoons \mathrm{AB_2} \tag{2}$$

with equilibrium constants given by

$$K_1 = \exp\left(\frac{\mu_{AB}^0 - \mu_A^0 - \mu_B^0}{RT}\right) = \frac{c_{AB}}{c_A c_B} \tag{3}$$

Table 1. Bilirubin - albumin binding constant (HSA human serum albumin, BSA bovine serum albumin)

	pH	T (°C)	High affinity sites (M^{-1})	Low affinity sites (M^{-1})	Reference
HSA	7.4	4	$4 \cdot 10^9$	-	Gray et al., 1978
HSA	7.4	37	$1.4 \cdot 10^8$	$5 \cdot 10^{5(a)}$	Jacobsen et al.,1977
HSA	8.5	24	$3 \cdot 10^6$	$1.4 \cdot 10^{5(b)}$	Beaven et al.,1973
HSA	7.4	-	$9.5 \cdot 10^7$	-	Petersen et al.,2000
HSA	7.4	-	$6.8 \cdot 10^7$	$4.5 \cdot 10^6 \div 5 \cdot 10^{6(c)}$	Ostrow et al., 1994
HSA	-	5	$1.3 \cdot 10^8$	$2.4 \cdot 10^{6(d)}$	Berde et al.,1979
HSA	-	23	$1.2 \cdot 10^8$	$4.9 \cdot 10^{6(d)}$	Berde et al. 1979
HSA	-	37	$6.4 \cdot 10^7$	$2.6 \cdot 10^{6(d)}$	Berde et al. 1979
BSA	7.4	25	$5 \cdot 10^7 - 7 \cdot 10^7$	$4 \cdot 10^6 - 5 \cdot 10^6$	Blauer et al.,1997

(a) Two sites at low affinity
(b) A third site is considered for high values of ionic strength
(c) A lot of low affinity sites in the range reported
(d) A third site is also considered

$$K_2 = \exp\left(\frac{\mu^0_{AB_2} - \mu^0_{AB} - \mu^0_B}{RT}\right) = \frac{c_{AB_2}}{c_{AB}c_B} \tag{4}$$

where c_B, c_A are the free bilirubin and albumin concentration, c_{AB} and c_{AB_2} are the 1:1 and 1:2 albumin-bilirubin complex concentrations and μ^0_i is the chemical potential of pure component *i* at the temperature and pressure of the system. Values of the binding constants reported in the literature (see Tab. 1) vary in a quite wide range; in any case, K_1 values in the range $10^7 - 10^8$ M^{-1} and K_2 values at least one order of magnitude lower are widely accepted. On this basis, the formation of complexes involving secondary sites can be neglected and only binding reaction (1) will henceforth be considered.

With this assumption, the fraction of unbound bilirubin can be calculated from total albumin and bilirubin concentrations (c_{alb} and c_{bil}, respectively) by coupling equation (3) with the following mass balance equations

$$c_{bil} = c_{AB} + c_B \tag{5}$$

$$c_{alb} = c_{AB} + c_A \tag{6}$$

In clinically interesting conditions $K_1 c_{bil} \gg 1$ and, as far as $c_{alb} > c_{bil}$, the unbound fraction of bilirubin can be approximated as follows:

$$\frac{c_B}{c_{bil}} \approx \frac{1}{K_1\left(c_{alb} - c_{bil}\right)} \ll 1 \tag{7}$$

Therefore, virtually all the bilirubin is present in the solution as AB complex (for example, less than 1% of bilirubin is in the unbound form if the difference $c_{alb} - c_{bil}$ is greater than 10^{-6}M), so that it may further be assumed that

$$c_{AB} \approx c_{bil} \quad ; \quad c_A \approx c_{alb} - c_{bil} \tag{8}$$

2.2. Bilirubin Phase-Partition

Binding of bilirubin to albumin is a fundamental phenomenon to be considered in describing bilirubin phase partition equilibria. In this section, bilirubin partition between two albumin-containing aqueous phases and bilirubin solubility in a polymer membrane will be considered, accounting for the effect of bilirubin-albumin binding. Bilirubin adsorption on solid phases is discussed in a following section. It is worth noting that the following analysis is referred to bilirubin; nevertheless, the same conclusions can be drawn for any tightly albumin-bound compound.

Firstly, let us consider two aqueous phases with different albumin concentrations. Bilirubin transfer between the two solutions, indicated as α and β, may be viewed as an exchange reaction

$$\mathrm{AB}^{\alpha} + \mathrm{A}^{\beta} \rightleftharpoons \mathrm{AB}^{\beta} + \mathrm{A}^{\alpha} \tag{9}$$

As a consequence, equilibrium conditions correspond to

$$\mu_{AB}^{\alpha} + \mu_{A}^{\beta} = \mu_{AB}^{\beta} + \mu_{A}^{\alpha} \tag{10}$$

or, assuming ideal solutions

$$\frac{c_{AB}^{\alpha}}{c_{AB}^{\beta}} = \frac{c_{A}^{\alpha}}{c_{A}^{\beta}} \tag{11}$$

If conditions (7) and (8) hold, the equilibrium condition (11) may be also rewritten in terms of total bilirubin and albumin concentrations as

$$\frac{c_{bil}^{\alpha}}{c_{bil}^{\beta}} = \frac{c_{alb}^{\alpha}}{c_{alb}^{\beta}} \tag{12}$$

or as

$$\frac{c_{alb}^{\alpha} - c_{bil}^{\alpha}}{c_{alb}^{\alpha}} = \frac{c_{alb}^{\beta} - c_{bil}^{\beta}}{c_{alb}^{\beta}} \tag{13}$$

Condition (12) corresponds to a bilirubin partition coefficient equal to the ratio of albumin concentrations in the two phases, whereas condition (13) shows that, at equilibrium, equal fractions of available albumin for bilirubin binding are present in both phases.

These simple equilibrium conditions apply to a dialysis process on an albumin rejecting membrane and provide a rationale for some empirical findings reported in the literature. For instance, Steiner et al. (2004) carried out albumin dialysis of human plasma spiked with bromosulphthalein (BSP, a strongly albumin-bound compound, with a binding constant as high as $2{\cdot}10^{7}\mathrm{M}^{-1}$) and observed that, at equilibrium, the concentration of BSP in each liquid phase was proportional to albumin concentration (see Fig. 2). The same authors extrapolated a similar result also for bilirubin, by the analysis of the time course of bilirubin concentration in a non-equilibrium experiment.

The applicability of equilibrium condition (12) to albumin-dialysis of strongly albumin-bound toxins is also confirmed by data obtained by the authors of this chapter and reported in Fig. 2. These data were collected at the end of albumin dialysis experiments of bilirubin-containing solutions in a two compartment closed loop system, and refer to equilibrium conditions.

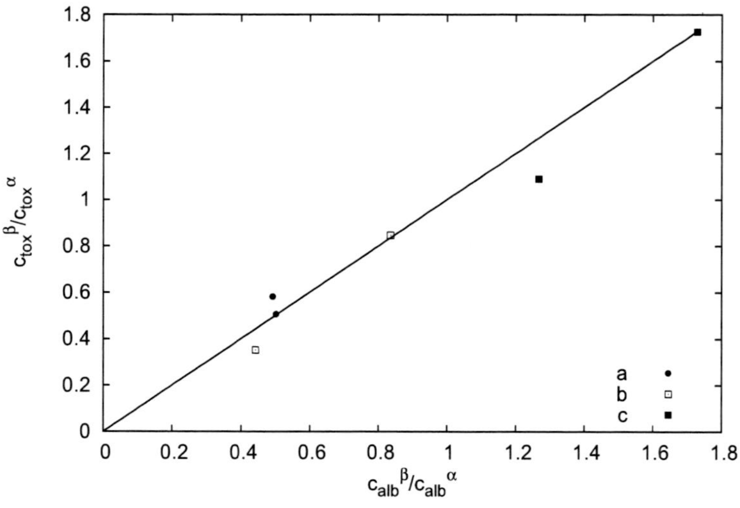

Figure 2. Strongly albumin-bound toxin partition between two albumin-containing solutions. Data were collected at equilibrium in a double-compartment, closed loop dialysis system, using different membrane modules and operating conditions. (a) Bilirubin partition using a F40 module (Fresenius Medical Care, Bad Homburg, Germany); (b) Bilirubin partition using a FX40 module (Fresenius Medical Care, Bad Homburg, Germany); (c) BSP partition, data reported by Steiner et al. (2004)

A similar chemical approach can be used to describe partition of bilirubin between an albumin-containing aqueous solution and an hydrophobic phase, like a polymeric membrane, in which free bilirubin is soluble: this phenomenon can be represented by the following reaction

$$\mathrm{AB}^{\alpha} \rightleftharpoons \mathrm{B}^{m} + \mathrm{A}^{\alpha} \tag{14}$$

where superscript m refers to the membrane phase. As a consequence, bilirubin solubility is given by

$$c_{bil}^{m} = \exp\left(\frac{\mu_{AB}^{0,\alpha} - \mu_{A}^{0,\alpha} - \mu_{B}^{0,m}}{RT}\right)\frac{c_{AB}^{\alpha}}{c_{A}^{\alpha}} = S_{bil}^{m}\frac{c_{AB}^{\alpha}}{c_{A}^{\alpha}} \tag{15}$$

or, if conditions (7) and (8) hold, in the simplified form

$$c_{bil}^{m} = S_{bil}^{m}\frac{c_{bil}^{\alpha}}{c_{alb}^{\alpha} - c_{bil}^{\alpha}} \tag{16}$$

3. Dialysis Process

Although the bilirubin MW is far below the cut-off of standard dialysis membranes, due to its extremely low water solubility at neutral pH, this toxin cannot be removed by conventional dialysis against an aqueous buffer solution, but a binder is required in the dialysate.

Albumin itself is used in the dialysate as binder for bilirubin, as well as other protein-bound toxins, in the "albumin dialysis" process used in extracorporeal therapy, both in single pass (SPAD) or closed loop with albumin regeneration (MARS). The beneficial effect of albumin in the dialysate for bilirubin removal in conventional renal replacement therapy has also been suggested (Chawla et al., 2005).

Starting from the first work of Stange and Mitzner (1993) that suggested the use of albumin dialysis for the treatment of liver failure, several studies have been devoted to the in-vitro and in-vivo investigation of this detoxification process (for instance, Dammer et al., 2007 and other papers in the same collection). Despite the big amount of literature on this subject, only few papers report data that can be used for a quantitative analysis of the process and for design purposes. Experimental data on albumin-dialysis of bilirubin in a two-compartment closed-loop system are reported by Stange et al. (1993, 1996): these authors compare the behavior of a standard polysulphone and a polyammide membrane pre-treated with albumin. Abe et al. (2004) compare the performance of different membrane modules for the removal of bilirubin. The rationale of albumin bound toxin removal by albumin dialysis has been discussed byPatzer (2006), while Meyer et al. (2004) report a theoretical model to evaluate albumin-bound toxin clearance in conventional hemodialysis. In vitro transport of bilirubin and other protein bound compounds has been analyzed by Raff et al. (2006).

In order to evaluate the bilirubin clearance obtained by an albumin dialysis device, a mathematical model is presented here that combines the bilirubin transfer rate across the membrane with the bilirubin mass balance in the feed and dialysate solution.

As for bilirubin transfer, diffusion through the membrane is considered as the rate-controlling step. In the framework of a diffusion-solution model, bilirubin transmembrane flux is given by

$$J_{bil} = \frac{\mathcal{D}_{bil}^{m}}{\delta}\left(c_{bil}^{m,\alpha} - c_{bil}^{m,\beta}\right) \tag{17}$$

where superscript m refers to the membrane phase; $c_{bil}^{m,\alpha}$ and $c_{bil}^{m,b}$ are the bilirubin concentrations in the membrane at the interface with phase α and β, respectively; $\mathcal{D}_{bil}^{m}$ is bilirubin diffusivity inside the membrane and δ is the membrane thickness. Assuming equilibrium conditions (equations 15 or 16) at the interfaces, the bilirubin flux may be rewritten in terms of bilirubin and albumin concentrations in the liquid phases as

$$J_{bil} = K_0\left(\frac{c_{AB}^{\alpha}}{c_{A}^{\alpha}} - \frac{c_{AB}^{\beta}}{c_{A}^{\beta}}\right) = K_0\left(\frac{c_{bil}^{\alpha}}{c_{alb}^{\alpha} - c_{bil}^{\alpha}} - \frac{c_{bil}^{\beta}}{c_{alb}^{\beta} - c_{bil}^{\beta}}\right) \tag{18}$$

where $K_0 = S_{bil}^{m}\mathcal{D}_{bil}^{m}/\delta$ is a characteristic transport coefficient of the membrane. It is worth noting that the driving force for bilirubin transfer is the difference between bilirubin to free albumin ratios in the two phases and, obviously, the flux becomes zero when equilibrium is achieved. At low bilirubin concentration ($c_{alb} \gg c_{bil}$) the driving force reduces to the difference between bilirubin to albumin molar ratios, as suggested by Steiner et al. (2004) and Dammeir et al. (2008) on the basis of experimental evidence.

The model of the albumin dialysis module is then obtained by combining the mass transfer rate equation (18) with bilirubin mass balances in the feed (α) and dialysate (β); for a counter-current hollow fiber module, the result is

$$\frac{dc_{bil}^{\alpha}}{dz} = \frac{K_0 A}{LQ^{\alpha}}\left(\frac{c_{bil}^{\alpha}}{c_{alb}^{\alpha} - c_{bil}^{\alpha}} - \frac{c_{bil}^{\beta}}{c_{alb}^{\beta} - c_{bil}^{\beta}}\right) \tag{19}$$

$$\frac{dc_{bil}^{\beta}}{dz} = \frac{K_0 A}{LQ^{\beta}} \left(\frac{c_{bil}^{\alpha}}{c_{alb}^{\alpha} - c_{bil}^{\alpha}} - \frac{c_{alb}^{\beta}}{c_{alb}^{\beta} - c_{bil}^{\beta}} \right) \tag{20}$$

where A is the membrane area, L the module length and Q is the volumetric flow rate. Equations (19) and (20) can be integrated with the boundary conditions:

$$z = 0 \quad c_{bil}^{\alpha} = c_b^{\alpha,in}; \quad z = L \quad c_{bil}^{\alpha} = c_b^{\beta,in} \tag{21}$$

Assuming that no albumin transfer occurs through the membrane, the total albumin concentration in the each phase is constant and equations (19)-(21) can be rewritten in terms of dimensionless variables $x = c_{bil}/c_{alb}$ and $\zeta = z/L$:

$$\frac{dx^{\alpha}}{d\zeta} = -\frac{K_0 A}{Q^{\alpha} c_{alb}^{\alpha}} \left(\frac{x^{\alpha}}{1 - x^{\alpha}} - \frac{x^{\beta}}{1 - x^{\beta}} \right) \tag{22}$$

$$\frac{dx^{\beta}}{d\zeta} = -\frac{K_0 A}{Q^{\beta} c_{alb}^{\beta}} \left(\frac{x^{\alpha}}{1 - x^{\alpha}} - \frac{x^{\beta}}{1 - x^{\beta}} \right) \tag{23}$$

with the boundary conditions:

$$\zeta = 0 \quad x^{\alpha} = x^{\alpha,in}; \quad \zeta = 1 \quad x^{\beta} = x^{\beta,in} \tag{24}$$

In SPAD or for a perfectly efficient dialysate regeneration system, the influent dialysate is bilirubin-free ($x_0^{\beta} = 0$). In this case, the above equations show that, for an assigned bilirubin-to-albumin molar ratio in the feed, the fraction of bilirubin removed $\left(x^{\alpha,in} - x^{\alpha,out}\right)/x^{\alpha,in}$ depends only on two dimensionless parameters: $\kappa = K_0 A/Q^{\alpha} c_{alb}^{\alpha}$, and $1/Z = Q^{\beta} c_{alb}^{\beta}/Q^{\alpha} c_{alb}^{\alpha}$. Fig. 3 reports a plot of module dimensionless clearance

$$\frac{CL}{Q^{\alpha}} = \frac{c^{\alpha,in} - c^{\alpha,out}}{c^{\alpha,in}} \tag{25}$$

Vs. $1/Z$ for different κ and $x^{\alpha,in}$ values. Fig. 3 clearly shows that:

- the higher the albumin concentration in the solution to be dialyzed, the lower the clearance that can be obtained, since both the values of $1/Z$ and κ decrease.
- clearance increases with increasing $1/Z$, i.e. increasing the dialysate flow rate or its albumin concentration; nevertheless a clearance limiting value, CL_{∞}, is obtained for $1/Z \gg 1$. CL_{∞} can be determined by the solution of the following equation

$$\frac{CL_{\infty}}{Q^{\alpha}} x^{\alpha,in} + \ln\left(1 - \frac{CL_{\infty}}{Q^{\alpha}}\right) = -\kappa \tag{26}$$

  From a practical point of view, if $1/Z$ is above 1-1.5, a further increase of this parameter should produce a negligible improvement of module clearance.

- substantial improvement can be obtained with larger $K_0 A$ values, i.e with larger modules or more permeable membranes.

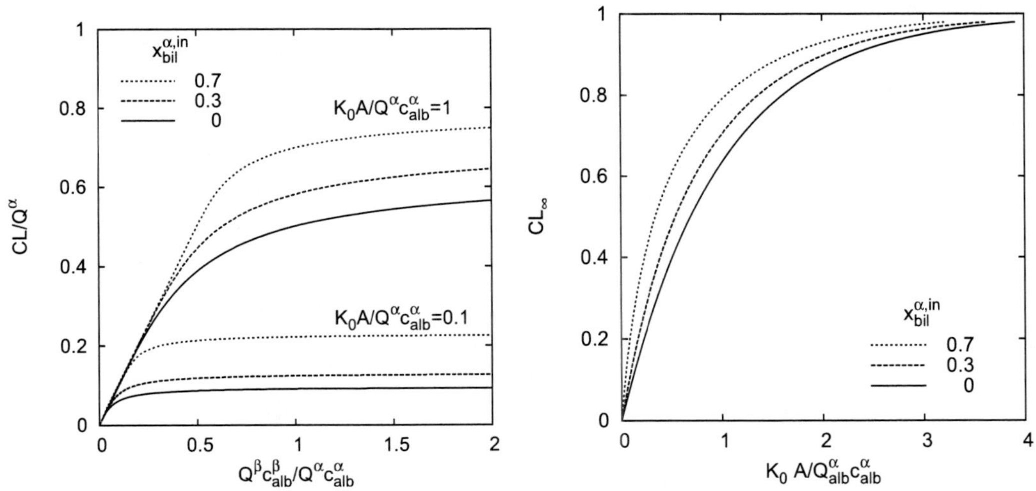

Figure 3. Clearance of strongly albumin-bound toxins obtained by albumin dialysis with fresh dialysate ($x^{\beta,in} = 0$). On the left: predicted values with different operating conditions. On the right: limiting values ($1/Z \to \infty$) for different inlet toxin concentrations.

- if the solution to be detoxified has a very low bilirubin-to-albumin molar ratio, i.e. in the limit of $x^{\alpha,in} \to 0$, equations (22) and (23) can be analytically integrated to give:

$$\frac{CL}{Q^\alpha} = \frac{1 - \exp\left[\kappa\left(1 - Z\right)\right]}{Z - \exp\left[\kappa\left(1 - Z\right)\right]} \tag{27}$$

  This limiting expression for the module clearance underestimates the exact solution when $x^{\alpha,in} > 0$ and can therefore be considered as a conservative approximation. It is worth noting that, in these conditions, the driving force for toxin transfer is $x^\alpha - x^\beta$, coherently with the empirical findings reported by Dammeir et al. (2007).

The order of magnitude of K_0A for the membrane modules used in albumin dialysis LSDs can be estimated by the data reported by Stange and Mitzner (1996) and referring to the time course of bilirubin concentration in a two-compartment, closed-loop dialysis system. The model presented in this section, was coupled with mass balances for the two compartments and fit against these experimental data. The optimal value found for K_0A was 2.61 μmol/min.

4. Adsorption Process

In recirculating albumin dialysis devices such as MARS, adsorption is one of the fundamental steps in dialysate regeneration. Fixed bed units are normally used for this task, and the design of such units requires information both on toxin adsorption equilibrium and kinetics.

The most common adsorptive media used in LSDs are activated carbon, non-ionic polymeric resins and anionic resins. Bilirubin adsorption equilibrium on these media from albumin-containing aqueous solutions has been previously investigated (Annesini et al., 2005, 2008, submitted) and expressions for the adsorption isotherms, accounting for the effect of albumin concentration, are available. It is worth remarking some of the main results of those papers. Firstly, on all of the adsorbent media tested, bilirubin molar uptake was largely greater than albumin molar uptake (which was, at least for anionic resins, negligible), indicating that bilirubin is mainly, if not only, adsorbed in its unbound form. Furthermore, a significant reduction of bilirubin adsorbed amount was always observed when albumin concentration was increased in the solution. Finally, comparing all the results, it can be stated that the adsorption capacity for bilirubin of the three different media increases in the following order

polymeric non ionic resin < activated carbon <anionic resin

The general expression of the bilirubin adsorption isotherm obtained by Annesini et al. (2005, 2008, submitted) was derived accounting for albumin-toxin association in the liquid phase and free toxin adsorption on the adsorbent solid. The same expression will be derived here in a simpler way, by the chemical thermodynamic approach followed throughout this chapter for the description of bilirubin phase partition.

For tightly albumin-bound toxins like bilirubin, adsorption on a solid medium can be viewed as an exchange reaction involving the adsorption sites on the solid:

$$\mathrm{AB} + \mathrm{S} \rightleftharpoons \mathrm{A} + \mathrm{BS} \tag{28}$$

at thermodynamic equilibrium:

$$K_{ads} = \frac{c_{AB}\left(1-\theta\right)}{c_A \theta} \tag{29}$$

where θ is the fraction of saturated binding sites or the ratio between the bilirubin adsorbed amount and the maximun adsorption capacity of the adsorbent for the free toxin ($\theta = n_{bil}/n_{m,B}$). With the approximations (8), the adsorption isotherm may be written in an apparent Langmuir form as

$$n_{bil} = \bar{n}_{bil} \frac{c_{bil}}{\bar{k} + c_{bil}} \tag{30}$$

with the following apparent constants:

$$\bar{n}_{bil} = \frac{n_{m,B}}{1 - K_{ads}} \tag{31}$$

$$\bar{k} = \frac{K_{ads}}{1 - K_{ads}} c_{alb} = m_{bil} c_{alb} \tag{32}$$

According to this model, the apparent Langmuir constant $\bar{k}$ should be an increasing function of albumin concentration, whereas the maximum adsorption capacity should be independent of it. Furthermore, if $c_{bil}/c_{alb} \ll m_{bil}$ a linear adsorption isotherm is obtained

$$n_{bil} = \frac{\bar{n}_{bil}}{m_{bil} c_{alb}} c_{bil} \tag{33}$$

Table 2. Parameters of the bilirubin adsorption isotherm on different adsorbent (Annesini et al., 2005, 2008, submitted)

Adsorbent Material	Eq.	$\bar{n}_{bil}$ μmol/g	m_{bil}	$\bar{n}_{bil}/m_{bil}$ μmol/g
Activated carbon (05112, Fluka)	(33)			43.4 ± 3.2
Polymeric non-ionic resin (Lewatit1064 MDPH, Bayer)	(30)	0.482 ± 0.082	0.34 ± 0.13	
Anionic resin (IRA 400, Sigma Aldric)	(30)	$a \exp\left(-c_{alb}/c^*\right)$ $a = 48.9 \pm 3.4$ $c^* = 1245 \pm 200\ \mu$M	0.24 ± 0.03	

It is worth noting that the slope of the linear isotherm (33) is a decreasing function of albumin concentration.

Since the free bilirubin adsorption isotherm from aqueous neutral solutions is not experimentally accessible (bilirubin is almost water insoluble at pH 7.4), $\bar{n}_{bil}$ and m_{bil} or, in the case of linear isotherm, the ratio $\bar{n}_{bil}/m_{bil}$, may be considered as adjustable parameters and determined by fitting of data obtained with albumin-containing solutions.

The experimental data of Annesini et al. (2005, 2008), showed that the chemical model proposed is suitable to describe bilirubin adsorption isotherms onto activated carbon and polymeric non-ionic resin (in the case of activated carbon, the linear form of the model (33) can be used). On the other hand, for adsorption on anionic resin, Annesini et al. (submitted) observed a clear reduction of the maximum bilirubin adsorbed amount as the concentration of albumin in the solution was raised. Such a behavior, which is not predicted by the simple chemical model presented here, suggests that, in this case, the effect of albumin on bilirubin adsorption is not limited to binding in the liquid solution. Different other phenomena that involve albumin and may cause the observed reduction of $\bar{n}_{bil}$ can be hypothesized; in particular, albumin may compete with bilirubin for adsorption sites or cause steric hindrance to its diffusion in the sorbent pores. Regardless of the phenomena actually involved, Annesini et al. (submitted) accounted for these non-chemical effects of albumin with an empirical relation between $\bar{n}_{bil}$ and c_{alb}. Tab. 2 reports a summary of the adsorption isotherm parameters for bilirubin on the different adsorptive media.

A comparison between bilirubin adsorption onto activated carbon and anionic resin at constant albumin concentration (120 μmol/l) is reported in Fig. 4 (specific adsorption on polymeric non-ionic resin is much lower and therefore not reported on the same plot). For this albumin concentration, the slope of the bilirubin adsorption isotherm at $c_{bil} = 0$ is 1.54, 0.36 and 0.012 l/g for anionic resin, activated carbon and polymeric resin, respectively. Fig. 4 clearly shows that the anionic resin is the best sorptive media for the bilirubin among those tested.

Since the anionic resin showed the higher adsorption capacity for bilirubin, only this adsorbent will be considered henceforth. In order to characterize bilirubin fixed-bed ad-

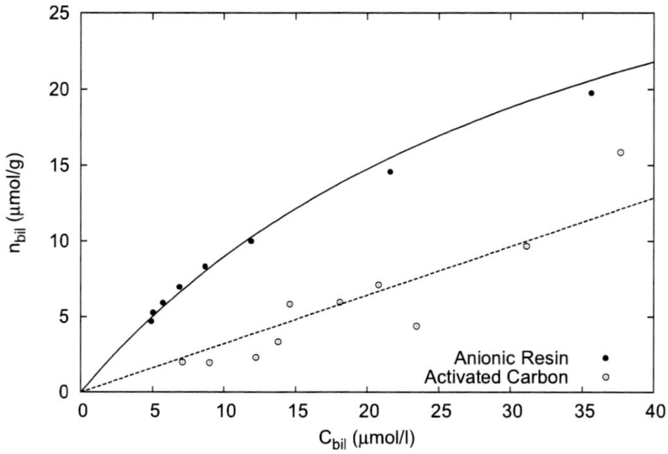

Figure 4. Typical bilirubin adsorption isotherms. lines: model fitting; circles: experimental data.

sorption kinetics, breakthrough curves of bilirubin-albumin solutions on the same anionic resin used by Annesini et al. (submitted) were experimentally determined. Fig. 5 reports some typical results obtained with different albumin concentrations in the feed. It can be clearly seen that the bilirubin adsorption rate is low, so that the column effluent contains a non-negligible bilirubin concentration even at early operating times, when the sorbent is far from saturation. This finding, also confirmed by some clinical data obtained during a MARS treatment (unpublished results), has a fundamental relevance in the design of an albumin regeneration system, because it suggests that, in the present configurations of devices such as MARS, the albumin is never completely regenerated.

Furthermore, the curves reported in Fig. 5 confirm the negative effect of albumin over bilirubin adsorption.

In order to design the adsorption units of a LSD, a model that provides a quantitative description of bilirubin adsorption in a fixed bed column is required. Such a model can be obtained by coupling the differential unsteady bilirubin mass balance in the column, with mass transfer kinetics from the liquid to the adsorbed phase.

Considering both solute convection and axial dispersion in the liquid phase, bilirubin mass balance in the column can be written as follows

$$\varepsilon\frac{\partial c_{bil}}{\partial t} + (1-\varepsilon)\rho_c\frac{\partial n_{bil}}{\partial t} = D_z\frac{\partial^2 c_{bil}}{\partial z^2} - v\frac{\partial c_{bil}}{\partial z} \tag{34}$$

were c_{bil} is the bilirubin concentration in the liquid phase, n_{bil} the bilirubin adsorbed amount per unit sorbent mass, ε is the bed porosity, ρ_c is the intrinsic density of the solid adsorbent, v the liquid superficial velocity and D_z the bilirubin axial dispersion coefficient. Assuming linear driving force (LDF) mass transfer kinetics (Glueckauf and Coates, 1947), the bilirubin mass balance in the adsorbed phase may be written as

$$\frac{\partial n_{bil}}{\partial t} = \frac{3}{R}k_c(n^*_{bil} - n_{bil}) \tag{35}$$

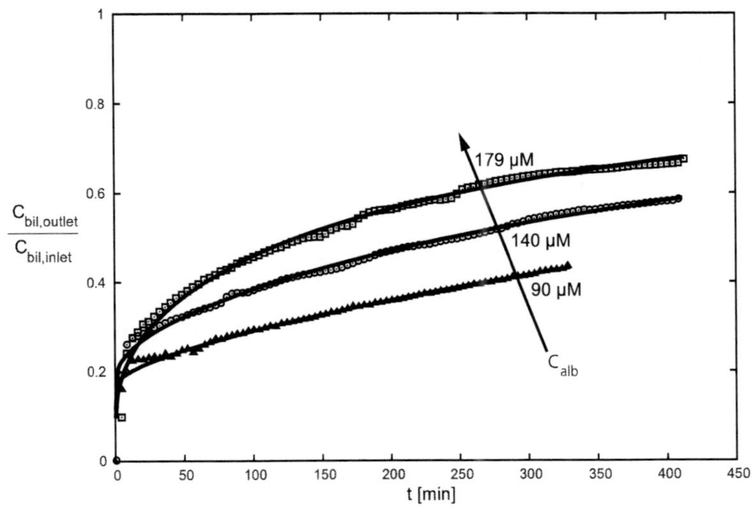

Figure 5. Typical bilirubin breakthrough curves on anionic resin. c_{alb}: albumin concentration in the solution; lines: model fitting; circles: experimental data. Operating conditions: feed flow-rate 1 ml/min ; resin mass 3 g ; inlet bilirubin concentration $c_{bil,inlet}$=100 μM.

where n^*_{bil} is the solute adsorbed amount at equilibrium with the liquid phase, given by equation (30), k_c is the LDF mass transfer coefficient and R is the radius of adsorbent particles.

Equations (34) and (35) can be integrated with the following initial and boundary conditions:

$$t = 0 \quad 0 \leq z \leq H \quad c_{bil} = 0 \; ; \; n_{bil} = 0 \tag{36}$$

$$t > 0, \quad z = 0, \;\; v\, c^{in}_{bil} = -D_z \frac{\partial c_{bil}}{\partial z} + v\, c_{bil} \tag{37}$$

$$t > 0 \quad z = 1 \quad \frac{\partial c_{bil}}{\partial z} = 0 \tag{38}$$

where c^{in}_{bil} is the bilirubin concentration in the inlet solution.

The system of equations (34) to (38) allows to predict bilirubin breakthrough curves, once the axial dispersion coefficient, equilibrium conditions and mass transfer coefficient are known. As for the dispersion coefficient D_z, the correlation proposed by Chung and Wen (1968) was used. The LDF mass transfer coefficient k_c is a lumped parameter that, from a general point of view, accounts for the diffusional resistance offered by adsorbent particles and the convective resistance in the liquid film surrounding the particles. Due to the intrinsic semi-empirical nature of the LDF model, k_c must be regarded as an adjustable parameter that can be obtained by fitting of experimental data. In this case, in order to improve the model adaptability to the experimental data, a modified form of LDF kinetics was considered, assuming k_c to be a decreasing function of time

$$k_c(t) = \frac{k^0_c}{1 + a t^b} \tag{39}$$

Table 3. Parameters used in the simulations.

Parameter	Value
K_0A	2.61 μmol/min
Q^α	180 ml/min
Q^β	180 ml/min
c^α_{alb}	606 μmol/l
$c^\alpha_{0,bil}$	240 μmol/l
resin mass	240 g
column liquid holdup	120 ml
total dialysate circuit volume	600 ml
k^0_c	$2.13 \cdot 10^{-7}$cm/s
a	0.043
b	0.54

Superscripts α and β refer to blood (fiber-side of membrane module) and albumin dialysate, respectively.

This time dependence of k_c can be related to the formation of an albumin skin layer on the resin surface, that results in a further, time-increasing diffusional barrier to bilirubin transfer to the solid phase.

The data reported in Fig. 5 were fitted using the model presented in this section. The optimal values found for the parameters of equation (39) are: $k^0_c = 2.13 \cdot 10^{-7}$ cm/s, $b = 0.54$, $a = 0.043$ min$^{-0.54}$.

5. Simulation of a Recirculating Albumin Dialysis Liver Support Device

The models presented in the previous sections can be combined to simulate a complete LSD, including different units. In this section, an example of this application is presented, focusing the analysis on bilirubin detoxification.

An albumin dialysis device with recirculation and regeneration of the albumin dialysate by adsorption on one anionic resin column is considered. This regeneration circuit is equivalent to that of MARS (see Fig. 1.): actually, MARS includes also a conventional dialysis module and an activated carbon column in the regeneration circuit, but these units have only a minor effect on bilirubin removal.

The simulations performed were aimed at evaluating the bilirubin clearance obtained by the device under different operating conditions. Since the analysis is focused on the device performance, a constant-composition feed was considered for the albumin dialysis module.

Fig. 6 reports time courses of bilirubin clearance obtained with the operating conditions reported in Tab. 3 and different albumin concentrations in the dialysate, ranging from 50 to 1500 μM. As the albumin concentration in the dialysate is increased, an initial improvement of the clearance is obtained, while at longer operating times the performance of the device is worsened.

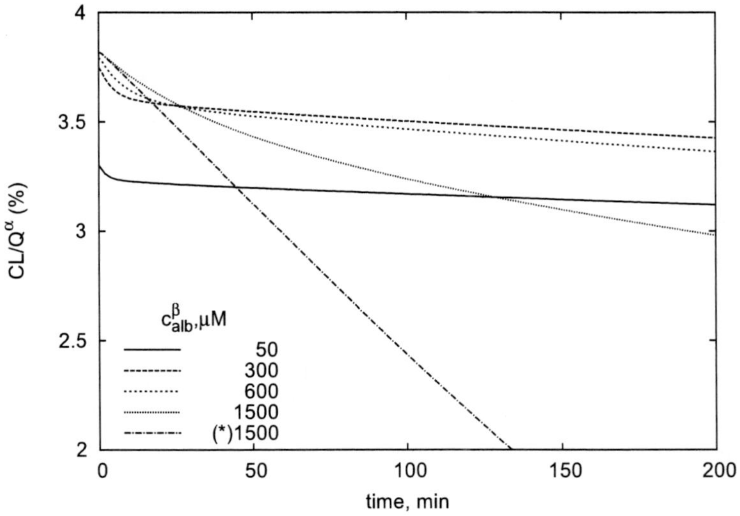

Figure 6. Simulated overall device clearances for bilirubin, as a function of time. The operating conditions considered are reported in Tab. 3. The different lines refer to different albumin concentration in the dialysate. The line marked with (*) was obtained setting to zero the column adsorption capacity for bilirubin.
Superscripts α and β refer to blood (fiber-side of membrane module) and albumin dialysate, respectively.

This result is clearly due to the two-fold effect of albumin in the dialysate on the overall detoxification process: on one hand, an higher albumin concentration enhances bilirubin transfer to the dialysate in the membrane module, but, on the other hand, impairs the regeneration of dialysate by adsorption. The first, positive effect is dominant at earlier operating times, when the adsorption column is far from saturation and the overall performance is determined by the dialysis process, while, at longer operating times, this condition is reversed and adsorption becomes the controlling process.

Anyway, it must be pointed out that, with the operating conditions considered in the simulations, a minor, if not negligible, initial improvement of the clearance is obtained increasing the albumin concentration in the dialysate from 300 μM to 1500 μM. This is not surprising, because such concentrations correspond to values of $1/Z$ in the range $0.5 \div 2.5$, where the membrane clearance is quite insensitive to changes in $1/Z$ (see Fig. 3). Furthermore, the overall clearances obtained are, in any case, as low as 3-4% and are limited by the low performance of the dialysis module (the value of κ calculated with the parameters reported in Tab. 3 is as low as $2.34 \cdot 10^{-2}$). Bilirubin clearances of a few percent of the feed flow-rate were also reported by Peszynski et al. (2002).

Finally, in order to highlight the importance of the regeneration efficiency for the prolongation of the treatment, Fig. 6 also shows a curve obtained by setting to zero the resin adsorption capacity for bilirubin. It is clear that, in a system without dialysate regeneration, the detoxification process becomes rapidly ineffective, due to the build-up of bilirubin concentration in the dialysate.

6. Conclusions

In this chapter, the application of an engineering analysis to albumin dialysis liver support devices was presented. Starting from the physico-chemical description of the elementary processes occurring in the device units, a mathematical model of a recirculating albumin dialysis liver support device was built. The model was subsequently used to calculate bilirubin clearances obtained by the device with different operating conditions. The analysis was focused on bilirubin removal, since this toxin is an important marker of the clinical state of liver failure patients and can also be considered representative of a wide class of strongly albumin-bound toxins.

This approach proved to be helpful in gaining a deeper insight into the working principles of liver support devices, providing a rationale for some empirical findings and leading to the detection of design issues.

In particular, the possible existence of an optimum dialysate albumin concentration emerged. The optimal value should be determined as a trade-off between two opposing effects of an increase of this parameter: the improvement of albumin dialysis efficiency on one hand, and the impairment of dialysate regeneration by adsorption on the other hand.

The simulations, based partially on parameters determined by in vitro experiments, showed that, for a device operating in conditions similar to those of MARS, the main effect of a further increase of dialysate albumin concentration would be a reduction of dialysate regeneration efficiency and, therefore, a more rapid reduction of the the device performance with time. Furthermore, for such a device, overall bilirubin clearances didn't exceed 4% of the blood flow-rate fed to the membrane module (in accordance with some literature data), this poor performance being limited by the slow bilirubin mass transfer across the membrane.

The information presented in this chapter can be helpful for the optimization of existing liver support devices and for the design of new ones; nevertheless, for a complete assessment of the performance of a given device, a similar analysis should be extended to the clearance of other toxins and some of the model parameters should be also checked against clinical data.

Acknowledgments

The Authors thank prof. Luigi Marrelli for his insightful comments, prof. Gilnardo Novelli and dott. Vincenzo Morabito for the useful discussion about the medical aspects. This work was financially supported by University of Rome "La Sapienza"

References

Abe, T; Shono, M; Kodama, T; Kita, Y; Fukagawa, M; Akizawa, T. Extracorporeal albumin dialysis. *Theapeutic Apheresis and Dialysis*, 2004, 8, 217– 222.

Annesini, MC; Di Paola, L; Marrelli, L; Piemonte, V; Turchetti, L. Bilirubin removal from albumin-containing solution by adsorption on polymer resin. *The International Journal of Artificial Organs*, 2005, 28, 686–693.

Annesini, MC; di Carlo, C; Piemonte, V; Turchetti, L. Bilirubin and tryptophan adsorption in albumin-containing solutions I. Equilibrium isotherms on activated carbon. *Biochemical Engineering Journal*, 2008 40 205-210.

Annesini, MC; Piemonte, V; Turchetti, L. Adsorption Equilibrium of albumin-bounbd toxins on anionic resin. *The International Journal of Artificial Organs*, (submitted).

Beaven, GH; d'Albis, A; Gratzer, WB. A Spectroscopic Study of the Haemin–Human-Serum-Albumin System. *European Journal of Biochemistry*, 1973, 33, 500-510.

Berde, CB; Hudson, BS; Simoni, RD; Sklar, LA. Human seum albumin. Spectroscopic studies of binding and proximity relationships for fatty acids and biliruin. *The Journal of Biological Chemistry*, 1979, 254, 391-400.

Blauer, G; Lavie, E; Silfen J. Relative affinities of bilirubin for serum albumins from different species. *Biochemistry and Biophysics Acta,* 1997, 42, 64-69.

Chawla, LS; Georgescu, F; Abell, B; Seneff, MG; Kimmel, PL. Modification of continuous venovenous hemodiafiltration with single-pass albumin dialysate allows for removal of serum albumin. *American Journal of Kidney Diseases*, 2005, 45, 51-56.

Chung, SF; Wen, CY. Longitudinal dispersion of liquid flowing through fixed and fluidized beds. *AIChE Journal,*1968, 14, 857-866.

Dammeir, N; Baumann, A; Suraj, M; Weiss, R; Mitzner, S; Schmidt, R; Stange, J.Capacity of different adsorbent combinations to mantain a high toxin/albumin gradient during albumin dialysis. *International Symposium on Albumin Dialysis in Liver Disease. Book of Abstracts 2002-2008*, Rostock (Germany), September 2008.

Evenepoel, P; Laleman, W; Wilmer, A; Claes, K; Maes, B; Kuypers, D; Bammens, B; Nevens, F; Vanrenterghem, Y. Detoxifyng capacity and kinetics of Prometheus- A new extracorporeal system for the treatment of liver Failure. *Blood Purification,* 2005, 23, 349-358.

Evenepoel, P; Laleman, W; Wilmer, A; Claes, K; Kuypers, D; Bammens, B; Nevens, F; Vanrenterghem, Y. Prometheus versus adsorbent recirculating systems: comparison of efficincy on two different liver detoxification decices. *Artificial Organs,* 2006, 30, 276-284.

Falkenaghen, D; Strobl, W; Vogt, G; Schrefl, A; Linsberger, I; Gerner, FJ; Schoenhofen, M. Fractional plasma separation and adsorption system: a novel system for blood purification to remove albumin bound substances. *Artificial Organs*, 1999, 23, 81-86.

Gray, RD; Stroupe D. Kinetics and mechanism of bilirubin binding to human serum albumin. *The Journal of Biological Chemistry,* 1978, 253, 4370-4377.

Glueckauf E; Coates J. Theory of Chromatography. Part IV. The Influences of Incomplete Equilibrium of the Front Boundary of Chromatograms and the Effectiveness of Separation. *Journal of Chemical Society*, 1947, 1315-1321.

Jacobsen, J. Studies of the affinity of human serum albumin for binding of bilirubin at different temperatures and ionic strength. *International Journal of Peptide Protein Research*, 1977, 9, 112-114.

Meyer, TW; Leeper, EC; Bartlett, DW; Depner, TA; Lit, YZ; Robertson, CR; Hostetter, TH. Increasing dialysate flow and dialyzer mass transfer area coefficient to increase the clearance of protein-bound solutes. *Journal of the American Society of Nephrology*, 2004, 15, 1927-1935.

Mitzner, S; Klammt, S; Stange, J; Schmidt, R. Albumin regeneration in liver support—comparison of different methods. *Therapeutic Apheresis and Dialysis*, 2006, 10, 108–117.

Ostrow, JD; Mukerjee, P; Tiribelli, C. Structure and binding of unconjugated bilirubin: relevance for physiological and pathophysiological function. *Journal of Lipid Research*, 1994, 3, 1715-1737.

Patzer, J. Principles of Bound Solute Dialysis. *Therapeutic Apheresis and Dialysis*, 2006, 10, 118-124.

Petersen, CE; Ha, CE; Haroalli, K; Feix, JB; Bhagavan, NV. *Journal of Biological Chemistry*, 2000, 275, 20985-20995.

Peszynski, P; Klammt, S; Peteters, E; Mitzner, S; Stange, J; Schmidt, R. Albumin dialysis: single pass Vs. recirculation (MARS). *Liver*, 2002, 22, 40-42.

Raff, M; Ertl, T; Krause, B; Storr, M; Goehl, H. Mass Transfer in artificial liver mambrane devices. *Desalination,*2006, 199, 234-235.

Rifai, K; Ernst, T; Kretshmer, U; Hafer, C; Haller, H; Manns, MP; Fliser, D. The Prometheus device for extracorporeal support of combined liver and renal failure. *Journal of Hepatology*, 2005, 39, 298-302.

Rozga, J. Liver support technology—an update. *Xenotransplantation,* 2006, 13, 380–389.

Sauer, IM; Goetz, M; Steffen, I; Walter, G; Kehr, DC; Schwartlander, R; Hwong, YJ; Neuhaus, P. In vitro comparison of the molecular adsorbent recirculation system (MARS) and single-pass albumin dialysis (SPAD). *Hepatology,* 2004, 39, 1408-1414.

Stange, J; Ramlow, W; Mitzner, S; Schmidt, R. Dialysis against a recycled albumin solution enables the removal of albumin-bound toxins. *Artificial Organs*, 1993, 17, 809–813.

Stange, J; Mitzner, S. A carrier-mediated transport of toxins in a hybrud membrane. Safety barrier between a patients blood and bioartificial liver. *The International Journal of Artificial Organs*, 1996, 19, 677-691.

Stange, J; Mitzner, S; Ramlow, W; Gliesche, T; Hickstein, H; Schmidt, R. A new Procedure for the removal of protein bound drugs and toxins. *ASAJO Journal,* 1993, 39, 621-625.

Stegmayr, BG. A survey of blood purification techniques. *Transfusion Apheresis Science,* 2005, 32, 209–220.

Steiner, C; Sen, S; Stange, J; Williams, R; Jalan, R. Binding of bilirubin and bromosulphthalein to albumin: implications for understanding the pathophysiology of liver failure and its management. *Liver Transplantation*, 2004, 10, 1531–1538.

INDEX

B

C

D

E

F

G

H

I

J

K

L

M

N

O

P

Q

R

S

T